# UNITEXT for Physics

**Series Editors**

Michele Cini, University of Rome Tor Vergata, Roma, Italy

Stefano Forte, University of Milan, Milan, Italy

Guido Montagna, University of Pavia, Pavia, Italy

Oreste Nicrosini, University of Pavia, Pavia, Italy

Luca Peliti, University of Napoli, Naples, Italy

Alberto Rotondi, Pavia, Italy

Paolo Biscari, Politecnico di Milano, Milan, Italy

Nicola Manini, University of Milan, Milan, Italy

Morten Hjorth-Jensen, Department of Physics and Astronomy, University of Oslo, Oslo, Norway

Alessandro De Angelis, Physics and Astronomy, INFN Sezione di Padova, Padova, Italy

UNITEXT for Physics series publishes textbooks in physics and astronomy, characterized by a didactic style and comprehensiveness. The books are addressed to upper-undergraduate and graduate students, but also to scientists and researchers as important resources for their education, knowledge, and teaching.

Roberto Piazza

# An Invitation to Probability and Data Analysis for Physicists

 Springer

Roberto Piazza
Chimica, Materiali e Ingegneria Chimica
Politecnico di Milano
Milan, Italy

ISSN 2198-7882 ISSN 2198-7890 (electronic)
UNITEXT for Physics
ISBN 978-3-031-83858-3 ISBN 978-3-031-83856-9 (eBook)
https://doi.org/10.1007/978-3-031-83856-9



This Springer imprint is published by the registered company Springer Nature Switzerland AG
The registered company address is: Gewerbestrasse 11, 6330 Cham, Switzerland

If disposing of this product, please recycle the paper.

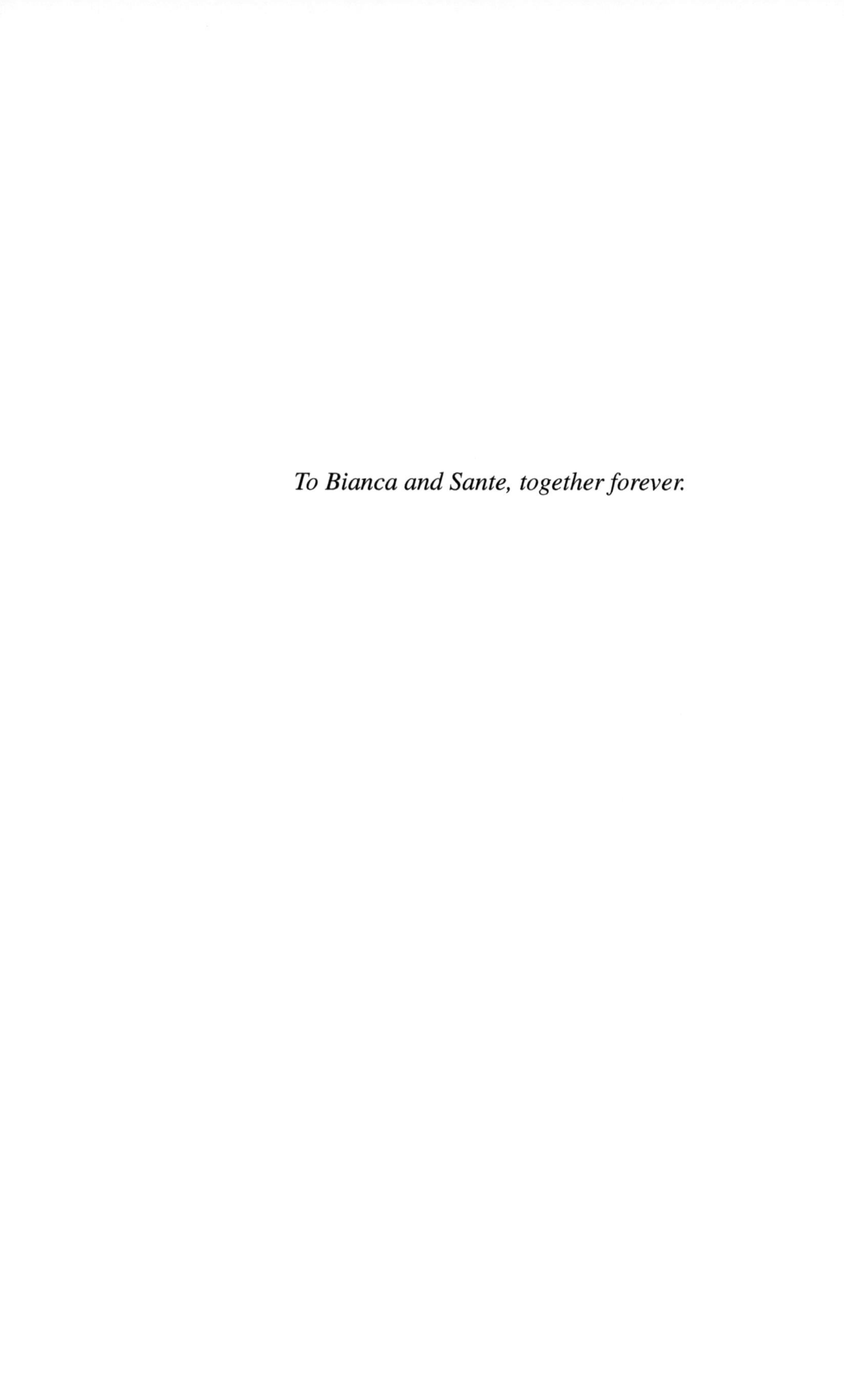

*To Bianca and Sante, together forever.*

# Preface

<br><br>

*To my 25 ± 5 readers*
*A. Manzoni*
*(updated for fluctuations)*

If any of my readers think they have just acquired a reference book covering all applications of probability and statistics in physics, I am sorry to disappoint them. This is not a standard textbook. Neither it is a technical manual. And, without a doubt, a 400-page tome is not a compendium either. In a way, it is a self-portrait, an account of my personal journey in physics.

Let me explain. The first encounter, or rather clash, with data analysis I had as a freshman was that kind of hasty introduction to laboratory courses that went under the name of "error theory". Practically, it consisted of giving a dignified look to the data collected during several tedious afternoons spent in the lab. Upon becoming familiar with statistical physics, dynamical systems, and of course quantum mechanics, I felt a greater affinity for probability, but I never really pondered over its basic ideas, which I always found a bit shaky. Surely, I was more captivated by the elegance of rigorous theoretical results.

Many years after, I can say that I was totally misled. My entire career as a physicist taught me that probability is fascinating, and crucial to capture the nature of the physical reality. That scrutinizing experimental data is more an art than a mechanical and sterile application of consolidated techniques. Sharing some of this late fascination with young physicists is my sole desire.

But what is the point of writing another book when there are already countless publications in probability and statistics available on the editorial market? My tentative answer is that my experience as a scientist has instilled in me a rather unconventional view of probability and statistics. These two fields are traditionally distinct and operated by distinct professionals, with statistics often seen as a kind of inverse probability. For example, while probability tells us that in the roll of three dice 11 is a slightly better total score than 12, statistics allows us to inquire whether the dice are loaded, if in a limited number of tosses, we observe the opposite.

However, such a Great Divide is somewhat blurred if you deal, as I do, with interdisciplinary subjects like complex systems or soft matter. Probability and statistics rather engage in a sort of ping-pong game, where a probability model indicates which tests are worth performing, or a clever scrutiny of the experimental results suggests a novel interpretation of the data statistics.

In this mutual exchange, probabilistic insights have the upper hand. Written by a lab scientist these words may sound odd, but this is what experience taught me. Regrettably, the majority of textbooks on probability theory are either too basic or too advanced for an aspiring physicist. Therefore, I have decided to primarily focus this book on probability, attempting to reconcile a graded approach with a precise but not excessively rigorous account that emphasizes applications to stimulating or intriguing physical issues. Besides, while most books on data analysis gloss over the meaning of probability, mostly considered as stuff for idle philosophers, I firmly believe that "what are we talking about?" is a question physicists *must* deal with. This also leads to reflect upon the meaning of randomness, complexity, and the baffling nature of quantum reality.

Less space is reserved for methods and techniques of data analysis, both because in this case there are already excellent manuals for physicists, and because I honestly find them to be less exciting. Nevertheless, I tried to delve into the nontrivial assumptions on which even the most common techniques of data testing are based on. Besides, I have taken a rather unconventional approach, mostly aimed at those young scientists who will have the privilege of experiencing physics like a kind of artisan business. That is, working in a small-scale lab where they are crucially involved in all phases of the investigation of a scientific puzzle, even if it is not a "fundamental" question of physics. And where, as I always say to my students, the most valuable instrument is still within their heads. In this kind of business, talking in abstract about experimental errors and accuracy is pointless, unless you are fully acquainted with the instrumental origins of signal fluctuations—a useful lesson for young theorists too. When dealing with new experimental evidence, moreover, digging into the data, looking at them from an original perspective, and presenting them in a form that facilitates modeling, can be more useful than a dry technical analysis. An attitude that I tried to convey right from Chap. 1.

In the end, stimulating curiosity has been the leading beacon of my efforts. It is not my job to judge whether and how much I have achieved this goal, but I hope that I have at least succeeded in a more modest intent, which was close to my heart too. Namely, to create that book that, in the shoes of a young physicist, I would have wanted to have at hand many years ago.

Milan, Italy                                                      Roberto Piazza
November 2024

**Competing Interests** The author has no competing interests to declare that are relevant to the content of this manuscript.

# Contents

# Chapter 1
# Digging Into Data: a Primer in Statistics

Tell the truth, nothing but the truth<br>
but not the WHOLE truth<br>
M. Kac

The primary purpose of this chapter is to review some basic tools that provide a quantitative picture of data that show statistical fluctuations, whatever the origin of this apparent randomness. But I also hope that the examples we will discuss may train you in the subtle art of "digging" into data. The latter may not be so essential in time-honored fields like high energy or nuclear physics, where the experimentalist benefits from consolidated methods of data analysis and the theorist may not even *need* to go through experiments. But scrutinizing, perusing, dissecting data becomes crucial if one deals with uncharted or frontier subjects like the physics of complex systems or biophysics. In these fields, taking an original look to data is often the pathway to new insights, both for experimentalists and theoreticians. Thus, I will try and show you how unexpected features may arise from selecting the quantities to be plotted, choosing a suitable range, changing the axes, scaling the data, focusing on specific details. To make these ideas clearer, however, we better start delving into the matter.

## 1.1 An (Apparent) Oxymoron, to Begin with

As a first approach to the quantitative analysis of experimental data, let us do a little mathematical experiment. Actually, an expression like "mathematical experiment" sounds as a kind of oxymoron that the majority of the mathematicians may find

© The Author(s), under exclusive license to Springer Nature Switzerland AG 2025
R. Piazza, *An Invitation to Probability and Data Analysis for Physicists*, UNITEXT for Physics, https://doi.org/10.1007/978-3-031-83856-9_1

offensive. But if you are not too fussy and have a taste for surprise, just follow me in taking a look at one of the most celebrated numbers, $\pi$. We know that $\pi$ is an irrational number, and therefore can be written as an infinite non-periodic sequence of decimals, the first 1000 of which are shown in Table 1.1. But how many times does a given digit (for example, 1,4, or 7) appear if we consider a number of consecutive decimals of $\pi$? In other words, if I consider $N$ digits of the sequence of decimals of $\pi$ and reckon how many times $n_k$ a certain digit $k$ appears, what can I expect? If there is no preference among the various digits, I can assume to find approximately $n_k \simeq N/10$ for each digit $k$. This sort of impartiality among the digits is satisfied with those numbers that in mathematics are said to be *simply normal* in a given integer base (in the case we are considering, base 10). It can be shown that almost all real numbers are simply normal. In fact, actually, much more can be shown. As a matter of fact, every possible pair, or triplet, or $n$-tuple of digits appears the same number of times in the distribution of decimals of almost all real numbers, which is expressed by saying that almost all real numbers are *normal* in *any* integer base greater than

**Table 1.1**  The first 1000 digits of $\pi$: $\pi = 3$.

```
1 4 1 5 9 2 6 5 3 5 8 9 7 9 3 2 3 8 4 6 2 6 4 3 3 8 3 2 7 9 5 0 2 8 8 4 1 9 7 1
6 9 3 9 9 3 7 5 1 0 5 8 2 0 9 7 4 9 4 4 5 9 2 3 0 7 8 1 6 4 0 6 2 8 6 2 0 8 9 9
8 6 2 8 0 3 4 8 2 5 3 4 2 1 1 7 0 6 7 9 8 2 1 4 8 0 8 6 5 1 3 2 8 2 3 0 6 6 4 7
0 9 3 8 4 4 6 0 9 5 5 0 5 8 2 2 3 1 7 2 5 3 5 9 4 0 8 1 2 8 4 8 1 1 1 7 4 5 0 2
8 4 1 0 2 7 0 1 9 3 8 5 2 1 1 0 5 5 5 9 6 4 4 6 2 2 9 4 8 9 5 4 9 3 0 3 8 1 9 6
4 4 2 8 8 1 0 9 7 5 6 6 5 9 3 3 4 4 6 1 2 8 4 7 5 6 4 8 2 3 3 7 8 6 7 8 3 1 6 5
2 7 1 2 0 1 9 0 9 1 4 5 6 4 8 5 6 6 9 2 3 4 6 0 3 4 8 6 1 0 4 5 4 3 2 6 6 4 8 2
1 3 3 9 3 6 0 7 2 6 0 2 4 9 1 4 1 2 7 3 7 2 4 5 8 7 0 0 6 6 0 6 3 1 5 5 8 8 1 7
4 8 8 1 5 2 0 9 2 0 9 6 2 8 2 9 2 5 4 0 9 1 7 1 5 3 6 4 3 6 7 8 9 2 5 9 0 3 6 0
0 1 1 3 3 0 5 3 0 5 4 8 8 2 0 4 6 6 5 2 1 3 8 4 1 4 6 9 5 1 9 4 1 5 1 1 6 0 9 4
3 3 0 5 7 2 7 0 3 6 5 7 5 9 5 9 1 9 5 3 0 9 2 1 8 6 1 1 7 3 8 1 9 3 2 6 1 1 7 9
3 1 0 5 1 1 8 5 4 8 0 7 4 4 6 2 3 7 9 9 6 2 7 4 9 5 6 7 3 5 1 8 8 5 7 5 2 7 2 4
8 9 1 2 2 7 9 3 8 1 8 3 0 1 1 9 4 9 1 2 9 8 3 3 6 7 3 3 6 2 4 4 0 6 5 6 6 4 3 0
8 6 0 2 1 3 9 4 9 4 6 3 9 5 2 2 4 7 3 7 1 9 0 7 0 2 1 7 9 8 6 0 9 4 3 7 0 2 7 7
0 5 3 9 2 1 7 1 7 6 2 9 3 1 7 6 7 5 2 3 8 4 6 7 4 8 1 8 4 6 7 6 6 9 4 0 5 1 3 2
0 0 0 5 6 8 1 2 7 1 4 5 2 6 3 5 6 0 8 2 7 7 8 5 7 7 1 3 4 2 7 5 7 7 8 9 6 0 9 1
7 3 6 3 7 1 7 8 7 2 1 4 6 8 4 4 0 9 0 1 2 2 4 9 5 3 4 3 0 1 4 6 5 4 9 5 8 5 3 7
1 0 5 0 7 9 2 2 7 9 6 8 9 2 5 8 9 2 3 5 4 2 0 1 9 9 5 6 1 1 2 1 2 9 0 2 1 9 6 0
8 6 4 0 3 4 4 1 8 1 5 9 8 1 3 6 2 9 7 7 4 7 7 1 3 0 9 9 6 0 5 1 8 7 0 7 2 1 1 3
4 9 9 9 9 9 9 8 3 7 2 9 7 8 0 4 9 9 5 1 0 5 9 7 3 1 7 3 2 8 1 6 0 9 6 3 1 8 5 9
5 0 2 4 4 5 9 4 5 5 3 4 6 9 0 8 3 0 2 6 4 2 5 2 2 3 0 8 2 5 3 3 4 4 6 8 5 0 3 5
2 6 1 9 3 1 1 8 8 1 7 1 0 1 0 0 0 3 1 3 7 8 3 8 7 5 2 8 8 6 5 8 7 5 3 3 2 0 8 3
8 1 4 2 0 6 1 7 1 7 7 6 6 9 1 4 7 3 0 3 5 9 8 2 5 3 4 9 0 4 2 8 7 5 5 4 6 8 7 3
1 1 5 9 5 6 2 8 6 3 8 8 2 3 5 3 7 8 7 5 9 3 7 5 1 9 5 7 7 8 1 8 5 7 7 8 0 5 3 2
1 7 1 2 2 6 8 0 6 6 1 3 0 0 1 9 2 7 8 7 6 6 1 1 1 9 5 9 0 9 2 1 6 4 2 0 1 9 8 9...
```

or equal to 2. However, it is almost impossible to rigorously prove that a *particular* number like $\pi$ is normal. For our purposes, then, the sequence of digits of $\pi$ is an uncharted territory that we want to investigate "experimentally".

At first glance, the distribution of the digits of $\pi$ does not resemble what we are used to considering random. For example, no zero appears in the first thirty decimals, which instead contain as many as six threes, and in the twentieth row there is even a sequence of six consecutive nines. Table 1.1, however, is only the initial part of the entire group of the first 10000 decimals of $\pi$ that I analyzed and that we will now discuss more accurately, first asking ourselves whether it is actually plausible to believe that each digit appears the same number of times in the sequence of decimals. Panel A in Fig. 1.1 shows the trend of the deviation $\Delta_6(N) = n_6(N) - N/10$, namely the number $n_6(N)$ of sixes found as a function of the number $N$ of examined decimals, wherefrom we subtract the number $N/10$ of results that we expect to find expect if $\pi$ were a normal number. In fact, things do not seem to be going too well. The deviation from the prediction, although with a somewhat oscillating trend, seems to *grow* progressively as $N$ increases. However, if, as in Panel B, we consider the *fraction* $f_6 = n_6/N$ of 6s that we obtain compared to the total number of decimals examined, we realize that $f_6$ tends to quickly settle around a value $f_6 \simeq 0.1$.

Where does this apparent contradiction come from? From Panel A of Fig. 1.1 we see that, as $N$ grows, it becomes increasingly rare for $n_6$ to be *exactly* equal to $N/10$. But from Panel B we also conclude that the deviation from the predicted value, even if it grows in absolute terms, becomes smaller and smaller relative to $N$, that is, it grows less rapidly than $N$. The behavior of the other digits does not differ qualitatively from what we have observed for the digit 6. If we then define in an analogous way for each digit $k$ the ratio $f_k = n_k/N$, we obtain, as the number $N$ of examined decimals increases, Table 1.2.

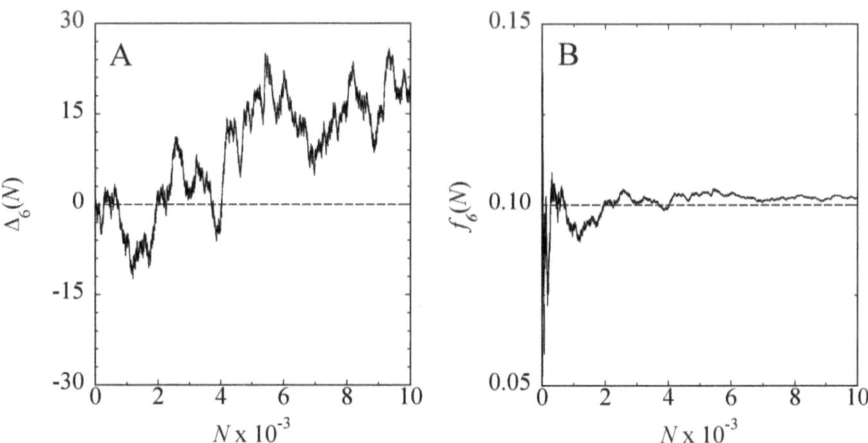

**Fig. 1.1**   Panel A: Deviation of the number $n_6$ of the digit 6 from the value $N/10$ in the first 10000 decimals of $\pi$. Panel B: Relative frequency $f_6(N)$ of the digit 6 in the distribution of the decimals of $\pi$

**Table 1.2** Distribution of the frequencies of the digits $k$ in $\pi$ as a function of the number N of considered decimals

| $N$ | $f_0$ | $f_1$ | $f_2$ | $f_3$ | $f_4$ | $f_5$ | $f_6$ | $f_7$ | $f_8$ | $f_9$ | $\Delta_f$ |
|---|---|---|---|---|---|---|---|---|---|---|---|
| 30 | 0.000 | 0.067 | 0.133 | 0.200 | 0.100 | 0.100 | 0.100 | 0.067 | 0.100 | 0.133 | 0.1563 |
| 50 | 0.040 | 0.100 | 0.100 | 0.160 | 0.080 | 0.100 | 0.080 | 0.080 | 0.100 | 0.160 | 0.1095 |
| 100 | 0.080 | 0.080 | 0.120 | 0.110 | 0.100 | 0.080 | 0.090 | 0.080 | 0.120 | 0.140 | 0.0648 |
| 300 | 0.087 | 0.100 | 0.117 | 0.103 | 0.123 | 0.090 | 0.103 | 0.063 | 0.113 | 0.100 | 0.0514 |
| 500 | 0.090 | 0.118 | 0.108 | 0.100 | 0.106 | 0.100 | 0.096 | 0.072 | 0.106 | 0.104 | 0.0371 |
| 1000 | 0.093 | 0.116 | 0.103 | 0.102 | 0.093 | 0.097 | 0.094 | 0.095 | 0.101 | 0.105 | 0.0218 |
| 3000 | 0.086 | 0.103 | 0.101 | 0.088 | 0.106 | 0.105 | 0.101 | 0.096 | 0.103 | 0.111 | 0.0232 |
| 5000 | 0.093 | 0.106 | 0.099 | 0.092 | 0.102 | 0.105 | 0.103 | 0.098 | 0.098 | 0.104 | 0.0147 |
| 10000 | 0.097 | 0.103 | 0.102 | 0.097 | 0.101 | 0.105 | 0.102 | 0.097 | 0.095 | 0.101 | 0.0097 |

As you can see, as $N$ increases all $f_k$ values quickly approach 0.1. This is better appreciated if we evaluate the overall deviation from the value 0.1 for all digits. However, it is not very useful to consider the simple deviations $f_k - 0.1$. The sum of these quantities is always zero, since positive and negative deviations exactly balance,

$$\sum_{k=0}^{9} (f_k - 0.1) = \sum_{k=0}^{9} \frac{n_k}{N} - 1 = 0.$$

This can be avoided by considering the sum of the *squares* of the deviations

$$\Delta_f^2 = \sum_{k=0}^{9} (f_k - 0.1)^2,$$

which is certainly greater than or equal to zero. The last column of the table shows that, by increasing the number of considered decimals by a factor of one hundred, $\Delta_f = \sqrt{\Delta_f^2}$ decreases by about an order of magnitude. Indeed, the plot in Fig. 1.2 shows that the values of $\Delta_f(N)$ are reasonably interpolated by a straight line, which in a double-log scale corresponds to $\Delta_f(N) = AN^{-1/2}$, where $A$ is a constant. We shall later see that there are good reason to expect that $\Delta_f$ is inversely proportional to $\sqrt{N}$. As a conclusion, we can then say that, from an "experimental" point of view, $\pi$ resembles a simply normal number, or in other words, that the distribution of the decimals of $\pi$ may have a property—the equanimity between all digits—that we surely expect from a number that we regard as "random".

Are you convinced? Good, then we can use $\pi$ to play a little game. From what we have seen, we expect that approximately half of the time a particular digit in the sequence is less than 5, and in the other half greater than or equal to this value. We can then think of the sequence of decimals of $\pi$ as the sequence of flips of a coin, stating that a particular flip results in "heads" if the corresponding digit in the sequence is

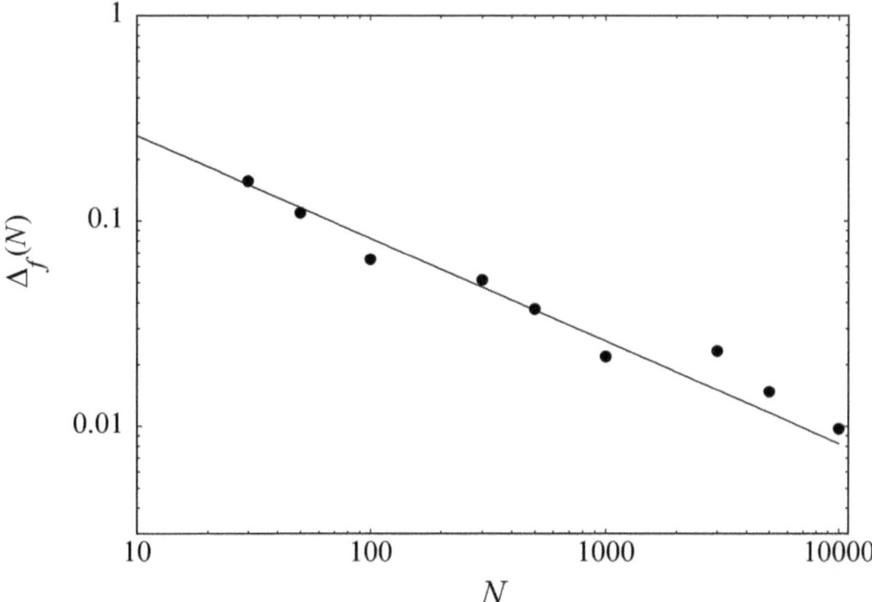

**Fig. 1.2** Dependence of $\Delta_f$ on the considered number of decimals of $\pi$. The straight line on the double-log plot is the function $\Delta_f(N) = 0.823N^{-1/2}$

less than 5, and "tails" otherwise. Suppose I choose tails and you choose heads. I want to analyze how my winnings (or losses) behave over the course of 10000 of these virtual flips, whose results are fixed by the values of the corresponding decimal of $\pi$. My gain after $N$ flips will be given by the quantity

$S(N)$= [number of tails in N flips] - [number of heads in N flips].

Figure 1.3 shows that the game ends more or less evenly, and that $S(N)$ displays the same irregular behavior found in Panel A of 1.1.

Yet, this particular sequence of $10^4$ flips, one of the many possible ones, has some additional surprises in store. Intuitively, one may indeed expect that I were ahead for about half the time, and that the same happened for you. Furthermore, we expect the "leader", i.e., the player who is momentarily ahead, to change often during the game. But the results blatantly contradict these predictions. According to the plot, even if the game ends up being slightly unfavorable for me, I am ahead for a good part (about 85%) of the time. Changes in leadership are actually very rare. We might think that these are quirks of the heads-or-tails game, or of $\pi$, but we will see that this is not the case. In particular, the long prevalence of a leader is a typical characteristic of all even games, in which the trend of fluctuations is far from intuitive. So do not complain too much if your favorite team stays at the top of the standings for most of the season, only to be finally overtaken by the direct pursuers!

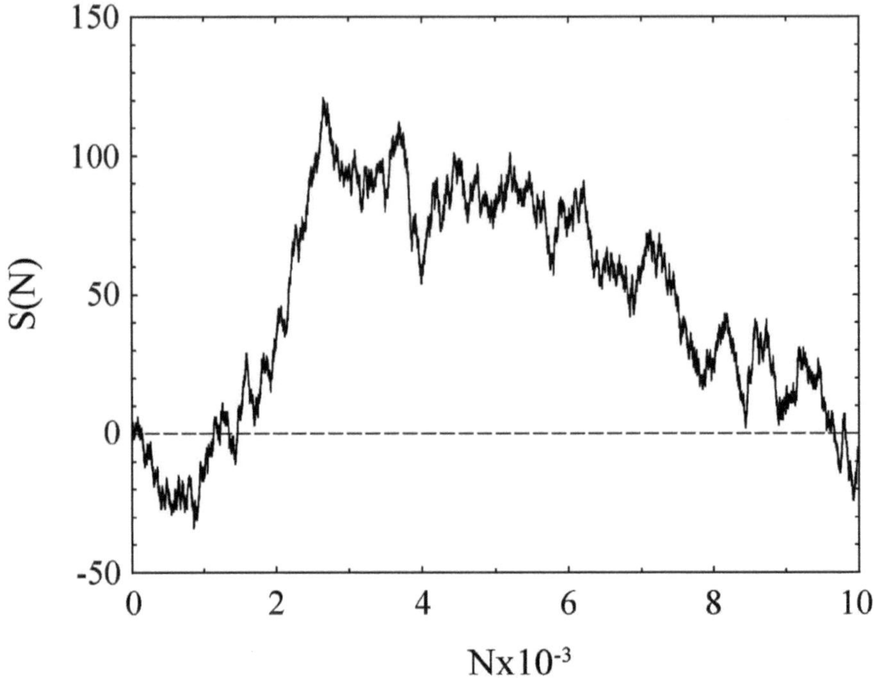

**Fig. 1.3** Playing "heads or tails" with the decimals of $\pi$

### *1.1.1   A Kick-Off Meeting with Randomness*

So far, we have given the word "random" a sort of intuitive meaning. As a matter of fact, randomness is probably one of the hardest concepts to be defined. Consider for instance the example we have just discussed. Evidently, a random number *must* be normal (in any base). But is that sufficient? Can any normal number be regarded as truly random? Well, surely not. One of the few real numbers that are known to be normal in base 10 is the so-called *Champernowne constant*,

123456789101112131415161718192021222324252627282930313233343536373 8 . . .

obtained by concatenating the natural numbers one after the other in order. I am pretty sure, however, that no one of my readers will consider this number "random", for I have just given you a precise rule to make it. If instead we construct a number less than one by linking in a string all prime numbers,

0.235711131719232931374143475359616771737983899710110310710911312 7 . . .

we obtain another normal number in base 10, known as the *Copeland–Erdös constant*. In this case, the rule is a bit more subtle, for we do not know a priori which numbers are prime, but it is still a rule, like "after the last prime you wrote, find and attach the next one" (even it may not be easy). What about $\pi$? Do we really believe that the most celebrated number of mathematics is *random*? Because of course there are *rules* to obtain $\pi$, the simplest one being "$\pi$ is the length of a circumference of unit diameter". Or, if you wish to be a bit fancier, "$\pi$ is that real number $r$ such that $\exp(ir) = -1$". So, the question is not whether there is a rule, but probably how much effort does that rule requires to be implemented. The problem is even thornier because of the natural tendency of our brain to look for patterns and find them even when they do not exist. For example, presented with a sequence of digits starting with 112358132133 some—I hope many—of you will continue it as 11235813213**354**, for this looks very much as the sequence of Fibonacci's numbers. Actually, however the correct answer is 11235813213**353**, because the former appears as a sub-sequence of the decimal expansion of $\pi$ (I don't know where, but it must surely be somewhere, if $\pi$ is a normal number!).[1]

The best example I know that highlights the difficulty of distinguishing between a random sequence and one that is constructed with a simple rule has once again to do with $\pi$, and is a memorable "Galilean dialogue" by Josef-Maria Jauch[2] It goes as follows:

SALVIATI:    Suppose I give you two sequences of numbers, such as

$$7853981633974483096156684\ldots$$

and

$$1, -1/3, +1/5, -1/7, +1/9, -1/11, +1/13, -1/15\ldots$$

If I asked you, Simplicio, what the next number of the first sequence is, what would you say?

SIMPLICIO:    I could not tell you. I think it is a random sequence and that there is no law in it.

SALVIATI:    And for the second sequence?

SIMPLICIO:    That would be easy. It must be $+1/17$.

SALVIATI:    Right. But what would you say if I told you that the first sequence is also constructed by a law and this law is in fact identical with the one you have just discovered for the second sequence?

SIMPLICIO:    This does not seem probable to me.

---

[1] If you wish to see which complex patterns can a numerologist find in $\pi$, just take a look at the 6th chapter of "The Magic Numbers of Dr. Matrix", by Martin Gardner (Prometheus Books, New York, 1985). An even more impressive story on what you can do with the first 10000 decimals of $\pi$ is told in "Not a Wake", by Michael Keith (Vinculum Press, 2010). Its content is almost unbelievable...but I don't want to spoil the surprise.

[2] J. M. Jauch, *Are Quanta Real? A Galilean Dialogue*, Indiana University Press (1990).

SALVIATI:    But it is indeed so, since the first sequence is simply the beginning of the decimal fraction of the sum of the second. Its value is $\pi/4$.

SIMPLICIO:    You are full of such mathematical tricks …

Salviati's "trick" just consists in using the formula for $\pi$ usually[3] attributed to Leibnitz, which you can also find from the series expansion for the inverse tangent,

$$\arctan(x) = x - \frac{x^3}{3} + \frac{x^5}{5} - \frac{x^7}{7} + \ldots = \sum_{k=0}^{\infty} \frac{(-1)^k x^{2k+1}}{2k+1},$$

evaluated for $x = 1$. This is yet another demonstration of how deceptive can be a sequence that looks random.[4] We will come back to the thorny problem of randomness in Chap. 6.

## 1.2  The Passwords of Statistics

### Statistical Samples

Our simple "mathematical experiment" allows us to introduce some key concepts of statistics on which we will try to build the quantitative description of the data. First off, doing statistics means dealing with a quantity $S$ that can take a certain number of values, and a statistical test consists in determining how often $S$ takes each of the possible values. In the example we considered, the statistical quantity is the digit that corresponds to each particular decimal in the sequence, which has possible values from 0 to 9. Unable to examine all the decimals of $\pi$, we only examined the first 10000 digits. In a real test, we are indeed bounded to consider only a *statistical sample*, namely, a limited collection of objects of any nature for which we determine the value of $S$.

For example, if the quantity $S$ of interest were the length of the nose of individuals, a statistical sample could be represented by the first hundred people we meet after

---

[3] And erroneously, since the expansion of $\arctan(x)$ was already known to the Indian mathematician Madhava of Sangamagrama in the 14th–15th century, and was then independently rediscovered by James Gregory in 1671.

[4] The comment by Salviati that follows, fully endorsed by Sagredo, the third character of Galileo's dialogues, is stimulating too:

SALVIATI: …It is exactly in this manner that laws of nature are discovered. Nature presents us with a host of phenomena which appear mostly as chaotic randomness until we select some significant events, and abstract from their particular, irrelevant circumstances so that they become idealized. Only then can they exhibit their true structure in full splendor.

SAGREDO: This is a marvelous idea! It suggests that when we try to understand nature, we should look at the phenomena as if they were messages to be understood. Except that each message appears to be random until we establish a code to read it.

leaving home in the morning. Or the sample could be made up of the molecules that escape in a fixed time interval from a small hole in a container filled with gas, and the statistical quantity the speed of the individual molecules that we measure with some technique.

## Population

The examination of the sample of $\pi$ we considered had however the purpose of drawing conclusions about the *entire* sequence of decimals. It is often useful to think of our statistical sample as a subset of what we will call *population*. The concept of population has a very concrete meaning both in the case of measuring the length of the nose (for example, all the people living in our neighborhood, or in the city, or in the whole planet), and in that of determining molecular speeds (the set of gas molecules inside a large container).

Yet, it is not always like this. For example, when we analyze the accuracy of a series of experimental measurements, the population will be only an abstract concept, which refers to an in principle unlimited repetition of the same experiment. In reality, we always deal with statistical samples. Nevertheless, the sample versus population distinction we introduced is convenient anyway, because it allows us to operationally separate a first phase of *description* of the sample data, followed by the *development of a model* of the population and a final phase of *comparison* between data and predictions.

The relationship between a sample and the population wherefrom it is extracted is the real nightmare of those who deal with statistics applied to social and economic sciences. First, we have already seen that a sample, in order to provide significant information about the population, must be as extensive as possible. A story that circulated at the beginning of the last century is that at Harvard University, one in three female students married a professor. Which was true. The only thing that was not specified is that the data referred to an academic year in which the number of women enrolled at Harvard was equal to three. This may seem like just a joke, but remember it when you read in some newspaper that one in five people has breakfast with Simpsons Donuts.

The main question, however, is whether the sample is a fair representation of the population. For example, suppose that I want to conduct a survey on how young people spend their holidays, and that to do this I email a detailed questionnaire to a number of youngsters chosen at random. Would I get an accurate picture of youth preferences? Obviously not, since the method I used to conduct the survey has the effect of excluding a subset of the population, those youngsters who use their mobile mainly to post on socials and consider emails just as stuff for old junk like me, that is likely not that small. This is, of course, an extreme example, and those who deal with surveys do not certainly make such mistakes (unless they do it on purpose to be able to reach some desired conclusion). But selecting a sample randomly is certainly the hardest problem in experimental statistics.

If you think you are on the safe side because you deal with "exact" sciences, you are deadly wrong. Later, we will analyze in depth how the number of data in a sample affects the statistical conclusions we draw, but the problem of the representativeness of the sample is present in experimental physics too. An example that immediately comes to mind, since it is related to things I usually do in my laboratory, is determining how the sizes of small particles dispersed in a fluid are distributed, for example water droplets suspended in the air, a meteorological situation well known in the region where I leave commonly called fog. A very efficient technique for doing this is to send a beam of light through the dispersing medium and analyze the properties of the light scattered by the particles. The trouble is that the amount of scattered light grows much faster than the radius $R$ of the particle—as $R^6$, for small particles. The presence of tiny particles is then masked by the predominant contribution to the intensity of the scattered light from the larger ones. Hence, this experimental technique tends to favor the detection of the latter, and if you do not take this into account you may obtain a completely wrong distribution of radii.

There are even thornier situations. The issue of sample representativeness is for example a fundamental problem in cosmology. Many of the conclusions that can be drawn for this strange science, which has the serious problem of being able to analyze a *single* experiment—precisely the real Universe, among the many imaginable universes—are based on the so-called "hypothesis of large-scale homogeneity", that is, on the fact that the statistical properties of the objects (the sample) that we observe in the region of the Cosmos (astronomically) close to our Galaxy reflect those of any region chosen at random in the Universe.

A second no less relevant problem lies in the *way* we ask questions. For example, suppose you wish to decide whether it is safer to travel by car or by plane. What would you compare? The number of plane crashes per year with the number of road accidents in the same period? Or the number of people killed in plane or road accidents compared to the number of people transported? Or again, the number of people killed per unit distance traveled either by plane or by car? As you can see, it is not immediately clear which is the correct question, or rather each answer is meaningful only in relation to the question we asked. The trouble is that many statistical statements we find in newspapers make no reference to how questions were asked.

More generally, what we are trying to do is extracting indirect information about a statistical quantity (for example, safety in travel) from the measurement of *another* quantity—the percentage of accidents in a certain period, or per unit of travel. The question is well posed only if there is a precise functional dependence between these quantities, and not just a certain more or less vague relationship, relying on subjective interpretation. Actually, the *indirect* determination of physical quantities is extremely common, and it will therefore be our task to thoroughly analyze the problem.

**Relative Frequencies**

When we talk about values of the quantity $S$, we do not necessarily refer to numbers. If, for example, we draw a sample from a box containing balls of different colors, and the quantity we consider is the color of the drawn ball, the values of $S$ are colors like red, or blue, or green. Very often, however, it is possible to associate a numerical value with each of the different results of a measurement of $S$. Thus, the length of the nose or the magnitude of the molecular velocities are statistical quantities that can be assumed to take in principle any numerical value in the interval $[0, +\infty)$, at least if we disregard the theory of relativity and some biological problems.

Luckily, in physics we almost always deal with quantities to which we can associate numbers, but we must make an important distinction about the set of values that $S$ can take. As for the description of the data, the simplest case is that of quantities that can only take a *finite* number of values, like the ten digits in the case of the sequence of decimals of $\pi$. Slightly different is the case of quantities that can only take *discrete* values but, at least in principle, an infinite number of them, such as the number of stars $N$ that make up a star cluster. Actually, there is a physical limit to the maximum size of a cluster, and saying that four stars in a row make up a cluster is somewhat arbitrary. But the range of values is so wide that in practice it is convenient to think of $N$ as a quantity that can take any integer value. Since we always analyze a finite number of data, in this case most of these values will not be represented in our sample. Finally, the most delicate (and most common) situation is that of quantities that can be assumed to take a *continuous* set of values, for example an entire interval of the real axis, as in the case of the length of the nose or molecular velocities. The problem in this case is that it is not possible to label the individual values assumed by $S$.

**Remark 1.1** Before we proceed, I have to give a crucial warning. In the last paragraph, I write "quantities that can be *assumed* to take", and not "quantities that *take*", on purpose. Indeed, I am willing to make the following statement, regardless of how disturbing or shocking it may sound,

Nothing is truly continuous!

Or, at least, no continuous quantity or property really exists in the physical world.[5] Even in classical physics assuming continuity down to the lowest limit leads to baffling problems (for example, a string that is truly continuous would host waves that are so high-frequency as to cause an energy catastrophe), and one of the main messages of quantum mechanics is that (at least in a finite volume) all physical quantities are discrete and have a lower bound.

Of course, no physicist in their right mind would ever give up treating properties and quantities as continuous, when possible. Calculus gave a tremendous boost to the development of science, and no field of physics can be investigated without the power of differential equations. But this is just a matter of convenience. In the following, we will see that adhering to a strict assumption of continuity can similarly lead to delicate issues in the definition of probability, misunderstandings in its interpretation, baffling questions about important concepts such as statistical entropy. So, never forget that continuity is just an extremely valuable illusion.

After my little warning, let us go back to our discussion. We have seen that the best way to analyze how often a quantity $S$, which can take a finite number of values, takes a particular value is to relate it to the total number of cases examined, that

---

[5] Mathematicians may have a different opinion, for most of them have a different concept of existence (which is perfectly legitimate).

is, to the size of the sample. Let us then consider a statistical sample consisting of $N$ elements, and label with an index $k = 1, 2, \ldots, r$ the values that the statistical variable $S$ we are analyzing can assume. If $n_k$ is the number of elements of the sample for which the $k$th value of $S$ is found, we will call *relative frequency* $f_k$ the quantity

$$f_k = \frac{n_k}{N}. \tag{1.1}$$

Note that the sum of the relative frequencies over all possible $r$ values for $S$ is always equal to one,

$$\sum_{k=1}^{r} f_k = \frac{1}{N} \sum_{k=1}^{r} n_k = 1. \tag{1.2}$$

Very often in statistics $n_k$ is also called the frequency of the value $k$. However, since, as we will see, $f_k$ plays a much more important role than $n_k$, to avoid confusion we shall not use this naming convention. So, even when we refer to $f_k$ simply as "frequencies", we always mean relative frequencies. By allowing $k$ to take any integer value, we can talk about relative frequencies even for quantities that admit an infinite but countable number of values. Of course, in this case, most of the relative frequencies for an experimental sample will be zero. We will deal with continuous-valued quantities later.

Sample, population, frequency are then the passwords that will allow us to enter the world of statistical description. But the last and most important keyword, which will grant us access to the quantitative analysis of statistical data, is the subject of the next paragraph.


## 1.3  Frequency Distributions

Let us go back to our sequence of decimals of $\pi$. So far we limited ourselves to considering only the behavior of each individual digit, with the implicit conviction that "one digit is worth the other", that is, that all digits are somehow equivalent. In fact, it is of little use to compare the behavior of different digits, since, as the size of the sample of decimals examined increases, all frequencies tend to flatten on a constant value equal to 0.1. But the case we were considering is trivial. Actually, what interests us most is analyzing how the relative frequency varies as a function of the value assumed by the statistical quantity $S$, which we can do by making a graph that has the values assumed by $S$ on the $x$-axis and the relative frequencies on the $y$-axis. Such a graph is called the *frequency distribution* of the investigated quantity for the sample we are analyzing. Of course, the concept of frequency distribution truly makes sense for statistical quantities with numerical values, as clarified in the following example.

**Example 1.2** (*In the footsteps of E. A. Poe*)  As a first example of a frequency
distribution, let us consider statistical data related to a quantity to which *no* numerical
value is associated. Some of you may have read a magnificent short novel by E. A. Poe
entitled *The Gold-Bug*. In the story, the protagonist William Legrand manages to find
the place where a treasure is hidden from a message written in mysterious characters
on the back of a beetle-shaped object. The technique he follows is to associate these
characters with the letters of the alphabet, and compare the frequencies with which
each character appears in the message with those of the distribution of letters in the
English language.[6]

To decipher a secret message encrypted in this simple way, the first step is to
evaluate the distribution of individual letters in a written text. Figure 1.4 compares the
relative frequencies of the letters in all the English words with their values obtained
by considering only those words (about 9000) made of only 5 letters.[7] The graph
allows some interesting difference to be spotted, such as the significant increase of
the frequency of the $s$'s likely due to the weight of plurals, also included in the word
set, a corresponding decrease of the frequency of $t$'s, or the different values of the
vowels in the two distributions.

But what is the actual use of a graph like the one in the figure, if not to make it
easier to compare at a glance the frequencies of the various letters? The particular
distribution of values along the horizontal axis depends only on having chosen the
alphabetical order to arrange the data, and would have been completely different
had we changed the arrangement criterion. The *shape* of the distribution is therefore
completely arbitrary, and this is precisely because we do not have a numerical criterion for ordering the values. A table might have been less immediate to read, but it
would have had the same information content.

## Ranking

As a matter of fact, a simple criterion to arrange the distribution of letters does exist,
and consists in *ranking* them according to the value of their frequency, as shown in
the inset of Fig. 1.4. Ranking surely makes some features of the distribution clearer.
For example, besides the impressive score ($f > 0.12$) of the letter $e$, we can identify
a group of "dominant" letters ($t, a, o, i, n, s, h, r$) that score between 6 and 10%,
while the almost all the other letters appear with a frequency below 0.03.

Yet, the ranking of a series of values is not at all equivalent to a distribution with
numerical values, since no linear one-to-one correspondence exist between the ranks
and the integers. For instance Bayern Munich won 32 times the German football
*Bundesliga*, followed by Borussia Dortmund and Borussia Mönchengladbach, equal

---

[6] Actually, since the message is short (namely, the sample is small), to solve the puzzle he needs a
finer analysis and some tricks. Read the story to find out more.

[7] The latter are of particular interest for fans of Scrabble or (like me) of the word game *Wordle*©
published daily by the New York Times.

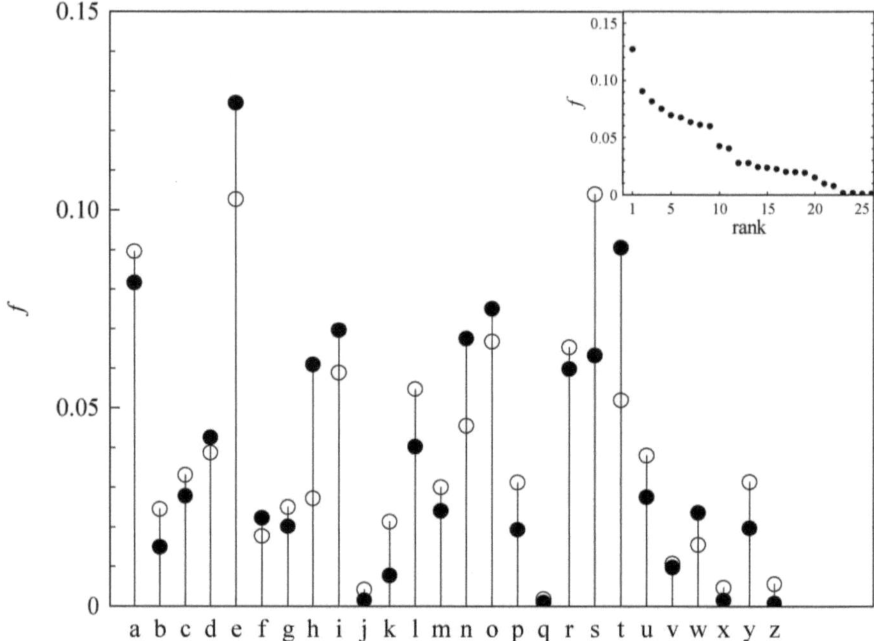


**Fig. 1.4** Comparison between the distribution of relative frequencies of the letters in English (•) and the same distribution, but limited to words of only 5 letters (o). The inset shows the data (for words of any length) ranked according to their score

winners of 5 titles. Evidently, just stating that Bayern ranks first and the two Borussias second on equal footing does not catch at all the disproportionate score of Bayern.[8]

Although it is often the sole approach to deal with several topics in economics or social sciences ranking is a nontrivial task, which sometimes can be ambiguous. A first requirement for a ranking to be unquestionable is transitivity, namely that when A ranks better than B, and B ranks better than C, A ranks better than C. Suppose however you have to choose what shape you should form with your hand in a game of Rock-Paper-Scissors. How would you rank each of these choices? Similarly, you may choose to have lasagna rather than a cheeseburger and a lobster rather than lasagna, but a cheeseburger rather than a lobster. It just depends on how you balance gluttony and parsimony when you compare these entries.[9]

---

[8] Something similar holds for the Italian Serie A, but you will never convince a loyal supporter of Inter FC like me to be more explicit….

[9] Individual intransitivity without incoherence may arise when there is more then one criterion for choosing between different options. For instance, suppose that both your liking ($L$) and the price ($P$) of cheeseburger, lasagna, and lobster are in the ratio 1:2:4. You can get a "circular intransitivity" of the kind we discussed if you compare any two of these options ($a$, $b$) by choosing that $a$ is better then $b$ if $[L(b) - L(a)] + 0.4[P(b) - P(a)]^2 > 0$, i.e., by attributing a super-linear (quadratic) weight to prices.

Even when individual choices are transitive, *collective* evaluations may be not. For example, suppose that three young students, Clair, Ellen, and Lucy apply for a PhD position in quantum optics and are evaluated by a committee of three professors named Roy, Arthur, and Donna. The panel members give these rankings:

|   | Roy | Arthur | Donna |
|---|-----|--------|-------|
| 1 | Clair | Ellen | Lucy |
| 2 | Ellen | Lucy | Clair |
| 3 | Lucy | Clair | Ellen |

Clearly, there is no way to select one of the candidates, either by the number of top positions (one each), or by contrasting them in pairs (two of three members prefer Clair to Ellen, but the same is true for Ellen vs. Lucy and Lucy vs. Clair). Even if the committee determines to compare the sum of the individual rankings, they all get 6 points.

But lack of transitivity is not the only hindrance to fair ranking. Another natural requirement is that, if we have ranked a number of options and we add a new one, the new option does not change the *relative* ranking between the others. Thus, if I prefer apricots to cherries, my preference will not overturn if I consider apples too, regardless of how I rank apples in comparison to other fruit. Once again, however, this is not a foregone conclusion for *group* ranking. For example, suppose that Snow White set up a prize for the Top Physicist of All Times. Taking into consideration all the denizens of the Physicists Heaven,[10] the award jury, composed by the Seven Dwarfs Doc (Dc), Grumpy (Gr), Sleepy (Sl), Bashful(Bf), Happy (Hp), Sneezy (Sn), and Dopey (Dp), preselects three guys named James (J), Isaac (I), and Albert (A). Having evaluated their CVs, the members of the jury rank them as follows

|   | DC | Gr | Sl | Bf | Hp | Sn | Dp |
|---|----|----|----|----|----|----|----|
| 1 | I | I | A | J | I | A | I |
| 2 | J | A | J | A | J | J | J |
| 3 | A | J | I | I | A | I | A |

and decide that the winner is the candidate with the least overall ranking, namely Isaac, whose positions in the ranking sum up to 13, while both James and Albert follow close behind with 14 and 15 points, respectively.

However, Snow White reprimands the jury, maintaining that they should have also considered a guy named Michael (M). Complying with the directive of the Boss, the Seven Dwarfs carefully deem this additional candidate, judging however

----

[10] The Physicists Heaven is a place where the Higgs field never existed.

that Michael may have some ability in the abstract, but is definitely not suited to theoretical reasoning. So, their final ranking is the following

|   | DC | Gr | Sl | Bf | Hp | Sn | Dp |
|---|----|----|----|----|----|----|----|
| 1 | I | I | A | J | I | A | I |
| 2 | J | A | J | A | J | J | J |
| 3 | M | M | M | M | M | M | M |
| 4 | A | J | I | I | A | I | A |

However, now it is *James* who leads and wins with 15 points, while Isaac trails with 16.[11] This fictional story is an illustration of the consequences of an "impossibility" theorem due Kenneth Arrow, 1972 Nobel laureate in economic sciences, stating that a ranking system that always satisfies a few apparently unassailable criteria, including that no new irrelevant option can upset the ranking, does not exist. Arrow's theorem has obvious consequences in attempts to devise a fair voting systems.[12]

**Example 1.3** (*Qualifying for university*)  As a second example of frequency distribution, consider instead the results of the admission test to the Italian Schools of Engineering.[13] Figure 1.5 shows the frequency distributions of the number of correct answers for the 30 multiple choice questions in mathematics, physics and chemistry included in the test. Consider first the results obtained in 2016 and 2018, which show a remarkably similar distribution. At variance with Example 1.2, we can pick up some specific features of these distributions:

1. Both distributions have a *maximum* value $f_k^{max} \simeq 0.066$ which is obtained for 9-10 correct answers approximately.
2. They have a certain *width*. A first way to estimate this width is by finding which values of $k$ have a frequency greater than $f_k^{max}/2$, which yields for both distributions an interval of values roughly ranging between 4 and 19 correct answers.
3. The distributions are however asymmetrical, showing a longer "tail" towards high values. This suggests that a student taken at random typically tends to answer correctly to a number of questions a bit *larger* than 10. Thus, $f_k^{max}$ may not be the best parameter to characterize the "typical value" of a distribution.

The distribution for the year 2022, however, shows substantial differences with the other two. Specifically, its maximum is definitely larger and the distribution is wider, flatter, and almost symmetric. At first glance, we can guess two different explanations for this rather striking change:

Guess 1 →    The Covid-19 pandemic (or the vaccination against it) has increased the students' skill in math and basic science.

---

[11] This is not the only case in the history of physics when a guy named James owes a lot to some Michael.

[12] A brilliant discussion of intransitivity in ranking and on the consequences of Arrow's theorem can be found in the splendid book H. W. Lewis mentioned in the selected bibliography.

[13] I thank CISIA, the University Consortium for the Admission to the Italian Universities, for the kind concession of the data.

Guess 2$\rightarrow$    Something in the test has changed.

Admittedly, the first guess sounds rather improbable—vaccines are great to save your life, but no brain booster has been put on the market so far—so we better lean towards the second one. Which is indeed the correct answer. Indeed, maybe because the committee that drew up the test feared that the long lockdown periods may have *weakened* the skill of the applicants, in 2022 the test has been recalibrated to be conceivably easier.[14]

The example we have just examined allows us to introduce the important concept of *cumulative distribution* $F_k$ as the sum of all relative frequencies $f_j$ up to the value $j = k$

$$F_k = \sum_{j=1}^{k} f_j, \tag{1.3}$$

which monotonically increases from 0 to 1 when $k$ increases from its minimum to its maximum value. The cumulative distributions for the number of correct answers, plotted in the inset of Fig. 1.5, show that for all three years the fraction of students scoring at least half of the maximum value is about 80%.

### 1.3.1  Histograms

So far we have considered statistical quantities with a discrete and finite set of possible values. But what can we say when we consider properties that can take a continuous set of values, such as the height of individuals? No matter how large the sample, and even assuming that we can measure height with arbitrary precision, it is quite hard finding someone who is *exactly* 6 feet high. Indeed, any statistical sample we may consider consists of a finite amount of experimental measures scattered in a continuum of possible values. Thus, as the sampled data increases, the experimental frequencies for values which are just points on the real axis become smaller and smaller. All we can do is splitting the overall range of possible values of the continuous variable $X$[15] into subintervals of a given width and collect together the data that fall within each subinterval. Of course, the number of individuals included in a given interval will increase as the width of the subintervals gets larger. We can then picture the data through a *histogram*,[16] constructed as follows:

---

[14] Was this a good move? You should first know that this is a *comparative* test. Namely, it does not matter that much your *absolute* score but rather your *relative* position in the final ranking of all participants. Hence, shifting the whole distribution to higher vales of $k$ does not mean at all providing a better selection.

[15] It is often convenient to indicate a statistical *variable X* with a capital letter, reserving lower-case letters for its *values* $x_i$.

[16] Note that a histogram is not a simple bar chart, which is just a pictorial representation of a *discrete* variable using bars to compare different categories of data, where bars do not need to be adjacent.

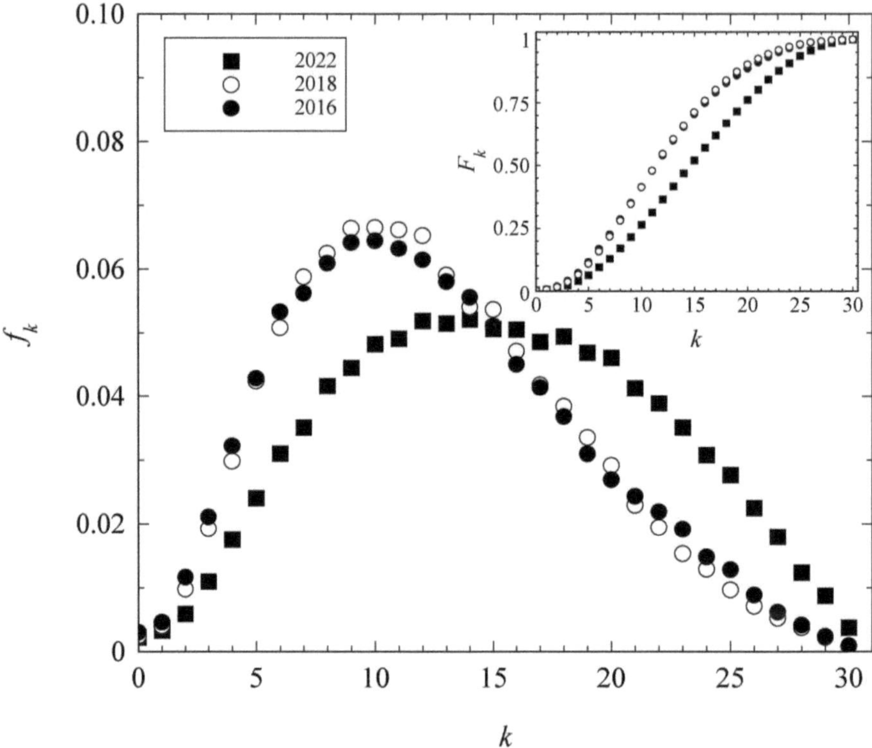



**Fig. 1.5** Number of correct answers in the math and science sections of the CISIA admission test to the Italian Schools of Engineering for the three years shown in the legend. The corresponding cumulative distributions are plotted in the inset

1. We fix the overall range $L$ of values to be considered as the difference between the maximum and minimum value obtained for the sample, and divide it into $r$ subintervals of width $\ell = L/r$.
2. We evaluate the number $n_k$ of data falling within the interval of values $(k-1)\ell \leq x < kl$, with $k = 1 \ldots r$.[17]
3. We arrange on the $x$-axis rectangles of base $\ell$ and height $n_k/\ell$ centered at $x_k = (k-1/2)\ell$.

We can also introduce a normalized *frequency histogram* by attributing to all those values of the variable $x \in [(k-1)\ell, kl)$ the same *normalized* relative frequency

$$f(x) = \frac{n_k}{N\ell}$$

---

[17] Considering half-open intervals avoids counting twice the data that lies at the ends of the subintervals.

namely, the relative frequency divided by the length of the interval and attributed to the value $x = k - 1/2)\ell$ in the center of that interval. It is easy to show that the total area under the rectangles of a frequency histogram is then always equal to one.

Sometimes, however, the number of data that falls within a certain interval can span a wide range. In this case, it helps varying the width of the intervals so that the number of data $n_k$ that fall within each interval $\ell_k$ is comparable, retaining anyway the condition $\sum_{k=1}^{r} \ell_k = L$. A similar problem can actually arise for variables taking discrete values too when the sample size is not very large compared to the number of possible values. For example, the extraction of a number in the Italian bingo ("tombola") can take 90 values. To get an idea of the distribution of results with a sample of only 100 extractions it may be convenient to collect the data in intervals from 1 to 10, from 10 to 20 and so on, and draw a histogram. Here, however, the choice is only practical, we can very well calculate the frequencies for each individual number, even if approximately 1/3 of these, as we shall see, will usually vanish.

There is a certain degree of arbitrariness in drawing a histogram, since its shape depends in part on the width we choose for the subintervals. Choosing wide subintervals a regular but less detailed trend is obtained, while narrower intervals emphasize details at the expense of regularity. Sorting out details that have a real meaning from those that just reflect noise due the small sample size is not trivial, so there is no golden rule for choosing the width of the subintervals. The optimal number of subintervals, however, grows significantly slower than the total number of data, approximately as $N^{1/3}$. As a rule, therefore, for samples with a a size between a few dozen and a few thousand data the sensible number of subintervals varies only between 5 and 20.

Deep down, however, a histogram just allows us to get an overview of the distribution of the data. For a quantitative analysis, there is actually no need of data binning, an action that actually erases many details thus throwing away information. Therefore, histograms are mainly a graphic aid, as the following demographic statistics clarify.

**Example 1.4** (*The height of our great grandfathers*)  As a first example of a histogram, let us consider the distribution of the height of people. We may expect accurate data on this anthropometric parameter of high social and economic value to be easily available, but I conversely found this task rather exhausting. Fortunately, at least as far as Italian males are concerned, we have at our disposal a surprising collection of data prepared for the Ministry of War by General Federico Torre, first Director General of the levy. Between 1860 and 1905, Torre collected with dedication the stature of over *twenty-one million* young Italians called to arms, obtaining tables of relative frequencies for a height between 125 and 199 cm divided into 1 cm intervals. Figure 1.6 shows Torre's original data relating to the 1900 conscription class, corresponding to about half a million conscripts.[18] As we can see, such a large and in first approximation homogeneous sample yields a very regular distribution,

---

[18] Torre's data have been carefully reanalyzed in B. A'Hearn, F. Peracchi and G. Vecchi, *Demography* **46**, 1 (2009). I am particularly grateful to Franco Peracchi for having provided me with both the original data and the authors' analysis.

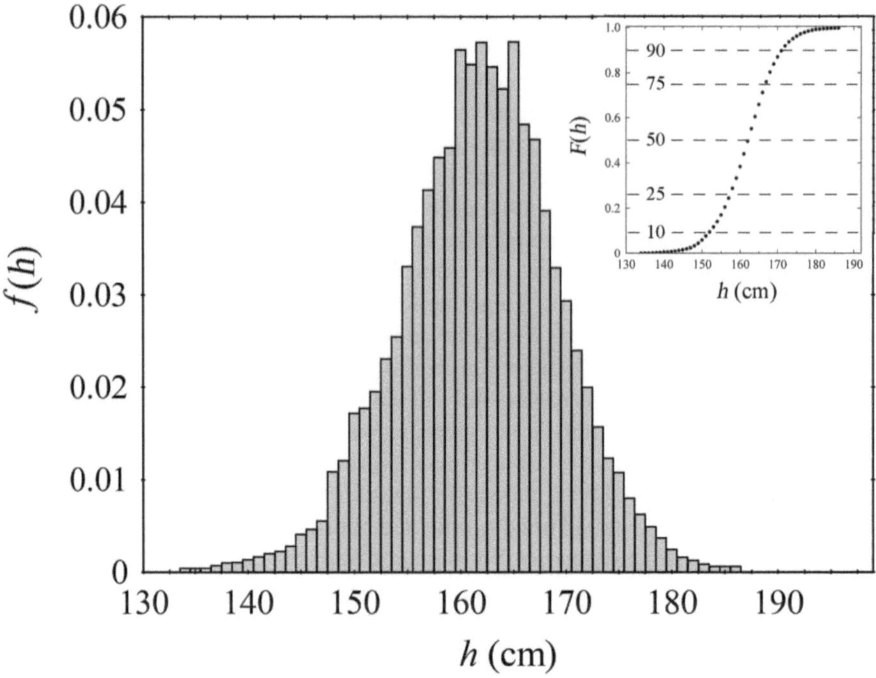

**Fig. 1.6** Distribution $f(h)$ of the height of Italian soldiers for the 1900 conscription class. The cumulative distribution $F(h)$ is shown in the inset together with the upper limits for five selected cumulants

with a typical "bell-shaped" form that we will encounter again. In particular, the distribution is remarkably symmetrical with respect to the maximum.

The figure inset shows the cumulative distribution, together with the values of few selected "percentiles", i.e., the scores below which a given percentage of the investigated subjects fall. So, for example the 75th percentile is the height (about 166 cm) below which 75% of the 1900 conscripts lie.

In Chap. 4 we will discover that the remarkable symmetry of the height distribution originates from a fundamental general result, and is not necessarily shared by other anthropometric quantities, for example the adults' body weight. Finding detailed data for the distribution of the body weight is even harder than for stature, firstly because the available data usually refer to a relatively small sample, and secondly because these data, like for most other anthropometric quantities, are usually reported as tables of percentiles. The latter are surely very useful for pediatricians and nutritionists, but not so much for our purposes. The values for the percentiles of the distribution of weight of about 5000 male adults in the US,[19] plotted in the inset of Fig. 1.7, show for example that half of the subjects weigh less or equal to 87.5 kg, while only about 15% of them exceed 110 kg. By considering differences in percentiles it is easy to obtain

---

[19] CDC Vital and Health Statistics, Series 3, Number 46 (2021).

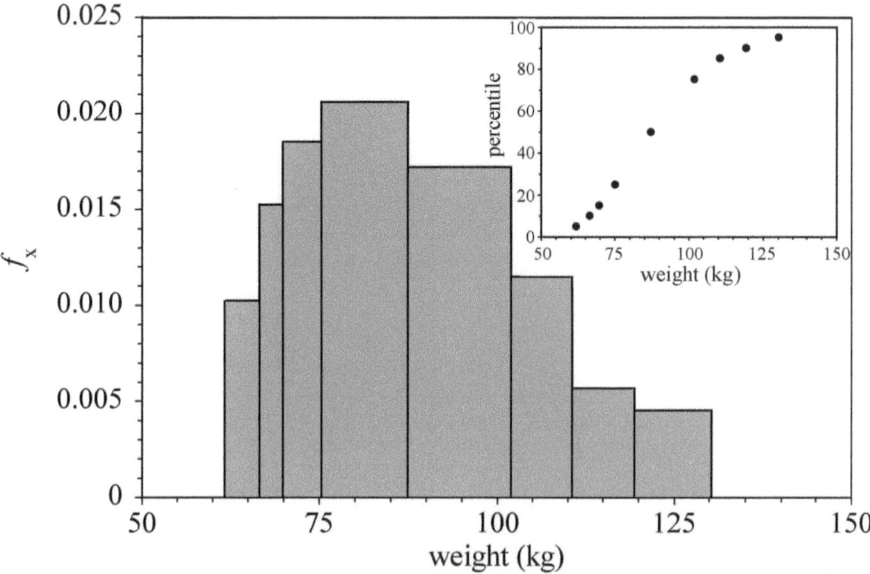

**Fig. 1.7** Distribution of the body weight of a sample of US males with an age $\geq 20$ years. The inset shows percentile values

an histogram, which however can be plotted only for 90% of the sample (4577 tested subjects), since neither a lower limit for the weight of the 5% lighter men, nor an upper limit for the heavier 5% can be fixed. Nevertheless, the histogram, displayed in the figure body, clearly show that the distribution is asymmetric, with a rather pronounced tail for large values of the weight (a dead giveaway of the problem of obesity in rich countries).

**Example 1.5** (*An ageing country*) Most western countries are ageing, and Italy is a top player in this game. Figure 1.8 takes advantage of a extensively used graphical illustration of the distribution of a population, dubbed the "population pyramid" to highlight the age distribution changes of Italian residents that took place in the past 20 years.[20] Clearly, from 2002 to 2022 the whole histogram has shifted to higher values. Indeed, inset A shows that in the same period the average population age has increases almost linearly from about 42 to almost 47 years. Inset B shows that, for what concerns birth rate, Italy actually ranks very bad among the world countries (223/228). Notice that the latter histogram is much more readable if one chooses to use bars of different width.

---

[20] The population pyramid usually made by comparing the distribution of male and females for the same year. Here I adapted it to compare the total population for two year to avoid the unavoidable mess deriving from plotting two sets of bars on the same $x$-axis, flattering at the same time my gender–sensitive readers.

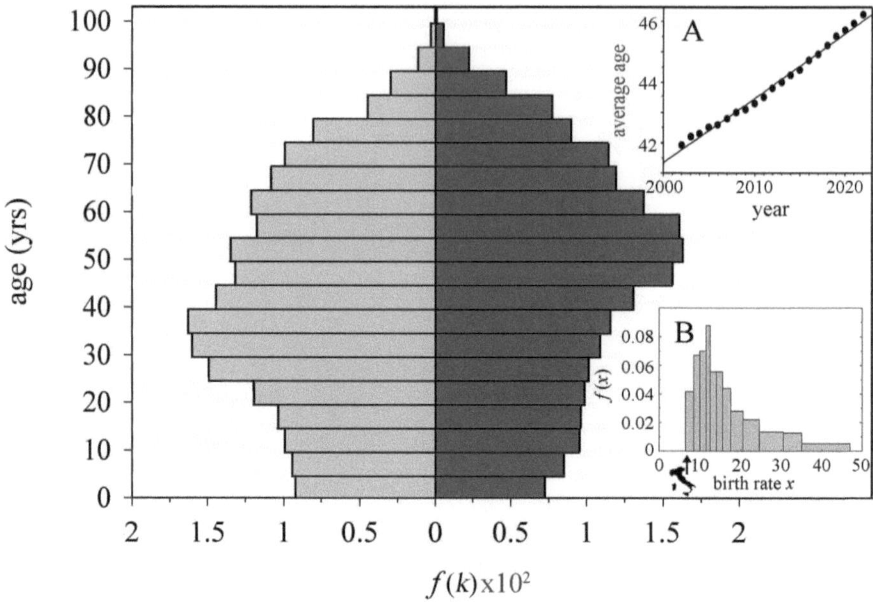


**Fig. 1.8** Body: Comparison of the Italian population pyramids for the years 2002 (light grey, left) and 2022 (dark grey, right). Inset A shows the almost linear increase in the population average in the past 20 years (ISTAT, Italian National Institute of Statistics). Inset B shows an histogram for the annual birth rate per 1000 inhabitants for 228 world countries. (*Source* CIA, Central Intelligence Agency, *The World Factbook*)

The aging of the Italian population has a precise origin. Figure 1.9 compares the specific fertility rate, i.e., the ratio of annual births to Italian women at a given age to the population of women of the same age for the years 1955, 2002, and 2022. Here, the total area under the histograms[21] is the number of babies born in a certain year compared to the total female population, i.e., the average number $\bar{n}$ of children per woman, which decreases from $\bar{n} = 1.27$ to $\bar{n} = 1.1.24$ even in the limited two-decades range considered in Fig. 1.8. A comparison with fertility rate at the dawn of the era of the "boomers" I belong to, when $\bar{n} = 2.33$, strikingly justify the drastic ageing of the Italian population we discussed.

The comparison also shows a significant shift towards older ages in the distribution, whose maximum increases from about 26 years in 1955 to 30 years in 2002, to 32 years in 2022. Thus, not only Italian women make fewer children, but they also make them when they are older. While the former feature may be regarded as a typical trend of all wealthy countries, the latter nozzle a clear alarm about the status

---

[21] The histograms are shown here just by symbols placed in the middle of the corresponding intervals, replacing the standard bars with a simpler graphical display that we will often use in what follows.

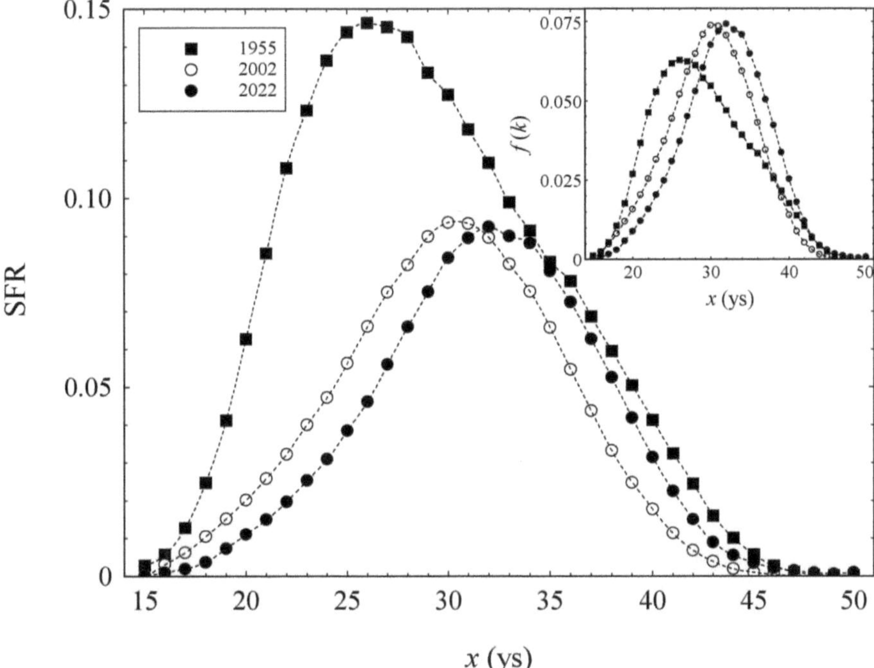



**Fig. 1.9** Specific fertility rates (SFR) for the years 1955, 2002, and 2022, obtained by relating for each woman of childbearing age of age $x$ the number of live births to the amount of female population of the same age. The insert shows the frequency distributions obtained by normalizing the area subtended by each curve. (*Source Italian National Institute of Statistics,* https://demo.istat.it/)

of welfare in my country. No attempt to increase the birth rate can work without adequate investments in policies to support the young mothers.

To correctly compare the shape of the three distributions, it is however better to scale each of them by the area it subtends, so to obtain frequency distributions. In the figure inset we can for example appreciate how, compared to 1955, the frequency distributions for 2002 and 2022 look narrower. Moreover, while in 1955 the distribution showed a significant tail towards older ages, nowadays the asymmetry of the curve is less pronounced and even inverted.

## 1.4  Outlining a Distribution

The description of the data of a statistical sample is therefore made by determining the distribution of relative frequencies, which implicitly contains all the information that we draw from the sample. Sometimes, however, we may be content of getting an idea of some general properties of a frequency distribution, such as what value it is centered around, or how wide it is, or whether it is symmetric or not. These properties are often referred to as the *characteristics* of a distribution. For small statistical samples, this is often all we can say about a distribution. Trying to give a finer description would be arbitrary, since the details of the distribution are generally only "accidents" due to the small number of data we consider. To do this, we can introduce parameters that quantitatively describe the characteristic features of a distribution.

### 1.4.1  Mean

Let us first try and find a parameter indicating the typical value around which a distribution is centered. We actually several options, like choosing for example the value $x_{max}$ for which there is a maximum of relative frequency, which is called in statistics the *mode* of the distribution, but we have seen that for an asymmetric distribution the value assumed more frequently does not coincide with what we have in mind for a typical value. A subtler reason that makes the mode a rather weak estimate is that its definition is based on the value of the distribution at a *single* point.

A sounder strategy is looking for a global parameter, related to the *entire* distribution of the data, like the value $x_m$ such that the total frequency with which $x < x_m$ is equal to the total frequency with which $x > x_m$, $f(x < x_m) = f(x > x_m) = 0.5$. This value that "splits in two" the distribution is called the *median*. If, for example, we were to analyze the distribution of incomes of a population, it is particularly useful to know that half of the individuals earns more than a certain amount, and the other half less. In this case, we are mainly interested in finding the "watershed" of the distribution, while we are relatively uninterested in knowing whether the upper band is all composed of people with an income just above the median, or whether among them there may be both small wealthy people and billionaires. The median can obviously be evaluated very easily, but its limit is that it separates the data into two equal groups without taking into account the values of the individual data.

To account more effectively for the specific distribution of data, let us ask ourselves if we can find a value $\bar{x}$, which we will call *mean* or *average value*, wherefrom the set of data deviates as much in excess as in deficit. Considering a sample of $N$ data $x_i$ ($i = 1...N$) of a statistical property x, we wish the sum of the deviations $\delta_i = x_i - \bar{x}$ with respect to the mean to vanish,

$$\sum_{i=1}^{N} \delta_i = \sum_{i=1}^{N} (x_i - \bar{x}) = \sum_{i=1}^{N} x_i - N\bar{x} = 0.$$

Thus the average value must be what in mathematics we are used to calling arithmetic mean,

$$\bar{x} = \frac{1}{N} \sum_{i=1}^{N} x_i, \tag{1.4}$$

Note that the mean does not in general coincide with either the median or the mode. For this to happen, the distribution must be symmetrical with respect to its maximum value, which in this case represents both the mode and, by symmetry, the median and the mean. The mean is not always the most sensible choice for the typical value of a distribution. For example, the presence of some billionaires can push the average income quite high. It would be a bit misleading, however, to say that this makes the population as a whole richer, or at least this statement should be taken with great care. As far as we are concerned, however, the mean will be the value we will assume as an indicator of the typical value of a distribution, even if the reason for introducing this "favoritism" with respect to the median or the mode will be clear only later.

If $X$ takes only discrete values, we can write the expression for the mean in terms of relative frequencies. Indeed

$$\bar{x} = \frac{1}{N} \sum_{j=1}^{r} n_j x_j = \sum_{j=1}^{r} f_j x_j \tag{1.5}$$

where $x_1, x_2, \ldots, x_r$ are the $r$ *values* that $X$ takes $n_1, n_2, \ldots, n_r$ times. This way of writing the mean allows, as we will see, a more direct comparison of the sample data with the parameters of a theoretical distribution for the entire population.

We would be tempted to do the same for a continuous variable, using the frequencies of the intervals of a histogram, but this does not work. In collecting the data in subintervals, we lose information about the actual position occupied by a data point within the subinterval. The mean calculated from the frequencies of a histogram will not therefore not coincide with that obtained directly, except when, for large samples, very narrow subintervals are chosen.

**Example 1.6** (*Back to* $\pi$) Let us examine the distribution of the frequencies of the digits $k$ in $\pi$ as a function of the number $N$ of considered decimals, given in table 1.2. Within the first 100 decimals the maximum frequency is obtained for $k = 9$, but this value jumps to $k = 1$ or $k = 5$ if we analyze 1000 or 10000 decimals, respectively. As you see, for such a "flat" distribution the mode makes no sense.

Evaluating the median is a bit more challenging. From the table we see that, if we consider for example 1000 decimals, the sum of all frequencies up to $k = 3$ is equal to 0.414, and to 0.507 if we add the frequencies up to $k = 4$. This just tells us that the median is between 3 and 4, and much closer to 4 than to 3, but does not provide us with a precise value. This problem arises every time we have to deal with discrete values. The best thing we can do is interpolating linearly between the two limits for $x_m$.

If we instead calculate the average value $\bar{k}$ of the distribution of digits using Eq. 1.5, we obtain $\bar{k} \simeq 4.77$ ($N = 100$), $\bar{k} \simeq 4.47$ ($N = 1000$) and $\bar{k} \simeq 4.49$ $N = 10000$). Hence, as $N$ increases, the mean approximates better the value

$$\bar{k}_{teo} = 0.1(0 + 1 + 2 + 3 + 4 + 5 + 6 + 7 + 8 + 9) = 4.5$$

which would be obtained if all digits were distributed with the same frequency $f_k = 0.1$.

### 1.4.2  Moments of a Distribution

In physics we often encounter quantities, called *moments*, defined as the product of a quantity times the distance from a point or an axis. For example, the position of the center of mass of a system is nothing more than the sum of the moments of the individual masses about the origin, divided by the total mass. Sometimes it is also convenient to introduce quantities that "weigh" the values of a quantity with the square of the distance from something. For example, we define the moment of inertia of a rigid body by weighing the individual mass elements with the square of the distance from an axis. An operation of this type introduces a kind of ranking between the masses, so that, for the same mass, a body has a greater moment of inertia if its mass is distributed more outward. The moment of inertia thus gives us additional information about the distribution of masses compared to that constituted by the position of the center of mass. To specify the fact that the weight we attribute to each mass is related to the square of a distance, we say that, while the position of the center of mass is related to a *first* moment of the mass distribution, the moment of inertia is calculated as a *second* moment.

We can exploit these ideas for a statistical distribution too by associating to each value $x_i$ taken by the variable $X$ a "mass" equal to the relative frequency with which that value appears. The total mass of this one-dimensional distribution is obviously unitary, since this is the value of the sum of the frequencies. What are the moments of the distribution about the origin? The position of the "center of mass" of the system, that is the first moment, is given by $x_{cm} = \sum_{j=1}^{r} f_j x_j$. But this is just the mean of the distribution, as much as the center of mass of a system is the point where the mass is concentrated on average. The mean is therefore the first moment of a frequency distribution with respect to the origin. We can now define an analogue of the moment of inertia, i.e., a second moment with about the origin, as the sum of the squares of the values taken by the quantity, multiplied by the frequencies associated with them, $\sum_{j=1}^{r} f_j$. This is equivalent to evaluating the average of the *square* of $X$,

$$\overline{x^2} = \sum_{j=1}^{r} f_j x_j^2 = \frac{\sum_{i=1}^{N} x_i^2}{N}, \tag{1.6}$$

where, we recall, the first summation is made on the $r$ *values* that the variable can take, while the second on the $N$ *data* of the sample. Writing $\overline{x^2}$, and *not* $\bar{x}^2$ is crucial, because the average of the square is generally greater than the squared average,

$$\overline{x^2} = \frac{\sum_{i=1}^{N} x_i^2}{N} \geq \left( \frac{\sum_{i=1}^{N} x_i}{N} \right)^2 = \bar{x}^2. \tag{1.7}$$

At this point nothing prevents us from going further, so by analogy we define the $k$th moment $M_0^k$ of a distribution about the origin the average of $x^k$,

$$M_0^k = \overline{x^k} = \sum_{j=1}^{r} f_j x_j^k = \frac{\sum_{i=1}^{N} x_i^k}{N}. \tag{1.8}$$

How much information do the moments yield about the distribution of $N$ experimental data? It is clear that knowing only the first, second, and perhaps third moment of a distribution, we have less information than that contained in the original data (in principle, to know everything about the distribution, we should know all the first $N$ moments). But the moments give us a completely different type of information, related to the whole distribution, not to a single point. Because of this the definition of the shape parameters of a distribution can be expressed in terms of moments, as we are going to see.

### *1.4.3  Standard Deviation and Skewness*

We look for a parameter that tells us how wide a distribution is, i.e., a quantity gauging to what extent the data deviates from the mean, the typical value of the distribution. However, a simple sum of the deviations from the mean vanish because positive and negative contributions exactly balance, as you can easily check. To avoid this inconvenience, we can instead consider the *squares* of the deviations with respect to the mean, which are certainly positive, or at most zero.

Physical quantities, however, have *dimensions*. The mean obviously has the same dimensions as the variable $X$, but the square of deviations have the same units of $X^2$. Thus, to have a quantity with the same dimensions as $X$, we introduce the *standard deviation $s_x$* as

$$s_x = \sqrt{\frac{\sum_{i=1}^{N} (x_i - \bar{x})^2}{N}} = \left[ \overline{(x - \bar{x})^2} \right]^{1/2}, \tag{1.9}$$

that is, as the root of the average of the squared deviations or *mean squared deviation*, which is therefore the parameter we will adopt to gauge the width of a distribution. Incidentally, we could have bypassed the problem of the alternating signs by considering the absolute values instead of the squares of the deviations, but the different choice we made, besides making calculations easier, will prove more opportune.

It is easy to see that for a discrete variable, the standard deviation can be written in terms of frequencies as

$$s_x = \left[ \sum_{j=1}^{r} f_j \left( x_j - \bar{x} \right)^2 \right]^{1/2}. \tag{1.10}$$

From the definition of standard deviation, we have

$$s_x^2 = \frac{1}{N} \sum_{i=1}^{N} \left( x_i^2 - 2\bar{x}x + \bar{x}^2 \right) = \frac{1}{N} \left[ \sum_{i=1}^{N} x_i^2 - 2\bar{x} \sum_{i=1}^{N} x_i + N\bar{x}^2 \right] = \overline{x^2} - 2\bar{x}^2 + \bar{x}^2$$

and therefore

$$s_x^2 = \overline{x^2} - \bar{x}^2, \tag{1.11}$$

which tells us that the square of the standard deviation is also the difference between the second moment and the square of the first moment (with respect to the origin). A distribution with a high average value will generally have a greater standard deviation than a distribution of similar shape, but with a lower mean. However, it is often useful comparing the shape of two distributions regardless of the absolute numerical values that the two statistical variables assume. More than the absolute width of a distribution, it is then useful to estimate its width in relation to the average value. To this aim, we will use the *relative* standard deviation, given by $s_x/\bar{x}$. While the standard deviation has the same dimensions as $x$ and therefore depends on the units of measurement we choose, the relative standard deviation has the advantage of being *dimensionless*. For example, the absolute and relative standard deviations of the distribution of the digits of $\pi$ for the values of $N$ used in Example 1.6 are given by

| $N$ | $s_k$ | $s_k/\bar{k}$ |
|---|---|---|
| 100 | 2.92 | 0.619 |
| 1000 | 2.90 | 0.649 |
| 10000 | 2.86 | 0.637 |

Note that the standard deviation is still a second moment of the distribution, but obtained by taking the *mean* as the origin. Why do we need to consider a moment with respect to the mean? If the average value is high, $\overline{x^2}$ is likely to be quite large, but this is just because the whole distribution is strongly shifted from the origin and has nothing to do with its width. We can then read Eq. (1.11) as a correction that removes the spurious contribution related to the value of the average.[22]

---

[22] What we have said has a mechanical equivalent in the Steiner Theorem, according to which the moment of inertia about a reference axis can be separated into the sum of the moment about an axis passing through the center of mass, a proper contribution of the considered body, plus a "transport" term, which depends only on where we have fixed the reference axis.

Our definition of $s_x$ actually has a little problem. It obviously makes no sense talking about the width of the distribution if we consider a statistical sample consisting of a *single* value $x_1$. Yet, according to (1.10), $s_x$ would instead vanish, rather suggesting that the distribution is infinitely narrow! We will see in Chap. 5 that there are good reasons to slightly modify Eq. (1.10) through a corrective factor that, in addition to having a precise theoretical meaning, makes in this case $s_x$ completely undetermined.

If for a very large number $N$ all the relative frequencies became about 0.1, we would expect an absolute standard deviation

$$
s_k = \sqrt{\sum_{k=0}^{9} 0.1 \left(k - \frac{9}{2}\right)^2} \simeq 2.87
$$

and a relative standard deviation $s_k/\overline{k} \simeq 0.638$. So, as the sample size increases, these results still seem to support the hypothesis of a uniform distribution of digits.

By extending the ideas we have just developed, we can define the moments of a distribution with respect to any value $x_0$ as

$$
M_k\left(x_0\right) = \overline{(x - x_0)^k} = \frac{1}{N} \sum_{i=1}^{N} (x_i - x_0)^k = \sum_{j=1}^{r} f_j\left(x_j - x_0\right)^k \tag{1.12}
$$

and in particular the moments about the mean, or *central moments*:

$$
M_k\left(\bar{x}\right) = \overline{(x - \bar{x})^k} = \frac{1}{N} \sum_{i=1}^{N} (x_i - \bar{x})^k = \sum_{j=1}^{r} f_j\left(x_j - \bar{x}\right)^k. \tag{1.13}
$$

The mean is actually that value $x_0$ about which the second moment, the mean square deviation, is minimal,

$$
\frac{\mathrm{d}}{\mathrm{d}x_0} \sum_{i=1}^{N} (x - x_0)^2 = 0 \Longrightarrow \sum_{i=1}^{N} x_i - N x_0 = 0 \Longrightarrow x_0 = \frac{1}{N} \sum_{i=1}^{N} x_i.
$$

It is easy to show that the value about which the sum of the absolute values of the deviations is minimal is instead the median.

Finally, we can define a parameter that allows us to evaluate how much a distribution is *symmetrical* about the mean, that is, whether and how much the distribution shows a longer "tail" towards one extreme or the other of the range of values of $x$. Here the algebraic signs of the deviations about the mean, which we have tried to get rid of by defining the standard deviation, are conversely important. We already know that a simple average of deviations does not work, since it always vanishes. A non-necessarily zero quantity that takes into account the sign of the deviations is the average of the *cubes* of the deviations, namely, the third moment about the mean, whose physical units are the cube of those of $X$. It is however more interesting to define a dimensionless quantity. To do this, we observe that a lack of symmetry is more easily spotted for a narrow distribution than for a very wide one (one of the few advantages of being chubby), and that it is therefore convenient to relate the absolute asymmetry to the standard deviation of the distribution. We then introduce the *skewness* $\gamma_x$ of a distribution as

$$\gamma_x = \frac{1}{Ns_x^3} \sum_{i=1}^{N} (x - \bar{x})^3 = \frac{M_3(\bar{x})}{s_x^3}. \tag{1.14}$$

Suppose for instance that we have obtained from two experimental samples of the physical quantities $P$ and $Q$ these two simple distributions

$$f_P = \begin{cases} 0.75 \text{ for } P = 1 \\ 0.25 \text{ for } P = 5 \end{cases} \qquad f_Q = \begin{cases} 0.25 \text{ for } Q = 1 \\ 0.75 \text{ for } Q = 5 \end{cases}$$

For the distribution of $P$ we have $\bar{P} = 2$ and (if the number of data is very large) $s_P = \sqrt{3}$. For $Q$ we have instead $\bar{Q} = 4$ and $s_Q = \sqrt{3}$. The standard deviation is therefore the same for both distributions, but or the asymmetry we have

$$\gamma_P = +\frac{2}{\sqrt{3}}; \ \gamma_Q = -\frac{2}{\sqrt{3}}.$$

In general, therefore, $\gamma_x > 0$ implies a tail for high values, while the opposite is true for $\gamma_x < 0$.

Further observations on the statistical parameters of a distribution can be drawn by analyzing the distributions discussed in Examples 1.4 and 1.5:

a) The mean, standard deviation, and asymmetry for the height distribution in Fig. 1.6 are given by:

| $\bar{h}$(cm) | $s_h$(cm) | $s_h/\bar{h}$ | $\gamma$ |
|---|---|---|---|
| 161.5 | 7.6 | 0.047 | -0.025 |

Stating the mode is instead quite arbitrary, since the values for $h = 162$ and $h = 165$ cm are almost equal, while the median $h_m \simeq 162$ cm is very close to the mean because of the very low value of $\gamma_h$. Observing Fig. 1.6, we can notice how, for this particular bell-shaped distribution, at least 2/3 of the data fall within an interval $\bar{h} - s_h < h < \bar{h} + s_h$. The curve is also quite narrow, with a relative standard deviation $s_h/\bar{h} \simeq 0.05$.

b) Extracting precise values for the parameters of the distribution of body weight $w$ is impossible because, as we mentioned, the histogram in Fig. 1.7 refers only to the tested subjects with a weight above the 5th and below the 95th percentile. Nevertheless for these subjects (which anyway correspond to 90% of the sample) one obtains $\bar{w} \simeq 89$ kg, $s_w \simeq 16$ kg, and $\gamma_w \simeq 0.45$. Compared with that of the height, the distribution of the body weight is surely wider, with a relative standard deviation of about 18%, and skewed towards higher values.

c) Finally, for the three frequency distributions for the fertility rate in the inset of Fig. 1.9 one obtains:

|      | $\bar{x}$ (ys) | $s_x$ (ys) | $s_x/\bar{x}$ | $\gamma$ |
|------|------|------|------|------|
| 1995 | 29.1 | 6.0 | 0.21 | +0.33 |
| 2002 | 30.1 | 5.3 | 0.18 | -0.09 |
| 2020 | 31.9 | 5.4 | 0.17 | -0.10 |

The mean age then increases in time, but not as much as the mode due to the increasing symmetry of the distribution.

### 1.4.4  Statistics by Classes and Weighted Parameters

Quite often, in particular in social sciences and economics, a sample is split into "classes" according to some, non necessarily numerical, criterion. For example, people can be divided into classes according to the country where they live, or to their gender, or to the sport they like to practice. When the sample is partitioned into classes we often know only the average value within each class of some quantity, so evaluating the mean for the whole sample requires some care. Consider for instance the following table, which gives the gross domestic product per capita (GDP-PC) in those world countries whose name starts with the letter $N$.

| Country | GDP-PC ($) | Pop (MM) | % Pop. |
|---|---|---|---|
| Namibia | 4911 | 26.4 | 0.848 |
| Nauru | 10648 | 0.12 | 0.004 |
| Nepal | 1337 | 29.2 | 9.362 |
| Netherlands | 55985 | 18.0 | 5.766 |
| New Caledonia | 36668 | 0.27 | 0.086 |
| New Zealand | 48249 | 5.2 | 1.677 |
| Nicaragua | 2255 | 6.7 | 2.162 |
| Niger | 533 | 25.4 | 8.144 |
| Nigeria | 2184 | 216.8 | 69.588 |
| North Macedonia | 6592 | 1.8 | 0.588 |
| Norway | 106149 | 5.5 | 1.776 |

If with nonchalance we take the mean of the data in the second column, we get an average GDP-PC among all these countries of about 25 k$. But this is ludicrous, since we put on equal footing the wealthy Norwegians and the rather poor Nigerians, although there are almost 40 Nigerians for each Norwegian. We better put on the scales also the population size of each country, and define a *weighted average*

$$\bar{x}_w = \frac{\sum_i w_i x_i}{\sum_i w_i} \tag{1.15}$$

where, in the case we are considering, $w_i$ is the population of each country weighing the value of its GDP-PC. Evaluating the weighted average of the GDP-PC's, we obtain $\bar{x}_w = 7.8\,\mathrm{k\$}$, which is less than 1/3 of the simple mean! Note that we can also write $\bar{x}_w = \sum_j \hat{w}_j x_j$, where the normalized weights

$$\hat{w}_j = \frac{w_j}{\sum_i w_i}$$

play the same role as the relative frequencies in Eq. (1.5) for the simple mean.

When dealing with data aggregated into classes, the concept of weighted average can naturally be extended to other parameters. So, one can define a weighted variance as

$$\sigma_w^2 = \sum_j \hat{w}_j \sigma_j^2 \tag{1.16}$$

## 1.5  A First Encounter with Brownian Motion

Do atoms really exist? Today this question may sound ludicrous, but till the end of the 19th century it was not. Indeed, several top scientists like Pierre Duhem, Wilhelm Ostwald, and Ernst Mach thought that the atomic theory was useful to model chemical reactions, but that atoms lack any physical reality. The main reason of their skepticism was that back then no one had taken a picture of an atom, so they stick to the old idiom "what the eye doesn't see, the heart doesn't grieve over".

Yet, a landmark experiment had already paved the way to the evidence that atoms *must* exist. To learn about this investigation, we have to go back to 1827 and peep through the shutters of a lab in London where Robert Brown, a Scottish botanist, is observing under the microscope some particles or granules suspended in water, with a size of a few microns, contained in the pollen of *Clarkia pulchella*, a plant akin to primrose. The trouble is that these wretched specks are not keen at all to keep still under observation, but rather seem to suffer from a kind of Saint Vitus' dance, stirring and jiggling madly about before poor Robert's eyes. Besides, the paths followed by the pollen particles are extremely irregular and look remarkably unlike one to the other.

At the time, it might have seemed easy to account for this. Most naturalists, in disagreement with physicists, believed that biological stuff possessed a kind of *élan vital* that made it superior to inanimate objects, and *Brownian motion*, as we shall call the effect observed by our hero, could have been a direct manifestation of this elusive life spirit. But Brown, who was not a physicist but no fool either, carefully avoided jumping to this conclusion. And he was right, since it was easy for him to show that even specks of humble and totally inanimate dust displayed the same frantic behavior. In fact all kinds of particles having a size lower than, say, a few micrometers, which are called *colloidal particles*, share, to a greater or lesser extent (the smaller they are, the more frantically they move), this hectic behavior.

Just a curious oddity in the early 19th century, Brownian motion became a truly puzzling enigma a few decades later, after Clausius and Kelvin had formulated the Second Law that put an end to any dreams of designing a perpetual motion machine of the "second kind", i.e., a machine supposed to *completely* turn heat into work, violating the Second Law of thermodynamics. Whereas in our everyday experience all kinds of motion eventually dissipate as heat, these tiny particles never stop, like an endless musical canon—a *Perpetuum Mobile*, indeed. The origin of Brownian motion remained a mystery till 1905, when Einstein[23] gave it a full and brilliant explanation, which is also a fundamental proof of the atomic nature of matter. Even more, Einstein's explanation of Brownian motion not only *requires* the existence of atoms, but also provides us with a way to estimate a quantity of primary importance in atomic theory, the Avogadro number.

The only assumption made by Einstein is that atoms and molecules are in perpetual motion too, with a kinetic energy proportional to temperature. Hence, a colloidal particle immersed in a fluid is unceasingly bombarded by these shooting nano-bullets that transfer a little of their own energy to it, so that, in a very short time, the kinetic energy of the particle will also be equal to thermal energy (neither more, otherwise the particle would give it back to the molecules, nor less, because then it would go on absorbing energy). This means that the particle must be moving too. But how? Each collision is a kind of "kick" (a feeble one, but still a kick) to the particle, kicks that come from all sides. On average, all these nudges even out, so there is no a preferred direction for the particle to move. Yet, if we look at a very short time interval this is not exactly true. There will always be a tiny loss of balance, causing the particle to perform an extremely irregular zigzag motion, similar to the staggering walk of a drunkard, which is called a *random walk*.

A random walk is a fundamental example of a *random* or *stochastic process*, namely, of a physical process that requires a statistical description. Later we shall extensively deal with random walks. For the moment, let us try and make a little computer simulation of a RW along a straight line, i.e., in one dimension. For example, let us imagine that we drank a little too much and went out into the night along the road facing the pub we visited, and of which we took full advantage. We don't remember well whether to go home we should go right or left, so we take a first step in a random direction, let us say to the right. Then we stop to think and as a consequence decide to retrace our steps, or to take another step in the same direction, and so on at each step.

Where will we be, after having taken a certain number $N$ of steps? For all intents and purposes, the problem is completely identical to that of a head or tail game that we described in Sect. 1.2. After all, since we have no idea how to get home, we could decide each time which side to go by tossing a coin. A *single* realization of an RW will therefore have a statistically similar appearance to that shown in Fig. 1.3. While wandering here and there, therefore, we expect not to move far from the starting point, even if we rarely pass in front of the pub, but we cannot say much more.

---

[23] And independently Marian Smoluchowski, although with a more tortuous approach.

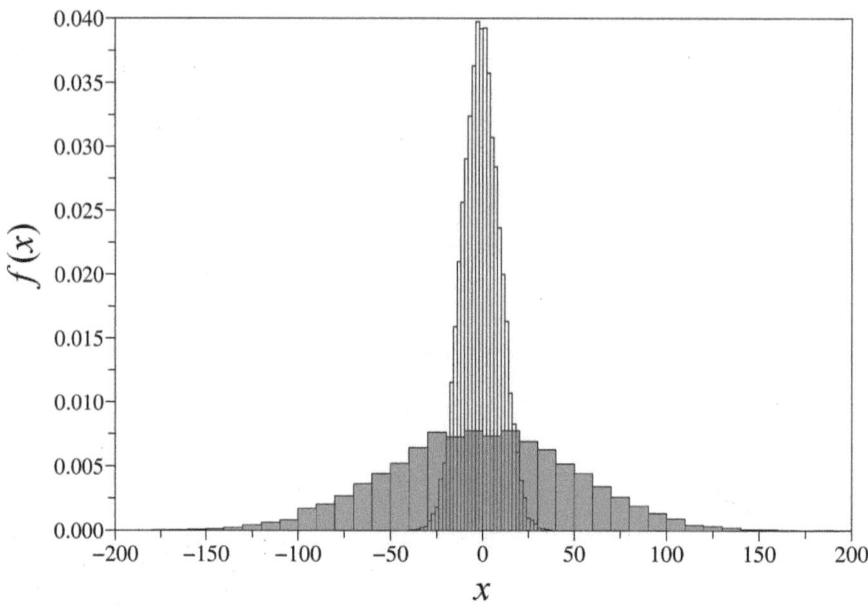

**Fig. 1.10** Distributions of the final position obtained from the simulation of $10^4$ RWs of 100 (narrower histogram) and 2500 (wider histogram) unit steps each

To gather the statistical properties of a RW, the only way to proceed is to repeat our experiment many times. Figure 1.10 shows two distributions of the final position $x$ reached by a similar drunkard, obtained by simulating 10000 distinct RWs, each consisting of $N = 100$ and $N = 2500$ unit-length steps, respectively. The similarity of both distributions to the "bell-shaped" distribution in Fig. 1.6 is quite remarkable, and leads us to begin to think that this type of distribution is, for some obscure reason, rather common. For both distributions, we have $\bar{x} \simeq 0$, as we could expect. The standard deviations are respectively equal to $s_x \simeq 10.05$ and $s_x \simeq 50.1$, values that coincide with good approximation with the *square root* of $N$.

A better visualization of the spreading of the random walkers is obtained by considering a 2-dimensional RW—the drunkard this time wanders around a field. To do this, I have simulated 2500 RWs of 1600 steps, each of unit length but directed at an angle $\vartheta_r$ with respect to the direction of the $x$-axis chosen randomly in $[0, 2\pi]$, thus corresponding to displacements $\Delta x = \cos \vartheta_r$ and $\Delta y = \sin \vartheta_r$. Panel A in Fig. 1.11 provides a clear graphical impression of the distribution of final positions, while the distributions for the components of the displacement along $x$ and $y$ show a trend very similar to that in Fig. 1.10.

It is also interesting to analyze the behavior of the distance $r$ from the origin, that is, of the root mean square displacement (RMSD) $r_{rms} = \sqrt{\langle x^2 + y^2 \rangle}$, which is clearly a positive quantity. Panel B shows that the frequency distribution for $r$ grows rapidly and has a maximum for $r \simeq 25$, while the value of the standard deviation for

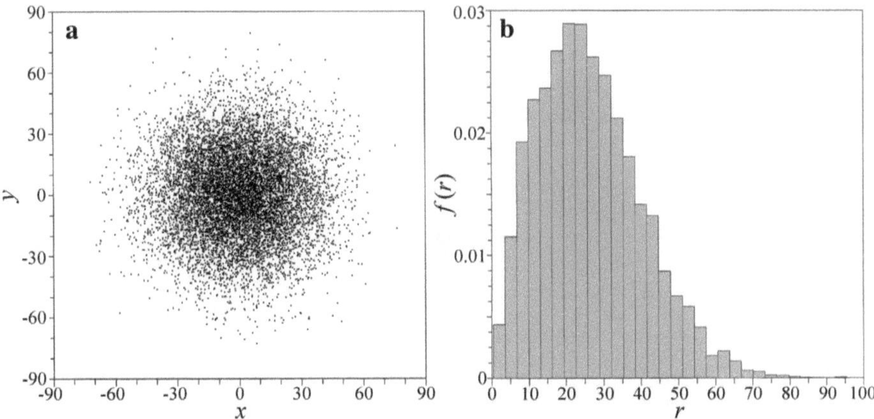

**Fig. 1.11** Panel A: Simulation of final positions for 2500 RWs in two dimensions, each of 1600 steps. The corresponding distribution of the mean square displacement $r$ is shown in Panel B

the distributions of both components, $s_x \simeq s_y \simeq 40$, corresponds to the RMSD for which the distribution decreases to about half of its maximum value. We will account for these results in the following chapters.

## 1.6 Correlations

We consider now data concerning *two* statistical properties $X$ and $Y$, obtained by measuring the values they assume under the same conditions, such as on the same object, at the same time, in the same place. Our sample is then made up of *pairs* of values $\{x_i, y_i\}$. Of course, this is what we do when we look for a physical law connecting $X$ and $Y$. For the moment, however, we will just consider a simpler question, namely, if we can spot some "affinity" between the way $X$ and $Y$ vary. If we do find some kind of connection, we may say that $X$ and $Y$ are, to a given extent, *correlated*.

Correlation is a concept of great importance in all fields of science, but it plays a paramount role in modern physics, so much that one is tempted to state with David Mermin that

*the proper subject of physics* is *correlation and* only *correlation*[24]

However, spelling out properly what correlation means is subtle and rather thorny matter, so we better proceed with a lot of caution and a bit of intuition. Both $X$ and $Y$ fluctuate, taking values in excess or defect of their average values $\bar{x}$ and $\bar{y}$. Yet, the fluctuations in $Y$ can be related to those of $X$ in several ways, which are typified by

---

[24] N. D. Mermin, Am. J. Phys. **66**, 753 (1998).

the trends for the three quantities $Y_1$, $Y_2$, and $Y_3$ shown below, where the horizontal axis could simply represent a series of subsequent measurements of each quantity.

We observe that when $X$ exceeds its mean, $Y_1$ typically tends to be in excess too, so we may say that $X$ and $Y_1$ are indeed correlated. For $Y_2$, the opposite occurs. This does not mean that there is no relation between the fluctuations of $X$ and $Y_2$, rather that are still correlated, but in the opposite direction or, as we shall say, *anti*correlated. Conversely, $X$ and $Y_3$, whose deviations from the mean do not show any evident relationship, are close to what we mean by *uncorrelated* variables.

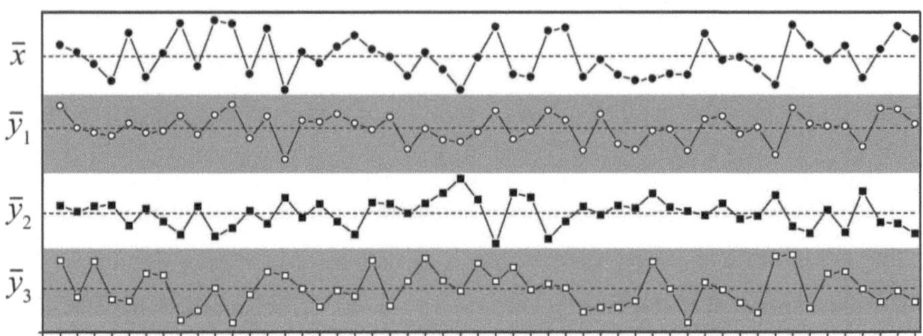

Note that the product of the deviations of $X$ and $Y_1$ from their means is generally positive, while it is predominantly negative for $X$ and $Y_2$. To quantify the degree of correlation, it is then natural to consider the average value of the product of the deviations,

$$s_{xy} = \frac{1}{N} \sum_{i=1}^{N} (x - \bar{x})(y - \bar{y}) = \overline{xy} - \bar{x}\bar{y}. \tag{1.17}$$

When $X$ and $Y$ are actually the *same* variable, we have $s_{xy} = s_x^2$, while for $Y = -X$ we have $s_{xy} = -s_x^2$. We can then regard $s_{xy}$ as a kind of "mutual" standard deviation of $X$ and $Y$, which, unlike $s_x$ and $s_y$ can also be negative. Actually we will later see that, as much as for the standard deviation, the definition of $s_{xy}$ requires a correction, significant however only when the sample is small. Since the physical dimensions of $s_{xy}$ are given by product of the dimensions of $X$ and $Y$, it is also convenient in this case to define a dimensionless quantity $r_{xy}$ which is called *linear correlation coefficient* between $x$ and $y$,

$$r_{xy} = \frac{s_{xy}}{s_x s_y}. \tag{1.18}$$

The reason for specifying "linear" will soon be evident.

When we consider a large data set, the correlation coefficient will approach $+1$ if the two variables are fully correlated, $-1$ if the are fully uncorrelated, and $0$ for uncorrelated variable. For example, the correlation coefficients of $Y_1$, $Y_2$, and $Y_3$ with $X$ are

$$r_{xy_1} \simeq +0.82 \; r_{xy_2} \simeq -0.81 \; r_{xy_3} \simeq -0.03.$$

The correlation coefficient $r$ has illustrious fathers like Francis Galton, Darwin's cousin who also introduced the concept of standard deviation, Auguste Bravais, the father of crystallography who first wrote Eq. (1.18), and Karl Pearson, a pioneer of mathematical statistics, discipline where $r$ plays a key role. Yet, as I anticipated, it should be handled with care. The following examples will help clarifying several issues.

**Example 1.7** (*Why "linear"?*) We start by considering two quantities that we *do* expect to be correlated, the height $h$ and body weight $w$ of people. To be more confident, I have selected a homogeneous sample by collecting anthropometric data for a total of almost 1,000 players of the major football leagues in Italy (season 2016–17) and England (season 2014–2015), all young and healthy individuals with enviable physical stamina. From the data sample one obtains

$$\bar{h} = 1.825\,\text{m} \,;\; \bar{w} = 76.10\,\text{kg} \,;\; \overline{hw} = 139.18\,\text{kgm} \,;\; s_h \simeq 0.06\,\text{m} \,;\; s_w \simeq 6.7\,\text{kg},$$

which gives[25] $r \simeq 0.74$, confirming a high correlation between height and weight. If we now plot $w$ versus $h$ as in Fig. 1.12, we see that the data are arranged in a point cloud around the straight line shown in the figure.[26]

It is indeed rather easy to show that two variables $X$ and $Y$ which are linearly related, are fully correlated or anticorrelated. Writing $Y = aX + b$, you should not find it too difficult to show (do it!) that

$$\bar{y} = a\bar{x} + b \,;\; \overline{xy} = a\overline{x^2} - b\bar{x} \,;\; s_y = |a|s_x$$

so that

$$r_{xy} = \frac{a\overline{x^2} - b\bar{x} - \bar{x}(a\bar{x} + b}{|a|s_x^2} = \frac{a(\overline{x^2} - \bar{x}^2)}{|a|s_x^2} = \text{sgn}(a)$$

Hence, $r = \pm 1$ depending only on the sign of $a$.

However, let me be a little naughty…Many of you have probably heard of the expression "Body Mass Index" (BMI), a quantity defined as the ratio between a person's weight in kilogram and the square of her/his height in meters. Physicians state that a healthy adult should have a value of the BMI in the range $18.5\,\text{kg/m}^2 \leq$ BMI $< 25\,\text{kg/m}^2$. Indeed, the BMI of the considered football players has an average value of $22.8\,\text{kg/m}^2$ and a standard deviation of only $1.3\,\text{kg/m}^2$, yielding a *relative* standard deviation of less than 6%. But, wait a minute! This means that the players' BMI is approximately constant, or in other words that their weight is proportional to the *square* of their height! Indeed, the points in the inset of Fig. 1.12 seems to cluster around a straight line as much as those in the figure body.

---

[25] Note that I stated $\bar{h}$, $\bar{w}r$, and $\overline{hw}$ with rather high precision. The reason is that the difference $\overline{hw} - \bar{h}\bar{w}$ is *very* small, so a poorer approximation leads to rather serious mistakes. This often happens when evaluating $r$, so be careful!.

[26] This is actually the "best" straight line that fits all data, a concept that will be discussed in a following chapter.

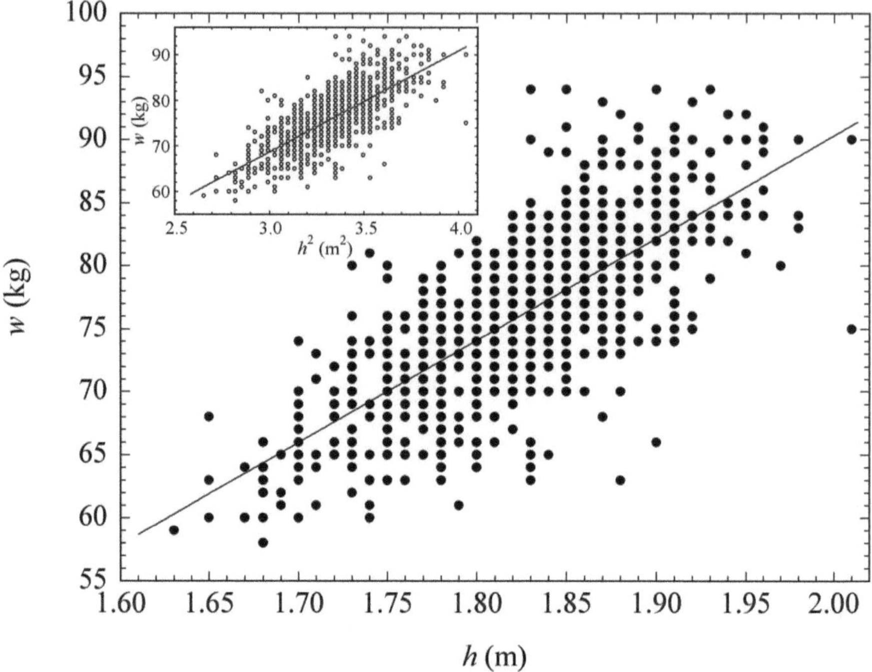

**Fig. 1.12** Body weight $w$ versus height $h$ for 540+420 players of the major Italy and England football leagues. The straight line is a linear fit to the data. The inset shows the same data, but plotted as $w$ versus $h^2$

So, does $w$ scales with $h$ or with $h^2$? The answer is that our data do not yield *any* conclusive answers. The main reason is very likely the very small range of $h$, which varies by just over 3% between all the players.[27] With such a limited range of variation of one of the two variables, no discrepancy from a linear relation can be easily detected.

After all, what does the BMI, not a dimensionless parameter but a mass per length square, indicate? Actually, by multiplying the BMI by $g$ one gets a pressure that, for a healthy subject, is in the range of $180 - 250\,\mathrm{Pa}$. It is hard to give these values, which are less than 0.3 % of the atmospheric pressure, any sound physical meaning. Besides, why should the weight of an individual scale with the square of the height, when the weight of an object with reasonably uniform density scales with its volume? The BMI was explicitly introduced by physiologists in 1972, but actually its idea goes back 1830, when Adolphe Quetelet, a Belgian mathematician and astronomer who first tried to apply statistics to social sciences, used $h^2$ as a correction factor to make the distribution of body weight more symmetric and similar to the bell-shaped

---

[27] Correspondingly, $h^2$ varies by less than 7%.

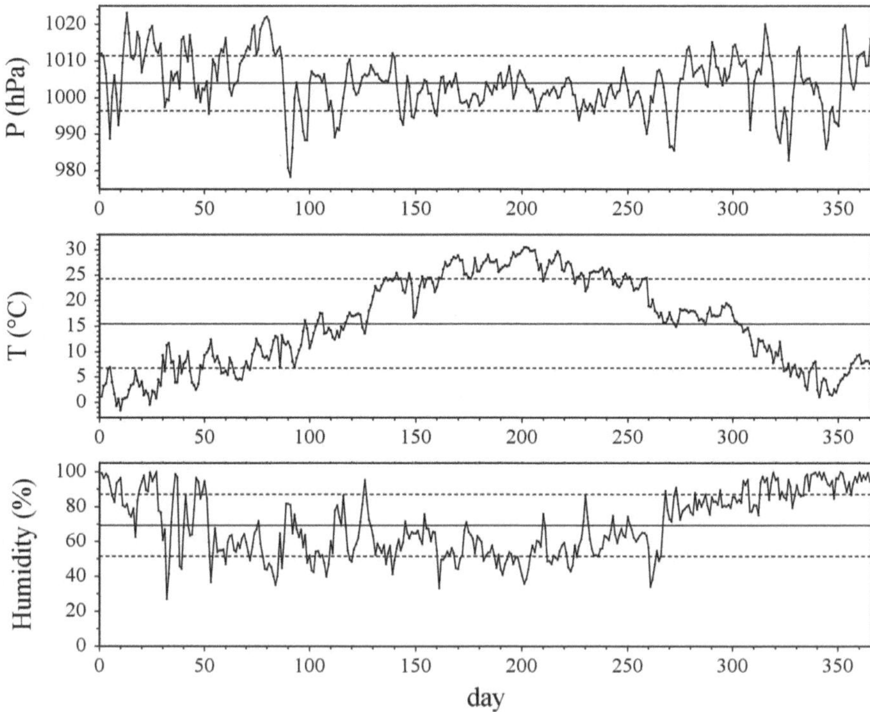

**Fig. 1.13** Daily atmospheric pressure, temperature, and relative humidity in Milan during the whole year 2022

distribution for the height—which indeed works. But the reason of the popularity of this parameter to assess the health risks associated with excessive weight is, at least to me, a mystery.

Coming back to the role of sample breadth to get a reliable estimate of correlation, let us consider the weather data in Fig. 1.13, showing the values of the average daily pressure $P$, temperature $T$, and relative humidity $H$ in my home town during 2022. One finds a negative and rather large correlation coefficient $r_{TH} \simeq -0.6$ between $T$ and $H$, which is what one expects—it rains ($H \geq 100\%$) more often on cold days.[28] Similarly, there is no physical reason to expect pressure and relative humidity to be correlated, and indeed $r_{PH} \simeq 0$. However, one may expect high pressure to be more typical of hot summer days. Yet, one find $r_{PT} \simeq -0.25$, which is rather small and surprisingly *negative*. Again, however, the range of variation of $P$ all over the year is *less than* 0.8%! My personal advise: Never trust correlations with a quantity varying in such a ludicrously small range!

---

[28] This does not mean that on hot days in Milan you won't feel miserably wet. This is relative, not *absolute* humidity!.

**Example 1.8** (*Lack of correlation versus independence*) Finding uncorrelated variables is not too difficult. The close frequencies of the single digits is a first clue of the randomness of the distribution of the decimals of $\pi$, but surely this is not enough. One basic criterion should be that, if we know the value of a decimal, we have no way to predict the next one. Indeed, by evaluating the correlation coefficient between the (9,999) successive pairs of decimals in $\pi$, one obtains $r_{n,n+1} \simeq -0.01$. This result can actually be extended to the subsequent followers, obtaining that $-0.021 \le r_{n,n+k} \le 0.018$ for $k = 1$ to 9. This lack of correlation between subsequent digits is surely a stronger indication of the randomness of the distribution.

However, stating that two quantities are not correlated does not mean at all that they are "independent". We will properly define the concept of independence of two or more variables only later, but right from now we can safely state that two quantities $X$ and $Y$ which are connected by functional relation $f(X, Y) = 0$ are surely *not* independent. Consider for example two variables connected by a *nonlinear* relation as simple as $Y = X^2$, where $X$ takes only the values $\pm 1$. Then, whatever the value $x_i$ of $X$, we always have $y_i = \bar{y} = 1$, so that $r_{xy} = 0$. Hence two mutually dependent variables are not necessarily correlated (unless of course they are *linearly* related). In the next chapter we will instead see that independence *does* imply absence of correlation.

**Example 1.9** (*Stock markets and small samples*) The concept of correlation plays a primary role in economics and finance too. Stating whether two quantities are correlated or not, for example, allows us to assess how the trend observed for a certain financial indicator influences another one. The most significant financial indicators are obviously the global stock market indices, that is, those that summarize the average trend of all stocks or the most significant ones. Figure 1.14 shows that the detailed trends of four important stock market indices in the whole decade 2013–2022 are remarkably similar.[29] Some differences can however be spotted. For example, the sudden drop of the securities between February and March 2020, due to the outbreak of the pandemic, is largely absent in the Shanghai index, where conversely the so-called "Chinese stock market turbulence", namely the popping of the stock market bubble between June 2015 and February 2016, is very noticeable.

---

[29] These indices are defined in very different ways and are obviously calculated in local currency. For example, the Nikkei 300 (NK) reflects the average value in yen of the 300 most significant Tokyo Stock Exchange securities, while the Dow Jones Industrial (DJ) is limited to considering the value in dollars on Wall Street of the 30 largest companies securities.

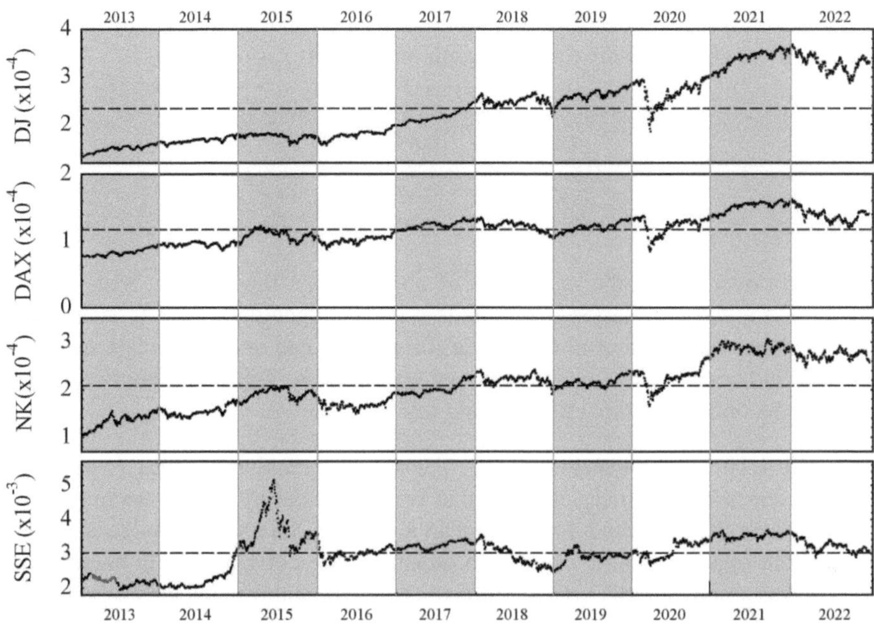

**Fig. 1.14** Daily closure index for the DJ (New York), DAX (Frankfurt), NK (Nikkei Tokyo) and SSE (Shanghai) stock markets for the decade 2013–2022

By considering all pairs of indices $I_j$, $I_k$, we can evaluate the *correlation matrix* $\mathbf{R}(j, k)$, having for elements the correlation coefficients $r_{jk}$[30]

|     | DJ   | DAX  | NK   | SSE  |
| --- | ---- | ---- | ---- | ---- |
| DJ  | 1.00 | 0.92 | 0.96 | 0.66 |
| DAX | 0.92 | 1.00 | 0.95 | 0.66 |
| NK  | 0.96 | 0.95 | 1.00 | 0.63 |
| SSE | 0.96 | 0.66 | 0.63 | 1.00 |

which is of course a symmetric matrix. As it could be expected, the correlation coefficients between the DJ, DAX, and NK are close to 1, while those with the SSE are lower, but still very significant.

---

[30] The correlation coefficients were calculated by considering only the days (about 2,200) when the four stock exchanges were simultaneously open.

However, if we narrow down our study to a much shorter period, things change. Indeed, if we consider only the last month of 2022, $\mathbf{R}(j, k)$ becomes

|      | DJ   | DAX  | NK   | SSE  |
|------|------|------|------|------|
| DJ   | 1.00 | 0.65 | 0.80 | 0.33 |
| DAX  | 0.65 | 1.00 | 0.67 | 0.26 |
| NK   | 0.80 | 0.67 | 1.00 | 0.54 |
| SSE  | 0.33 | 0.26 | 0.54 | 1.00 |

showing a decrease of the $r_{j,k}$ values of about 1/3 on the average, with peaks of 50−60% for the correlation coefficients with the Shanghai market. Therefore, a much smaller sample does not convey a correct estimate of the correlation between the four indices. So, my second personal advise is: be sceptical about correlation coefficients obtained from a limited data set!

**Example 1.10** (*Half-siblings and false friends*) Correlation has little or nothing to do with *causation*. Namely, the fact that two quantities $X$ and $Y$ are moderately, or even strongly correlated does not mean at all that $X$ induces, produces, leads to $Y$. Nor the opposite. Quite often, two quantities are correlated only because they share a "common parent". For example, the following table gives the average height of women ($h$) in 18 European countries and the share of women in each of these countries who eats at least a portion of fruit a day ($fr$) or profess the Roman Catholic religion ($rc$).[31]

|             | $h$ (m) | $fr$ (%) | $rc$ (%) |
|-------------|---------|----------|----------|
| Austria     | 1.66    | 58.7     | 72.1     |
| Belgium     | 1.64    | 61.4     | 75.5     |
| Croatia     | 1.67    | 62.3     | 73.5     |
| Czechia     | 1.68    | 54.1     | 34.2     |
| Denmark     | 1.69    | 53.4     | 0.64     |
| France      | 1.64    | 64.9     | 75.5     |
| Germany     | 1.66    | 65.8     | 31.8     |
| Greece      | 1.65    | 52.2     | 1.04     |
| Hungary     | 1.62    | 60.8     | 58.2     |
| Ireland     | 1.64    | 70.3     | 76.1     |
| Italy       | 1.61    | 72.2     | 96.6     |
| Netherlands | 1.70    | 49.5     | 31.3     |
| Poland      | 1.65    | 60.9     | 94.3     |
| Portugal    | 1.61    | 70.2     | 90.4     |
| Slovakia    | 1.67    | 60.0     | 73.1     |
| Slovenia    | 1.67    | 63.6     | 81.3     |
| Spain       | 1.62    | 71.5     | 87.8     |
| Sweden      | 1.67    | 55.3     | 1.60     |

---

[31] *Sources Eurostat, Catholic-Hierarchy.org.*

From this data set we obtain the correlation matrix which surely sounds disconcerting:

|      | $h$   | $fr$  | $rc$  |
|------|-------|-------|-------|
| $h$  | 1.00  | -0.78 | -0.59 |
| $fr$ | -0.78 | 1.00  | 0.75  |
| $rc$ | -0.59 | 0.75  | 1.00  |

Indeed, what does it mean such a large negative correlation coefficient between fruit consumption and the stature of adult women? That the good habit of eating fruit inhibits growth? And should the Vatican promote fruit consumption (see $r_{fr,rc}$), of course concealing the evidence that daughters raised in the Catholic faith stay rather short (see $r_{h,rc}$)? Clearly, there is no direct causal relation between these three variables. The roots of these strong correlations are rather that countries like Italy, Spain, Portugal, and France, where fruit consumption is high, are also mostly Catholic and populated by women that on the average cannot compete in terms of stature with their Dutch or Danish counterpart.

There are however many situations where the origin of high values of $r$ cannot be easily traced. The Internet is full of weird examples, but surely the most hilarious web site is Spurious Correlations, a project developed and kept by Tyler Vigen.[32] that is surely worth visiting. The apparent correlations brought to light are surely bizarre, but should be taken with a grain of salt, because the sample size is tiny (10–13 data pairs) and the range of one or both the compared quantities is often rather limited. Nevertheless, they point further out, if needed, that correlation does not necessarily imply causation, a concept that should be crystal clear to anyone dealing with statistics.

Alas! This is not always the case in several fields, medicine above all, where the caveat we raised are often ignored. Unfortunately, sometimes this is also true for question of the highest priority like global warming. For instance, no mentally healthy scientist can deny that there is a strong correlation between $CO_2$ emissions and planetary temperature rise. But stating that $CO_2$ emissions are the (only) cause of the observed global warming is another story. Likely true, but another story, requiring further evidence to be fully supported.


**Example 1.11** (*Correlation besides concurrency*) Besides being strictly adequate for linear relationships alone, the $r$ coefficient does not capture many facets of what we intuitively mean by "correlation". So far, we have considered only sets of data for two or more properties obtained in concurrent conditions. However, there are quantities that, although weakly related when measured at the same time or in the same place, display a much higher degree of correlation when we abandon the requirement of space-time coincidence.

Consider for instance Fig. 1.15, showing the results of one of the first detailed experimental studies of predator-prey dynamics in biological systems performed on

---

[32] https://www.tylervigen.com/spurious-correlations.

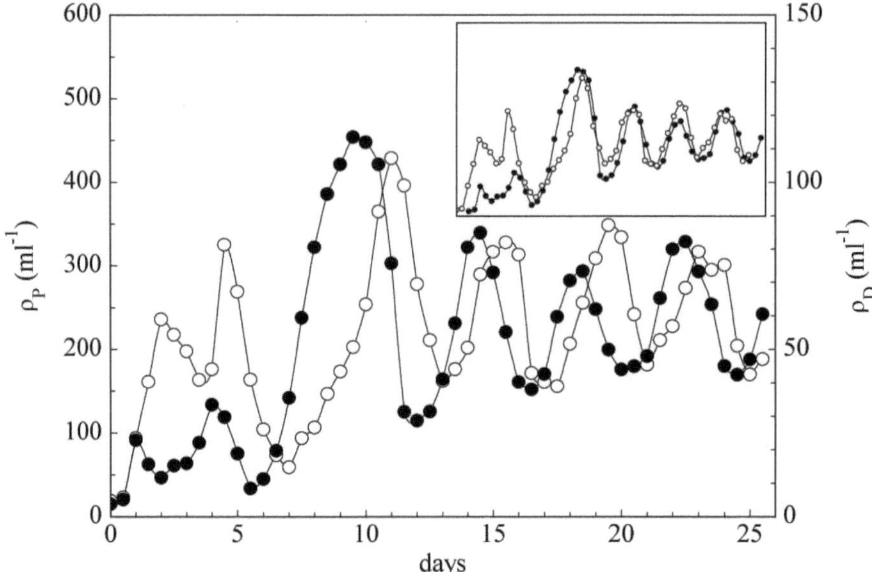

**Fig. 1.15** Time dependence of the population of *Paramecium* (full dots) and *Didinium* (open dots). In the inset, the correlation between the two trends are compared by forward shifting the time behavior for *Paramecium* by 1.2 days

the system Didinium-Paramecium, two unicellular ciliates.[33] *Didinium nasutum* is an amazingly voracious predator and its favourite meal is by far *Paramecium*.[34] The figure body shows that the concentrations $\rho_P(t)$ and $\rho_D(t)$ of the two species in a culture medium oscillate in time, with a similar period but definitely out of phase, with the growth or decay of the predator concentration lagging behind those of its prey. The correlation coefficient between the two sets of data is indeed rather modest, $r \simeq 0.32$.

However, if we introduce a forward shift of slightly more than one day in the time axis of $\rho_P(t)$, as shown in the inset, the correlation coefficient becomes as large as $\rho \simeq 0.74$. This means that the data, when compared with a suitable time difference, are indeed *strongly* correlated. This broader view of the link between two fluctuating quantities will became clearer in Chap. 5, where we introduce the concepts of correlation function of a random process.

---

[33] B. G. Veilleux, PhD thesis, University of Alberta (1976).

[34] Dididium is not particularly afraid of the fact that its prey can be much larger than itself. Once it has captured and paralyzed *Paramecium* using its "nose" (wherefrom *nasutum*), it slowly but obstinately engulfs all this big burger, sometimes to the point of bursting because of its greed.

## 1.7  Characteristic Scales and Scale Invariance

Physics is not (only) math. In mathematics, expressions like $y = \exp(-x)$ or $x = \sin(t)$ make perfect sense. Physical quantities, however, have *dimensions*, and must be measured with an *algebraic* combination of basic units. Consider now $x = \exp(t)$, and suppose that $t$ is a time. In what should we measure $x$? In obscure "sines of seconds"? Evidently, for non-algebraic functions like $\sin \alpha$ the only way out is making its argument $\alpha$ *dimensionless* writing $x = x_0 \sin(\omega t)$ where $\omega$ is a frequency and $x_0$ is a constant with the same units of $x$. Similarly, if $x$ is a length, we better write an exponential decay as $y = y_0 \exp(-x/x_0)$ where $x_0$ is a length.

Such a requirement on the argument of non-algebraic (transcendental) functions introduce therefore a *characteristic scale* (of length, time, frequency, energy...), which can often be identified from the physics of the investigated problem. Thus in the former example, $x_0$ is the characteristic length scale at which $y$ decreases to $1/e$ of its value in the origin. Identifying a characteristic scale (of length, time, energy...) in a problem is very often a winning strategy for developing models of physical phenomena. Thus, for example, all thermodynamic phenomena are characterized by the natural energy scale $k_B T$, given by the product of the Boltzmann constant by the absolute temperature. Often, the analysis of a specific problem brings out new characteristic scales that guide the solution. In the next two examples, we discuss two frequency distributions that show an exponential decay whose origin we will extensively discuss in the following chapters.

**Example 1.12** (*Radioactive decays*) The totally random nature of radioactive decay has been a baffling enigma in the early days of nuclear physics. We can indeed measure the half-life of a radioactive source, i.e., the time required on average for half of the radioactive atoms to decay but, when we observe a radioactive emission from the sample, we have no way to predict when the next one will occur. What we can do, however, is measuring the statistical distribution of the waiting time $\tau$ between two subsequent decays. Figure 1.16a shows one of the first measurement of the distribution of waiting times for the $\alpha$-decay from a very thin layer of uranium oxide by Marsden and Barrat in 1910.[35] Notwithstanding the limited number of measured intervals (around 300), the plot shows that the logarithm of the relative frequency is rather accurately fit by a straight line, showing that the length of the time intervals is exponentially distributed with a decay constant $\gamma \simeq 0.53\,\mathrm{s}^{-1}$. In Chap. 3 we will see where such distribution originates from.

**Example 1.13** (*Speckle patterns*) When an expanded laser beam is scattered off a rough surface or by an inhomogeneous medium, it generates on a detector or on an observation screen a speckle file, i.e., a pattern of bright and dark spots, originating from the interference of the radiation scattered from different points. If the speckle pattern is imaged on a photographic plate or a digital sensor with a lens system, the speckle size and structure depends not only on the laser wavelength and on the

---

[35] E Marsden and T Barratt *Proc. Phys. Soc. London* **23** 367 (1910).

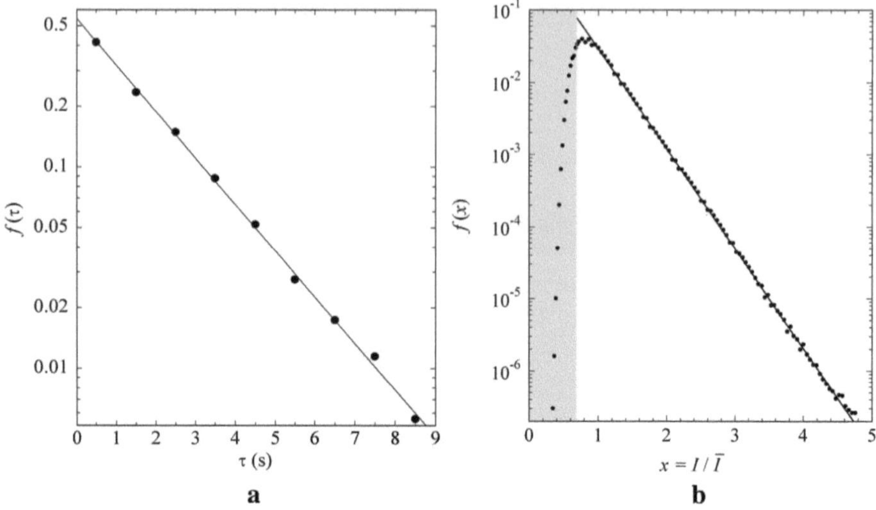


**Fig. 1.16**  **a** Distribution of time intervals between two successive $\alpha$-decays from a layer of uranium oxide obtained by Marsden and Barratt in 1910. **b** Distribution of the intensity of laser speckles generated by a scattering medium. The grey region corresponds to the background noise

distance from the scattering medium, but also on the properties of the imaging optics. Regardless of the detailed structure of the speckle field, however, the distribution of the light intensity on the image plane displays some very general features that we will extensively discuss in the following.

As a sneak peek, Fig. 1.16b shows the distribution $f(x)$ of the intensity $I$ of the speckle field scaled to its average value $\bar{I}$, generated by light scattering from a colloidal suspension.[36] The figure shows that $f(x)$ is anything but uniform, decreasing exponentially by almost six orders of magnitude when $I$ is above a minimum value $I_n \simeq 0.7\bar{I}$ due to background noise. Actually, by subtracting out the noise and writing $I' = I - I_n$, $\bar{I}' = \bar{I} - I_n$, the distribution decays almost exactly as $\exp(-I'/\bar{I}')$. We shall later inquire about the origin of this simple functional form.

There are however problems that are *scale-free*. For example, if two quantities are related to each other by a power law, $y = Cx^\alpha$, we do not need to scale $x$ by any characteristic value $x_0$, as long as the dimensions of the prefactor $C$ (which is *not* a physical quantity, but simply a constant that couples quantities of different nature) are $[C] = [y][x]^{-\alpha}$. Note that if we scale the units of measurements of $x$ by a factor $k$ (e.g., from miles to kilometers, if $x$ is a length), so that $x \to x' = kx$, $y$ simply scales as $y \to y' = k^\alpha y$, namely, $y$ does not change provided that we gauge it with a suitably scaled yardstick. This feature, which is not shared by functions with an intrinsic scale like an exponential, originates the so-called *self-similarity* of

---

[36] The distribution was obtained by averaging a large number of 16-bit images on a CMOS sensor, with each images made of $220 \times 40$ pixels.

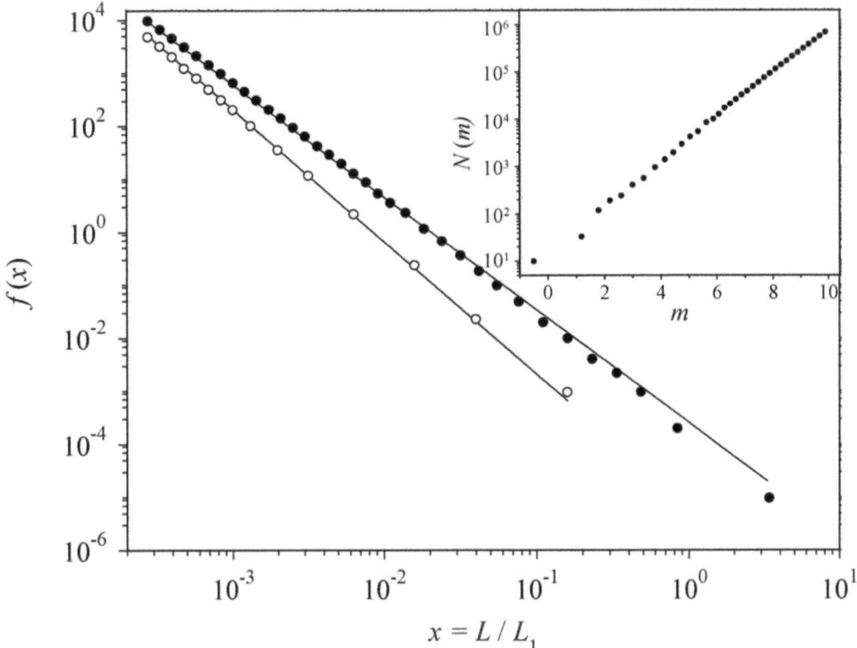

**Fig. 1.17**  Inset: Number of stars $N(m)$ versus their visual magnitude $m$. Body: Frequency distribution for the number of stars with magnitude $m < 10$ as a function of apparent luminosity $L$ relative to that of first magnitude stars $L_1(\bullet)$. The straight line on the double–logarithmic graph corresponds to a power law $f = A(L/L0)^{\alpha}$, with $\alpha \simeq 2.15$. The graph also shows the frequencies related to only stars with spectral characteristics similar to the Sun ($\circ$), interpolated by a power law with exponent $\alpha = 5/2$

quantities described by power laws. While a large number of physical phenomena have long been known to display an exponential behaviour, many recent advances in statistical physics came from considering quantities that, on the contrary, show scale invariance and therefore self-similarity. In the next examples we discuss two scale-invariant distributions showing a power-law behavior of particular interest for astronomy and geophysics.

**Example 1.14**  (*The brightness of stars*) The apparent (or visual) magnitude $m$ is an index that quantifies the brightness of visible stars, chosen so that an increase of $+1$ in the value of $m$ corresponds to a reduction of $10^{2/5} \simeq 2.512$ in the apparent brightness of a star. Thus first magnitude stars ($m = 1$), are 100 times brighter than stars of sixth magnitude, the weakest to be perceived by the naked eye.[37] How does the number of visible stars vary with $m$? The inset in Fig. 1.17 shows the distribution of $N(m)$

---

[37] Some particularly bright celestial bodies have a negative apparent magnitude. For example, Sirius, the brightest star, has $m \simeq -1.5$, Venus at its maximum brightness has $m = -4.4$, and for the Sun, we have $m = -26.7$.

for over 36000 stars with $m < 10$, i.e., all the stars visible with a small amateur telescope, taken from the Tycho astronomical catalog[38] The semi-logarithmic scale shows that the number of stars grows exponentially with $m$.

Yet, $m$ is only an index chosen for convenience according to a *logarithmic* scale of the visual intensity. If we respectively denote with $L$ and $L_1$, the apparent brightness of a star of apparent magnitude $m$ and of a first magnitude star, we have $m = 1 + 2.5 \log(L_1/L)$. It is then more meaningful to plot the histogram of the frequencies versus the normalized apparent brightness $L/L_1$. If we wish to keep unitary the area under the histogram, however, this change of variable requires some care, since equal intervals $\Delta m$ for $m$ do not correspond to equal intervals $\Delta y$ for $y = L/L_1$. What we want is keeping the same *area* for each rectangle, so we must have $f(y)\Delta y = f(x)\Delta x$. Hence, if the width of the intervals is very small, one can simply rescale the relative frequencies as $f(y) = (\mathrm{d}x/\mathrm{d}y)f(m)$, but in general one has to explicitly find the start and end points of each interval for $L/L_1$. This is what I actually did to get the graph in the body of Fig. 1.17, which shows that the frequency distribution for the apparent luminosity is well described by a power law with an exponent $\alpha \simeq 2.15 - 2.18$. Thus, the apparent brightness of stars shows scale invariance.

Can we account for this result? The apparent brightness of a star depends both on its distance $R$ from Earth (as for a candle, a light bulb, or any source that emits radiation isotropically, the apparent intensity decreases as $R^{-2}$) and, obviously, on its *absolute* brightness $L_{abs}$, namely the total power it radiates. The latter can vary by many orders of magnitude and, in addition, the emitted radiation may have very different spectral properties. There are indeed "red dwarfs", which emit a much smaller amount of light than our Sun, and "blue supergiants" as bright as thousands of suns. The distribution of $L$ is therefore closely linked to that for $L_{abs}$, which in turn is determined by both the mechanisms of star formation and those of stellar evolution, since the luminosity and spectral class of a star change over time.[39] Therefore, it is therefore not at all simple to predict the distribution of $L_{abs}$ that we observe at the present time of cosmic evolution. It is surprising, however, how the combined effect of such complex mechanisms translates into a simple power law behaviour, which should be predicted, at least qualitatively, by any good theoretical model.

If all stars had the *same* absolute luminosity, things would be much simpler. Since we can write $L \propto L_{abs}/R^2$, the number of stars $N(L)$ with an apparent luminosity greater than a given value $L$ will be equal to the number of stars of this kind contained in a sphere of radius $(L_{abs}/L))^{1/2}$,

$$N(L) = An \left( \frac{L_{abs}}{L} \right)^{3/2},$$

---

[38] To be precise, this is the star photovisual brightness $V$. For the Tycho catalog, see https://www.cosmos.esa.int/web/hipparcos/tycho-2..

[39] Actually, between spectral emission and absolute luminosity there is a deep connection quantified by the Herzprung-Russell diagram, one of the cornerstones of stellar astrophysics.

where $A$ is a constant and $n$ is the number of stars per unit volume. However, the vast majority of stars with $m \leq 10$ lies in a small region (in an astronomical sense, of course!) of the Galaxy close to us, where $n$ can be assumed to be approximately constant. Then the number of stars $n(L)\mathrm{d}L$ with apparent luminosity between $L$ and $L + \mathrm{d}L$ is obtained by taking the derivative of the previous expression, which yields

$$n(L) = CL^{-5/2},$$

where $C$ is another constant. We therefore expect that the relative frequency of stars with a given apparent luminosity scales as $L^{-2.5}$. To see if this works, let's consider only those stars, among those with $m < 10$, with spectral properties alike those of our Sun.[40] Figure 1.17 shows that this simple prediction is in good agreement with observational data.

**Example 1.15** (*The violence of earthquakes*) Earthquakes, like stars, can be enormously diverse, from small tremors detectable only by the pens of seismographs, to cataclysmic events that can even change the landscape. A scale like the one devised at the end of the 19th century by Giuseppe Mercalli, which ranks earthquakes according to their destructive effects, is scarcely useful for studying the geophysics of seismic events, in addition to being somewhat anthropomorphic. The modern Richter scale uses a single index, still called magnitude $m$, to quantify the intensity of an earthquake, set up as for the stellar magnitude on a logarithmic scale based on the amplitude of the maximum displacement of the seismograph pens. An increase of one degree of magnitude thus corresponds to an increase of a factor of $10^{3/2} \simeq 31.6$ of the quake energy, which can be taken either as the energy actually released or the fraction that reaches the Earth's surface.[41] Already in 1954, Beno Gutenberg and Charles Richter himself observed a remarkable relationship between the number of observed earthquakes in a given period and their magnitude, which can be written as $N(m) \propto 10^{-bm}$, where $b \simeq 1$. Among data archives, those supervised by Caltech about seismic events in California, a notoriously high-risk area, are particularly accurate and extensive[42] The inset of Fig. 1.18, which shows the frequency distribution of about 6500 quakes with magnitude $3 \leq m < 6$ occurring in southern California from the beginning of this millennium to the time I am writing, fully confirms the Gutenberg–Richter empirical law.

Once again, however, it is more useful to consider the frequency distribution of a more meaningful physical quantity, such as the released energy scaled to its

---

[40] For experts and sticklers, I selected from the Tycho catalog only those stars with a B-V color index 0.6 and 0.7, which corresponds to a surface temperature $T$ between 5750 and 6100 K.

[41] Note that even earthquakes with $m < 3$, which are usually detected at the surface only by seismographs, can release an underground energy comparable to an aerial bombardment.

[42] See https://scedc.caltech.edu.

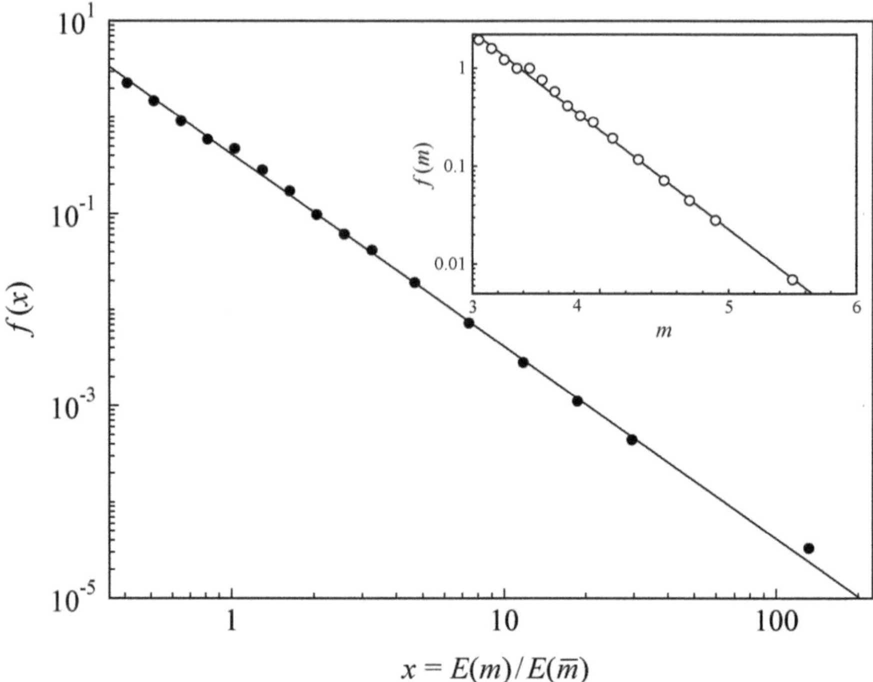

$$x = E(m)/E(\overline{m})$$

**Fig. 1.18** Inset: Relative fraction of earthquakes with Richter magnitude $3 \leq m \leq 6$ detected in South California between January 2001 and November 2023, which have an average magnitude $\bar{m} \simeq 3.44$ The straight line in the semi-log plot has slope -1. Body: Frequency distribution for the energy $x = E(m)/E(\bar{m})$ released by these seismic events normalized to its value for $m = \bar{m}$, fitted as $f(x) = Ax^{-2}$

value $\overline{E} = E(\bar{m})$ corresponding to the average magnitude $\bar{m}$.[43] As shown in the body of Fig. 1.18, we obtain again a power law $f(E/\overline{E}) = A(E/\overline{E})^{-\alpha}$ with an exponent $\alpha \simeq 2$. While it is not easy to determine the trend of the brightness of stars, predicting the intensity of earthquakes seems to be almost mission impossible. However, attempts to justify the Gutenberg-Richter law and to frame it in a more general context have generated many interesting theoretical ideas.

**Example 1.16** (*On companies, towns, and income*) Scale-free distributions are ubiquitous in economic and social data too. Here I will discuss two specific examples concerning the country where I live. Let us first try to classify the companies operating in Italy according to the total number of employees $N$ working in each individual company. Here we mean the term "companies" in the broad sense, considering industries of all productive sectors, to which belong companies that can have tens of

---

[43] Like in the example of the distribution of star luminosity, this requires to take correctly into account the effect of the change of variable on the interval width, so to keep a unit area under the histogram.

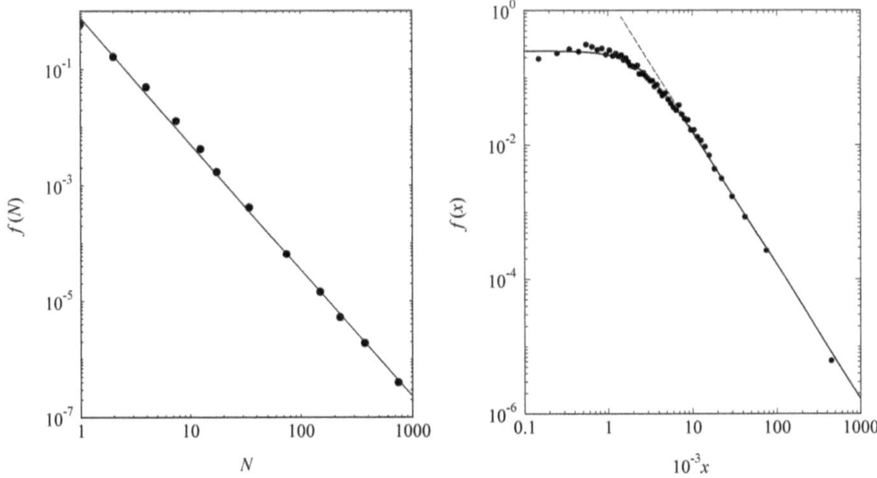

**Fig. 1.19 a** Frequency distribution of the working units in Italy as a function of the number $N$ of employees fitted as $f(N) = AN^{-\alpha}$, with $\alpha \simeq 2.2$. **b** Distribution of the Italian municipalities as a function of the number $x$ of residents. The full line is a fit with a FB-like function as described in the text

thousands of employees, businesses such as small family-run shops, even independent contractors and freelances. More properly, then, we should speak of "working units".

The question is, can we identify a standard size, i.e., a typical number of employees for a working unit? Fig. 1.19a, which shows the distribution of working units with a number of employees between 1 and $1000^{44}$ as a function of $N$, gives us a negative answer, for the relative frequencies follow once again a power law distribution $f(N) \propto N^{-\alpha}$, with $\alpha \simeq 2.2$. I personally find this result quite surprising. Such a fast decrease in the number of working units with $N$ means in particular that more than half of the Italians work in units with less than ten employees, a fact that has obvious social interest.

As a second example, I have considered the population of all Italian towns and cities with less than 1,000,000 residents, which excludes only Rome and Milan. Figure 1.19b shows that the frequency distribution, approximately constant for small towns, decays as a power law too, with an exponent very close to $-2$ for all municipalities with more than 5–10 thousands residents. However, the *whole* distribution is nicely fitted as

$$f(x) = f(0)\left[1 + \left(\frac{x}{x_c}\right)^2\right]^{-d}, \tag{1.19}$$

---

[44] ISTAT, *9th General Census of Industry, Services and Non-profit Institutions*, 2011 (unfortunately, detailed data on larger companies are not provided). I found that a similar power-law trend is shared by the data for many other countries.

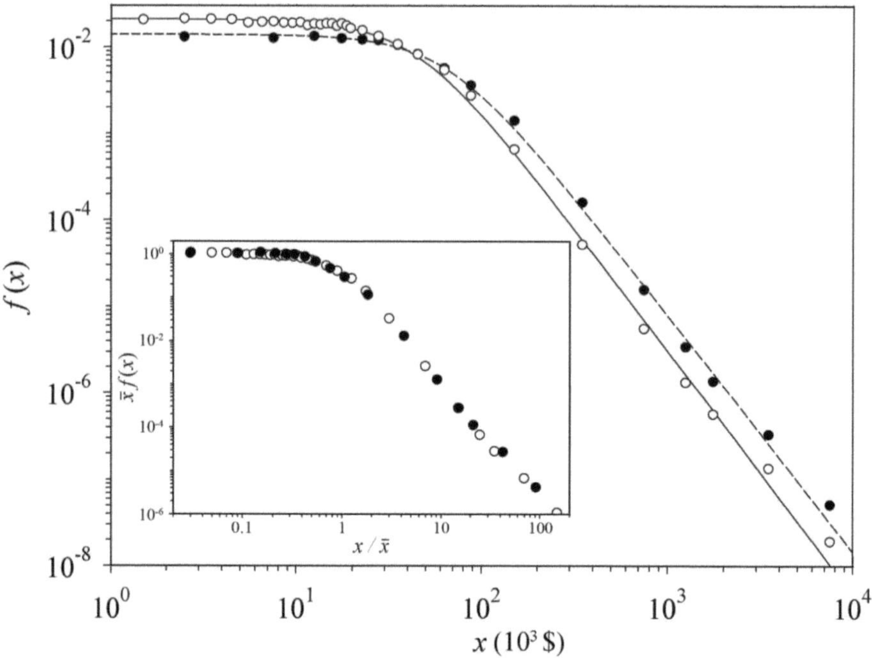

**Fig. 1.20** Distribution of the gross incomes in thousands of dollars of US residents, as reported in tax return forms for the years 2000 (○) and 2020 (●), fitted with Eq. (1.19). The inset shows that the two distributions, when scaled to their average values, superimpose almost perfectly (*Source: US Internal Revenue Service*)

with $d = 1$ and $x_0 \simeq 6500$, which curiously resembles the Fisher-Burford (FB) function[45] used to fit the scattering of light from a fluid close to the critical point or from fractal particle aggregates (where $2d$ is the fractal dimension). In this regard, the "roll-off" value $x_0$ plays the role of the minimum town size above which scale invariance holds.

Actually, Eq. (1.19) is apparently capable of describing other data sets showing a power-law decay but saturating to a maximum frequency for small values of the variable. For example, Fig. 1.20 shows that the same function rather accurately fit the distributions for the gross income of the US residents for the years 2000 and 2020 with a similar exponent $d \simeq 1.4$. In fact, if the gross income is scaled to its average value for the two years, $\bar{x}_{2000} \simeq 50.2$ and $\bar{x}_{2020} \simeq 82$ (and $f(x)$ is multiplied by $\bar{x}$, so to keep the same area under the distribution) the two distributions nicely superimpose. Note that the increase of about 63% in two decades is not much larger than the cumulative rate of inflation of about 54%, so that the Americans have not become much richer (or at least, not *all* of them)!

---

[45] M. E. Fisher, R. J. Burford, Phys. Rev. **A156**, 583 (1967).

Personal income, which according to Fig. 1.20 varies over several orders of magnitude,[46] is particularly suited to highlight why quantities distributed as a power-law $f(x) \propto x^{-\alpha}$ can be a nightmare for statisticians. Indeed, the long tails of these distributions, much longer than those of an exponential whatever the value of the exponent $\alpha$, poses serious problems on representativeness. Suppose for instance that you want to estimate the distribution of the body weight on a sample of 1,000 individuals. This is not a large sample but, paying some attention on individual selection, you could at least get a reasonable approximation of the mean or standard deviation of the distribution. Anyway, the result would not change much if you select Mr. A, a man in his sixties with a balanced diet and in decent physical shape, instead of Mr. B, a Class III obese addicted to fast food. After all, given the distribution in Fig. 1.7, it is almost impossible that the weight of these two individuals differs by more than a factor of three. But suppose you want to analyze the distribution of the *income* of the same individuals, and that Mr. A goes by the first name of Elon or Jeff…As a matter of fact, outcomes that occur very rarely, but that completely distort the values obtained for the parameters of a distribution, can also occur in apparently less extreme situations.

**Example 1.17** (*Scale invariance and Benford's law*) The town population data in Fig. 1.19b also allow us to illustrate a rather puzzling distribution that often occurs in statistics, particularly when dealing with scale-invariant properties. Suppose we consider the frequency distribution of the *first* digit of the number of people living in each one of the Italian municipalities we discussed. Christened by the decimals of $\pi$, we may naively expect it to be a rather uniform distribution from 1 to 9. Nothing could be further from the truth! Fig. 1.21 shows that the digit 1 appears in about 30% of the cases, while the frequency of the larger digits decreases monotonically, until the towns with a number of residents starting with 9 is only 4%. This very surprising distribution is actually rather common, occurring in a variety of situations. First pointed out by the American astronomer Simon Newcomb, it was thoroughly investigated by Frank Benford of the General Electric Company, who assembled thousands of numbers from diverse sources ranging from the area of lakes and the length of rivers to molecular weights and, precisely, population size, showing that the frequency for the first or leading digit of these numbers followed the distribution already suggested by Newcomb, who for the digit $d$ predicted that

$$f(d) = \log\left(1 + \frac{1}{d}\right) \tag{1.20}$$

There have been a large number of diverse attempts to account for this ubiquitous law, but no one truly conclusive. Indeed, quoting Rachel Fewster, who wrote a nice and accessible paper about Benford's law[47]

---

[46] At least 4, but remember that Fig. 1.20 includes only individuals with a gross income up to 10 million dollars. IRS data show that in 2020 there were about 26,500 US residents with a larger income (no upper limit is stated).

[47] R. M Fewster, *The American Statistician*, **63**, 26 (2009).

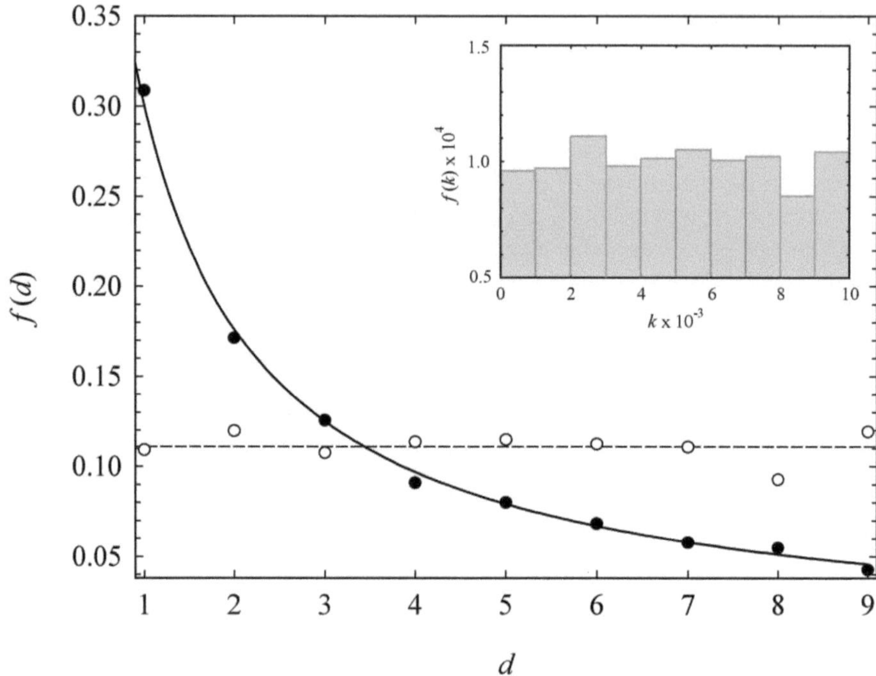

**Fig. 1.21** Distribution of first digit $d$ of the number of residents on Italian municipalities ($\bullet$) with a population larger than 1,000 inhabitants, compared with the corresponding leading digit distribution for the group of four decimals of $\pi$ ($\circ$), whose histogram is shown in the inset. The full line is the Benford law discussed in the text

Benford's Law, also known as the first-digit law, has long been seen as a tantalizing and mysterious law of nature. Attempts to explain it range from the supernatural to the measure-theoretical, and applications range from fraud detection to computer disk space allocation. Publications on the topic have escalated in recent years, largely covering investigation of the law in different data sources, applications in fraud and computer science, and new probability theorems. The underlying reason why Benford's Law occurs is, however, elusive.

It is however worth noticing that Benford's law typically holds only for distributions covering several order of magnitudes in frequency. Indeed, let us contrast what we have found for the data in Fig. 1.19b with the frequencies of the first digit for a much flatter distribution. To this aim I subdivided the decimals of $\pi$ in groups of four digits, obtaining then 2,500 integers ranging between[48] 1 and 9,999. As expected for a normal number, their frequency histogram, shown in the inset of Fig. 1.21, is almost

---

[48] Of course, initial zeros are not considered.

uniform. At variance with the town population data, however, also the distribution of the first digit of these numbers is approximately constant. We may then try to find a simple argument to explain the strong affinity between distribution extending on a wide frequency range and Benford's law. Yet, this first requires to cut our teeth with some basic concepts in probability, which is what we will do in the next chapter.

# Chapter 2
# Probability Basics

> *Comment oser parler des lois du hasard?*
> *Le hasard n'est-il pas l'antithèse de toute loi?*
> *J. Bertrand*

As Julius Caesar would have said, the moment has come to "roll the dice"—in a non figurative sense as well—and ask ourselves if we develop a theoretical framework describing the behavior of a statistical variable. Just as relative frequency is the basic concept to describe statistical data, the keystone of theoretical modeling is the idea of probability.

The task we set ourselves seems at first glance easy, since the concept of probability is well rooted in common sense and language. The trouble is that the meaning we associate with it changes depending on the situation. Consider, for example, the following statements:

(A)  *The probability that a particular digit in the sequence of decimals of $\pi$ is equal to "seven" is 10%;*

(B)  *The probability of getting "four" by rolling a fair die is 1/6;*

(C)  *The probability that tomorrow a rainstorm spoils our trip is more than 20%.*

These sentences imply quite different readings of the concept of probability.

(A)  Since the relative frequency with which we obtain a particular digit tends to settle, as the number of considered digits increases, around a value $p \simeq 0.1$, we are stating that it is plausible to find approximately $Np$ sevens within a group of $N$ digits. Here by "probability" we therefore mean some kind of limit of the relative frequencies as the sample size increases. Since we have no prior indication of how the digits are distributed, the natural way to define $p$ is an "experimental" approach.



R. Piazza, *An Invitation to Probability and Data Analysis for Physicists*, UNITEXT for Physics, https://doi.org/10.1007/978-3-031-83856-9_2

(B)    We could think of 1/6 as a limit frequency here too, but to give this estimate, none of us really feels the need to start rolling a die furiously. When we think of a "fair" die, we assume *a priori* that each face will approximately appear with the same frequency. Perhaps even in this case our response comes from experience, but it is difficult to appeal to background knowledge when, for example, we speculate that in an ideal gas each atom moves along any direction with the same probability.[1] This *equiprobability* assumption is often a simple first guess, but further considerations based on the *symmetry* of the problem may lead us to make more complex assumptions.

(C)    Here we cannot surely evaluate the degree of probability by analyzing a sequence of identical events (there is only one tomorrow!), nor can we even identify a class of equiprobable events. What we do is use information we already have, like the value of atmospheric pressure, the height at which swallows fly, or the infallible nose of our grandfather, to attribute a degree of probability to a future event. We are thus *inferring* our subjective degree of certainty from other facts we are aware of.

Without discussing for now the solidity and effectiveness of these diverse *interpretations* of probability, let me just anticipate that each of them does not seem to be able to fully account for all the situations in which we use probability concepts. Fortunately, it is possible to develop an axiomatic theory of probability that abstracts from the particular interpretation we give it, wherefrom rigorous calculation rules can be deduced. This general and robust approach, which works in almost all situations of interest for natural sciences, was pioneered by Andrej Nikolaevič Kolmogorov. In his own words[2]

> The theory of probability, as a mathematical discipline, can and should be developed from axioms in exactly the same way as Geometry and Algebra. This means that after we have defined the elements to be studied and their basic relations, and have stated the axioms by which these relations are to be governed, all further exposition must be based exclusively on these axioms, independent of the usual concrete meaning of these elements and their relations.

Here we will not delve much into the abstract theory of probability, which requires fairly complex mathematical tools, contenting ourselves with an intuitive non rigorous approach. The basic idea is of introducing calculation rules derived from what we mean in practical terms by "measuring", thus looking at probability as a particular *measure* associated with the subsets of a given set of possible outcomes.

---

[1] Note however that stating what "equal probability" means is in this case far more complicated than for the results of rolling a die. While the latter are a finite number, the possible direction of motion of an atom are not only infinite, but also continuous. Defining probability for infinite or even continuous possible outcomes is far more challenging, as we will discover throughout this whole book!.

[2] A. N. Kolmogorov, *Foundations of the Probability Theory*, Chelsea Publ. Comp., New York (1956). It is worth noticing, however, that the first English edition was published only 33 years after the original one (in German).

## 2.1 Rules of Calculation

Structure of the sample space
A statistical experiment generates not just one, but several possible outcomes. The set of all possible outcomes, usually denoted with $\Omega$, is called the *sample space*, and its elements (or points) *elementary events*. For example, when we roll a die, $\Omega = \{1, 2, 3, 4, 5, 6\}$ and a particular result like $\{2\}$ is an elementary event. However, if we were to attribute a probability only to elementary events, the game would end soon. For instance, when we roll a die we want to state a probability to the generic occurrence of an odd score too, regardless of its specific value. We will then call *event* any subset of $\Omega$. Thus, the "odd" event corresponds to the union of the elementary events $\{1\}$, $\{3\}$, $\{5\}$, and therefore to the subset $\{1, 3, 5\}$.

This simple definition works if the sample space is *discrete*, i.e., it contains a finite or an infinite but countable number of points. However, when the elements of the sample space are *uncountable*, for example because they map onto a continuous set[3] like the height or weight of people we consider in the previous chapter, it is often impossible to find a "metrics" that provides a consistent measure of *all* subsets of $\Omega$. In this case, we look for a nontrivial (and hopefully large) collection of subsets $\Sigma$ of the sample space that can constitute the domain of a set function with numerical values satisfying what we intuitively expect from a measure. This can be achieved if $\Sigma$ is a $\sigma$-*algebra* (or $\sigma$-*field*), i.e., a nonempty collection of subsets including $\Omega$ and closed under complement and countable union[4]:

- $\Omega \in \Sigma$ (the whole sample space is an event).
- $E \in \Sigma \implies \overline{E} \in \Sigma$ (if $E$ is an event, then its complement $\overline{E}$ is an event too).
- $\forall i : E_i \in \Sigma \implies \bigcup_{i=1}^{\infty} E_i \in \Sigma$ (if $E_1, E_2, \ldots$ is a finite or numerable collection of events, then their countable union is an event too).[5]

We have defined events in the framework of set theory, but it is useful to discuss them in the context of Boole algebra, where events are actually *logical propositions*. Let us then recall some equivalences between the operation on events in these descriptions:

---

[3] Never forget, however, the important warning about continuity I previously gave!.

[4] It is worth pointing out that one can consider a restricted $\sigma$-algebra of events $\Sigma \subset \Omega$ also for sample spaces with a finite number of points. For example if $\Omega = \{\omega_1, \omega_2, \omega_3\}$ we could take for $\Sigma$ only the subsets $\{\omega_1, \omega_2\}$ and $\{\omega_3\}$ (plus of course $\{\emptyset\}$ and $\{\Omega\}$). This may not be particularly useful in classical problems, but it is sometimes mandatory in quantum mechanics, where the properties $\omega_1$ and $\omega_2$, though physically distinct, cannot be obtained separately.

[5] Recall that $\forall$ is the logic operator "for all".

| Set operation | | Boole algebra |
|---|---|---|
| complement: $\overline{E}$ | negation (NOT): | $\neg E$ |
| union: $E_1 \cup E_2$ | conjunction (AND): | $E_1 \wedge E_2$ |
| intersection $E_1 \cap E_2$ | disjunction (OR): | $E_1 \vee E_2$ |
| identity for $\cup$: $\emptyset$ | identity for $\vee$: | 0 |
| identity for $\cap$: $\Omega$ | identity for $\wedge$: | 1 |

The union and the intersection as set operations (or AND/OR as logical connectives) have the same properties of, respectively, addition and multiplication in elementary algebra, which include[6]:

**Associativity:**   $A \cup (B \cup C) = (A \cup B) \cup C \Longleftrightarrow A \wedge (B \wedge C) = (A \wedge B) \wedge C$
**Commutativity:**   $A \cup B = B \cup A, A \cap B = B \cap A \Longleftrightarrow A \wedge B = B \wedge A, A \vee B = B \vee A$
**Distributivity:**   $A \cap (B \cup C) = (A \cap B) \cup (A \cap C) \Longleftrightarrow A \vee (B \wedge C) = (A \vee B) \wedge (A \vee C)$

Probability space

We now want to associate with each event $E \in \Sigma$ a probability. In current use, probabilities are often given as percentages from 0, if an event is almost impossible, to 100 if it is almost certain, but it is more convenient to assume that probabilities are real numbers between 0 and 1. I mentioned "almost impossible" and "almost certain" events because any coherent scheme requires that events with zero probability can occur, and that an event with unit probability may not happen. For example, if we throw an arrow it is natural that the larger the target, the more easily we hit it. Even if the target shrinks down to a point, we cannot exclude that a lucky shot hits the mark: We can only say that the ratio between the number successful shots and the total number of throws vanishes as the number of attempts increases.[7]

To introduce a "composition rule" for probabilities that agrees with our usual concept of measurements, we just need to note that, when we measure surfaces, the total area delimited by two figures is equal to the sum of the two areas, as long as the two figures do not overlap. Two events that do not "overlap", i.e., such that $E_1 \cap E_2 = \emptyset$ so that when $E_1$ occurs $E_2$ does not and viceversa, are called *mutually exclusive*.

We could then simply introduce a function $P : \Sigma \to \mathbb{R}$ with the property that, for any two mutually exclusive events $E_1$ and $E_2$, $P(E_1 \cup E_2) = P(E_1) + P(E_2)$. If the number of events is not finite, however, we have to extend this property to a countable union of mutually exclusive events. Kolmogorov's axioms can then be stated as follows:

---

**K1**:   $\forall E \in \Sigma :$   $P(E) \geq 0$
**K2**:   $P(\Omega) = 1$
**K3**:   For a collection $\{E_i\}$ of mutually exclusive events: $P\left(\bigcup_i E_i\right) = \sum_i P(E_i)$

---

[6] This is why $\Sigma$ is structured as an algebra (or field).

[7] This is somehow equivalent to what we do in statistical mechanics when we say that some quantity is "negligible in the thermodynamic limit", i.e., for a macroscopic system.

Using axioms K1 and K2, which together provide the "yardstick" of probability, one can easily show that $P : \Sigma \to [0, 1]$, while axiom K3 extends our intuitive rule of measure to a possibly infinite collection of events. The triplet $\{\Omega, \Sigma, P\}$ is called a *probability space*.

Immediate consequences of the axioms are

$$\boxed{P(\emptyset) = 0} \tag{2.1}$$

which is simply obtained from K3 by observing that, for any $E$, $E \cap \emptyset = \emptyset$ and $E \cup \emptyset = A$. Moreover, calculating the probability of both sides of the identity $\Omega = E \cup \overline{E}$ using K2 and K3, we obtain

$$\boxed{P(\bar{A}) = 1 - P(A)} \tag{2.2}$$

Kolmogorov's axiom K3 also implies that probability is *monotonic*, i.e., for any two events $E_1$, $E_2$ such that $E_1$ is a subset of $E_2$

$$\boxed{E_1 \subseteq E_2 \implies P(E_1) \leq P(E_2)}. \tag{2.3}$$

writing $E_1$ as $(E_1 \cap E_2) \cup (E_1 \cap \overline{E}_2)$,[8] the union of two mutually independent events, and noticing that $(E_1 \cap E_2) = E_2$, one has indeed

$$P(E_1) = P(E_2) + P(E_1 \cap \overline{E}_2) \leq P(E_2).$$

With a little effort the monotonicity of $P$ can be extended to a countable decreasing sequence of events $E_1 \subseteq E_2 \subseteq \ldots \subseteq E_n \subseteq \ldots$ to obtain the limit property

$$\boxed{\lim_{n \to \infty} P(E_n) = P\left(\bigcup_{n=1}^{\infty} E_n\right)}. \tag{2.4}$$

Finally, for *any* events $E_1$ and $E_2$

$$\boxed{P(E_1 \cup E_2) = P(E_1) + P(E_2) - P(E_2 \cap E_2)}, \tag{2.5}$$

which intuitively means that we have to subtract out the probability of $E_1 \cap E_2$ to avoid counting twice the points in common between the two events. Formally, Eq. (2.5) is obtained by writing

$$\begin{cases} E_1 \cup E_2 = E_1 \cup (E_2 \cap \overline{E}_1) \\ E_2 = (E_1 \cap E_2) \cup (\overline{E_1 \cup E_2}), \end{cases}$$

---

[8] $E_1 \cap \overline{E}_2$ is often called the *set difference* $E_1 \setminus E_2$.

two relations that are easily checked where the r.h.s. are unions of mutually exclusive events. Then Eq. (2.5) results from applying K3 to both relations and eliminating $P(E_2 \cap \overline{E}_1)$. Note that Eq. (2.4) implies that, for a pair of events, probability is generally *subadditive*, i.e., $P(E_1 \cup E_2) \leq P(E_1) + P(E_2)$.

To keep up a close parallel between probability and measure, when we deal with a small number of events we can visualize the sample space as a square of unit area, associating to each event an area equal to its probability and obtaining the probability of other events by composing them like surfaces. For example, Fig. 2.1 easily allows to check Eqs. (2.2) and (2.3).

**Example 2.1** (*A poor definition of sample space*) Two written exams in Calculus and Modern Physics take place at the same time on the same day. You have a 45% chance of passing the former and a 65% chance of passing the latter. Since the exams take place simultaneously, the events MAT= {"pass Calculus"} and PHY= {"pass Modern Physics"} are obviously mutually exclusive. Therefore, since we have $P(\text{MAT}) = 0.45$ and $P(\text{PHY}) = 0.65$, according to K3 the probability of passing either MAT or PHY is given by $P(MAT \cup PHY) = 1.1$. This is too good to be true, and indeed it is wrong!   The fact is, the previous chances of passing the two exams make sense only if you *take* them, thus the true elementary events are:

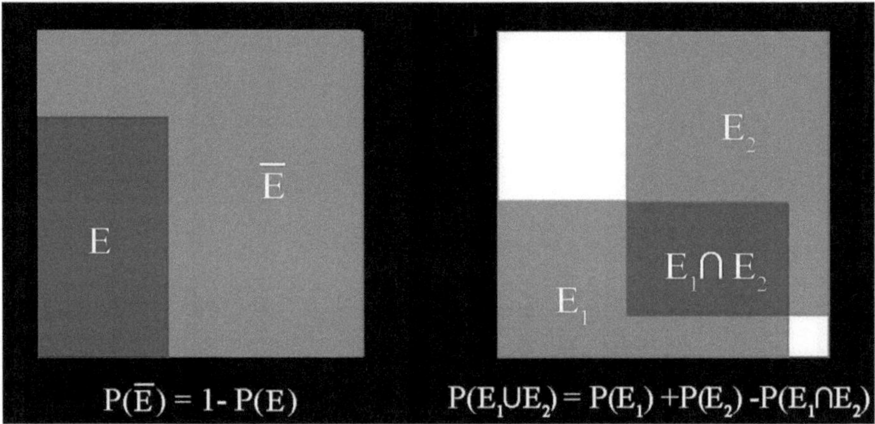



**Fig. 2.1** Using "normalized Venn diagrams" (squares with a unitary surface area) to visualize and check Eqs. (2.2) and (2.3)

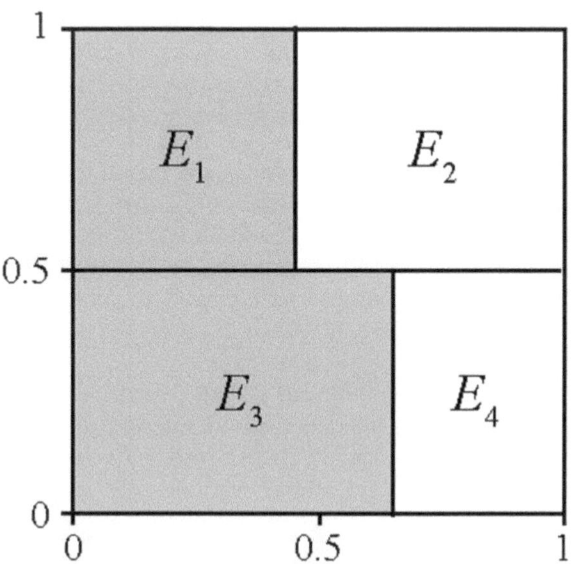

**Fig. 2.2**  The true events in Example 2.1. The grey area corresponds to $P(E_1 \cup E_3)$

$$
\begin{aligned}
E_1 &= \quad \{\text{"take Calculus and } pass \text{ it"}\} \\
E_2 &= \quad \{\text{"take Calculus and } fail\text{"}\} \\
E_3 &= \quad \{\text{"take Modern Physics and } pass \text{ it"}\} \\
E_4 &= \quad \{\text{"take Modern Physics and } fail\text{"}\}
\end{aligned}
$$

Of course, to give probabilities to these events we also have to know how likely you will take each one of them. Assuming that make your choice by flipping a coin we have (see the diagram in Fig. 2.2)

$$
P(E_1) = 0.225; \ P(E_2) = 0.275; \ P(E_3) = 0.325; \ P(E_4) = 0.175,
$$

and the chance of passing one or the other is $P(E_1 \cup E_3) = 0.55$, a much less exciting prospect.[9]

The example we made may sound a bit like a joke, but you will possibly change your mind when we discuss probability in quantum mechanics, where a proper definition of the sample space is anything but trivial.

The axioms we have introduced allow us to combine probabilities of distinct events, but they do not tell us how to assign a probability value to elementary events. In some cases, like in the previous example, it is sufficient to assume that these values are assigned at the beginning of the problem, without asking how. However, a situation in which it is possible to unambiguously assign probability values is when we are dealing with a finite number of elementary events that we can consider as

---

[9] Of course, if you are smart, you better choose to put back the coin in your pocket and take Modern Physics without hesitation.

*equiprobable*, as in the case of the results of rolling a die. In this case, the graphical interpretation we have introduced is particularly simple, since we can obtain the probability of each event by composing "tiles" all with an area equal to the probability of an elementary event. This type of problem is discussed in Example 2.2.

**Example 2.2** (*Rolling classical and quantum dice*) The word "hazard" was originally used to indicate a medieval game of chance played with two or three dice, which may have been brought to Europe by the knights returning from the Crusades, and possibly named from a castle in Palestine or from the word *az-zahr* (meaning "dice") in ancient Arabic. As we shall see, comparing the chances of getting different results when rolling three dice has been studied by the founding father of physics, Galileo.

For the moment, we start with the simpler problem of finding the probability that the sum of the results obtained by rolling two dice is a prime number. Let us first ask ourselves how many distinct results we can get. Since for each value obtained for the first die there are six possible ones for the second, we have a total of 36 pairs of possible results that, assuming that both dice are fair, we can consider as equiprobable. For each pair, we will therefore assign a probability $p = 1/36$. The prime numbers among the possible values of the sum are 2, 3, 5, 7, and 11. However, we must consider in how many ways it is possible to obtain each of these values. A sum of 2 can just be obtained with the pair $(1, 1)$, 3 with $(1, 2)$ and $(2, 1)$, and 5 with the pairs $(1, 4)$, $(2, 3)$, $(3, 2)$, $(4, 1)$. Including the 6 pairs giving 7 and the 2 pairs giving 11, we find that there are 15 possible pairs of values whose sum is a prime number. The probability of obtaining a prime number when rolling two dice will then be equal to

$$P = \frac{15}{36} = \frac{5}{12} \simeq 0.417.$$

Are we sure that this is *always* the correct solution? Well, let's be a bit careful, because we implicity made a crucial assumption. Namely, that we can somehow *label* the two dice differently, for instance because we roll them in sequence, so that we can distinguish the first die we rolled from the second, or because they have a different color or size. But suppose that the two dice are *identical*, and that we roll them simultaneously. How could we find the probability for the different sums? Using an "experimental" approach, we could roll the dice and then take a picture of them resting on the game board, repeating these steps many times so to evaluate the probability of a given result by extrapolating its relative frequency of occurrence. Doing so, however, there is no way to distinguish a results like $(2, 3)$ from $(3, 2)$. With a little effort, you can find that in this case there are 8 pairs corresponding to a prime number over a total of 21 pairs, hence

$$P = \frac{8}{21} \simeq 0.381.$$

But let me be a bit picky by introducing the rule that results where the two (identical) dice show the same face are banned. Then the total number of "legal" results decreases to 15, and those corresponding to primes to 7, so we have

$$P = \frac{7}{15} \simeq 0.467,$$

a different value once again. The three different games we discussed are shown in Fig. 2.3. You may say we can always tell the difference, albeit minimal, between two real dice, and that they are the nit-picking authors who should be banned. As we will see, however, this is not true in the quantum world. And Nature is *very* picky!

Another rather surprising result where the radical difference between classical and quantum probabilities peeps out is the following example.

**Example 2.3** (*Rich, smart, and handsome people*) Do you trust me if I say that the number of people who are rich and handsome is very likely less than the sum of those who are rich and smart and those who are handsome and dull? No? Well, you should. Indeed, it is enough to prove that for *any* triple of events $E_1$, $E_2$, and $E_3$

$$P(E_1 \cap E_2) \leq P(E_1 \cap E_3) + P(E_2 \cap \overline{E_3}), \tag{2.6}$$

and then put

$$
\begin{aligned}
E_1 &= \quad \{\text{"being rich"}\} \\
E_2 &= \quad \{\text{"being handsome"}\} \\
E_3 &= \quad \{\text{"being smart"}\} \text{ (so that } \overline{E_3} = \{\text{"being dull"}\}
\end{aligned}
$$

Eq. (2.5) can be proved as follows. Write

I) $\quad P(E_1 \cap E_3) = P(E_1 \cap E_2 \cap E_3) + P(E_1 \cap \overline{E_2} \cap E_3)$

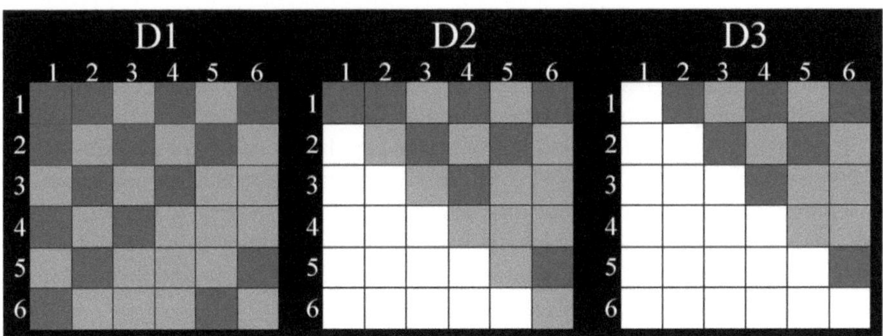

**Fig. 2.3** Graphical interpretation of the three game of chances discussed in Example 2.2 using two distinguishable dice (D1), or indistinguishable dice that can (D2) or cannot (D3) show the same face. All valid results are shown in shades of grey while those corresponding to a prime number are just those in dark grey

II)    $P(E_2 \cap \overline{E_3}) = P(E_1 \cap E_2 \cap \overline{E_3}) + P(\overline{E_1} \cap E_2 \cap \overline{E_3})$

III)    $P(E_1 \cap E_2) = P(E_1 \cap E_2 \cap E_3) + P(E_1 \cap E_2 \cap \overline{E_3}).$

From I) and II) we have

$$P(E_1 \cap E_3) + P(E_2 \cap \overline{E_3}) = P(E_1 \cap E_2 \cap E_3) + P(E_1 \cap E_2 \cap \overline{E_3}) +$$
$$+ P(E_1 \cap \overline{E_2} \cap E_3) + P(\overline{E_1} \cap E_2 \cap \overline{E_3})$$

Since the sum of the first two terms at the r.h.s. is just $P(E_1 \cap E_2)$, and the last 2 are necessarily positive, we surely have

$$P(E_1 \cap E_3) + P(E_2 \cap \overline{E_3}) \geq P(E_1 \cap E_2).$$

The general result we have just found may sound surprising but is unquestionable. At least in the classical world. As we shall see, however, the inequality (2.5) *can* be violated in a quantum system. Actually, it is at the roots of a remarkable and rather disturbing theoretical result by John Bell that has passed countless experimental tests, opening the road to applications of the mysterious feature of quantum entanglement.

Assigning probability values unambiguously is also possible when we can picture the sample space as a domain in $\mathbb{R}^n$, such as a segment, a surface, a volume, so that the probability of each single event can be put in correspondence with the length, area, volume of a subset of that domain: In this case one speaks of *geometric probability*. The simple problems that follow exemplify this method.

**Example 2.4** (*The slide rule*) When I was young, all engineers had to own a slide rule. This fantastic gadget enabled those who could use it to do multiplications and divisions, taking roots, calculating trigonometric functions as quickly and almost as precisely as pocket calculators, at that time in their infancy, with no need for a battery. I won't spend time to explain you in detail how a slide rule works—although any physicist would love it—but let me just reveal the secret of its operating mechanism: it exploits logarithms. Looking at the picture of a slide rule of those times in Fig. 2.4, you can see that the marks on the graduated scale are spaced logarithmically (for example, the spacing between 1 and 2 is exactly equal to the spacing between 5 and 10), so that multiplications become sums and divisions differences.

If we choose a point at random on a standard linear ruler, the probability that it lies between marks $n$ and $n + 1$ is clearly constant, i.e., it does not depend on $n$. Things are however different on a logarithmic ruler. Indeed, if the ruler covers the interval [1, 10], the probability $p$ that the point falls between two consecutive integers is given by the length of interval between the two marks and the total length of the ruler, that we can safely assume to be one. Thus

$$p = \log(n + 1) - \log(n) = \log\left(1 + \frac{1}{n}\right),$$

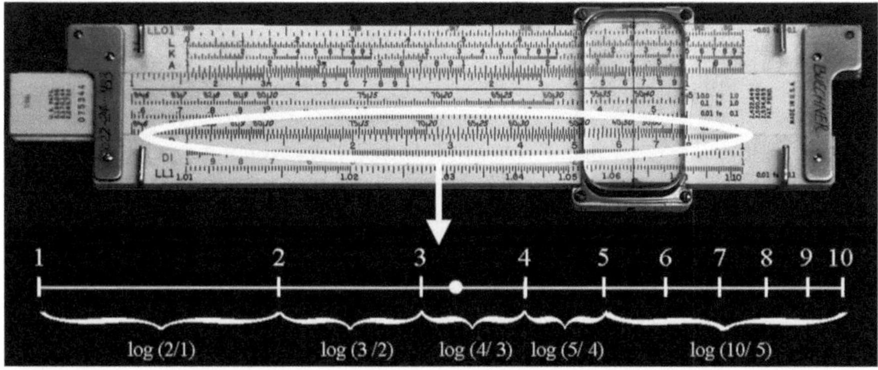


**Fig. 2.4**  A slide rule (top) and one of its logarithmic scales with a randomly placed dot. In the point is placed as in the figure, the probability of being between marks 3 and 4, is $\log 4 - \log 3 = \log(4/3) \simeq 0.125$

which coincides with the Benford law we discussed in the previous chapter, suggesting that the distribution of the first digit must be somehow related to a logarithmic scale.

We can learn more from this example, by considering a question posed by a group of Florentine gentlemen to Galileo.[10]

> The real value of a horse is 100 scudi (*Scudo* was the name of a coin used in various Italian states during and after the Renaissance). A guy estimates it worths 1000 scudi, another one just 10. The question is, who gave the better estimate and made the less extravagant guess?

The query had already be presented to a parish priest, Tolomeo Nozzolini, who thought that guessing 1000 scudi was definitely much worse. Galileo, on the contrary, regarded both estimates as equivalent, reasoning as follows:

> The deviations of the estimates from the right value must be judged according to the *geometric* proportion...Therefore, they depart equally from what is correct two individuals who speculate the double versus the half, or *ten times versus one tenth*, of the correct guess.

Besides pointing out for the first time the significance of what we now call the *relative* standard deviation, Galileo recognized that there are quantities for which errors are the same if they are of the same *order of magnitude*. Since on a logarithmic ruler two decades (e.g. 1 to 10, or 100 to 1000) take on the same space, these are quantities which are equiprobable not on a linear, but on a log scale. In other words, their possible values are equally spaced in *order of magnitude*. We will meet again similar subtle problems in defining equiprobability for continuous quantities.

---

[10] Galileo Galilei, *Lettere intorno alla stima di un cavallo*, Opere, Firenze, Vol.14 (1855) (my loose and prosaic translation).

**Example 2.5** (*"Squaring" a coin*) A coin with a diameter $d = 2\,\text{cm}$ is thrown onto a chessboard made up of squares with sides measuring $\ell = 3\,\text{cm}$. What is the probability that the coin is completely enclosed in a square $S$ without touching its edges?

The solution is readily found if we notice that this will not happen it and only if the center of the coins lies at a distance larger than 1 cm from any sides of the square, and therefore inside a square $S'$ with a side of 1 cm. The probability $p$ we are looking for is then given by the ratio between the areas of $S'$ and $S$, that is,

$$p = \frac{1}{9}.$$

**Example 2.6** (*Catching the connection*) Coming from the beautiful town of Bergamo, you reach the Central Station in Milan with a local train $x$ to catch a connection with an Eurostar $y$ to Rome. However, since we are not in Switzerland, both $x$ and $y$ usually arrive in Milan a bit "capriciously", say, at a random time between 8:00 ($t_i$) and 8:15 ($t_f$). While $x$ stops at the station for 5 min, $y$ stops for only 3 min.

(a)   What is the probability $p$ that you catch the connection? Calling $t_x$ e $t_y$ the arrival times of the two trains, we must have $t_x < t_y + 3$. Since the arrival of both trains is equally likely between $t_i$ and $t_f$, the probability we are looking is the ratio between the area in white and the total area of the square in Fig. 2.5A. Thus we have

$$p = \frac{225 - 144/2}{225} = 0.68.$$

(b)   What is the probability $q$ that you do not have to wait for the Eurostar on the platform? For this to happen, when $x$ arrives $y$ has already to be on the

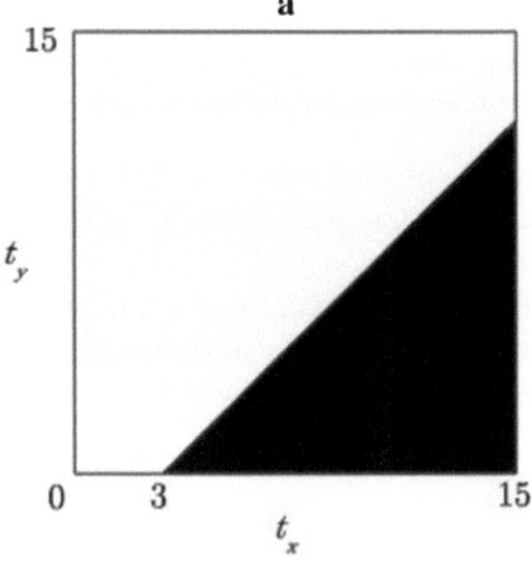


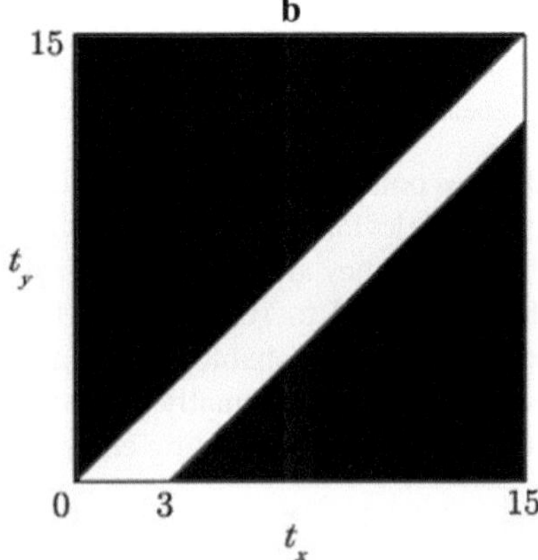


**Fig. 2.5** Combined time line for the local train $x$ and the Eurostar $y$. Both the axes indicate minutes after 8:00

departure track. Therefore, we must have (Fig. 2.5B) $t_y < t_x < t_y + 3$, thus $q = p - 1/2 = 0.18$.

**Example 2.7** (*From a stick to a triangle*) We want to break a stick of unit length in three pieces and use them to make a triangle. What is the chance that we succeed, if we choose the two breakpoints at random?

To be able to form a triangle, the sum of any two of the pieces must of course be larger than the third one. Let us then call $x$ and $y$, with $y > x$, the distance of the two breakpoints from one end of the stick, and $A = x$, $B = y - x$, $C = 1 - y$ the length of the three resulting segments. Since $A + B + C = 1$ we must have:

$$\begin{cases} A < B + C \Longrightarrow 2A < 1 \Longrightarrow x < 1/2 \\ B < A + C \Longrightarrow 2B < 1 \Longrightarrow y - x < 1/2 \\ C < A + B \Longrightarrow 2C < 1 \Longrightarrow y < 1/2 \end{cases}$$

Figure 2.6 shows that the previous conditions are satisfied within the white triangle inscribed in the grey triangle, which is the region where, by assumption. $x < y$. The chance $p$ of making a triangle with the three sticks is then

$$p = \frac{\text{area of the white triangle}}{\text{area of the grey triangle}} = \frac{1}{4}.$$

As it often happens with problems in probability, however, with a bit of insight we can reach the same conclusion in a much easier manner. To make a triangle, all the three pieces $A$, $B$, and $C$ must be shorter than half the original stick. If you think a bit, this means that a triangle forms if and only if the two breakpoints are on either

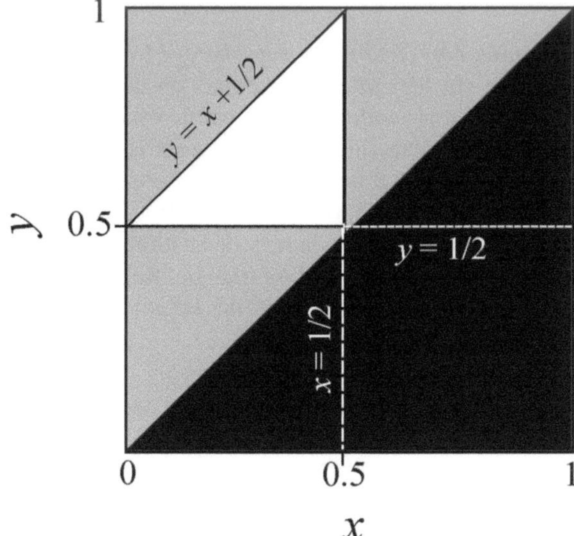


**Fig. 2.6** Building a triangle from a segment

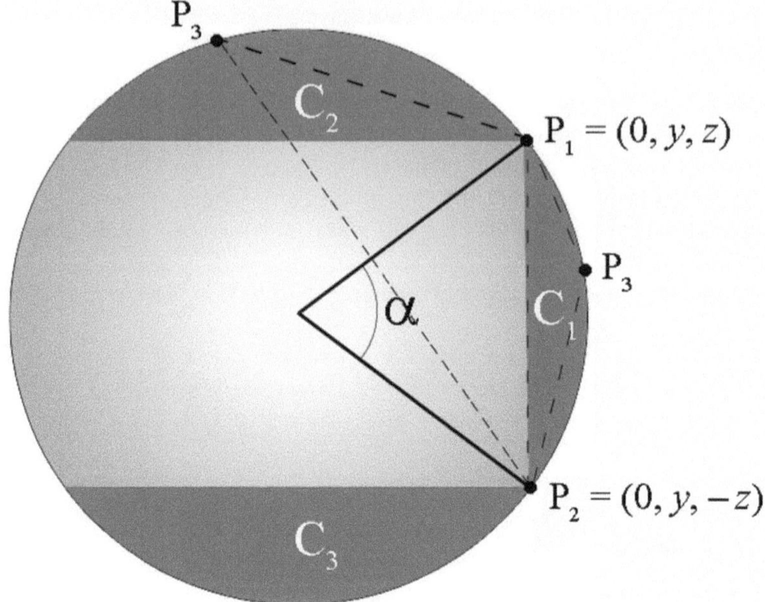



**Fig. 2.7** Building an acute triangle from three points on a spherical surface. The third point has to lie

side of the midpoint of the stick. Since they are placed randomly and independently, the probability that one of the breakpoints is in right and the other one in the left half is simply

$$p(R \cap L) = p(R)p(L) = 1/2 \times 1/2 = 1/4.$$

**Example 2.8** (*Triangles on a sphere*)[11] Like people, triangles can be acute or obtuse. Suppose we first select two points on the surface of a sphere or radius $r$. If we then choose at random a third point, what is the chance $p$ that the plane triangle formed by connecting the three points is acute? Suppose we choose the first two points with central angle $\alpha$. With no less of generality, we may assume they have coordinates $P_1 = (0, y, z)$ and $P_2 = (0, y, -z)$. The section through the $x = 0$ plane in Fig. 2.7 shows that, when $P_3$ lies within one of the three spherical cap $C_1$, $C_2$, $C_3$, we will instead get an *obtuse* triangle. Recalling that the surface area of a spherical cap is $A = 2\pi r h$, where $h$ is the height of the dome, the surface area where we should put $P_3$ is

$$4\pi r^2 - A(C_1) - 2A(C_2) = 4\pi r^2 \sin \frac{\alpha}{2} - 2\pi r^2 \left(1 - \cos \frac{\alpha}{2}\right).$$

---

[11] Adapted from https://www.cut-the-knot.org/Probability/GoldenTriangleOnSphere.shtml.

Thus, dividing by the total area of the sphere,

$$p = \sin\frac{\alpha}{2} - \frac{1}{2}\left(1 - \cos\frac{\alpha}{2}\right)$$

Those we discussed are only very simple examples of probability problems that can be solved using a geometric approach. Actually, the investigation of other questions, related for example to the possible distributions of lines or segments on a $2d$ surface, are often much more complicated, highlighting, as we will see in Chap. 6, a crucial difference between continuous and discrete random quantities.


## 2.2  Independent Events


A concept that we will find extremely useful is that of *independent events*. For example, suppose that in my bedroom there are two drawers, one containing 10 pairs of socks, of which 2 are blue, and the other 12 shirts, of which 3 are blue. I get up early in the morning and, to avoid waking my wife out of a deep sleep, I pick up a shirt and a pair of socks at random from the two drawers without turning the light on. What is the probability that I get both blue socks and a blue shirt? The probability of the event $A = \{$"Taking out a blue shirt"$\}$ is $p_A = 1/4$ and of the event $B = \{$"Taking out a pair of blue socks"$\}$ is $p_B = 1/5$. Since these two events are in no way linked, because drawing out a blue shirt does not does not affect the chance of picking out blue socks, and vice versa, we expect $P(A \cap B) = p_A p_B = 1/20$. We generalize this observation by saying that two events $A$ and $B$ are independent if and only if[12]

$$\boxed{P(A \cap B) = P(A)P(B)} \tag{2.7}$$

Obviously, two mutually exclusive events are *never* independent since the occurrence of one of them implies that the other one does not occur, except in the trivial case where one of the two events has zero probability—any event with zero probability is independent even from itself!

Digging deeper into the concept of independence, note that in the example we discuss A and B belong to *different* sample spaces $\Omega_A$ and $\Omega_B$—we have two distinct drawers, containing different things. We can therefore formalize the concept of independent event by two different sample spaces $\Omega_A$ and $\Omega_B$, and define a *compound event* as an element of the product set[13] $\Omega_A \times \Omega_B$, i.e., an event of the form $E = A \cap B$ with $A \in \Sigma_A$ and $B \in \Sigma_B$.[14] This is actually the most general and rigorous definition of independence, in the sense that any pair of independent events

---

[12] Note that the property of independence is *symmetric*, i.e., if $A$ does not depend on $B$, $B$ does not depend on $A$, which does not fully correspond to its common sense meaning.

[13] Also called "Cartesian product".

[14] More formally, the $\sigma$-algebra of $\Omega_A \times \Omega_B$ is the minimal $\sigma$-algebra containing both $\Sigma_A$ and $\Sigma_B$.

$A$, $B$ can eventually be regarded as a event $E \subseteq (\Omega_A \times \Omega_B)$, although this may not be immediately apparent. For example, the probability of drawing a two of spades from a deck of 40 cards is obviously 1/40, which is the product of the probabilities of extracting a two, 1/10 and of extracting a spades, 1/4. Here, the first $\sigma$-algebra is made up of all the 10 card values, while the second is made up of the 4 possible suits.

The concept of independence is really *the most distinctive feature of probability*, which distinguishes it from a generic function $\mu : \Sigma \to R$ that can be considered a "measure" in mathematics. For example, in the case of Euclidean surface geometry there is nothing like "two independent surfaces", such that the area of their common region is the product of their areas!

**Example 2.9** (*Pólya's proofreading*) Let's assume that this book has at least three readers, Alice, Bob, and myself (as I guess I am supposed to be). After reading the book, both Alice and Bob send me a file detailing all the misprints and mistakes they discovered. Alice finds $A$ errors and Bob $B$, with $C$ errors found by both of them. They are both careful readers (or, at least, much better than I am), but I still wonder whether there are other, undiscovered errors. Can I give an estimate for the number of mistakes that went unnoticed?

Well, if I trust the interpretation of probability as a limit frequency, yes. Indeed, if the (unknown) total number of errors $M$ is relatively large (I bet it will be), calling $P_A/P_B$ the probabilities that Alice/Bob finds an error we may estimate

$$A \simeq P_A M , \quad B \simeq P_B M.$$

Now, Alice and Bob are not acquainted with each other,[15] so they can be consider as independent proofreaders. Therefore,

$$P(C) = P(A \cap B) = P_A P_B \implies C \sim P_A P_B M$$

From the trivial identity $M = (M P_A)(M P_B)/(M P_A P_B)$ we can then appraise

$$M \simeq \frac{AB}{C}.$$

So, the number of misprints and mistakes $M_u$ that remained undiscovered can be predicted to be approximately as

$$M_u = M - (A + B - C) \simeq \frac{(A - C)(B - C)}{C},$$

---

[15] As a consequence of their interactions with the author, they may later meet, join their efforts, and eventually fall in love with each other. Proofreading may provide a viable alternative to dating apps....

which, rather surprisingly, does *not* depend on the total number $M$ of errors. This very simple but remarkable application of independence has been suggested by the Hungarian[16] mathematician George Pólya, probably one of the top experts in probability of all times.

**Example 2.10** (*Looking for ET*) In our galaxy, there are approximately $N = 10^{11}$ stars, many of which have a planetary system. The data collected by the Kepler space telescope lead to estimate[17] that the probability $p$ that around one of these stars orbits a potentially habitable planet is of the order of $3 \times 10^{-3}$. However, being "potentially" habitable does not mean that all conditions for the appearance of life are satisfied: Life requires water, several organic compounds that are essential building blocks for the biomolecules, a suitable atmosphere with a screen protecting from high levels of cosmic rays. As a matter of fact, we do not yet have enough information to establish all the conditions that make a planet actually suited to host life, but let us pessimistically assume that the chance that these conditions are fully met and then life does develop is as low as $q = 10^{-7}$.

We then ask ourselves what is the probability that around a star in the Galaxy orbits another inhabited planet. We must pay some attention to the different sample spaces we consider. For $p$, the set $\Omega_p$ is obviously the set of all the stars in the Galaxy, but for $q$ we refer to the sample space $\Omega_q \subset \Omega_p$ consisting only of those stars with (at least) a planet that hosts life. Since each star in $\Omega_p$ has a probability $q$ of being also in $\Omega_q$, the probability of the event $E=$ {"a star in the Galaxy has a planet that hosts life"} is then

$$P(E) = pq = 3 \times 10^{-11}$$

which looks like a very small value so much that in the whole Galaxy we expect to find only a bunch of planets, say 2 to 4, hosting life.

But let us ask instead what is the probability $P_1$ that *at least* one planet in the Galaxy hosts life. This is given by $P_1 = 1 - P_0$, where $P_0$ is the probability that *no* planet in the Galaxy hosts life. To evaluate $P_0$, we proceed as follows. We have seen that the probability that a star does not have a planet hosting life is $1 - pq$. The probability that another star has not got such a planet too is still $1 - pq$, since the two events are clearly independent. The probability that no star has a planet hosting life is then given by

$$P_0 = (1 - pq)^N.$$

Recalling that $\ln(1 - x) \approx -x$ for $x \ll 1$, we have $\ln(P_0) = N \ln(1 - pq) \approx -Npq$, so

$$P_0 \approx e^{-Npq} = e^{-3} \simeq 0.05 \implies P_1 \simeq 0.95$$

Therefore, despite our low expectations regarding the number of inhabited planets, if our assumptions hold it is likely that at least one planet hosts life!

---

[16] Or possibly Martian, according to Leo Szilard.

[17] https://exoplanets.nasa.gov/news/1664/about-half-of-sun-like-stars-could-host-rocky-potentially-habitable-planets.

What we have just found leads us to reflect about very low probability values. In common language, a statement that something is 99 percent likely is usually interpreted as the event being inevitable. Consider however the next example.

**Example 2.11** (*Rain on the pyramids (timeworn and contemporary)*) Cairo and Las Vegas are known for their dry climates, with an average of 14 rainy days in a year for Cairo and 25 for Las Vegas. Of course, the number of rainy days fluctuates throughout the year, but let us focus only the wetter months of December and January, when it typically rains 7 d in Cairo and 6 in the Sin City, so that the probability that it rains on a day of this period on the pyramids of Giza, near Cairo, and of the Luxor Casino Hotel, in Las Vegas, are $p_c = 7/62$ and $p_v = 6/62$, respectively. The probability $p$ that it rains *on the same day* in both cities is then

$$p = p_c p_v = \frac{7}{62} \times \frac{6}{62} \simeq 0.01,$$

since these are clearly two independent events. So, it is 99% likely that, on a December or January day, it won't rain simultaneously in the two desert cities. Reasoning as before, however, the probability that this *never* happens in any of the $N = 62$ days of these two months is

$$P_0 \approx \exp(-N p_c p_v) = \exp(-21/31) \simeq 0.5,$$

so there is still a 50% chance that this will actually happen.

Surprised? Well, you will soon learn that Lady Luck holds *a lot* of surprises that can baffle our intuition. Like in the next example.

**Example 2.12** (*The judoka dilemma*) You are a judo green belt. The test you must take consists of facing two opponents in three fights and winning two consecutive ones, with the rule that you cannot face the same opponent in two consecutive matches. The opponents are your friend $G$, who has recently started practicing this sport and is only a yellow belt, and your instructor $N$, who is obviously a black belt. The possible sequences of matches you can face are clearly $GNG$ and $NGN$. Which one suits you? At first glance, the first sequence seems more tempting, since you fight twice against the weaker opponent, but it is not like that. Call $p$ and $q$ the chances of defeating $N$ and $G$ respectively, with $p < q$. If you choose the sequence $GNG$, you pass the test if you win in the first and second of the three fights (event $A$), or in the second and third (event $B$). Assuming you are tireless and that you are not discouraged by a defeat, victories in two different fights may approximately be regarded as independent events, hence $P(A) = P(B) = pq$. Thus the probability of winning all three fights, which is the event $A \cap B$, is $pq^2$ and your overall probability of victory is:

$$P(A \cup B) = P(A) + P(B) - P(A \cap B) = pq(2 - q).$$

If you instead choose the sequence $NGN$ the reasoning is similar, with the difference $P(A \cap B) = p^2 q$, so we have

$$P(A \cup B) = pq(2 - p)$$

and, since $p < q$, you better choose the sequence $NGN$.

What can we say for three or more events? If they are independent pairwise, we might conclude that they are "totally" independent, i.e., independent as a whole, but this wrong. See the following example.

**Example 2.13** (*Dressing in solid blue and white*) Suppose you have two shirts, say blue and white, and two pairs of pants, also blue and white. Consider the events $A =$ {"wear the blue shirt"}, $B=$ {"wear the white pants"} and $C=$ {"be dressed in plain colors"}. It is easy to show that

$$P(A \cup B) = P(A)P(B)$$
$$P(A \cup C) = P(A)P(C),$$
$$P(B \cup C) = P(B)P(C)$$

but if any two of these events occur, the third necessarily does not. Therefore, the three events are not totally independent.[18]

Three events $A$, $B$, $C$ are said to be totally independent if they are pairwise independent and if *in addition* we have $P(A \cup B \cup C) = P(A)P(B)P(C)$. Generalizing to an arbitrary number of events, we shall say that $E_1, E_2, \ldots, E_n$ are totally independent if and only if

$$\boxed{P(E_{j_1} \cap E_{j_2} \cap \ldots \cap E_{j_m}) = P(E_{j_1})P(E_{j_2}) \ldots P(E_{jm})} \qquad (2.8)$$

for every index subset $\{j_1, j_2, \ldots, j_m\} \subseteq \{1, 2, \ldots, n\}$.

## 2.3  Conditional Probability

If two events $A$ and $B$ are *not* independent, we expect that the probability that $A$ takes place is modified by the occurrence of the event $B$, and vice versa. For example, the probability that the result of rolling a die is the event $A=$ {"two"} is $1/6$. But if we already know for sure that the result of the roll is even, that is, if event $B=$ {"even"} occurs, the probability of the occurrence of $A$ becomes $1/3$, because there are only three even numbers.

---

[18] How would you write the event $C$ using $A$ and $B$?.

We will then call *conditional probability* $P(A|B)$ of $A$ "given" $B$ the probability of obtaining $A$ when the event $B$ occurs with certainty. Of course, for two independent events we want to have $P(A|B) = P(A)$. We then define, for $P(B) \neq 0$,

$$\boxed{P(A|B) = \frac{P(A \cap B)}{P(B)}}\,, \tag{2.9}$$

which because of Eq. (2.5) satisfies our requirement. The definition corresponds to stating that the probability that both $A$ and $B$ occur is equal to the product of the probability that $B$ occurs times the probability that $A$ occurs given $B$. Note that if we instead evaluate $P(B|A)$ we get:

$$P(B|A) = \frac{P(B \cap A)}{P(A)} \implies \boxed{P(B|A) = P(A|B)\frac{P(B)}{P(A)}}\,, \tag{2.10}$$

which is *different* from $P(A|B)$ unless $P(A) = P(B)$. For instance in the die rolling example, Eq. (2.10) obviously gives $P(B|A) = 1$.

Similarly, considering that $A \cap B$ and $A \cap \overline{B}$ are mutually exclusive events, one has

$$\boxed{P(A) = P(A|B)P(B) + P(A|\overline{B})P(\overline{B})}\,. \tag{2.11}$$

This apparently trivial "either-or" option turns out to be surprisingly useful for solving seemingly complex problems.

Expressions like those we used to introduce the concept of conditional probability ("if we know for certain that...") seem to imply something subjective, as if the probabilities of future events were modified by our degree of knowledge about the occurrence of other events. While comparing the different interpretations of probability, we will show how this view, when used with little caution, can lead to baffling conclusions. Expressions like the previous one just serve to make the idea of conditional probability more familiar to us. Let us instead read Eq. (2.9) in the light of our graphic scheme, as in Fig. 2.8. The occurrence of $B$ causes the sample space to "collapse" into the set of events that are compatible with $B$. Therefore, the probability of "$A$ given $B$" is nothing more than the total probability of $A$ in an event space restricted to the subset associated to the event $B$. This means that:

(a)   Only that part of $A$ which is contained in $B$ is considered.
(b)   We change yardstick, i.e., the probability of an event is no longer related to the area of $\Omega$ but to that of $B$.

The definition we have given of $P(A|B)$ operationally amounts to apply this procedure, which can be seen as a *projection operation* $\Omega \to B$. Note that, in this graphical interpretation, $A$ does not depend on $B$ if the area of $A \cap B$ is to the area of $B$ as the area of $A$ is to the area of the whole $\Omega$, i.e., if the "scaled area" of $A$ is not modified by a change of scale that turns $\Omega$ into $B$.

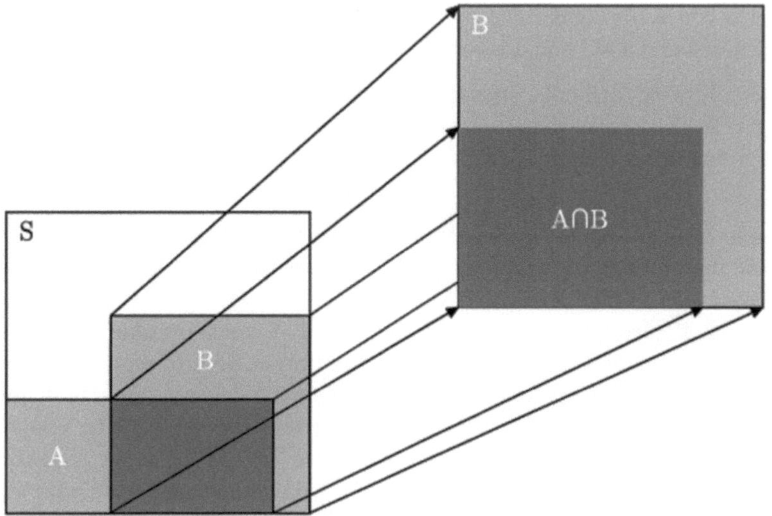


**Fig. 2.8** A geometric interpretation of conditional probability

One can actually take a reversed view, assuming $P(A|B)$ as a *primary* concept, and stating that two quantities are independent if and only if $P(A|B) = P(A)$ and $P(B|A) = P(B)$. As a matter of fact, some authors, in particular those who favor an "inductive" (or "Bayesian") interpretation of probability that we will later discuss, regard *all* probabilities as conditional. Thus, even when we simply write $P(A)$, we actually mean $P(A|\Omega)$, which is the probability that something happens whatever else happens. Note that, in our geometric interpretation, $P(A|\Omega)$ is a projection onto the whole sample space without any restriction.[19]

**Example 2.14** (*Drawing balls of two colors*) As we will see, a lot of questions in probability can be mapped onto an equivalent problem of drawing different objects from a container, usually called an "urn". Let us then consider a simple problem about conditional probability. An urn contains three red balls and two blue balls. What is the probability that in a series of two draws, a red ball is drawn first and then a blue ball? The probability that a red ball is extracted in the first draw is obviously $P(R) = 3/5$. However, if we do not put the ball back in the urn, the result of the second draw is not independent from the first, since now the number of balls in the urn has changed and so has the distribution of balls between the two colors. Since there are four balls left, two of which are blue, we will have $P(B|R) = 1/2$. Then, for the overall probability we have

$$P(R \cap B) = P(B|R)P(R) = 3/10.$$

---

[19] One of the advantages of this approach is that one then *defines* $P(A \cap B) = P(A|B)P(B)$, which holds even when $P(B) = 0$.

It is up to you verifying that the same result is obtained by assuming that a blue ball is drawn first and a red ball afterwards.

The following example stresses the crucial difference between $P(A|B)$ and $P(B|A)$. It also shows that it is often easier to evaluate one of these two conditional probabilities than the other, using then Eqs. (2.10) or (2.11) to find the harder one.

**Example 2.15** (*A scaring test*) Suppose that you have not been feeling well for some time and therefore go to the doctor for a check-up. Your doctor looks quite worried and advise you to take a test to check whether you have got a rare and incurable disease that affects only one in a thousand people. The test is rather reliable, but not completely, since it has given a correct answer only in 95% of the tested cases.

Unfortunately the test comes back positive, so you despair, thinking that you are affected with a probability $P = 0.95$ by this serious illness. Luckily, you are very wrong. Let us say $T$ the event "the test is positive" and $M$ the event "you are sick". The fact is, you are *not* looking for $P(T|M)$, which happens indeed with a chance of 95%, but rather to $P(T|M)$, which we can easily evaluate using Eqs. (2.10) and (2.11):

$$P(M|T) = \frac{P(T|M)P(M)}{P(T|M)P(M) + P(T|\overline{M})P(\overline{M})} = \frac{0.95 \times 0.001}{0.95 \times 0.001 + 0.05 \times 0.999},$$

that is

$$P(M|T) \simeq 0.02,$$

which is a much more reassuring prospect. The result may look paradoxical, but consider how much more frequently the test will give a positive result to a healthy person than to a sick one, as the tested individuals are overwhelmingly healthy.[20]

Chess players generally agree that the white, who moves first, begins the game with a certain advantage, and almost two centuries of statistics tend to support this idea. To compensate the advantage of the first move, in the Go game the Black (who plays first) is even given an initial penalty called *Komi*. Recently, the first-mover advantage has acquired noticeable—though sometimes questionable–relevance in marketing strategies. In the following example we will scrutinize the origin of this benefit in a much simpler game.

**Example 2.16** (*The first-mover advantage*) Tom and Jerry throw two dice, in a game where the winner is the first to get a sum of the dice values equal to 11. If Jerry throws before Tom, what is the probability $p$ that he wins?

Calling $J$ the event "Jerry wins" and $F$ the event "11 is obtained on the first throw", using the "either-or" option (2.11) we have

$$p = P(J) = P(J|F)P(F) + P(J|\overline{F})P(\overline{F}),$$

---

[20] See however Chap. 6.

where, since we can get 11 with (5,6) or (6,5) out of 36 possible results, $P(F) = 2/36 = 1/18$, $P(\overline{F}) = 17/18$ and, by definition, $P(J|F) = 1$. As for $P(J|\overline{F})$, which implies that 11 does *not* come out on the first throw, this must be equal to the probability that Jerry wins in a game where *Tom* throws first, and therefore it is equal to $1 - p$, because now it is Tom who has a winning probability $p$. Thus we obtain

$$p = \frac{1}{18} + \frac{17}{18}(1 - p) \implies p = \frac{18}{35}.$$

Since $p > 0.5$, Jerry is the favorite, as we could expect.

The next example stems from a personal memory. But don't ask me more...

**Example 2.17** (*Identical or fraternal?*)  We know that twins can be identical (monozygotic) or just fraternal (dizygotic). It is also known that monozygotic twins are about 1/3 of the twin pregnancies.

$Q_1$: What is the probability that a certain twin pregnancy is monozygotic, knowing that an ultrasound examination has shown that the two fetuses are in separate placentas, a fact that occurs for all dizygotic pregnancies but, on the average, for only 1/3 of the monozygotic ones?
Let $M$ be the event "monozygotic twins", $D$ the event "dizygotic twins", and $2P$ the event "distinct placentas". We have

$$P(M \cap 2P) = P(2P|M)P(M) = \frac{1}{3} \times \frac{1}{3} = \frac{1}{9},$$

and for the total probability that the two fetuses are in separate placentas

$$P(2P) = P(2P|M)P(M) + P(2P|D)P(D) = \frac{1}{3} \times \frac{1}{3} + 1 \times \frac{2}{3} = \frac{7}{9}.$$

Therefore
$$P(M|2P) = \frac{P(M \cap 2P)}{P(2P)} = \frac{1}{7}.$$

$Q_2$: How much does this probability changes if a chromosomal map of the amniotic fluid later shows that the twins are of the same sex?
Calling $S$ the event "twins are of the same sex" we have $P(S|D) = 1/2$ (one quarter of the pairs will consist of two males and one quarter of two females) and $P(S|M) = 1$ (all pairs of monozygotic twins are of the same sex!). Noticing that $\overline{D} = M$ and using (2.11) again

$$P(2P \cap S) = P(2P \cap S|D)P(D) + P(2P \cap S|M)P(M).$$

Since all dizygotic twins are in separate placentas, the first term coincides with $P(S|D)P(D) = 1/3$ and similarly, since all monozygotic twins are of the same sex, the second term coincides with $P(2P|M)P(M) = 1/9$. We then obtain $P(2P \cap S) = 4/9$. Noticing that

$$P(M \cap S \cap 2P) = P(M \cap 2P),$$

because the first event is completely contained in the second, we finally have

$$P(M|S \cap 2P) = \frac{P(M \cap S \cap 2P)}{P(2P \cap S)} = \frac{P(M \cap 2P)}{P(2P \cap S)} = \frac{1/9}{4/9} = \frac{1}{4}.$$

As an exercise, try and see how easily you can reach these results using the graphical scheme we introduced.

### 2.3.1  Bayes' Theorem

Under this rather high–sounding name passes a result that is actually a simple extension of the "either-or" option (2.11). Nevertheless, Bayes' theorem is particularly useful for "readjusting", using new data, probability values that had been estimated on the bases of what was previously known.

Consider a series of events $B_i$ that are mutually exclusive and at the same time "fill" the whole sample space,

$$\begin{cases} \forall i \neq j : B_i \cap B_j = \emptyset \\ \bigcup_i B_i = \Omega \end{cases}$$

in other words, the event collection $\{B_i\}$ is a *partition* of $\Omega$ in disjoint subsets. Then, an obvious extension of Eq. (2.11) is

$$P(A) = \sum_i P(A|B_i)P(B_i)$$

Note that it is not at all mandatory that the events $B_i$ have any connections with $A$. For example, if in the evening I eat pasta, risotto, soup or *couscous*, but never two of these entries at the same time, the probability that it will rain tomorrow in Milan is equal to the probability that it will rain if I eat pasta times the probability that I eat pasta, plus the probability that it will rain if I eat risotto times the probability that I eat risotto, plus…As we have seen in the previous examples, this "decomposition" is particularly useful when it is easier to calculate the probabilities of A conditioned by a specific event $B_i$.

Consequently, using (2.10) and the previous expression for $P(A)$, we can write the probability of one of the events $B_i$ conditioned by the occurrence of the event $A$ as

$$\boxed{P(B_i|A) = \frac{P(A|B_i)P(B_i)}{\sum_j P(A|B_j)P(B_j)}}$$ (2.12)

which is precisely Bayes' theorem. In other words, the occurrence of the event $A$ allows us to give a new and more reliable estimate for the probability of the event $B_i$. In particular, we note that when the condition $B_i$ makes the occurrence of $A$ much more plausible compared to the other mutually exclusive conditions $B_j$, i.e., $P(A|B_i) \gg P(A|B_{j \neq i})$, the occurrence of A considerably increases the change of being in the condition $B_i$. Equation (2.12), which we have already implicity used in Example 2.15, is at the roots of Bayesian estimation methods that we will discuss in Chap. 6. We'll soon discover how useful it is when we need to assess the degree of "information" associated with a collection of events that have a known probability.

## 2.4 The Art of Combinatorics

We have already introduced the concept of a compound event in Sect. 2.2 as an element of the product set between two elements belonging to distinct sample spaces. Elementary compound events are therefore pairs $\{\omega_1, \omega_2\}$, with $\omega_1$ and $\omega_2$ points of the two samples spaces.[21] Compound events can be regarded as the result of a sequence of experiments, also of different kind like flipping a coin and then drawing one of the 20 labelled balls contained in an urn, where the elementary compound events are $\{\text{head}, k\}$ or $\{\text{tail}, k\}$, with $k = 1, 2 \ldots 20$. More generally, we can consider sequences of $N$ experiments, where the elementary compound events are elements of $\Omega_1 \times \Omega_2 \times \ldots \times \Omega_N$.

The rolling of two dice that we discussed in Example 2.2 tells us that the problem of counting the total number of elementary events may be rather challenging. Indeed, when we considered two dice that can be labelled, for example because they have a different color or because they are rolled in sequence, the elementary events were of the form $\{n_1, n_2\}$ with $n_1$ and $n_2$ than could take all values between 1 and 6. However, when we considered two dice that cannot be distinguished,[22] the number of distinct events went down to 21. In that case, the elementary compound events cannot be considered as points of the Cartesian product $\Omega_1 \times \Omega_2$, where $\Omega_{1,2} = \{1, 2, 3, 4, 5, 6\}$, since in this product space an event like $\{1, 2\}$ is different from the event $\{2, 1\}$.[23] Rather, they are elements of the set $\Omega'$ made of all the pairs of results

---

[21] For uncountable sample spaces, $\omega_1$ and $\omega_2$ must of course belong to $\Sigma_1$ and $\Sigma_2$. In the following, however, we will deal only with finite sample spaces.

[22] Note that being *indistinguishable* is a more stringent condition than being *identical*. In other words, two object that are identical are not necessarily indistinguishable. For example, even if they are perfectly identical in size, color, material, two dice *can* be distinguished if I roll them in sequence or on different boards. Identity is a question of physical properties, while indistinguishability is a statistical concept.

[23] Which is equivalent to say that the Cartesian product of two sets is not commutative.

of rolling the two dice *regardless of the order* in which they were obtained. Thus, we consider as a single one two pairs differing only for a reversal of the events.[24] The last case we considered, where results with the two dice showing the same face are banned, is equivalent to choose a $\sigma$-algebra $\Sigma'$ that does not contains all the elementary events of $\Omega'$, even if their number is finite, excluding all pairs of equal results.

Counting "objects" that cannot be distinguished requires then different and often more elaborate methods. The fundamentals of probability we have discussed provide us with the guidelines to assess the chance of any event, no matter how complicated, but there are no "golden rules" that can solve every counting problem. Counting is actually the subject of the subtle art of combinatorics, whose elements we will use to develop some strategies for solving many recurring problems.

### *2.4.1  Counting Distinguishable Stuff*

Let us start by counting, distributing, reshuffling "objects" that can be distinguished because they are objectively different from one another or, if they are identical, because we can still find a way to "label" them.

#### 2.4.1.1  Simple Counting: Multiple Choices

Suppose that during my course I check the level of learning of the class with a multiple–choice test. The test is made of $k$ questions, which should be given a single answer among $r$ possible ones. You should immediately be able to see that, if the students answer all questions, there are $r^k$ different overall results for the test. We get the same answer if we calculate the total number of ways of putting $k$ balls in $r$ boxes, because we have $r$ possible choices for each ball, or for the number of different extractions of $k$ balls from a box that contains $r$ balls, if, after each drawing, we *put back* the ball in the box (this is called *sampling with replacement*).[25] Similarly, the total number of subsets of a set containing $n$ elements is $2^n$, because for each element, we can choose whether or not it belongs to the subset: here $r = 2$ and $k = n$. The same result is obtained for the total number of different paths in a 1-dimensional random walk of $n$ steps, since each step can either be forward or backward.

More generally, consider a sequence of $n$ identical trials, in which each trial has $k$ possible outcomes (like the number of answers to a given question, or of the boxes in which a ball can be tossed). A result of this compound experiment can be represented

---

[24] Recalling our definition of independence, this means that the outcomes of the two dice *cannot* be regarded as independent because, after rolling them, we go through all results "merging those" that should not be distinguished.

[25] Note that we are counting as distinct two results that differ only for the order of extraction. For example, an event that corresponds to the extraction of the ball number 2, followed by the ball number 7 and the ball number 3 is different from the extraction that leads to the sequence $\{3, 2, 7\}$.

as an ordered sequence $\{r_1, r_2, \ldots r_i, \ldots, r_k\}$, where $r_i$ is the specific outcome of trial $i$. Thus, there are $r^k$ possible results. We can also think of a slightly more complicated experiment, consisting in a sequence of $k$ diverse trials, where each trial has *its own number $r_j$* of possible outcomes, with $j = 1, \ldots, k$: generalizing, we easily see that the total number $R$ of possible results of this compound experiment is

$$R = r_1 \times r_2 \times \ldots \times r_k = \prod_{j=1}^{k} r_j \tag{2.13}$$

So, for example, if I have three suits, five shirts, seven ties, and two pairs of shoes, I can dress in 210 distinct ways, obviously ignoring any elementary criterion of aesthetics. Similarly, if we have to move from town A to town B, passing by town C, and there are 3 roads joining A and C and 2 roads from C to B, we can choose among 6 possible routes.

**Example 2.18** (*Au Bonheur des Dames*)[26] The previous results are very simple, but often a combinatorial problem looks simple only if you attack it from the right side. Suppose, for instance, that we visit the fantastic world of Ladies' Delight. On this planet, each man *must* marry a single woman, but, because the number $k$ of men is much larger than the number $r$ of women, a woman can take as many husbands as she wishes. In how many ways can ladies and gentlemen combine? Seen from the point of view of men, the problem is like those we just met: Each one of them has to make a single choice among $r$—provided, of course, that he tolerates to share his wife with other men—so there are $r^k$ possible outcomes.

Seen from the ladies' side, however, the story is more complicated, because each one of them can choose any number $m \leq k$ of husbands, no matter in which order, with the condition, however, that two women do not share any husbands. Thus they better let the men choose, at least as a sop. Yet, seeing the problem from this side teaches us something interesting. In fact, the ladies' selections split the ensemble of $k$ men into $r$ nonintersecting subsets. Hence, the number of possible ways to partition a set $E$ made of $k$ elements in $r$ subsets $E_i$, such that $E_i \cap E_j = \emptyset$ for all $i$ and $j$, is $r^k$. Suppose however that the former conclusion had been posed to you as a *question*: would you have found an easy answer? Now, if you wish, you can try to extend the result to $k \leq r$ (admittedly, an unfair condition for ladies!), assuming of course that some subsets can be empty.

### 2.4.1.2  Selecting and Rearranging: Permutations

Suppose that we select a number $k$ of students from a total of $n$ in a class, put them in a row, and take a picture of the whole row. Then, the students come back to their desks, and we repeat the whole operation by selecting again $k$ students (not

---

[26] If you wonder about the strange French title of this example, ask Émile Zola.

necessarily different from the first ones), putting them in a row, taking a picture…and so on and so forth. We wish to know how many different pictures $(n)_k$ we can get. Note that two pictures must be regarded as different even if they just differ by the *order* of the students: namely if two pictures show the same students, but comparing them we see that Alice and Bob have exchanged their place, these pictures are *not* the same. Unless we wish to spend a whole semester playing this rather boring game, we better find an easy way to evaluate $(n)_k$ in advance.

In general, what we want to find is the number of ways in which a subset of $k$ elements can be selected, *in a specific order*, from a set of $n$ elements. Aha, but this is just a simple application of the result (2.13) for a compound experiment, where in the first trial we have $n$ choices, in the second one $n - 1$ (a student has already been chosen), and so on, until at trial $k$ there are $n - (k - 1)$ choices. Hence,

$$(n)_k = n(n - 1)(n - 2) \ldots (n - k + 1) \tag{2.14}$$

Note that $(n)_k$ is also the number of different ways of drawing in an orderly way $k$ balls from a box that contains $n$ balls, in a *sampling without replacement* experiment, namely, an extraction in which we *do not* put back the balls in the box.

These "arrangements" of $k$ distinct objects, chosen among a total of $n$, are often called *k-permutations*. The special case $k = n$, which is simply named *permutations* of $n$ objects, is particularly interesting. From (2.14), we see that the number of permutation is just the product of all integers up to $n$: hence, we do not need any new symbols for permutations, because they are

$$n! = 1 \cdot 2 \cdot \ldots \cdot n \tag{2.15}$$

that is, the *factorials* of the integers.[27] It is very useful to include also the case $n = 0$, defining $0! = 1$. Factorials, in particular of *huge* numbers like the Avogadro number, are ubiquitous in statistical physics. We shall come back to them. For the moment, note that $(n)_k$ can be compactly written as

$$(n)_k = \frac{n!}{(n - k)!} \tag{2.16}$$

### 2.4.1.3 Selecting Without Ranking: Combinations

In a sampling without replacement experiment, we may however be interested only in which specific balls have been drawn, and not in the *order* of their extraction, which means that all the $k!$ permutations of the balls we draw count as a single one. This number, which is also the *number of different subsets of $k$ elements ("k-subsets")*

---

[27] For this reason the symbol $(n)_k$ for the $k$-permutations is also known as a *falling factorial*.

*in a set of n*, is called *k-combinations*, and is indicated with the special symbol $\binom{n}{k}$. Then, from Eq. (2.16), we have[28]

$$\boxed{\binom{n}{k} = \frac{(n)_k}{k!} = \frac{n!}{k!(n-k)!}} \tag{2.17}$$

For example, the probability of having a deck poker is $P \simeq 2.4 \times 10^{-4}$. There are indeed $13 \times 48$ hands that are worth a poker, because for each of the 13 groups of 4 cards of equal value there are 48 ways to choose the fifth card, and the probability of any hands is given by $p = \binom{52}{5}^{-1}$, so $P = 624p = 1/4165$.

The numbers $\binom{n}{k}$ are also call *binomial coefficients*, because they are directly involved with the expansion of the *n*th power of a binomial $(x + y)$,

$$(x + y)^n = \sum_{k=0}^{n} \binom{n}{k} x^k y^{n-k}. \tag{2.18}$$

Indeed, each term of degree $k$ in $x$ is the product of $n$ factors, $k$ of which are equal to $x$ and $(n - k)$ to $y$, and there are as many terms of this kind as the number of ways of selecting $k$ factors equal to $x$ out of $n$. This is the simplest form of the so-called *binomial theorem*, already known to the Persian poet and mathematician Omar Khayyam.

The binomial coefficients satisfy some useful relations. For instance,

$$\binom{n}{k} = \binom{n}{n-k}, \tag{2.19}$$

which is directly obtained from the definition (2.17). Moreover, applying the binomial theorem with $x = 1$ and $y = 1$, we get

$$\sum_{k=0}^{n} \binom{n}{k} = 2^n. \tag{2.20}$$

Actually this relation can also be obtained directly, noticing that the left–hand side is the total number of subsets of a set with $n$ elements, while at the right–hand side we first count the number of $k$-subsets, and then we sum over $k$ (which is "counting by classes"), obtaining of course the same results.

The second demonstration is a typical example of a brilliant route to obtain combinatorial results: we prove an identity by showing that both sides count, in a different way, the *same* objects. A beautiful example of this method is given by identity

---

[28] Note that $\binom{n}{0} = \binom{n}{n} = 1$.

$$\binom{n+1}{k+1} = \binom{n}{k} + \binom{n}{k+1}. \tag{2.21}$$

The left–hand side is the number of $(k+1)$-subsets in a set, $S_{n+1}$, of $n+1$ elements. Now, call $S_{k+1}$ one of these subsets, and label $a$ one of the elements of $S_{n+1}$. Subset $S_{k+1}$ either contains $a$, or it does not. If it does, then $S_k = S_{k+1} - a$ is a $k$-subset of $S_n = S_{n+1} - a$. If it does not, then $S_{k+1}$ is a $(k+1)$-subset of $S_n$. In fact, at the right–hand side we separately count all the subsets of the first and of the second kind, and finally sum them.

The last result allows us to build the Pascal triangle[29] which you are arguably familiar with. This construction is usually presented as an equilateral triangle, but I prefer to draw it as a right triangle (which is closer to the original version by Pascal),

$$
\begin{array}{llllllllll}
 & 1 \\
 & 1 & 1 \\
 & 1 & 2 & 1 \\
n & 1 & 3 & 3 & 1 \\
\downarrow & 1 & 4 & \mathbf{6} & \mathbf{4} & 1 \\
 & 1 & 5 & 10 & \mathbf{10} & 5 & 1 \\
 & 1 & 6 & 15 & 20 & 15 & 6 & 1 \\
 & 1 & 7 & 21 & 35 & 35 & 21 & 7 & 1 \\
 & \cdot & \cdot & \cdot & \cdot & \cdot & \cdot & \cdot & \cdot & \cdot \\
\end{array}
$$

$$k \rightarrow$$

where the rows correspond to $n = 0, 1, \ldots$ and the columns to $k = 0, 1, \ldots$, so that the element in position $(i, j)$ is $\binom{n}{k}$. Pascal's triangle is actually obtained by repeatedly applying Eq. (2.21)—look for instance at the numbers in boldface. You can directly check that Eqs. (2.19) and (2.20) hold. Moreover the sum with *alternate* signs of the numbers in a row vanishes,

$$\sum_{k=0}^{n} (-1)^k \binom{n}{k} = 0, \tag{2.22}$$

which can be demonstrated by choosing $x = -1$ and $y = 1$ in Eq. (2.18).

Another interesting consequence of the binomial theorem (2.18) is obtained by fixing $y = 1$ and taking the derivative of both sides with respect to $x$,

---

[29] My hands almost refused to write the expression "Pascal triangle". In fact, this useful construction was already known to the Chinese mathematicians of the 13th century, but was publicized in Europe only in 1556 by Niccolò Fontana, better known as "Tartaglia" (from the Italian verb "tartagliare", which means "to stutter") in his best–known work *Il General trattato di numeri et misure*. Only a century later Blaise Pascal dedicated a full book to the triangle that bears his name. At least outside Italy, for my fellow countrymen will forever dub it the *Tartaglia* triangle!.

$$n(x + 1)^{n-1} = \sum_{k=1}^{n} k \binom{n}{k} x^{k-1},$$

where the term for $k = 0$ can be omitted because it vanishes anyway. Choosing now $x = 1$, we see that *every* natural number $n$ can be written as

$$n = \frac{1}{2^{n-1}} \sum_{k=1}^{n} k \binom{n}{k}, \tag{2.23}$$

which is not so intuitive (at a first glance, it may not even obvious that the right–hand side is an *integer!*).

Equation (2.23) can be checked using the "standard" Pascal triangle too. When drawn as a right triangle, however, the latter has several other interesting features. For instance, the second column is the sequence of natural numbers, the third one the sequence of the so–called "triangular" numbers $\binom{n}{2} = \frac{n(n+1)}{2}$, which are also the sum of the first $n$ natural numbers, the fourth one the sequence of the "tetrahedral" numbers $\binom{n}{3} = \frac{n(n+1)(n+2)}{6}$, which are the sum of the first $n$ triangular numbers, and so on and so forth. Even more interesting is that the sums of the numbers along the diagonals that are perpendicular to the hypotenuse give the sequence of the *Fibonacci*[30] numbers $F_n$, which are defined by the recurrence relation $F_{n+2} = F_n + F_{n+1}$, with $F_1 = F_2 = 1$ (hence, the first $F_n$ are $1, 1, 2, 3, 5, 8, 13, 21, 34, \ldots$). This property of the binomial coefficients is slightly hard to prove. Instead, you should find it easy to show that, when a row of the triangle contains a prime number $p$, *all* the other numbers in that row, except the first and the last, are multiples of $p$.

We can also define the binomial coefficient for *negative* integers,

$$\binom{-r}{k} = \frac{(-r)(-r - 1) \ldots (-r - k - 1)}{k!}, \tag{2.24}$$

which are related to "standard" binomial coefficients by

$$\binom{-r}{k} = (-1)^k \binom{r + k - 1}{k}. \tag{2.25}$$

Indeed,

$$\binom{-r}{k} = \frac{1}{k!} \prod_{j=0}^{k-1} (-r - j) = \frac{(-1)^k}{k!} \prod_{j=0}^{k-1} (r + j) = (-1)^k \binom{r + k - 1}{k}.$$

---

[30] In this case, it is the Italian mathematician, Leonardo Pisano (called "Fibonacci" because he was the son—*filius*—of the rich merchant Guglielmo dei Bonacci, from Pisa), who gets credit for something he did not discover. In fact, this sequence, which he introduced in 1202 in his famous *Liber Abaci*, had already been known for centuries by Indian mathematicians.

Equation (2.25) allows the binomial theorem to be extended to obtain, for every $x, a$ such that $|x| < |a|$,

$$(x + a)^{-n} = \sum_{k=0}^{\infty} \binom{-n}{k} x^k a^{-n-k} = \sum_{k=0}^{\infty} (-1)^k \binom{r + k - 1}{k} x^k a^{-n-k}, \qquad (2.26)$$

where the r.h.s. is in fact the Maclaurin expansion of the l.h.s.[31]

### 2.4.1.4  Set Partitioning and Multinomial Coefficients

Binomial coefficients allow us to count the number of $k$-subsets of a set of $n$ elements. Equivalently, we can say that $\binom{n}{k}$ gives the number of ways this set can be split into *two parts*, respectively containing $k$ and $n - k$ elements. Let us generalize this concept, by evaluating in how many ways we can partition a set in a *generic* number $r$ of subsets. Specifically, we plan to split a set $S$ of $n$ elements into $r$ subsets such that subset $S_i$ contains $k_i$ elements, with the obvious condition $k_1 + k_2 + \ldots + k_r = n$. These numbers, which we shall indicate with $M_n^r$, are called *multinomial coefficients*.[32]

To evaluate $M_n^r$, we first extract $k_1$ elements from $S$, which can be done in $\binom{n}{k_1}$ ways. Then, we draw $k_2$ elements from the remaining $n - k_1$ in $\binom{n-k_1}{k_2}$ ways, and so on and so forth, till we extract $k_{r-1}$ elements (the last $k_r$ elements are then selected too!). Hence, noticing that $k_r = n - \sum_{i=1}^{r-1} k_i$, we have

$$M_n^r = \frac{n!}{k_1!(n - k_1)!} \times \frac{(n - k_1)!}{k_2!(n - k_1 - k_2)!} \times \ldots \times \frac{(n - k_1 - \ldots - k_{r-2})!}{k_{r-1}!k_r!}.$$

Simplifying all common factors at the numerators and denominators, we find

$$\boxed{M_n^r = \frac{n!}{k_1!k_2!\ldots k_r!}} \qquad (2.27)$$

Extending the line of reasoning we used to obtain Eq. (2.18), you should easily be convinced that

$$\left( \sum_{i=1}^{r} x_i \right)^n = \sum_{k_1+\ldots+k_r=n} M_n^r \, x_1^{k_1} x_2^{k_2} \ldots x_r^{k_r} = \sum_{k_1+\ldots+k_r=n} \frac{n!}{k_1!k_2!\ldots k_r!} \, x_1^{k_1} x_2^{k_2} \ldots x_r^{k_r},$$

---

[31] Equation (2.26) is a particular case of Newton's generalized theorem, which is valid for any complex value of $r$ within the open circle of radius $a$ in the complex plane.

[32] Because of their relation to binomial coefficients, the coefficients $M_n^r$ are sometimes indicated as

$$\binom{n}{n_1, n_2, \ldots, n_r}.$$

.

where the sum index emphasizes that $k_1, \ldots, k_r$ must satisfy $\sum_{i=1}^{r} k_i = n$. This *multinomial theorem* holds for all positive integers $n$ and $r$. Choosing all $x_i = 1$, we get a generalization of Eq. (2.20),

$$r^n = \sum_{k_1 + \ldots + k_r = n} \frac{n!}{k_1! k_2! \ldots k_r!} \tag{2.28}$$

Now, this result should remind you of the planet of Ladies' Delight! As we have already seen, at the left–side we have the total number of combinations, since each gentleman must choose one and only one of $r$ possible wives. At the right-hand side, we just see the problem from the *ladies'* side: each woman can get any number of husbands she manages (possibly none),[33] until they are "sold out". Correctly, these two ways of counting coincide.

Let us see a simple application of the previous concepts, which can also be regarded as an introduction to Sect. 2.4.2.

**Example 2.19** (*More on two-color urns*)

An urn contains $b$ blue balls and $r$ red balls. The balls are drawn out from the urn one by one, and aligned on the table in an row. The first question we ask is: how many different (by which we mean, *distinguishable*) alignments can we obtain, if we extract *all* the balls?

The number of permutations of $b + r$ balls is $(b + r)!$. However, our balls belong to two distinct "parties", the Reds and the Blues, and no change in the row can be detected if we exchange two balls of the same party. Thus, the number of different arrangements is

$$\frac{(b + r)!}{b!\, r!} = \binom{b + r}{r} = \binom{b + r}{b}$$

Now a slightly harder question: If we take out only $m$ balls, with $m \leq r, m \leq b$ (so that we neither not run out of red, nor of blue balls), how many ways we have of drawing exactly $k$ red balls? The ways of selecting $k$ red balls out of our initial "nest-egg" of $r$ is $\binom{r}{k}$. For each one of them, there are $\binom{b}{m-k}$ ways of selecting the remaining $m - k$ balls from the $b$ blue balls that we have at disposal, so the number of ways of drawing $k$ red balls out of $m$ is

$$\binom{r}{k}\binom{b}{m - k}.$$

Interestingly, since the total number of way of drawing $m$ balls, no matter their color, from an urn containing $r + b$ balls, is $\binom{r+b}{m}$, we have

---

[33] Even if it is named "Ladies' Delight", this planet is not necessarily a delight for *all* the ladies....

$$\sum_{k=0}^{m} \binom{r}{k}\binom{b}{m-k} = \binom{r+b}{m}, \tag{2.29}$$

which is called *Vandermonde's identity*.

The next example is so frequently discussed that I would gladly avoid taking it up again, but since we will have the opportunity to view it from a perspective of interest for physics, let me do it anyway.

**Example 2.20**  (*Birthday matching*)

Q: In a class consisting of $N$ students, what is the probability $P$ that at least two of them have their birthdays on the same day?

A: We first evaluate the probability $\bar{P} = 1 - P$ that all students were born on different days. Since for each student we have 365 possible choices, the total number of $N$-tuples that we can form with the birthdays of each student is given by $365^N$. Of these, there are

$$D_{365,N} = 365 \times (365 - 1) \times \ldots \times (365 - N + 1)$$

in which all birthdays are distinct, because it is in fact a sampling without replacement. Hence $\bar{P}$ is given by

$$\bar{P} = \frac{365 \times (365 - 1) \times \ldots \times (365 - N + 1)}{365^N} = 1 \times \left(1 - \frac{1}{365}\right) \times \ldots \times \left(1 - \frac{N-1}{365}\right).$$

This is a rather complicated expression, which can however be approximately evaluated if $N \ll 365$ by taking the logarithm of both members, remembering that, for small $x$, $\ln(1 - x) \simeq -x$, and taking into account that the sum of all integers up to $k$ is $k(k+1)/2$,

$$\ln \bar{P} \simeq 0 - \frac{1}{365} - \frac{2}{365} - \ldots - \frac{N-1}{365} = -\frac{N(N-1)}{730},$$

which yields

$$P \simeq 1 - \exp\left(\frac{N(N-1)}{730}\right).$$

This result is quite surprising: 23 students are enough for the probability of finding two of them who have their birthdays on the same day to be greater than 50%. And in a class of 40 students, this probability is almost 90%! Why? Simply because the number of *pairs* that we can form with $N$ objects is $N(N-1)/2$, thus for $N$ large it grows as $N^2$. So, even if the probability that two *selected* students were born on the same day is low, the total probability increases rapidly with $N$.

The rapid growth of the number of pairs has a strong relationship with a basic problem in physics, namely, the crucial role played by the range of interactions. For instance, the van der Waals theory of the liquid-gas transition, which is a landmark in condensed matter, assumes that molecules interact via a pair potential $u_{ij}(r)$ that vanishes for $r \to \infty$ faster than $r^{-3}$. If this condition is not met, the number of pairs of molecules involved gets so large that the average potential, on which this archetypal mean–field theory is based, cannot be properly defined. Similarly, a consistent theory of ferromagnetism would hardly be possible if the quantum exchange force were not very short-ranged at their core, so that they basically act only between of pairs of nearest neighbours.

The solution we found for the birthday matching problem can be extended to any situation where $k$ objects must be arranged in $n$ places, each of which can contain more than one object. If $k \ll n$, the probability of finding at least two objects in the same place is then

$$P = 1 - \exp\left[-\frac{k(k-1)}{2n}\right],$$

which means that $P$ becomes important when $k$ becomes of the order of $\sqrt{n}$. In the opposite limit, we note that if $n$ objects are randomly placed in $n$ places, the probability $P$ that each place contains one and only one object is

$$P = \frac{n!}{n^n}.$$

This value is extremely low even for small $n$. For example, for $n = 5$, we have $P \simeq 0.038$ and, for $n = 10$, $P \simeq 3.6 \times 10^{-4}$.

## 2.4.2 Counting Indistinguishable Objects

We move now to the slightly harder problem of counting objects that are "indistinguishable" in the strict sense of the word: namely, there is absolutely no way to set one of them apart from the others. Actually, we will see that we have *already* got some useful results.

### 2.4.2.1 From Anagrams to Bosons

Suppose we wish to count the number of distinct (but not necessarily meaningful) anagrams of the Italian word *ANAGRAMMA* (which, of course, means "anagram"). There are 9 letters, which give $9! = 362880$ permutations, but they are *not* all distinct, because two anagrams that differ just by the exchange of the two M's, or of any of the four A's, are of course identical. Then, the total number of distinct permutations is reduced to just (!) $9!/(2!4!) = 7560$.

More generally, what we plan to do is counting the distinct permutations of a set of $n$ elements, which can however be grouped in $r \le n$ "families" of identical objects (some of them possibly containing a *single* object). A specific element can then labeled with the name of the family it belongs to. Yet, this is clearly the same problem of splitting a set in a number $r$ of subsets, where we know the number of elements $k_i$ of each subset: for what we have just seen in the last section, this is exactly the *multinomial coefficient* $M_n^r$ given by Eq. (2.27). Hence, the number of distinct anagrams of a word made of $L$ letters is $L!/(r_1!r_2!\ldots r_\ell!)$, where $r_i$, with $i = 1, \ldots, \ell$, is the number of repetitions of each one of the $\ell$ distinct letters in that word.

An equivalent problem is calculating in how many ways we can put $n$ identical balls in $r$ "labeled" boxes, if we *assign* the number of balls that each of these boxes contains. This has an immediate counterpart in the distribution of identical quantum particles with integer spins, i.e., bosons. Here the "boxes" are the *single–particle states* (which *can* be labeled), and the prescribed number of balls in a given box is the *occupation number* of that state. The number of different ways to distribute $n$ non interacting bosons among $r$ different single-particle states with a set $\{n_\lambda\}$ of given occupation numbers is then

$$M_n^r = \frac{N!}{\prod_{k=1}^r n_k!} \tag{2.30}$$

When the number of states $r$ is much larger than the number of particles $N$, so large that is becomes extremely improbable that two or more particles are in the same state, so that the occupation numbers $n_k$ can only be 0 or 1, we find

$$\lim_{r/N \to \infty} M_n^r = N!,$$

which corresponds to the classical Maxwell-Boltzmann limit for distinguishable particles.

However, we may also want to know in how many ways $n$ identical balls can be distributed among $r$ boxes *without* pre–assigning the number of balls in each box, provided that the sum of these occupation numbers is $n$.[34] This problem does not look that easy, yet it can be solved with a simple trick. Suppose we have 10 balls and 7 boxes. One of the possible configurations may look like this:

$$\| \bullet \ \bullet \| \quad \| \ \bullet \ \| \bullet \bullet \bullet \| \quad \| \bullet \ \bullet \| \bullet \ \bullet \|$$

where | stands for a "separation wall" between two boxes, and we indicate with $\|$ the two walls at the extremes of the sequence of boxes. How can we generate all possible rearrangements of this configuration? Well, it is not difficult to see that this can be made by permuting in all possible ways *both* the 10 balls *and* the $7 - 1 = 6$ *internal* walls. However, we must take into account that any two permutations that just differ by the exchange of some balls, or of some walls, are equivalent. Thus, we have to divide the 16! permutations of balls plus walls by the 10! permutations of the balls and by the 6! permutations of the separation walls. In general, therefore, the number of distinct placements $\mathcal{P}_B(n, r)$ of $n$ identical balls in $r$ boxes is

---

[34] In Eq. (2.28), $\sum_{i=1}^r k_i$ is also constrained to be equal to $n$. Then, an equivalent problem is finding how many *addends* we have in that summation.

$$\mathcal{P}_B(n,r) = \frac{(n+r-1)!}{n!(r-1)!} = \binom{n+r-1}{n} = \binom{n+r-1}{r-1}, \qquad (2.31)$$

where the subscript $B$ reminds that this is also the number of ways of distributing $n$ independent bosons among $r$ states, for all possible values of the occupation numbers.[35] Then, $\mathcal{P}_B(n,r)$ can also be regarded as the number of ways of tagging $n$ identical objects with $r$ labels, provided that some of these labels can be the same.

It is up to you showing, using a similar "shuffling" of balls and walls, that the number of ways of placing $n$ bosons in $r \leq n$ states so that each state contains at least one boson is $\binom{n-1}{r-1}$.

**Example 2.21** (*"And yet it differs: Galileo on rolling dice"*) One of the oldest problems in probability was posed to Galileo by someone, probably the Grand Duke of Tuscany, who was a great fan of the game of dice. But let's leave the floor to Galileo himself:

> There is an obvious reason why in a game of *[three]* dice certain numbers are more advantageous than others, which is that they can be made up with more variety of numbers…Nevertheless, although 9 and 12 can be made up in as many ways as 10 and 11, and for this reason should be considered of equal value, it is nevertheless known that long observation has made players consider 10 and 11 to be more advantageous than 9 and 12…Now, to oblige him who has ordered me to produce whatever comes to my mind about such a problem, I will expound my ideas, in the hope not only of clarifying this question, but also of opening the way to clearly see the reason why all the details of the game have been arranged and adjusted with great care and judgment (Galileo Galilei, *Sopra le scoperte dei dadi*, Opere, Firenze, Barbera, Vol.8 (1898) (my loose and prosaic translation).)

Summing up, why in the roll of three dice 10 or 11 are better total scores than 9 or 12, if all these results are obtained with the same number of combinations? Galileo immediately points out the each of these combination can be obtained in several ways by permuting the dice. However, the number of permutations is not the same for all the (six) combinations. Let us compare for instance 9 and 10.

→   A sum of 9 can be obtained with all the distinct permutations of:
   $\{1,2,6\}, \{1,3,5\}, \{2,3,4\}, \{1,4,4\}, \{2,2,5\}, \{3,3,3\}$.

→   A sum of 10 can be obtained with all the distinct permutations of:
   $\{1,3,6\}, \{1,4,5\}, \{2,3,5\}, \{2,2,6\}, \{2,4,4\}, \{3,3,4\}$.

Of course, these distinct permutations of one of these triplet is nothing but the number of its "anagrams", which is $3! = 6$ if all numbers are different, $3!/2! = 3$ if two numbers are the same, and $3!/3! = 1$ if all numbers are the same. So we have

For 9 →   $3 \times 6 + 2 \times 3 + 1 \times 1 = 25$ distinct permutations.
For 10 →   $3 \times 6 + 3 \times 3 = 27$ distinct permutations.

---

[35] The last two equalities in (2.31) simply mean that this is also the number of ways to choose $n$ objects among a total of $n+r-1$, and name them "balls", or $r-1$ objects and dub them "separation walls".

Since the total number of possible results in the throw of three dice is simply $6^3 = 216$, the probability of obtaining a 10 is $p_{10} = 27/216 = 1/8 = 12.5\%$, slightly larger than the probability $p_9 = 25/216 \simeq 11.7\%$ of getting a 9.

It's up to you showing that exactly the same difference is obtained when comparing 11 with 12, and maybe scrutinize what the Florentine dice-players would have found using "boson" or "fermion" quantum dice.

#### 2.4.2.2   From Gentlemen's Delight to Fermions

As you know, here is another kind of identical quantum particles, fermions, which behave quite differently from bosons. Like the dice in Example 2.2 that are forbidden to show the same face, they simply refuse to share the same box. For them, however, the situation is much simpler. To see why, let us visit another planet, Gentlemen's Delight, where the situation is slightly less favorable for ladies, since the number $n$ of men is *smaller* than the number $r$ of women. On that planet, the wedding law in force states that each man *must* marry a woman, but two men cannot share the same wife.[36] Then, the first man can choose among $r$ ladies, the second one among $(r - 1)$, and so on. The order in which they choose does not matter, we only care about the couples they eventually form. Notice that this is like stating that, when they are bachelors, all the men are identical: They get tagged only by the name of the lady they marry. You should immediately be able to conclude that the total number of ways $\mathcal{P}_F(n, r)$ of marrying the $n$ men to the $r$ women is just the binomial coefficient

$$\mathcal{P}_F(n, r) = \binom{r}{n} = \frac{r!}{n!(r - n)!}, \tag{2.32}$$

which is of course also the number of distinct placements $\mathcal{P}_B(n, r)$ of $n$ identical balls (the gentlemen) in $r$ boxes (the ladies), with the condition that a box can contain at most one ball (which necessarily implies $r \geq n$). This is exactly the case of the distribution of $n$ independent fermions among $r$ possible single–particle states.

### 2.4.3   The Stirling Approximation

Let us say something more about $n!$, which, as we have seen, gives the number of permutations on $n$ distinct objects. In statistical physics, we actually have to consider factorials of *huge* numbers, typically of the order of the Avogadro number $\mathcal{N}_A$. This allows us to use a powerful approximation for "large" $n$ that gives $n!$ in terms of elementary functions,

$$\boxed{n! \simeq \sqrt{2\pi n}\, n^n\, \mathrm{e}^{-n}} \tag{2.33}$$

---

[36] On this planet there are surely some spinsters, but remember that this may happen on Lady's Delight too (where, however, there are no bachelors).

This formula is called *Stirling's approximation*, from the name of James Stirling, who proved it in his most famous work, *Methodus Differentialis sive Tractatus de Summatione et Interpolatione Serierum Infinitarum* (short titles were not that fashionable in those days) published in 1730, although it was probably known to Abraham de Moivre too, who introduced what we will call the Gaussian distribution. I put "large" in quotes, because this approximation is already quite good even for *small* $n$: for $n = 5$, it underestimates $n!$ by 2%, and by 0.8% for $n = 10$. We shall first account for the large-$n$ behavior of $n!$ given by Eq. (2.33) by a simple graphical method. Since $n!$ grows very fast with $n$, we better consider its *logarithm*,

$$\ln(n!) = \ln\left(\prod_{k=1}^{n} k\right) = \sum_{k=1}^{n} \ln(k).$$

The value of $\ln(n!)$ can then be regarded as the sum of the areas of $n$ rectangles with unit base centered about the natural numbers $k$, with $k = 1$ to $n$, whose height is $\ln(k)$. Figure 2.9, where these rectangles are plotted together with the continuous function $f(x) = \ln(x)$, shows that, to obtain their total area, we must first add to the area subtended by $f(x)$ half of the rectangle centered in $n$. We should then add all the curved triangles that lie above $f(x)$ (those shown in dark grey), and subtract all those lying below it (shown in light grey). By increasing $k$, however, the area of these little triangles decreases, and moreover two adjacent dark and light triangles become more and more similar to each other, because $d^2 f/dx^2$ rapidly decreases with $x$, so their contributions balance. Hence, they can be neglected for large $k$,

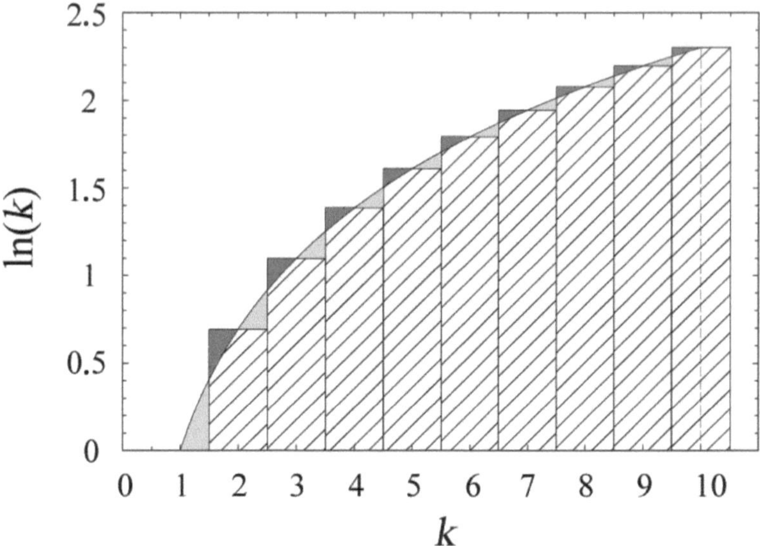

**Fig. 2.9**  Graphical interpretation of the Stirling approximation

while their contribution for the smallest values of $k$ can be accounted for by just adding a constant term $c$. We then write

$$\ln(n!) \approx [\text{area under } \ln(x)] + [\text{area of the last half–rectangle}] + \text{constant}$$

But the area under $\ln(x)$ is just

$$\int_1^n \ln(x)\, dx = [x(\ln(x) - 1)]_1^n = n[\ln(n) - 1] + 1,$$

thus we obtain $\ln(n!) \approx n[\ln(n) - 1] + (1/2)\ln(n) + c$, namely, writing $C = e^c$,

$$n! \approx C\, n^{n+(1/2)} e^{-n}.$$

In many practical cases, evaluating the constant $C$ is not imperative. Indeed, writing

$$\ln(n!) \simeq n[\ln(n) - 1] + \frac{1}{2}\ln(n) + \ln(C),$$

we see that, when $n$ is of the order of $\mathcal{N}_A$, the last term is surely negligible. In fact, even the *second* term, which is only of the order of $\ln(n)$, can be neglected. Hence, at leading order in $n$, we shall often be content to approximate

$$\ln(n!) \simeq n \ln(n) - n \tag{2.34}$$

An explicit evaluation of $C$ can however be instructive. Let us first introduce an interesting representation of the factorial,[37]

$$n! = \int_0^\infty x^n e^{-x}\, dx, \tag{2.35}$$

which, as we will see, is also the starting point for a generalization of the factorial to non integer numbers. The integrand $f(x) = x^n \exp(-x)$ (2.35) is the product of a rapidly increasing function of $x$ with a fast exponential decrease, hence it exhibits a sharp maximum for some value of $x$. Writing $f(x) = \exp[g(x)]$, $g(x) = x \ln x - x$ has its maximum, in $x = n$, where $g(n) = n \ln n - n$. Around the maximum we can then make a "parabolic" approximation

$$g(x) \simeq g(n) + \frac{1}{2}\left(\frac{d^2 g}{dx^2}\right)_n (x - n)^2 = n \ln n - n - \frac{1}{2n}(x - n)^2,$$

---

[37] Proving this equality is not hard. Show first that it holds for $n = 0$. Then prove it by mathematical induction, showing that, if it holds for $n$, it also holds for $n + 1$, which just requires an integration by parts.

so that, putting $t = x - n$,

$$n! \simeq n^n \mathrm{e}^{-n} \int_{-n}^{\infty} \mathrm{e}^{-t^2/2n}\, \mathrm{d}t.$$

Since the integrand is strongly peaked around $t = 0$, we can safely extend the lower limit of the integral to $t = -\infty$. In the next chapter, we will see that this integral is equal to $\sqrt{2\pi n}$, which yields the Stirling approximation (2.33).

This is just an example of a simple way to approximate the integral of a strongly peaked function $f(x)$: the integral is quadratically expanded around the maximum $\bar{x}$, obtaining the integral of $\exp(-ax^2)$, where $a = (\mathrm{d}^2 f/\mathrm{d}x^2)_{\bar{x}}$, which can be safely extended over the whole axis. Actually, this method can be generalized to evaluate integrals of a function of a *complex* variable, using an integration path in the complex plane. Since the appropriate extremum is in this case a saddle point, this method is known as *saddle point integration*. The simplified version we used here is anyway sufficient for our needs.

**Example 2.22** (*Zero crossings*) Consider a random walk of a point on a line. What is the probability that after a certain number of steps (of unit length) the point will be at the origin, i.e., at the starting point? For this to happen, the point must take as many steps in the positive direction as in the negative direction. If we indicate the total number of steps (which are necessarily even) with $2n$, we have as many "distinct paths" that bring us back to the origin as there are ways to choose $n$ steps in the positive direction out of $2n$, which are $\binom{2n}{n}$. Since the total number of possible paths is $2^{2n}$, and because all these paths are equiprobable, the probability $P_{0,2n}$ to come back to the origin after $2n$ steps is

$$P_{0,2n} = \binom{2n}{n} 2^{-2n}. \tag{2.36}$$

Using Stirling's approximation, it is easy to show that, if $n$ is large,

$$P_{0,2n} \approx \frac{1}{\sqrt{\pi n}}. \tag{2.37}$$

As you can see, the probability of returning to the origin after $2n$ steps decreases as the *square root* of $n$.

One can then show that the probability $\mathcal{P}_{2n}$ to return for the *first* time to the origin after $2n$ steps is simply given by the difference

$$\mathcal{P}_{2n} = P_{0,2(n-1)} - P_{0,2n}, \tag{2.38}$$

which, using (2.36), yields

$$\mathcal{P}_{2n} = \frac{1}{2n-1} P_{0,2n} = \frac{1}{2n-1} \binom{2n}{n} 2^{-2n}, \tag{2.39}$$

Note that $\mathcal{P}_{2n}$ decreases quite fast with $n$ (for large $n$, as $n^{-3/2}$). This is actually the origin of those slow oscillations and of the few changes of leader that we had met in the first chapter, both in the "coin flipping with $\pi$" and in random walk simulations. These and many more observations on the behavior of single paths in 1-dimensional random walks can be found in the first volume of the book by W. Feller cited in the bibliography.

### *2.4.4  Generalizing the Factorial: The Gamma Function*

Factorials are ubiquitous in combinatorics, but there is a very disconcerting aspect of them. Indeed, while there is a simple formula, $\sum_{k=1}^{n} k = \frac{n(n+1)}{2}$, giving the *sum* of the first $n$ integers, finding a function that returns the value $n!$ of their *product* is a much, much harder task! Many mathematicians, including Christian Goldbach and Daniel Bernoulli,[38] struggled in vain with this puzzle in the early 18th century, which however was successfully solved by Leonhard Euler in 1730. Using the symbol and expression that Legendre introduced 80 years later, which is equivalent and currently more popular, the required function is

$$\Gamma(n) = \int_0^\infty t^{n-1}\mathrm{e}^{-t}\mathrm{d}t,$$

which in the Appendix is shown to generate the factorial of any integer $n \geq 1$ as

$$\boxed{n! = \Gamma(n+1)} \qquad (2.40)$$

What is however even more interesting is that the recursive relation $\Gamma(x+1) = x\Gamma(x)$ is valid also for a *non integer* number $x > 0$. Therefore, we can define the Euler *gamma function*

$$\boxed{\Gamma(x) = \int_0^\infty t^{x-1}\mathrm{e}^{-t}\mathrm{d}t} \qquad (2.41)$$

which generalizes the factorial to any positive real number. The Appendix, where the properties of the gamma function are discussed, shows that $\Gamma(x)$ can be extended also for $x < 0$, except for negative integers, where $\Gamma(x)$ diverges. Besides, the Stirling approximation is valid for $\Gamma(x)$ too,

$$\Gamma(x+1) \simeq \sqrt{2\pi x}\, x^x \mathrm{e}^{-x} \quad (x \gg 1). \qquad (2.42)$$

---

[38] Goldbach is particularly known for his conjecture, one of the oldest unsolved problems in number theory, stating that every even natural number greater than 2 is the sum of two primes. About Daniel Bernoulli, we will hear a lot in the next chapters.

# Chapter 3
# Random Variables

*Everyone believes in the normal law, the experimenters because they imagine that it is a mathematical theorem, and the mathematicians because they think it is an experimental fact.*

*G. Lipmann (cited by H. Poincaré)*

The concept of *random* (or *stochastic*) *variable* (RW) naturally emerges from the discussion made in the last chapter, where we have seen that we can link to a $\sigma$-algebra of events a probability function with the properties of an abstract measure. It is therefore natural to define a random variable $X$ as a function from $\Sigma$ to the set $\mathbb{R}$ of real numbers that associate to each event $E$ its probability measure $P(E)$.

When we deal with a finite number of events, this formulation does not present any problems. Indeed, the probability of an event $E$ consisting of the points $x_1, x_2, \ldots, x_n$ in $\Omega$ is $P(E) = \sum_{i=1}^n P(x_i)$. Thank to Kolmogorov's axiom K3, this can be safely extended to a random variable defined on a sample space made of a countably infinite number of elementary events. However, for continuous random variables some caution is required.[1] Actually, we often deal with events that can be associated with subsets of $\mathbb{R}^n$ like segments, surfaces, volumes, and so on. Consequently, a continuous random variable could be misunderstood as a common function $f : \mathbb{R}^n \to \mathbb{R}$, and the events identified with intervals, rectangles, cuboids, and so on to which we can associate a probability measure proportional to their length, area, volume.

---

[1] Anyway, I will never tire of telling you, even at the cost of being pedantic, that continuity is just an extremely useful deception.



R. Piazza, *An Invitation to Probability and Data Analysis for Physicists*, UNITEXT for Physics, https://doi.org/10.1007/978-3-031-83856-9_3

Yet, not all events can be matched with geometrical sets, and even if they do, these are not necessarily simple boxes[2] in $\mathbb{R}^n$. For instance, what probability should we give at the event $E$ corresponding to the set of all points $x$ in $[0, 1]$ such that

$$P(x) = \begin{cases} 1 \text{ if } x \notin \mathbb{Q} \\ 0 \text{ if } x \in \mathbb{Q} \end{cases},$$

where $\mathbb{Q}$ is the set of rational numbers? For sure, we cannot sum "point by point", because a single point in a continuum has measure zero. Actually, this question can be solved only using integration methods that are definitely more sophisticated than those we learn studying elementary calculus. For the moment, therefore, we better stay on the safe side, and deal first with discrete random variables.

## 3.1   Probability Distributions for Discrete RVs

Consider a RV $k$ that can have $N$ discrete values $k_i$. We wish to give a precise meaning to the question:

What is the probability $P(k_i)$ that $k$ takes on a specific value $k_i$?

This goal can easily be achieved by selecting all those results matching the same value $k = k_i$ and stating that $P(k_i)$ is the sum of the probabilities of the single results, i.e. the sum of the probabilities of the elementary events associated to the same value $k_i$ of $k$. $P(k_i)$ is then a function of the value $k_i$ we consider, which we shall call the *probability distribution* for $k$. Obviously, because of the way it is defined, a probability distribution function can only take non-negative values. Because probabilities sum up to one, we must have

$$\sum_{i=1}^{N} P(k_i) = 1 \tag{3.1}$$

a condition that is expressed by saying that a probability distribution must be *normalized*. This is exactly what we found for the sum of the relative frequencies of data for a statistical sample, and is a natural choice if we regard probabilities as limits of relative frequencies. If the variable $k$ can assume an infinite number of discrete values—for instance all primes, or all even integers—the summation in (3.1) becomes an infinite series. For $P(k)$ to be a "healthy" distribution, the latter is then required to converge.[3]

---

[2] A box in $\mathbb{R}^n$ is the Cartesian product of $n$ intervals on the real axis, $I_1 \times I_2 \times \ldots \times I_n$.

[3] Since in this case all summed terms in (3.1) are *non-negative*, if the series converges, it converges absolutely. Hence, the order of summation is irrelevant.

**Example 3.1**  ((*Classical and quantum double-dice games*) The probability distribution for the outcomes of the roll of a single die is obviously constant, with $P(k) = 1/6$ for all the six values of $k$. In Example 2.2, however, we have seen that the sum of dice faces in a double-dice game does not always have the same number of outcomes. Besides, the possible outcomes depend on whether the dice are distinguishable or not. In the latter case, they differ for dice that show a "fermion-like" rather than a "boson-like" character. The following table, which you are invited to check, shows the number of outcomes for each value of the sum for "classical" ($n_c$), "boson-like" ($n_b$), and "fermion-like" ($n_f$) dice.

| $k$ | 2 | 3 | 4 | 5 | 6 | 7 | 8 | 9 | 10 | 11 | 12 |
|---|---|---|---|---|---|---|---|---|---|---|---|
| $n_c(k)$ | 1 | 2 | 3 | 4 | 5 | 6 | 5 | 4 | 3 | 2 | 1 |
| $n_b(k)$ | 1 | 1 | 2 | 2 | 3 | 3 | 3 | 2 | 2 | 1 | 1 |
| $n_f(k)$ | 0 | 1 | 1 | 2 | 2 | 3 | 2 | 2 | 1 | 1 | 0 |

Figure 3.1 shows the three corresponding probability distributions $P(k)$ obtained by dividing $n_c$, $n_b$, and $n_f$ by the total number of outcomes, which is $N_c = 36$, $N_b = 21$, $N_f = 15$, respectively.

In the next example we introduce a probability distribution of relevant practical interest where the random variable can assume in principle all possible positive integer values.

**Example 3.2**  (*The geometric distribution*) Suppose theorists predict that, in some high–energy physics experiments, a very short-lived particle, the "ephemeron" (E), is produced. We try to catch this event using a suitable detector D that, however, has a chance $p < 1$ to detect E in a single experiment, and therefore a probability $1 - p$ to miss the event. So, we repeat the experiment several times. We want to estimate the probability $P(k; p)$ that D detects *for the first time* the ephemeron at trial $k$.

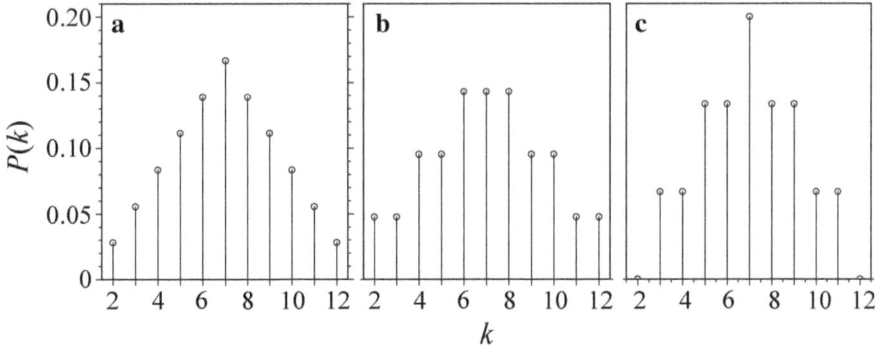

**Fig. 3.1** Probability distribution for the sum of the outcomes when rolling two "classical" (C), "boson-like" (B), and "fermion-like" (C) dice

We clearly have $P(1; p) = p$. However, for the positive outcome to happen at the second trial, D must also have failed to detect E at the first trial, which happens with probability $1 - p$. It we regard the trials as independent from each other (they may be performed on different days, or in different labs), then the probability that D does not detect E in the first experiment *and* detects it in second one is just the product $P(2; p) = (1 - p)p$. Similarly, the probability that the particle detection does not takes place in the first two trials, while it does at the third one, will be $P(3; p) = (1 - p)^2 p$. Generalizing this result, we have

$$P(k; p) = p\,(1 - p)^{k-1}, \tag{3.2}$$

which is called a *geometric distribution*. Using the sum of a geometrical series of ratio $(1 - p) < 1$, we readily find that

$$\sum_{k=1}^{\infty} P(k; p) = p \sum_{k=1}^{\infty} (1 - p)^{k-1} = p \sum_{k'=0}^{\infty} (1 - p)^{k'} = p\,\frac{1}{1 - (1 - p)} = 1,$$

hence the distribution is properly normalized.

The behavior of the $P(k; p)$ versus $k$ is better appreciated by putting $k_0 = -1/\ln(1 - p)$, which yields

$$P(k; p) = \frac{p}{1 - p}\,\exp\left(-\frac{k}{k_0}\right), \tag{3.3}$$

The geometrical distribution has then an exponentially decreasing trend, with a decay rate $(k_0)^{-1}$, which is shown in Fig. 3.2 for $p = 0.2$.

The geometric distribution plays an important role as the simplest "failure model" in safety, reliability, and quality control problems. Suppose for example that we add to a production plant an alarm that should go off if something goes wrong, but that there is a chance $p$ that this control system will *not* work when a potentially dangerous event takes place. We may expect several of these events to happen during the period of operation of the plant. Suppose that $p$ does not change in time—no "aging effects", which might be a strong assumption—so that whether the system works or not when an event takes place does not depend on its response to the previous events. Then, $P(k; p)$ gives the probability that the alarm fails to operate *for the first time* when the $k$th event takes place.

More generally, we can say that the geometric distribution $P(k; p)$ gives the probability of having the first success (or failure) at trial $k$ in a sequence of *independent* experimental trials, in which the probability of success (or failure) in a single trial is $p$. Or, if you like urns, $P(k; \frac{r}{r+b})$ is the probability of extracting for the first time at trial $k$ (with no replacement) a red ball from an urn that contains $r$ red and $b$ blue balls.

A slightly different question is asking how many *failures* are expected before the first success. Since this coincides with having the first success at the $k + 1$ trial, you

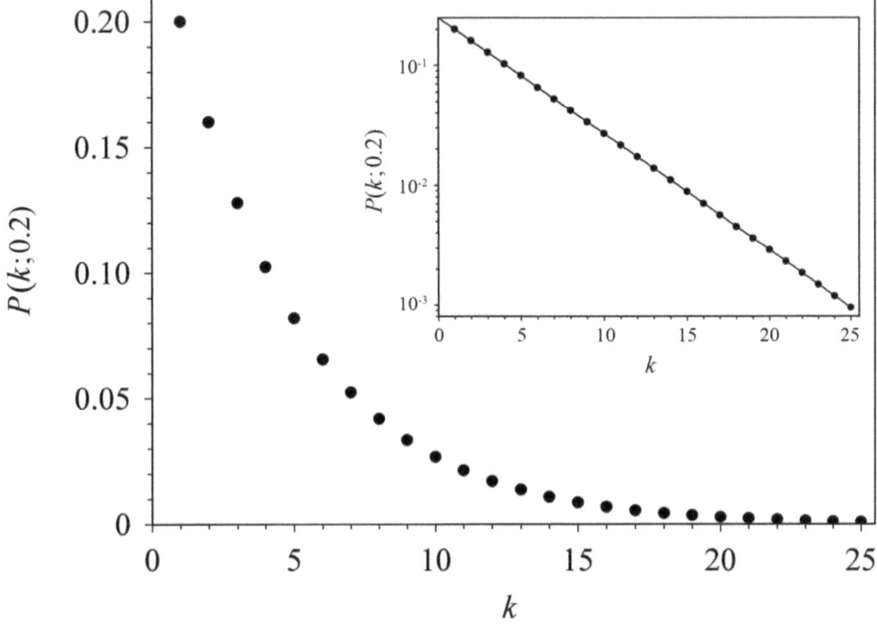

**Fig. 3.2** Geometric distribution for $p = 0.2$ compared in the semilog inset to the function $y = 0.8 \exp[x / \ln(0.8)]$ (full line)

should immediately see that the probability distribution for the number $k$ of failures is

$$P'(k) = p(1 - p)^k = p \exp\left(-\frac{k}{k_0}\right).$$ 

(3.4)

The parameter $k_0$ is then a characteristic scale for the decay of the geometric distribution. In Chap. 1, however, we have seen that several relevant frequency distribution decay as a power law. The next example shows a probability distribution with a scale-free behavior.

**Example 3.3** (*A random tour*) From Eq. (2.39), we see that in a 1D random walk, the probability $P(k)$ of returning to the origin for the first time after $2k$ steps is given by:

$$P(k) \equiv \mathcal{P}_{2k} = \frac{1}{2k - 1}\binom{2k}{k}2^{-2k},$$

where $k$ is a RV that takes on all positive integer values. Since every return to the origin is like starting the random walk again from the beginning, $P(k)$ also describes the distribution of *half* the number of steps required for a "random tour" starting and ending at the origin. Using Eq. (2.38), we can see that $P(k)$ is correctly normalized.

Indeed,

$$\sum_{k=1}^{\infty} P(k) = \sum_{k=1}^{\infty} [P_{0,2(k-1)} - P_{0,2k}] = P_{0,0} - P_{0,2} + P_{0,2} - P_{0,4} + \ldots = P_{0,0} = 1,$$

since of course the probability of being back at the origin after 0 steps, when we are already there, is 1!

Using (2.37), for large $k$ we have

$$P(k) \xrightarrow[k \to \infty]{} \frac{1}{\sqrt{\pi k}} \frac{1}{2k-1} \sim \frac{1}{2\sqrt{\pi}} k^{-3/2}, \tag{3.5}$$

which then shows a power-law decay $P(k) \propto k^{-\alpha}$ with exponent $\alpha = 1.5$.

This theoretical prediction can be checked numerically. The following table shows the results obtained from the simulation of 230 random walks of 1000 steps each, for a total of about 5000 passages through the origin. Since $P(k)$ decreases rapidly as the half-width $k$ of the interval between two passages increases, it is convenient to collect the data in classes of increasing width $\Delta k = k_{max} - k_{min}$ centered around $\bar{k} = (k_{max} + k_{min})/2$, evaluating of course the relative frequencies $f_k$ as for a continuous variable.

| $\bar{k}$ | 1 | 2 | 3 | 4.5 | 8 | 18 | 38 | 75 | 175 |
|---|---|---|---|---|---|---|---|---|---|
| $\Delta k$ | 0 | 0 | 0 | 1 | 4 | 14 | 24 | 50 | 100 |
| $f_k \times 10^2$ | 52.42 | 12.74 | 6.842 | 3.052 | 1.490 | 0.439 | 0.133 | 0.046 | 0.013 |

Figure 3.3, where $f(k)$ and $P(k)$ are compared shows that, except for very short intervals, where the Stirling approximation is worse, the agreement looks quite good. However, this is merely a qualitative assessment. In the last chapter we will learn how to make a better comparison between the experimental frequencies and a theoretical probability distribution.

## 3.2 Expectation, Variance and Higher Moments

In many cases, we are not interested in the entire probability distribution for a random variable (or maybe we are not even able to find it) but only in some of its general properties, such as the typical value around which it is centered, or its width. Like we did for the statistical distributions of experimental data, we then want to introduce parameters allowing us to characterize the key aspects of a distribution. Let us start by defining an analogue of the average value of a frequency distribution, which we

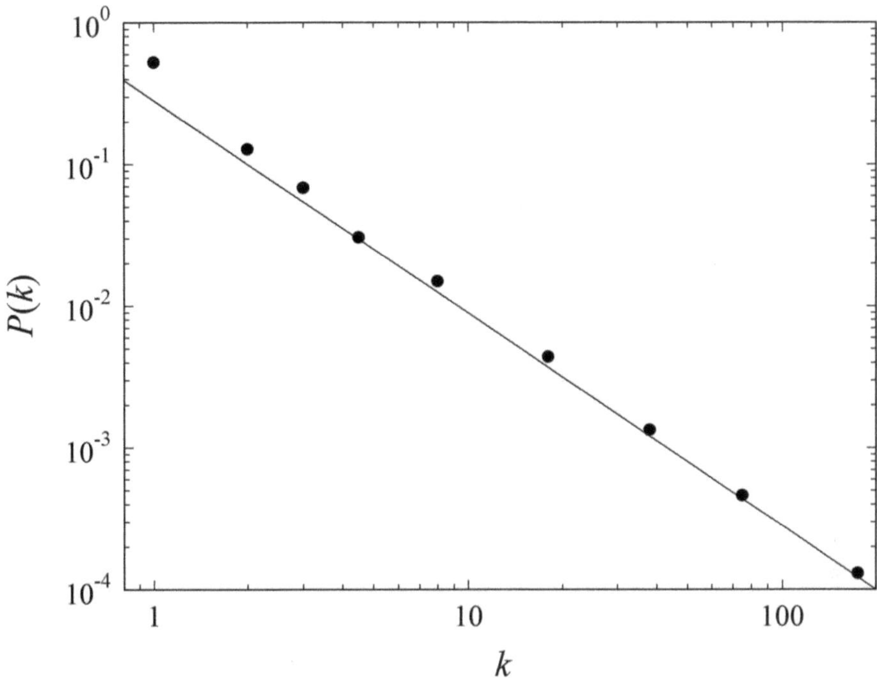

**Fig. 3.3** Comparison between the simulation data for the number of steps between two passages at the origin and Eq. (3.5)

will call *expectation*[4] $\langle k \rangle$, to emphasize that in some way it is the value that we most probably expect to find for the variable $k$, by setting

$$\langle k \rangle = \sum_{i=1}^{N} k_i P(k_i) \tag{3.6}$$

where $N$ may not be finite. As for the mean, $\langle k \rangle$ does not necessarily coincide with the maximum value of $P(k)$. If the distribution is symmetrical about a given value, however, that value coincides with expectation. Because of the way it is defined, the expectation value is sometimes called the "mean" of the RV. You may also do the same, if you like, but this is risky. Actually, while the mean of a distribution of experimental data is just a number that can always be calculated, the expectation of a probability distribution of a RV that may take an infinite number of values may not even exist, as we shall shortly see.

---

[4] Most books on probability denote the expectation as $E(k)$, but I am sure that any aspiring physicist will be much more at ease with the notation I choose to use.

If $k$ and $h$ are two RVs, their sum $z = k + h$ is a random variable too. We deal with sums of RVs in the next chapter, but since these sums happen frequently, it is appropriate to anticipate a particularly useful result: as for the mean, the expectation of the sum of two or more RVs, is equal to the sum of their expectation values. More generally, since the expectation is obtained with a *linear* operation, if $a$ and $b$ are constants, $\langle ak + bh \rangle = a \langle k \rangle + b \langle h \rangle$.

Going on with our "recycling", we call *moment of order r about the origin* of the variable $k$ the quantity

$$\boxed{\langle k^r \rangle = \sum_{i=1}^{N} (k_i)^r P(k_i)} \tag{3.7}$$

and *central moment* of order $r$

$$\boxed{\langle (k - \langle k \rangle)^r \rangle = \sum_{i=1}^{N} (k_i - \langle k \rangle)^r P(k_i).} \tag{3.8}$$

The symbols $\langle k^r \rangle$ e $\langle (k - \langle k \rangle^r \rangle$ recall that moments are calculated with the same kind of operation used to define the expectation. In the next chapter, we will delve much further into this analogy. As much as for $\langle k \rangle$, when $k$ admits an infinite number of values the moments may not exist. Specifically, it is possible to show that the moments of order $r > r_0$ do not exist if $P(k)$ decreases slower than $k^{-(r_0+1)}$.

The previous experience can guide us in defining a parameter that describes the width of a probability distribution, that is, how much the variable tends to deviate from its expectation value. So we introduce the expectation of the deviation with respect to $\langle k \rangle$ squared, i. e., the second moment with respect to the expectation, which we will call *variance* of the probability distribution,

$$\boxed{\sigma_k^2 = \langle (k - \langle k \rangle)^2 \rangle = \sum_{i=1}^{N} (k_i - \langle k \rangle)^2 P(k_i).} \tag{3.9}$$

The variance $\sigma_k^2$ is then the counterpart for a probability distribution of the *square* of the standard deviation for a frequency distribution. The width of a probability distribution will therefore be taken as the *square root* of the variance $\sigma_k = \sqrt{\sigma_k^2}$. Exactly as we did for $s^2$, it is easy to show that the variance is equal to the difference between the second moment and the square of the first moment,

$$\boxed{\sigma_k^2 = \langle k^2 \rangle - \langle k \rangle^2 .} \tag{3.10}$$

Since the expectation is a linear operation, it is then immediate to show that, if $k' = ak + b$,

$$\sigma_{k'}^2 = a^2 \sigma_k^2. \tag{3.11}$$

Regrettably, while the typical value of a probability distribution and of an experimental frequency distribution are named differently (expectation versus mean), this is not the case for the distribution width, which is called "standard deviation" in both cases (maybe distinguishing between them by specifying "of the population" instead of "of the sample"). To prevent confusion, I will keep using at least different symbols. ($\sigma_k$ vs. $s_k$).

We can finally introduce the skewness $\gamma$ of a probability distribution relating it to the third central moment,

$$\boxed{\gamma = \frac{1}{\sigma_k^3} \left\langle (k - \langle k \rangle)^3 \right\rangle,} \tag{3.12}$$

which then coincides with the third moment of the "standardized" variable $(k - \langle k \rangle)/\sigma$ (a variable transformation that we will often use in the following),

$$\boxed{\gamma = \left\langle \left( \frac{k - \langle k \rangle}{\sigma} \right)^3 \right\rangle} \tag{3.13}$$

As a first application, let us evaluate the expectation and variance of the distributions discussed in Example 3.1. The distribution for the sum of the values in the roll of two dice, either classical or not, is symmetric about the value $k = 7$, hence, necessarily $\langle k \rangle = 7$. For the variance, we have:

- Classical dice:

$$\sigma_k^2 = \frac{1}{36} \left[ 1 \times (2-7)^2 + 2 \times (3-7)^2 + 3 \times (4-7)^2 + \ldots \right] = \frac{35}{6} \simeq 5.83.$$

- Boson-like dice:

$$\sigma_k^2 = \frac{1}{21} \left[ 1 \times (2-7)^2 + 1 \times (3-7)^2 + 2 \times (4-7)^2 + \ldots \right] = \frac{20}{3} \simeq 6.67.$$

- Fermion-like dice:

$$\sigma_k^2 = \frac{1}{36} \left[ 0 \times (2-7)^2 + 1 \times (3-7)^2 + 1 \times (4-7)^2 + \ldots \right] = \frac{14}{3} \simeq 4.67.$$

Note that for the roll of a *single* die the expectation is $\langle k \rangle = \dfrac{\sum_{k=1}^{6} k}{6} = \dfrac{7}{2}$ and, for the variance, $\sigma_k^2 = \dfrac{\sum_{k=1}^{6}(k - 7/2)^2}{6} = \dfrac{35}{12}$. In a game with two classical dice $x$ and $y$, therefore, we have

$$\sigma_{x+y}^2 = \sigma_x^2 + \sigma_y^2,$$

namely, the variance of the sum of the outcomes is given by the sum of the (equal) variances of the single outcomes. We will see that this is a feature of the variance of the sum of two *uncorrelated* variables (as we would expect if the two outcomes were independent events). For "quantum" dice, however, this is not true: for boson-like dice $\sigma_{x+y}^2 > \sigma_x^2 + \sigma_y^2$, while the opposite happens for fermion-like dice. This indicate, as we will later discuss, that the former are positively and the latter negatively correlated.

It is also interesting to find the parameters of the geometric distribution (3.2). For the expectation, we have

$$\langle k \rangle = \sum_{k=0}^{\infty} kp\,(1 - p)^{k-1} = -p\,\frac{\mathrm{d}}{\mathrm{d}p} \sum_{k=0}^{\infty} (1 - p)^k = -p\,\frac{\mathrm{d}}{\mathrm{d}p}\left(\frac{1}{p}\right) = \frac{1}{p}, \quad (3.14)$$

where we have included the value $k = 0$ because it vanishes anyway, and we have exchanged the derivative with the summation because the series, made of positive terms, is absolutely convergent. This result is understandable: For instance, if the detection probability is 5%, we may expect that 20 trials are typically needed before the event is detected. Note that, when $p \ll 1$, $\langle k \rangle$ is almost equal to $k_0$, the reciprocal of the exponential decay rate. The variance can be evaluated using Eq. (3.10) which can be written as

$$\sigma_k^2 = \langle k^2 \rangle - \langle k \rangle^2 = \langle k(k - 1) \rangle + \langle k \rangle - \langle k \rangle^2 = \langle k(k - 1) \rangle + \frac{1}{p} - \frac{1}{p^2}.$$

The first term at the r.h.s. can be found with some effort (which I spare you, but you may still want to try) using a trick alike that used for $\langle k \rangle$, to obtain

$$\langle k(k - 1) \rangle = \frac{2(1 - p)}{p^2},$$

which finally yields

$$\sigma_k^2 = \frac{1 - p}{p^2} \tag{3.15}$$

When considering instead the geometric distribution for the number of failures before the first success one obtains, with a similar method,

$$\langle k \rangle = \frac{1 - p}{p} \;;\; \sigma_k^2 = \frac{1 - p}{p^2} \tag{3.16}$$

Substituting $p = 1/(\langle k \rangle + 1)$, the probability distribution (3.4) can be written in terms of its expectation $\langle k \rangle$ as

$$P'(k) = \frac{1}{\langle k \rangle + 1} \left( \frac{\langle k \rangle}{\langle k \rangle + 1} \right)^k, \tag{3.17}$$

a result which have an important analogous in physics that we will investigate in the next example.

**Example 3.4** (*Black body radiation*) Classical electromagnetic radiation of frequency $\omega$ and a wave-vector $\mathbf{k}$ is regarded in quantum physics as an ensemble of photons, zero rest-mass particles of energy $\epsilon = \hbar \omega$ and momentum $\mathbf{p} = \hbar \mathbf{k}$. Photons also have a spin (an intrinsic angular momentum) of magnitude $|\mathbf{S}| = \hbar$ that can be parallel or antiparallel to $\mathbf{p}$, which classically correspond to the two states of circular polarization. A *mode* of the electromagnetic field consists of photons with the same wave-number and in the same polarization state.

However, a field mode where the number of photons has an exact value, with no fluctuations, does not have a classical analogue. Classical fields are always described by a distribution of the number of photons. The average number $\langle n_\lambda \rangle$ of photons in a mode $\lambda$ of a radiation field at thermal equilibrium at temperature $T$ (known as *blackbody radiation*) is given by the Bose-Einstein statistics[5]

$$\langle n_\lambda \rangle = \frac{1}{e^{\beta \hbar \omega} - 1}, \tag{3.18}$$

where $\beta = (k_B T)^{-1}$ is the reciprocal of the thermal energy. Therefore

$$\exp(\beta \hbar \omega) = \frac{\langle n_\lambda \rangle}{\langle n_\lambda \rangle + 1} \tag{3.19}$$

But at thermal equilibrium the probability $P(n_\lambda)$ that the number in mode $\lambda$ is *exactly* equal to $n_\lambda$ is proportional to the Boltzmann factor $\exp(-\beta E)$ of the total energy $E = n_k \hbar \omega$. To find the proportionality constant $C$, we require that

$$\sum_{n_\lambda = 0} P(n_\lambda) = C \sum_{n_\lambda = 0} \left( e^{\beta \hbar \omega} \right)^{n_\lambda} = 1,$$

obtaining

$$P(n_\lambda) = \left( 1 - e^{\beta \hbar \omega} \right) e^{n_k \beta \hbar \omega}.$$

---

[5] More precisely, the Bose-Einstein statistics for zero mass particles, whose chemical potential vanishes.

Substituting from (3.19) we finally have

$$P(n_\lambda) = \frac{1}{\langle n_\lambda \rangle + 1} \left( \frac{\langle n_\lambda \rangle}{\langle n_\lambda \rangle + 1} \right)^{n_\lambda}, \tag{3.20}$$

which shows that the distribution of photons in a mode $\lambda$ is a geometrical distribution of the form (3.4). The variance of the photon number is therefore $\sigma_{n_\lambda}^2 = (1 - p)/p = \langle n_\lambda \rangle (\langle n_\lambda \rangle + 1)$, and the relative standard deviation (which gauges the strength of the fluctuations in the photon number)

$$\frac{\sigma_{n_\lambda}}{\langle n_\lambda \rangle} = \left( 1 + \frac{1}{n_\lambda} \right)^{1/2}. \tag{3.21}$$

Hence, for large $\langle n_\lambda \rangle$ the fluctuations become comparable with the average photon number.

The next example shows that probability distributions may not always have a finite expectation. It is known as the St. Petersburg paradox, from the place where it was proposed by Daniel Bernoulli in 1738 (after an idea by his cousin Nicholas).

**Example 3.5** (*The St. Petersburg paradox*) Consider a rather peculiar game of heads or tails. Suppose you choose "heads" and flip the coin. Should heads appear, the bank will pay you 1 €, and the game comes to an end. Conversely, if tails shows up, you flip the coin again and, if you get heads this second time, you win 2 €. Otherwise, you flip the coin again and again until you get heads. If this occurs at the $(n + 1)$th toss, you win $2^n$ €. How much should the bank make you bet, so as not to lose? It is clear that you should bet at least as much as what you expect to earn. It is evident that you should wager at least as much as you expect to earn.

The probability of getting heads for the first time at the $(n + 1)$-th flip is found as in Example 3.2 using $p = 1 - p = 1/2$. Thinking of your gain $g$ as a RV that takes all powers of two as values, the probability of earning $g = 2^n$ €will then be equal to $P(2^n) = 1/2^{n+1}$. We saw that this geometric distribution is correctly normalized. But how much you can expect to gain? We have

$$\langle g \rangle = \sum_g g P(g) = \sum_{n=0}^{\infty} 2^n \left( \frac{1}{2} \right)^{n+1} = \sum_{n=0}^{\infty} \frac{1}{2} = \infty,$$

which is certainly not a good prospect for the bank! Intuitively, the probability distribution we are considering does not have a finite expectation because it decreases too slowly as $n$ increases, that is, because it has very long "tails". We can indeed write $P(g) = (2g)^{-1}$, from which we see that the gain distribution is a scale-invariant power law with exponent $-1$.

**Example 3.6** (*Frustration of a sticker collector*) Some of you may have spent some part of their childhood and early adolescence collecting stickers, like I did. I don't know about you, but I have never been able to complete any sticker book. After an initial period of enthusiasm, when the pages filled up quickly, finding one of the few missing stickers always resulted in astronomical waiting times. So let us ask ourselves: how many stickers should we presumably buy to complete a collection made up of a total of $N$ stickers?

Suppose we have already collected $m$ stickers: How many attempts $k_m$ we have to make to find the next one? Since we still miss $N - m$ stickers, in each of these attempts we have $N - m$ chances to make a good buy out of a total of $N$, i. e., a chance of succeeding $p_m = (N - m)/N$. But this is nothing but the expectation of a geometric distribution with $p = p_m$, so that $\langle k_m \rangle = 1/p_m = N/(N - m)$. The total number of stickers that typically have to be purchased to complete the album is then given by

$$\langle k \rangle = \langle k_0 \rangle + \langle k_1 \rangle + \ldots + \langle k_{N-1} \rangle = N \left( \frac{1}{N} + \frac{1}{N - 1} + \ldots + \frac{1}{2} + 1 \right)$$

that is, by the product of $N$ times the sum of the reciprocals of the integers from 1 to $N$. If $N$ is very large, we can use a trick similar to the one used to derive Stirling's formula, considering each of the terms as the area of a rectangle centered on an integer $n$, of unit base and height $1/n$, and replacing the expression in brackets with the area enclosed by the function $y = 1/x$. Here too we must be careful with the integration limits and not neglect the area of the half-rectangle between 1/2 and 1. We can then write:

$$\langle k \rangle \approx N \int_{1/2}^{N} \frac{1}{x} dx = N[\ln(N) - 1/2] = N \ln(2N). \tag{3.22}$$

Therefore, to complete a collection of just 100 stickers, we must typically buy about 500 of them, which is the mathematical foundation of the practice of exchanging stickers.

To give a similar example, walking around a city of about 1,300,000 residents like Milan and assuming to meet a thousand randomly found people every day, it would take over fifty years before having met at least once each one of the citizens, which is much more than the four years or so needed if every time we meet a different person.

## 3.3  The Binomial Distribution

The problem we will address in this section is particularly interesting not only in itself, but also because it is a starting point for much of what we will say in the following. Indeed, the distribution that we will discuss is not only interesting in itself, but also because it generated in appropriate limits two other distributions of paramount importance in physics. We still deal with an experiment consisting of a sequence of independent trials, but this time we ask ourselves what is the probability that $k$ trials, over a total of $n$, are successful if the success probability in a single trial is $p$.[6] For instance, if we flip a (fair) coin $n$ times, we want to find the probability of obtaining exactly $k$ heads, with $p = 1/2$. We write this distribution as $B(k; n, p)$ to stress that this is a function of $k$, in which $n$ and $p$ appear as *parameters*.

Reasoning as we have done for the geometric distribution, we can easily see that, because in the sequences we are focusing on there are $k$ successes, but also $n - k$ failures (each one occurring with probability $1 - p$), and since the results of the single trials are independent events, the probability of any sequence of this kind is $p^k(1 - p)^{n-k}$. However, this is just the probability of a *single* sequence, and we have many of them. For instance, [**thttthhtht**] is a sequence of 10 tosses of a coin in which we obtain 4 heads, but the same number of successes is shared by [**thththththtt**], [**hhhh**tttttt], and by many other sequences too. How many? Well, this is just the number of ways of choosing four "h" over ten letters regardless of the order in which we choose them, which is the binomial coefficient $\binom{10}{4}$. Generalizing, the number of sequences with $k$ successes over $n$ trials is $\binom{n}{k}$. Hence we have

$$B(k; n, p) = \binom{n}{k} p^k(1 - p)^{n-k} = \frac{n!}{k!(n - k)!} p^k(1 - p)^{n-k}, \qquad (3.23)$$

which is called a *binomial distribution* (but, sometimes, a *Bernoulli distribution* too). Using again an urn analogy, we can say that $B(k; n, p)$, with $p = r/(r + b)$, is also the probability of getting $k$ red balls over a total of $n$ in an extraction *with replacement* from an urn containing $r$ red and $b$ blue balls. You may already guess that the binomial distribution applies to a wide range of situations in science, some of which we will later discuss.

Figure 3.4a shows the binomial distribution for some values of $p$, with $n = 20$ fixed. We notice that:

- The maximum of the distribution is obtained for a value $k \simeq np$ .
- For $p \neq 0.5$ the distribution is asymmetric, with a long tail for high ($p < 0.5$) or low ($p > 0.5$) values of $k$.

In Fig. 3.4b, where we instead consider the shape of the distribution as $n$ varies with $p = 0.1$ fixed, we can also observe that:

- As $n$ increases, the distribution becomes more symmetric for all values of $p$ and progressively turn into a "bell-shaped" curve.
- The width of the distribution, measured for instance at half the its maximum, grows with $n$, but as fast as the maximum increases.

**Expectation, variance, and skewness**

$$\langle k \rangle = \sum_{k=0}^{n} k \, \frac{n!}{k!(n - k)!} \, p^k(1 - p)^{n-k} = np \sum_{k=1}^{n} \frac{(n - 1)!}{(k - 1)!(n - k)!} \, p^{k-1}(1 - p)^{n-k},$$

---

[6] A single attempt of this kind is also called a *Bernoulli trial*, from Jakob Bernoulli, who was the first to investigate this question.

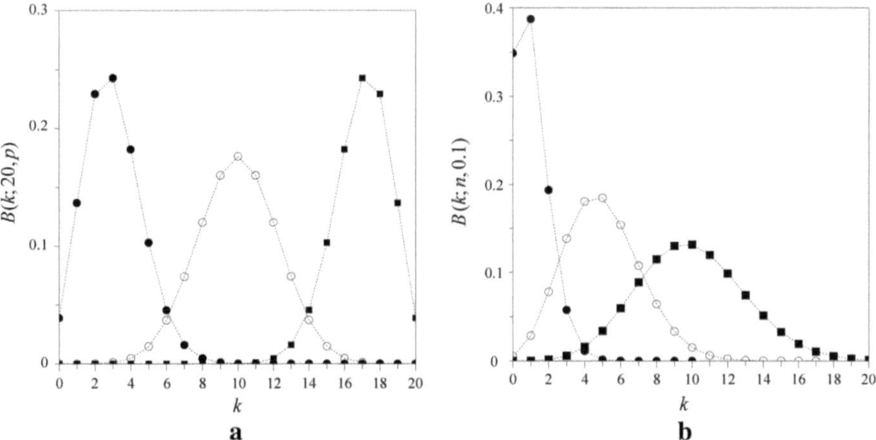

**Fig. 3.4** **a** Binomial distribution for $n = 20$ (fixed) and $p = 0.15$ ($\bullet$), $0.50$ ($\circ$), $0.85$ ($\blacksquare$). **b** Binomial distribution for $p = 0.1$ (fixed), and $n = 10$ ($\bullet$), $50$ ($\circ$), $100$ ($\blacksquare$)

where the first equality is obtained by observing that the term with $k = 0$ vanishes. Putting $k' = k - 1$ and $n' = n - 1$, and using once again the binomial theorem, we find

$$\langle k \rangle = np \sum_{k'=0}^{n'} \frac{n'!}{k'!(n'-k')!} \, p^{k'} (1-p)^{n'-k'} = np. \tag{3.24}$$

Thus, the expectation is exactly the number of trials we would guess, if we just knew that the probability of success in a single trial is $p$. Slightly less intuitive is the result for the variance of the distribution, which can be obtained like for the geometric distribution by first evaluating $\langle k(k-1) \rangle$ applying twice the trick used to find $\langle k \rangle$, and writing $\sigma_k^2 = \langle k(k-1) \rangle + \langle k \rangle - \langle k \rangle^2$. This yields

$$\sigma_k^2 = \langle k^2 \rangle - \langle k \rangle^2 = np\,(1-p). \tag{3.25}$$

Note that the width of the distribution grows as $\sqrt{n}$, and then as $\sqrt{\langle k \rangle}$ too, whereas the *relative* standard deviation $\sigma_k / \langle k \rangle$ *decreases* as $n^{-1/2}$. With a bit more effort, one can also find that the skewness of the binomial is

$$\gamma_k = \frac{1-2p}{\sigma_k^2} = \frac{1-2p}{np(1-p)}, \tag{3.26}$$

which vanishes for $p = 1/2$ and decreases as $n^{-1}$.

The expression for the expectation value can be obtained in a simpler way by remembering that the expectation is obtained by a linear operation. At the $i$-th trial we can then associate a RV $k_i$, which takes only the value 1, if the attempt is successful, and the value 0 otherwise. The value $k_i = 1$ has probability $p$, and $k_i = 0$

has probability $(1 - p)$ (this is called a *Bernoulli variable*). The expectation of each elementary variable $k_i$ is therefore $\langle k_i \rangle = p \cdot 1 + (1 - p) \cdot 0 = p$, while the total number of successes is given by $k = k_1 + k_2 + \ldots + k_n$, with expectation

$$\langle k \rangle = \langle k_1 \rangle + \langle k_2 \rangle + \ldots + \langle k_n \rangle = np.$$

**Example 3.7** (*Once more on coins & digits*)

(a) With ten coin flips we expect, for the number $k$ of heads,

$$\langle k \rangle = 5; \quad \sigma \simeq 1.6; \quad \frac{\sigma_k}{\langle k \rangle} \simeq 0.32,$$

with $\gamma_k = 0$, while for a thousand flips we get

$$\langle k \rangle = 500; \quad \sigma_k = \sqrt{250} \simeq 16; \quad \frac{\sigma_k}{\langle k \rangle} \simeq 0.03,$$

that is, the relative width decreases by a factor of 10 by increasing the number of trials by a factor of 100. For the same value of $n$, the maximum of the variance, and therefore of the width of the distribution, is obtained for $p = 0.5$.

(b) If we gather the decimals of $\pi$ in groups of 20, we expect a particular digit, for example 4, to appear on the average twice in each group. Yet, the number $k$ of fours should actually vary among the groups according to a binomial distribution with $p = 0.1$ and $n = 20$. The following table compares this prediction with the experimental frequencies $f(k)$ obtained by analyzing 500 groups of 20 digits (the meaning of the last column will be discussed later).

| $k$ | 0 | 1 | 2 | 3 | 4 | 5 | 6 | 7 | 8 |
|---|---|---|---|---|---|---|---|---|---|
| $f(k)$ | 0.122 | 0.246 | 0.310 | 0.186 | 0.092 | 0.034 | 0.008 | 0.002 | 0 |
| $B(k; 20, 0.1)$ | 0.122 | 0.270 | 0.285 | 0.190 | 0.090 | 0.032 | 0.009 | 0.002 | 0 |
| $P(k, 2)$ | 0.135 | 0.271 | 0.271 | 0.180 | 0.092 | 0.036 | 0.012 | 0.003 | 0.001 |

**Example 3.8** (*Displacement in a 1D random walk*) We have seen in Chap. 1 that a one-dimensional random walk of $N$ steps, where a step to the right or to the left are made with the same probability $p = 1/2$, is equivalent to a game where a fair coin is flipped $N$ times. Thus, the number $k$ of steps to the right out of a total of $N$ will be distributed according to a binomial distribution $B(k; N, 0.5)$. Starting at the origin and taking as positive a step to the right, the final position $x$ after $k$ step of equal length $L$ will then be given by

$$x = L[k - (N - k)] = L(2k - N)$$

Therefore:

(a) $\langle k \rangle = N/2$ implies $\langle x \rangle = 0$, i.e., on average the random walker is found at the starting point.
(b) Using (3.11), $\sigma_k = \sqrt{N}/2$ results in $\sigma_x = 2L\sigma_k = L\sqrt{N}$, that is, the size of the "explored" region (the root mean square displacement) increases as the square root of the number of steps.

Finding the entire probability distribution of the displacement is a bit more challenging, and is basically an extension of Example 2.22. It is easy to see that an odd or even number of steps respectively leads to a distance from the origin that is, in units of $L$, an odd or an even integer. Therefore, the probability of being $k$ steps away from the origin after $N$ steps always vanishes if $N + k$ is odd.[7] Suppose then that both $N$ and $k$ are both even or odd. Then, the probability $P_N(k)$ of being $k$ steps away from the origin after $N$ steps is equal to the probability of choosing $(N + k)/2$ steps out of $N$ (either in the positive or in the negative direction), i.e., by the binomial distribution

$$P_N(k) = B\left(k; \frac{N+k}{2}, \frac{1}{2}\right) = \frac{N!}{\left(\frac{N+k}{2}\right)! \left(\frac{N-k}{2}\right)!} \frac{1}{2^N},$$

This result can easily be extended to the case $N + k$ odd (which must vanish) by writing

$$P_N(k) = \frac{1 + (-1)^{N+k}}{2} B\left(k; \frac{N+k}{2}, \frac{1}{2}\right), \tag{3.27}$$

The probability distribution for the walker final position is then given by $L P_N(k)$, for all integer values of $k$ between $-N$ and $+N$.

**Example 3.9** (*Improving a warning system*) Suppose that we include in the control system of an experimental setup some device that gives a warning about events that require, for example, to suspend a data acquisition process. To verify the authenticity of an alarm, we actually insert into the setup three of these devices, assuming that the alarm threshold corresponds to a signal from at least two devices. These warning devices, however, are not perfect, so that, when an event occurs, they trigger with a probability of only 80%. The chance of detecting a risk event is then equal to the probability that at least two devices out of three goes off,

$$p(k = 2) = B(2; 3, 0.8) + B(3; 3, 0.8) \simeq 0.90,$$

significantly higher than the value $p = 0.64$ we would obtain using only two devices. If we then decide to introduce a fourth device, we have

$$p(k = 2) = 1 - p(k < 2) = 1 - B(0; 4, 0.8) - B(1; 4, 0.8) \simeq 0.97.$$

---

[7] $N$ odd $\rightarrow$ $k$ even $\rightarrow$ $(N + k)$ even, but also $N$ even $\rightarrow$ $k$ even $\rightarrow$ $(N + k)$ even.

## 3.4  The Poisson Distribution

As the number $n$ of trials increases calculating the coefficients $\binom{n}{k}$ in the binomial distribution becomes more and more difficult because of the rapid growth of the factorials with their argument. On the other hand, almost all applications of physical interest correspond precisely to situations where $n$ takes very large values. It is then useful to investigate what form the binomial distribution takes on when $n \to \infty$. Since the distribution is determined not only by the total number of trials, but also by the probability $p$ of success in a single trial, we can take the limit in two different ways:

1. We *fix* the probability of an event in a single trial and increase the number of trials, namely, $n \to \infty$ with $p = $ costant, so that $\langle k \rangle = np \to \infty$ too.
2. We increase the number of trials, but at the same time *decrease* the probability of success in a single trial, so that the *expectation* remains constant, namely, $n \to \infty$ and $p \to 0$ with $np = a$, where $a$ is a constant, which amounts to studying very improbable events that however, have a large number of chances to occur.

These distinct limits will lead us to introduce two probability distributions of extreme interest for physics, and in general for the analysis of statistical data. Note that the first distribution can also be thought of as a limiting case of the second, taking the further step $a \to \infty$. We start dealing with the second case.

Using the constant parameter $a = np$, we write the binomial distribution as

$$B(k; n, a) = \frac{n!}{k!(n-k)!} \left(\frac{a}{n}\right)^k \left(1 - \frac{a}{n}\right)^{n-k}.$$

We expect the probability of obtaining a number of successes $k \gg np$ to be very small and therefore, since $np$ is fixed can assume all those values of $k$ that have a significant probability to be much smaller than $n$. We can then approximate

$$\frac{n!}{(n-k)!} = n(n-1)...(n-k+1) \simeq n^k,$$

because all the factors at the r.h.s. are very close to $n$, and

$$\left(1 - \frac{a}{n}\right)^{n-k} \simeq \left(1 - \frac{a}{n}\right)^n \simeq e^{-a},$$

because this limit is indeed the *definition* of $e^{-a}$. This yields the *Poisson distribution*

$$\boxed{P(k; a) = \frac{a^k e^{-a}}{k!}.}$$
(3.28)

Therefore, by limiting the generality of the binomial distribution to the case $n \to \infty$ with constant $np$, we gain a lot in terms of simplicity:

- the Poisson distribution is determined by a *single* parameter $a$ (while to specify the binomial we need both $n$ and $p$.
- factorials of large numbers (of order $n$) are not needed, while familiar functions such as exponentials and powers appear.
- the calculation of the distribution as $k$ increases is particularly simple, since

$$P(k; a) = \left(\frac{a}{k}\right) P(k-1; a)$$

so that all terms can be obtained recursively from $P(0; a) = \mathrm{e}^{-a}$.

Figure 3.5 shows the Poisson distribution for some values of $a$ (which is not necessarily an integer). We note that the maximum of the distribution is for $k = a$, and that for small values of $a$ the distribution shows a marked asymmetry, similar to what we have seen for the binomial.

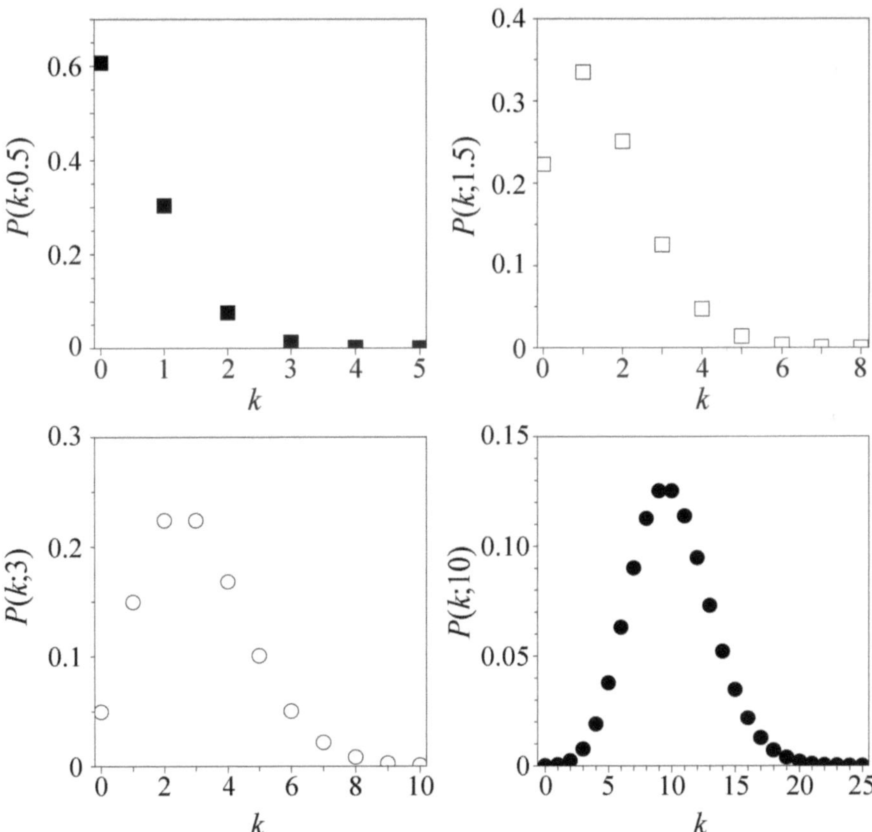

**Fig. 3.5** Poisson distribution for $a = 0.5$ (■), 1.5 (□), 3 (o) e 10 (•)

**Expectation, variance, and skewness.**

$$\sum_{k=0}^{\infty} P(k; a) = \mathrm{e}^{-a} \sum_{k=0}^{\infty} \frac{a^k}{k!} = \mathrm{e}^{-a}\mathrm{e}^{=}1$$

therefore the Poisson distribution is normalized. For the expectation we have

$$\boxed{\langle k \rangle = a} \tag{3.29}$$

which is not surprising since $a = np$ is the expectation of the binomial we started from. Indeed,

$$\langle k \rangle = \sum_{k=0}^{\infty} k P(k; a) = a\mathrm{e}^{-a} \sum_{k=1}^{\infty} \frac{a^{k-1}}{(k-1)!} = a\mathrm{e}^{-a} \sum_{k'=0}^{\infty} \frac{a^{k'}}{k'!} = a,$$

The variance can be obtained by first showing, with a similar method, that

$$\langle k(k-1)(k-2)\ldots(k-m+1) \rangle = a^m. \tag{3.30}$$

Then, from $\sigma_k^2 = \langle k^2 \rangle - \langle k \rangle^2 = \langle k(k-1) \rangle + \langle k \rangle - \langle k \rangle^2$, we have

$$\boxed{\sigma_k^2 = a} \tag{3.31}$$

The Poisson is then a distribution characterized by the distinctive property

$$\sigma_k^2 = \langle k \rangle,$$

while its relative standard deviation $\sigma_k / \langle k \rangle$ decreases as $\langle k \rangle^{-1/2}$, just like the binomial. Note that the variance of the Poisson is always larger than that of a binomial with the same expectation, and in particular it is twice the variance of a binomial with $p = 0.5$.

Using (3.30) one can also show that

$$\boxed{\gamma_k = \frac{1}{\sqrt{a}}} \tag{3.32}$$

hence the skewness decreases like the relative standard deviation.

The last row of the table in Example 3.7 shows the values of the Poisson distribution $P(k; 2)$, which corresponds to expecting on the average two "fours" in a sequence of 20 decimals of $\pi$. Although $n = 20$ is a moderately large value and $p = 0.1$ is not very small, the Poisson already approximates the binomial rather well, except for the tail at high values, where it significantly overestimates in percentage the values of $B(k; 20, 0.1)$.

**Example 3.10** (*An unlucky number*) On the wheel of a (French) roulette there are 37 (36 plus the zero) numbers, so the chance that in a spin of the wheel the ball stops on any of them is $p = 1/37 \simeq 0.027$. For reasons that I find hard to understand,[8] the number 13 seems to be the least–played number on the entire roulette board. The probability that 13 is *not* the winning number for $N$ spins, for the happiness of the superstitious bettors, is therefore $P = (1 - p)^N = \exp[N \ln(1 - p)]$ that, since $p$ is a rather small number, we can approximate as

$$P \simeq \exp(-Np).$$

The same result is obtained by noticing that, because $p \ll 1$, the number of times $k$ that the ball will stop on 13 follows approximately a Poisson distribution $P(k; a) = a^k e^{-a}/k!$, where $a = pN$. Thus, the chance that 13 will never appear is $P(0; a) = \exp(-Np)$ (which means that 13 has a chance of about 30% of coming out in just 13 spins).

**Example 3.11** (*Back to birthdays*) Let us reconsider the calculation we did in Example 2.20. There are $365 \times 365$ possible pairs $(d, d')$ of birthday dates. To two people who have their birthdays on the same day corresponds one of the 365 pairs of the type $(d, d)$, which has then a probability $p = 1/365$. With $N$ people we have $N(N - 1)/2$ possible pairs of birthday dates, thus we expect on the average $a = N(N - 1)/730$ pairs of people who have their birthday on the same day. If $a$ is not too large, we can assume that the number of pairs $k$ who have their birthdays on the same day is distributed according to the Poisson distribution $P(k; a)$, thus the chance that at least one pair has their birthday on the same day will then be equal to $1 - \exp(-a) = 1 - \exp[-N(N - 1)/730]$, which is the approximate result we previously found.

**Example 3.12** (*Radioactive decays*) In the first chapter we have seen that radioactive decay is one of the most striking examples of a random phenomenon in physics. In fact, the Poisson distribution, originally derived in a context that had little to do with science and almost ignored until the beginning of this century, took a key role in physics thanks to some landmark experiments on the $\alpha$-decay. It soon became evident that other physical phenomena, such as the emission of electrons from the heated metal filament of a thermionic valve (thermoelectric effect), or from a metal surface illuminated (photoelectric effect) share the same statistical properties.

The table below, taken from *Radiation from Radioactive Substances*, by E. Rutherford, J. Chadwick, and C.D. Ellis (1930), shows one of the first experimental results related to the statistical properties of radioactive decay, where the authors analyzed the number of counts measured in a time interval of $7.5\,$s by a small area detector placed at a distance from an intense radioactive source. Let us call $n(k)$ the number of intervals in which $k$ counts were measured and assume, anticipating what we shall

---

[8] After all, both in Italy and in Tibet, 13 is regarded (for different reasons) as a *lucky* number.

later prove, that the experimental average $\bar{k} = 3.87$, obtained on $N = 2608$ measurement intervals, is a good estimate of the expectation $\langle k \rangle$ of the probability distribution for $k$. If we compare the relative frequencies $n(k)/N$ with the Poisson distribution $P(k; \bar{k})$, or $n(k)$ with $N P(k; \bar{k})$, we find a quite reasonable agreement. It is useful to point out that the measurement intervals used in the experiment were very long compared to the average time between two emissions. In such a long time interval, the total number of $\alpha$ particles emitted from the source is almost constant, but the number of them that fall on the small detector are fluctuates significantly. An experiment of this type therefore probes the *spatial* rather then the temporal fluctuations of the emission.

| $k$ | 0 | 1 | 2 | 3 | 4 | 5 | 6 | 7 | 8 | 9 | 10 | 11 | 12 | 13 | 14 |
|---|---|---|---|---|---|---|---|---|---|---|---|---|---|---|---|
| $n(k)$ | 57 | 203 | 383 | 525 | 532 | 408 | 273 | 139 | 45 | 27 | 10 | 4 | 0 | 1 | 1 |
| $N P(k; \bar{k})$ | 54 | 210 | 407 | 525 | 508 | 394 | 254 | 140 | 68 | 29 | 11 | 4 | 1 | 1 | 1 |

**Poisson Distribution for Point Events in a Continuum**

We have obtained the Poisson distribution as a limiting case of the binomial, but it can actually be applied to a much wider and, at a first glance, apparently unrelated class of problems. If you have ever lain in the grass during a starry night in late August, looking at that meteor shower dubbed "the tears of St. Lawrence", you have surely noticed that these abrupt events come in "bunches": you wait a long time without catching sight of a single shooting star, and then, suddenly, two, three, or even more of them shoot within a short lapse of time. In other words, the number of meteors observed in a given time interval shows strong fluctuations. Looking for shooting stars is particularly annoying when you do it with your friends, for it is pretty sure that, as soon as one of them shouts "here it is!", it is already too late to catch sight of it. Meteors are indeed sudden, almost instantaneous events that irregularly break the peaceful flow of time. This is the kind of situation we wish to describe.

In general, we then consider a sequence of *point-like* events occurring at random in a continuum. This "continuum" can be time, in which case "point–like" means that these events take place almost instantaneously. But we may also refer to a *spatial* continuum: for instance, we may want to describe the distribution of specks of dust floating in a room, or deposited on a table. To illustrate the point, we shall stick to time, with no loss of generality. We assume that these events are totally *independent* one from the other, namely, that the chance that one of them takes place is not influenced by the previous occurrence of another event, even if the latter happened just a while before.

We wish to find the probability $P(k; t)$ that, in a time interval $t$, $k$ events take place, with the only information that $a$ events take place *on the average*. Hence, the typical frequency of the events is $\alpha = a/t$, and the typical time between two of them is $\tau = 1/\alpha$. Let us first evaluate the probability $P(0; t)$ that *no* event takes place in the time interval $t$. We divide $t$ in sub-intervals $\Delta t$. Choosing $\Delta t \ll \tau$, we can be sure that either no event, or just a *single* one of them, takes place in $\Delta t$ (no chance for two events, otherwise we choose a smaller $\Delta t$).[9] Then, since the probability that an event takes place in $\Delta t$ is $p = \alpha \Delta t$, the probability that it does *not* take place in $\Delta t$ is $P(0; \Delta t) = 1 - \alpha \Delta t$. Because the events are independent, for the whole interval $t$ we just have to take the product of all probabilities,

$$P(0; t) = [P(0; \Delta t)]^{t/\Delta t} = (1 - \alpha \Delta t)^{t/\Delta t} \xrightarrow[\Delta t \to 0]{} P(0; t) \simeq e^{-\alpha t} = e^{-a}.$$

To evaluate $P(k; t)$, we consider a slightly larger time interval $t + \Delta t$. Since at most one event can take place in $\Delta t$, there are only two ways for $k$ events to happen in a time interval $t + \Delta t$: either $k - 1$ events take place in $t$, and the last one happens during $\Delta t$, or $k$ events already took place at time $t$, and nothing happens in $\Delta t$. These two possibilities are mutually exclusive, hence $P(k; t + \Delta t)$ is just the sum of their probabilities,

$$P(k; t + \Delta t) = P(k - 1; t)p + P(k; t)(1 - p).$$

Substituting $p$ and rearranging this equation, we obtain

$$\frac{P(k; t + \Delta t) - P(k; t)}{\Delta t} + \alpha P(k; t) = \alpha P(k - 1; t),$$

which becomes, for $\Delta t \to 0$,

$$\frac{\mathrm{d}P(k; t)}{\mathrm{d}t} + \alpha P(k; t) = \alpha P(k - 1; t).$$

This is a recursive equation, which gives $P(k; t)$ once we know $P(k - 1; t)$. To solve it, observe that the function $f_k(t) = e^{\alpha t} P(k; t)$ satisfies $f_k'(t) = \alpha f_{k-1}(t)$. It is not difficult to see that a solution of this equation is $f_k(t) = (\alpha t)^k / k!$. Hence, recalling that $\alpha t = a$,

$$P(k; t) = \frac{(\alpha t)^k e^{-\alpha t}}{k!} \implies P(k; a) = \frac{a^k e^{-a}}{k!}$$

which satisfies $P(0; a) = e^{-a}$ and coincides with the Poisson distribution (3.28).

Therefore, points randomly arranged on a line, on a surface, or in a volume, are not distributed uniformly, and show denser and rarer regions such that the number of points in each sub-interval follows a Poisson distribution. For example, a random distribution of points on a surface could have the appearance of Fig. 3.6. I did not choose to represent the points with stars on a black background for purely aesthetic reasons. The distribution on the celestial vault of the stars visible to the naked eye, which are located in a nearby and therefore quite homogeneous region of the Galaxy, is in fact approximately a Poisson distribution, much to the chagrin of our habit of seeing bears, mythological hunters, or legendary queens of Ethiopia.

---

[9] In fact, we can choose $\Delta t$ as small as we like, since these events are just instants on the time axis, with zero duration.

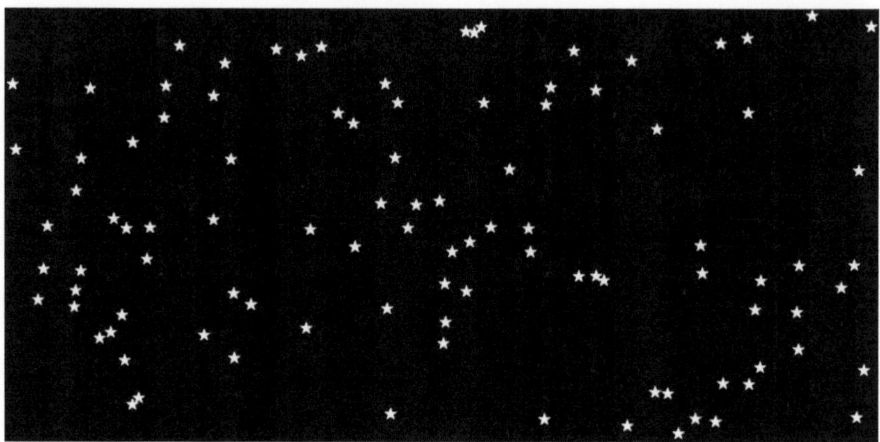

**Fig. 3.6** Starry night, according to Poisson

This alternative way of looking at the Poisson distribution is particularly useful in physics, in particular if we consider the distribution in *space* of noninteracting (namely, independent) particles, for instance the molecules of an ideal gas. Equation (3.28) tells us these particles will not be uniformly distributed: rather, if the average number of particles in a volume $V$ is $\langle N \rangle$, we expect fluctuations of this number that are of the order of $\Delta N = \sqrt{\langle N \rangle}$, which is the standard deviation of the Poisson. So, within a randomly chosen volume $V$, we may expect to find a number of particles $N = \langle N \rangle \pm \Delta N$. Of course, if $V$ (and then $\langle N \rangle$) becomes very large the *relative* fluctuations $\Delta N / \langle N \rangle$ decrease as $N^{-1/2}$. In a macroscopic volume $V$, these fluctuations can normally be neglected. For instance, 1 cm$^3$ of gas at room temperature and pressure contains/about $2.7 \times 10^{19}$ molecules, thus $\sqrt{\langle n \rangle} \simeq 5.2 \times 10^9$, giving a relative fluctuation $\Delta n / \langle n \rangle$ of about two parts per ten billions. But if we consider a small cube with a side of 100 nm, the relative fluctuations increase to about 0.6%. They are precisely the random density fluctuations on these microscopic scales that give rise to the diffusion of light by a gas, and among other things to the blue color of the sky.

As we mentioned in Chap. 1, the concept of fluctuations is one of the most important in physics and science in general. What we have just found is that, even if there is no force that induces *correlations* between the particle positions, a system is not perfectly uniform, but rather shows spontaneous and unavoidable number fluctuations, ruled by the Poisson distribution. Then, you may wonder what happens if *there are* forces between the particles that tend to correlate them. Actually, this is one of the main themes of statistical physics, but we can get a qualitative answer with a simple example. Suppose you fill a large room either with a large number of friendly people, or with a lot of rather unsociable fellows, who are not particularly eager to meet each other. Enjoying the company, the first kind of people will likely gather in lively groups, because they *attract* each other, in some sense. Conversely,

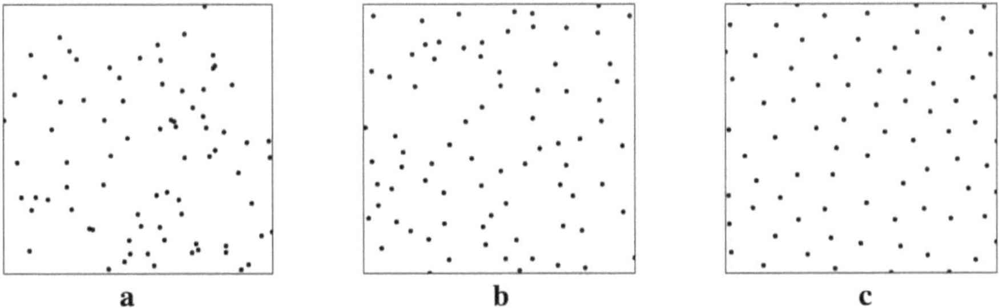

**Fig. 3.7** Random distribution on a square with side $L$ of the centers of disks with diameter $d = 0.01L$ (**a**), $0.05L$ (**b**) e $0.1L$ (**c**)

the other fellows, who care a lot about their privacy, will try to stay apart from the others as much as they can. Somehow, they *repel* each other. In the first case, we then expect number fluctuations to be *larger* than in the "noninteracting" case (which, in this example, corresponds to a room full of people that literally don't *see* each other). Unsociable people will conversely fight against fluctuations, even the little ones that would spontaneously arise if they just ignore each other. This is exactly the case of fluctuations in a system of particles that can freely move around, such as the molecules in a real fluid: attractions enhance fluctuations, whereas repulsions quench them.

In Fig. 3.7, for example, I have simulated three random distributions of about eighty points, but with the constraint that the distance between two of them cannot be less than a fixed fraction of the side of the square in which they are enclosed. Thus, these are actually distributions of "hard disks" of diameter $d = \alpha L$ that cannot overlap. As you can see, while for $\alpha = 0.01$ the distribution is qualitatively very similar to that of Fig. 3.6, spatial fluctuations tend to decrease as increases, until, for $\alpha = 0.1$, when the total area covered by the disks is about 63% of the square a nearly uniform distribution is obtained.[10]

A fascinating example of similar "repulsive interactions" in biology is described by Stephen J. Gould, a great paleontologist and theorist of evolution, in his popular science book *Bully for Brontosaurus*. On the walls of the Waitomo caves, which are a famous tourist attraction in New Zealand, a prodigious number of larvae of the insect *Arachnocampa luminosa* grow and live. Like common fireflies, these insects emit a greenish light, but while fireflies probably do this to confuse predators, these carnivorous larvae are excellent predators, which use light as a beacon to attract and eat other insects (mainly midges, but they do not disdain cannibalism). Therefore, it is much better for the larvae to stay as far away as possible from each other to maximize the hunting territory (and also to avoid unpleasant encounters with their own kind). In fact, the cave is dotted with a very uniform distribution of luminous points, much

---

[10] Actually, we could further increase $\alpha$ since the maximum *random* packing of disks (a quantity which is hard to rigorously define) is about 82% of the surface area. However, this can be obtained only by redistributing with complex strategies the disks that have already be placed. What fractional area could one fill with an ordered packing of disks on a triangular lattice?.

more similar to Fig. 3.7c than to Fig. 3.7a. To quote Gould, the glowworm grotto of Waitomo is an "ordered sky".

On the contrary, blackbody radiation is an important example of fluctuations that are strongly *enhanced* compared to the Poisson case. Indeed, in Example 3.4, we have shown that photon fluctuations do not decrease with the average photon number, but rather become comparable to it. This means that, with respect to the random fluctuations of a sequence of independent events, photons from a blackbody source reach the detector in clusters. In fact, photons from a radiation in thermal equilibrium are said to be "bunched".[11] However, are there nonequilibrium conditions where photon number fluctuations are Poisson-like? The answer is yes, and the peculiar nonequilibrium conditions we mentioned above characterize a class of light sources that have deeply affected our everyday life, as the next example shows.

**Example 3.13** (*Coherent radiation*) In Example 3.4, I mentioned in passing that a radiation field with no fluctuations in the photon number (a so-called *number* or *Fock state*) has no classical analogue. Let me be a bit clearer, by comparing the classical and quantum properties of a monochromatic radiation wave. Classically, the wave has precise values for the amplitude and the phase. In the quantum theory of the electromagnetic field, however, the product of intensity and phase indeterminacies cannot be smaller than a minimal value (this is just another uncertainty principle). A wave where the number of photons has an exact, non-fluctuating value $n$ has a constant amplitude, proportional to $\sqrt{n + 1/2}$, but its phase is *completely* undetermined, so that the field has a zero time-average everywhere.

Can we construct a probability distribution for the number of photons with the same frequency, such that the product of phase and amplitude indeterminacies of the associated field is minimal, so that is it the closest analogous of a classical monochromatic wave? A positive answer was found in 1963 by Roy Glauber, who showed that the simplest photon distribution with these properties is a *Poisson* distribution[12]

$$P(n; \langle n \rangle) = \frac{\langle n \rangle^n \, e^{-\langle n \rangle}}{n!}. \tag{3.33}$$

For these *coherent* (or *Glauber*) *states*, the relative fluctuations *decrease* with the photon number as $\langle n \rangle^{-1/2}$. But the most important point is that (3.33) is a very good approximation of the distribution of photons for the radiation emitted by a laser operating in a single (longitudinal) mode. You can surely appreciate why Glauber was awarded the 2005 Nobel Prize in Physics.

**Example 3.14** (*Visual threshold*) What is the minimum amount of light that our eye can detect? We should first spend a few words about the mechanism of vision. We

---

[11] This can be qualitatively explained because of the "bosonic" nature of photons, but the intensity fluctuations of thermal radiation can also be given a rigorous classical explanation that we will discuss in Chap. 5.

[12] For those of you acquainted with using raising and lowering operators $\hat{a}^{\dagger}$, $\hat{a}$ to calculate the stated of a quantum oscillator, the Glauber states are the eigenvectors of $\hat{a}_\lambda$ for a photon state $\lambda$.

"see" because light is absorbed by particular molecules located in the visual receptors, which are cone or rod structures of the retina. The chemical signal generated by the absorption is then turned into an electrical impulse that travels along the optic nerve. Physically, however, light cannot be absorbed in arbitrary quantities, but only as a integer number of photons with energy $\epsilon = \hbar\omega = hc/\lambda$, which ranges for $\lambda$ in the visible between 1.65 and 3.3 eV, or $(2.65 - 5.3) \times 10^{-19}$ J. We plan to find the minimum number of photons that can trigger a visual stimulus. To maximize the visual sensitivity of a subject, it is worth taking some precautions:

COLOR: The maximum sensitivity of the human eye is found in a region of the visible light spectrum that corresponds to blue–green.

DARK ADAPTATION: If you have ever observed the night sky, you will have noticed that the number of stars you see gradually increases. Indeed, in low light conditions the sensitivity of the eye steps up progressively, increasing by a factor of thousand after about half an hour spent in complete darkness.

WAY OF LOOKING: Those who have used a telescope know that viewing is better when looking slightly sideways into the eyepiece, which means focusing the image laterally with respect to the center of the retina. This is because the maximum density of rods, the most sensitive receptors, is off-axis by an angle of approximately $20°$ from the optical axis of the eye.

DURATION: In the presence of continuous exposure to light, the eye progressively loses sensitivity. To achieve maximum efficiency, it is better to expose the subject to light pulses with a duration not exceeding one tenth of a second. For pulses with a duration shorter than 10 ms, the minimum amount of light energy required to produce a visual stimulus, which is proportional to the product of intensity and exposure time, is almost constant.

To investigate the problem, let us consider a light pulse consisting of an average number $\langle n \rangle$ of photons. About half of these are reflected or absorbed before reaching the retina. Moreover, the receptors absorb at most 20% of the photons that reach the retina. The average number of photons actually absorbed will then be $\langle k \rangle = f \langle n \rangle$, where the loss factor $f \lesssim 0.1$. The absorption of a photon of light is a random process entirely analogous to radioactive emission, and the probability of absorbing $k$ photons is then given by the Poisson distribution $P(k; \langle k \rangle)$. A visual stimulus will then be generated if $k > k_0$, where $k_0$ is the minimum number of excitations required to "see". The overall probability of obtaining a stimulus will then be given by the cumulative probability for $k \geq k_0$,

$$P(k > k_0) = P(k_0; \langle k \rangle) + P(k_0 + 1; \langle k \rangle) + \ldots = \sum_{k=k_0}^{\infty} P(k_0; \langle k \rangle) \qquad (3.34)$$

which, for a fixed $\langle k \rangle$, is a function that depends on the parameter $k_0$.

Figure 3.8 shows $P(k > k0)$ numerically calculated for some values of $k_0$. Note that the curves become steeper and steeper by increasing $k_0$, and that their shape can

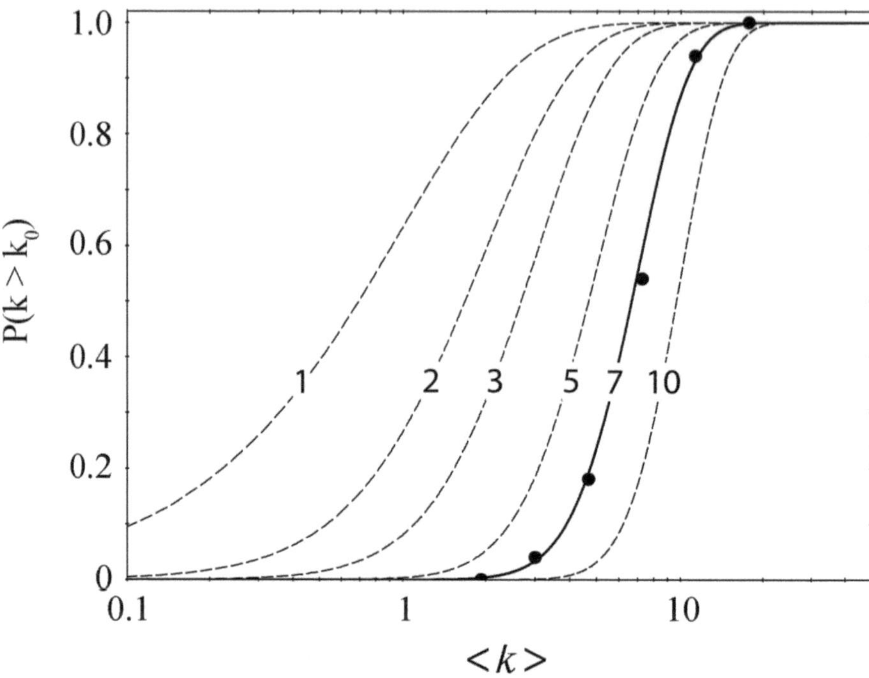



**Fig. 3.8** Cumulative probability (3.34) of a visual stimulus for several values of the parameter $k_0$

be compared even if the loss factor is not accurately known, since on a logarithmic scale changing $f$ just means translating the data. The figure also shows the results of a landmark experiment by Hecht, Shlaer and Pirenne.[13] The points correspond to the response frequencies of a subject (Shlaer himself) as a function of $\langle k \rangle$, determined assuming that $f \simeq 0.08$ (the quantity experimentally fixed by the intensity of the pulse is actually $\langle n \rangle$). For $k_0 = 7$ the agreement with the theory is truly remarkable. Even considering that the exact relation between $\langle k \rangle$ and $\langle n \rangle$ is not accurately known, we can still conclude that a light pulse consisting of only 50–100 photons, corresponding to an energy of the order of $10^{-17}$ J, can already be detected. This is a figure of merit typical of the best light detectors we make, the photomultipliers, which would however "burn out" immediately if exposed even to a small fraction of the light intensity that allows you to comfortably read this page. The eye, therefore, in addition to being an extremely sensitive detector, has also an extremely high *dynamic range*, namely, it can cover a huge range of light intensity values.[14]

---

[13] S. Hecht, S. Shlaer, and M. H. Pirenne, *Journal of General Physiology* **25**, 819 (1942).

[14] An extensive discussion of the physical limits of vision can be found in G.B. Benedek and F. M. H. Villars, *Physics with Illustrative Examples from Medicine and Biology: Vol. 2 (Statistical Physics)*, Springer-Verlag, Berlin (2000), wherefrom I took this example.

Reading the original paper by Hecht, Shlaer, and Pirenne, you will also appreciate how clear and detailed were papers in experimental physics at that time compared to today's.

**Example 3.15** (*Dead time in detection*)  The efficiency of a detector of short–duration physical events (such as a Geiger counter or a photomultiplier) is limited by the presence of a "dead time", that is, by the fact that for a time $t_m$ following an event the detection system is completely "blind" to the occurrence of a second signal. If events occur at a regular repetition rate, the detector can then count them all only when the time lapse between two of these events is greater than $t_m$, hence when the frequency of events is less than $\nu = 1/t_m$. But even if only one event takes place on the average in a time $t_m$, when the events follow a Poisson distribution the probability that two or more of them occur is

$$P = 1 - P(0; 1) - P(1; 1) = 1 - 2\mathrm{e}^{-1} \simeq 0.26.$$

Thus, there is more than a 26% chance of losing some counts along the way. It is easy to verify that to reduce this probability to an acceptable value, say less than 1%, the frequency of the events per second should not exceed $0.1\nu$. Compared to the counting of events that repeat with a precise period, the maximum acceptable frequency for completely random events is therefore reduced by an order of magnitude.

## 3.5  Probability Density for Continuous RVs

As we mentioned, providing a rigorous definition of a continuous random variable requires some care, so we will be a bit more formal then usual. It is first useful to recall which subsets of the real axis can be taken as measurable, if we wish to retain the notion that measure of an interval $(a, b) \in \mathbb{R}$ is simply its length $|b - a|$. This leads to define the *Borel field* on $\mathbb{R}$ as the smallest $\sigma$-algebra $\mathfrak{B}$ that contains all open intervals of $\mathbb{R}$.

Consider now a probability space $\Omega, \Sigma, P)$ where $\Omega = \{\omega\}$ is a set of elementary events, $\Sigma$ is a $\sigma$-algebra in $\Omega$, and $P$ is a probability measure defined for every event $E \in \Sigma$, and let $B \in \mathfrak{B}$ denote an element of the Borel $\sigma$-algebra of subsets of the real line $\mathbb{R}$. We then state that[15]

A one-to-one function $X : \Omega \to R$ is a random variable if and only if the inverse image $X^{-1}(B) = \{\omega : X(\omega) \in B\}$ of any Borel subset $B \in \mathfrak{B}$ belongs to $\Sigma$.

---

[15] Once again, we shall indicate the *variable* with a capital letter $X$, to distinguish it from its *values* $x$, although we shall use the value for the variable when there is no risk of confusion. Note that this was not a problem for a discrete variable $k$, whose values $k_i$ are specified by a subscript. In fact, this tells us that $x$ actually plays the role of a *continuous index* to indicate the values of $X$.

Therefore we can define the probability that the $X$ takes a value in any measurable subset $B$ of $R$ as the *probability $P\{X^{-1}(B)\}$ of the inverse image of $B$*, which is then a real number in $[0, 1]$. In particular, if $B = (-\infty, x)$, $F(x) = P(X \leq x)$ is the *cumulative distribution function* of $X$.

### 3.5.1  *Probability Density*

Note that the previous definition can be applied not only to a continuous, but also to a *discrete* RV. This only requires that there is a finite or countable set $B = \{x_1, x_2, \ldots, x_n, \ldots\}$ such that $P(X \in B) = 1$. However, while for a discrete RV we have a simple *prescription* to calculate the probability of a given set of outcomes, just by summing the probabilities of each single value, this does not work for a continuous RV, because the probability that $X$ takes on a single value $x$ is necessarily zero. For instance, we can try and find a model that predicts the probability $P$ for the height of an individual to be in the range $h = [1.77 - 1.78]$ m, but it does not make any sense to evaluate the probability that someone is *exactly* $\sqrt{\pi} = 1.777245385\ldots$ meters tall, up to all the infinite decimals of this irrational number! We can restrict of a factor of hundred the height interval $\Delta h$, still getting a value that can be tested on the population of a large country, but for sure $P$ *must* vanish when $\Delta h \to 0$.

Therefore, the only thing we can do with a continuous RVs is to give the probability $P(x; \Delta x)$ that it lies within a finite range $[x, x + \Delta x]$. However, if this range is small, we expect $P$ to be *proportional* to $\Delta x$. Hence, for an infinitesimal range $dx$, we can write

$$P(x; \Delta x) = f(x)dx, \tag{3.35}$$

where $f(x)$ is called the *probability density function* (p.d.f.) for $X$.[16]

**Remark 3.15.1** In physics, there is a crucial difference between a probability and a probability *density*: while $P(x; dx)$ is a pure number, $f(x)$ is *not* dimensionless, but rather has the units of the *inverse* of $X$, which we write $[f(x)] = [X]^{-1}$. For example, if $X$ is a time, $f(x)$ is a frequency.[17]

KEEP IT IN MIND!

The *cumulative probability distribution*, i.e., the total probability that $X$ is lower or equal than $x$, is then

$$F(x) = \int_{-\infty}^{x} f(x')dx'. \tag{3.36}$$

---

[16] By analogy with the pair of physical quantities {mass–mass density}, the distribution for a variable where probability is "lumped" in single values is often called in statistics the *mass distribution function*, m.d.f. As a physicist, I shun this denomination, but, after all, one cannot pretend that statisticians know number density, momentum density, energy density….

[17] In this case, we will usually denote the probability density and the cumulative distribution with $f(t)$ and $F(t)$, respectively.

and the probability that $X$ takes a value between $x_1$ and $x_2$

$$P(x_1 \leq x \leq x_2) = F(x_2) - F(x_1) = \int_{x_1}^{x_2} f(x)\mathrm{d}x. \tag{3.37}$$

Introducing probability densities allows us to generalize the properties and quantities we defined for a random variable with $N$ discrete values to a continuous variable $X$ defined in the interval $[a, b]$ of the real axis.[18] The following table compares the two cases.

|  | discrete variable $k$ | continuous variable $x$ |
|---|---|---|
| sample space | $N$ discrete values | continuous interval $[a, b]$ |
| normalization | $\sum_{i=1}^{N} P(k_i) = 1$ | $\int_a^b f(x)\mathrm{d}x = 1$ |
| expectation | $\langle k \rangle = \sum_{i=1}^{N} k_i P(k_i)$ | $\langle x \rangle = \int_a^b x f(x)\mathrm{d}x$ |
| central $r$-moment | $\langle k^r \rangle = \sum_{i=1}^{N} (k_i)^r P(k_i)$ | $< x^r >= \int_a^b x^r f(x)\mathrm{d}x$ |
| variance | $\sigma_k^2 = \langle k^2 \rangle - \langle k \rangle^2$ | $\sigma_x^2 = \langle x^2 \rangle - \langle x \rangle^2$ |
| skewness | $\gamma = \dfrac{1}{\sigma_k^3} \langle (k - \langle k \rangle)^3 \rangle$ | $\gamma = \dfrac{1}{\sigma_x^3} \langle (x - \langle x \rangle)^3 \rangle$ |

When one of the interval limits $a$ or $b$ is not finite, several issues that we already pointed out for a discrete variable with infinite values come out again, possibly amplified. For instance the normalization condition can be satisfied only if, for $x \to \pm\infty$, $f(x) \to 0$ faster than $x^{-1}$. Similarly, you can easily check that a moment of order $r$ exists only if $f(x)$ decreases faster than $x^{-r+1}$.

The study of a generic function of a continuous RV will be later addressed on detail. Right from now, however, it is worth anticipating that we can define the expectation of $y(x)$ as

$$\langle y(x) \rangle = \int_a^b y(x) f(x)\mathrm{d}x, \tag{3.38}$$

---

[18] We introduced the concept of probability density to express the probability that $X$ lies in a given interval $[a, b]$, which is the situation we will commonly encounter, but this definition can be extended to any measurable subset $B \subset \mathbb{R}$ by putting

$$P(X \in B) = \int_B f(x)\mathrm{d}x$$

However, this requires interpreting the integral in the sense of Lebesgue. Some of you may already be familiar with this more advanced and surely mich more powerful integration method. If you are not, for our purposes you just have to know that Lebesgue defined the integral that bears his name by partitioning the *range* of a function and summing up sets of $x$-coordinates belonging to given $y$-coordinates rather than, as done in the Riemann integral, by partitioning the *domain*. Graphically, while the Riemann integral considers the area under a curve as made out of vertical rectangles, the Lebesgue definition considers therefore horizontal "strips", summing the areas under the function that they intercept. The Lebesgue integral exists for a much larger class of functions than the Riemann integral, such as those that show an infinite number of discontinuity points such as the set of irrational numbers in $[0, 1]$ we discuss in the introduction.

provided of course that his integral exists. In general, $\langle y(x) \rangle \neq y(\langle x \rangle)$, as we have seen for moments, which are just particular cases. However, if $y(x)$ is a *convex* (concave upward) function over the whole domain $[a, b]$, we can say something more. In fact, this means that, for all $x_0 \in [a, b]$, one can find a straight line passing through the point $[x_0, y(x_0)]$ such that the whole curve lies above it,

$$y(x) \geq y(x_0) + m(x - x_0),$$

where $m$ is the slope of that line (convince yourself by tracing a graph).[19] Choosing in particular $x_0 = \langle x \rangle$, taking the expectation value of both sides, and observing that $\langle (x - \langle x \rangle) \rangle = 0$, we obtain

$$\langle y(x) \rangle \geq y(\langle x \rangle), \tag{3.39}$$

which is called *Jensen's inequality*. For a *concave* function, we have of course $\langle y(x) \rangle \leq y(\langle x \rangle)$. So, for instance, $\langle \ln(x) \rangle \leq \ln \langle x \rangle$.

**Example 3.16** (*The uniform distribution*)  As a simple example, consider a continuous variable $X$ that is *uniformly* distributed in the interval $[a, b]$, that is, with a constant probability density. For the p.d.f. to be normalized, we must have

$$f(x) = \begin{cases} \dfrac{1}{b - a} & \text{for } a \leq x \leq b \\[2mm] 0 & \text{otherwise,} \end{cases} \tag{3.40}$$

which is called a *boxcar function* and corresponds to a cumulative distribution

$$F(x) = \begin{cases} 0 & (x < a) \\[2mm] \dfrac{x - a}{b - a} & (a \leq x \leq b) \\[2mm] 1 & (x > b) \end{cases} \tag{3.41}$$

The expectation is

$$\langle x \rangle = \frac{1}{b - a} \int_a^b x \, dx = \frac{1}{b - a} \left[ \frac{x^2}{2} \right]_a^b = \frac{a + b}{2},$$

which is, rather obviously, the central point of the interval. Then, since

$$\langle x^2 \rangle = \frac{1}{b - a} \int_a^b x^2 \, dx = \frac{1}{b - a} \left[ \frac{x^3}{3} \right]_a^b = \frac{a^2 + ab + b^2}{3},$$

---

[19] Of course, when $y(x)$ is differentiable, one trivially has $m = y'(x_0)$.

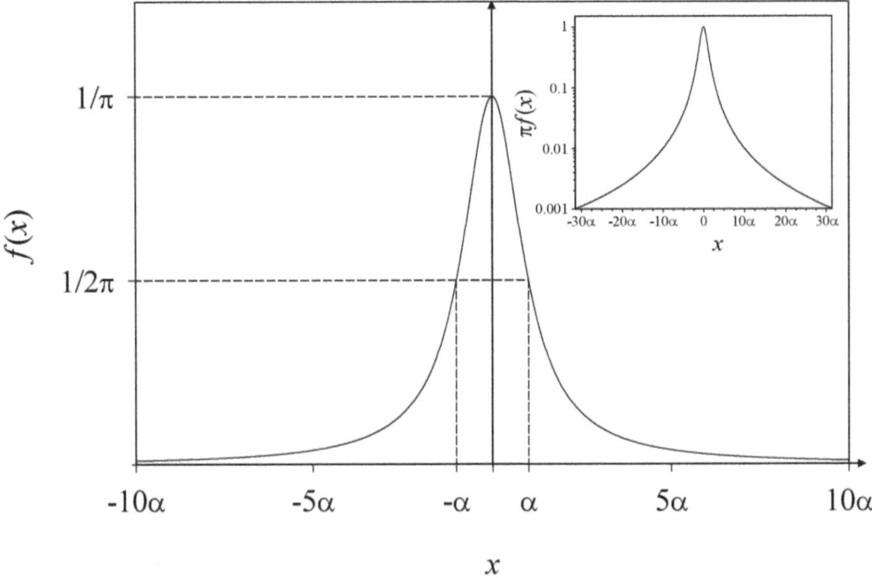

**Fig. 3.9** Cauchy distribution. The semi-logarithmic inset shows that the probability density decreases to a value $f(x) = 10^{-3}$ only for $|x| \gtrsim 30\alpha$

some elementary steps lead to

$$\sigma_x^2 = \langle x^2 \rangle - \langle x \rangle^2 = \frac{(a-b)^2}{12}.$$

It is quite interesting to see what happens when both $a$ and $b$ converge to the same value $x_0$, so that $f(x)$ becomes a very tall rectangle with a vanishing base, representing the probability density for a very 'localized' random variable. In this limit, $f(x)$ becomes a mathematical object you have surely already met, the Dirac delta $\delta(x - x_0)$, whose properties are summarized in the appendix. This is not a true p.d.f., because $\delta(x)$ is *not* a function,[20] but we will regard it as a kind of 'generalized' probability density that we will find useful in many situations. The corresponding cumulative density is the Heaviside 'step' function $H(x - x_0)$ discussed in the appendix.

**Example 3.17** (*The Cauchy distribution*)  As we mentioned. when one of the two extremes of the interval of definition is infinite, the moments of $f(x)$ may not necessarily be finite. A very interesting example of a "pathological" distribution of this kind is the *Cauchy distribution*, defined for all $x \in \mathbb{R}$ as

---

[20] Although you probably call it the '$\delta$-function', as I also use to do: but we physicists have the right to be a little careless at times....

$$\boxed{f(x) = \frac{\alpha}{\pi(x^2 + \alpha^2)}} \tag{3.42}$$

where $\alpha$ is a constant. The functional form of this p.d.f., which is shown in Fig. 3.9, is also a particular example of what in physics is called a *Lorentzian function*, or simply "Lorentzian", which is also the characteristic shape of lifetime–broadened[21] spectral lines. In high-energy physics, where it is better known as the (non-relativistic) Breit-Wigner distribution, it is often used to model resonances (unstable particles). But the Cauchy distribution finds also application in diverse fields, ranging from information theory to finance.

The cumulative distribution is obtained with a simple integration as

$$\boxed{F(x) = \frac{1}{2} + \frac{1}{\pi}\arctan\frac{x}{\alpha}} \tag{3.43}$$

which, taking $x \to \infty$, immediately shows that $f(x)$ is correctly normalized. However, neither the expectation, nor the variance, are definite because *all* moments

$$\langle x^r \rangle = \int_{-\infty}^{\infty} \frac{\alpha x^r}{\pi(x^2 + \alpha^2)}\,\mathrm{d}x$$

diverge (just look at the behavior for $x \to \pm\infty$). For the expectation, we may still think of using, as an indication of the "typical" value, $x = 0$, which is both its median and its mode (where it attains the maximum value $f(0) = 1/(\pi\alpha)$). However, this can be *very* dangerous.[22] Even more dangerous is considering $\alpha$ an equivalent of the standard deviation, even if gives an indication of the width of the distribution because $f(\alpha) = f(0)/2$. In fact, as we shall see, the Cauchy distribution is a prototypical example of a probability density that does not satisfy what is, arguably, the most important theorem of probability theory, the Central Limit Theorem.

It is finally worth mentioning a connection between the Cauchy distribution and the St. Petersburg paradox. To restore fairness towards the bank, let us indeed "symmetrize" the game. In this variant, when you get heads on the first throw you do not stop, but flip the coin again until tails come out. With a sequence of $n$ consecutive

---

[21] A Lorentzian line broadening is the results of describing the decay of an excited electronic, vibrational, or rotational state as a damped harmonic oscillator.

[22] One may think of defining the expectation value as the Cauchy principal value

$$\langle x \rangle = \lim_{a\to\infty} \int_{-a}^{a} x f(x)\,\mathrm{d}x,$$

which, in fact, vanishes for the Cauchy distribution (check). However, this is rather arbitrary, because, for instance (check this too),

$$\langle x \rangle = \lim_{a\to\infty} \int_{-2a}^{a} x f(x)\,\mathrm{d}x = \infty.$$

heads you now *lose* $2^n \text{\euro}$. We can show that, when the game is repeated many times, the distribution of your earnings can actually be fitted with a Lorentzian. This may seem strange at first glance, because in the original game the probability distribution of the earnings $P(g)$ decreases as $g^{-1}$, which does not match the $x^{-2}$ power law of the tails of the Cauchy distribution. However, when comparing these values to a continuous distribution we must take into account that the values obtainable for $g$ in a non-symmetrized game are *not* equally spaced. So, remembering that $g = 2^n$, to properly construct a histogram we must take into account that

$$P(2^{n-1} < g \leq 2^n) = (2^n - 2^{n-1})f(g) = \frac{g}{2}f(g) = \frac{1}{2g},$$

that is, the relative frequencies *normalized to the width of interval* decrease indeed as $f(g) = g^{-2}$.

**Example 3.18** (*On waiting times and lotteries*)  The Poisson is a distribution for the discrete variable $k$, but we can actually derive from it a *continuous* distribution $P(t; dt)$, which gives the probability that the first of a sequence of events, which take place randomly in time, happens between $t$ and $t + dt$. Indeed, $P(t; dt)$ is equal to the product of the probability that no event takes place until $t$, which is $P(0, t) = \exp(-\alpha t)$, times the probability that an event does take place in $dt$, which is $\alpha dt$,

$$P(t; dt) = e^{-\alpha t} \alpha dt.$$

Therefore, the function

$$f(t) = \alpha e^{-\alpha t}, \tag{3.44}$$

which has the dimensions of a frequency, can be regarded as the probability density that the first event takes place between $t$ and $t + dt$, or, setting $t = 0$ at the time when the last event has taken place, $f(t)$ can also be seen as the probability distribution of the time intervals between two events. When applied to radioactive decays, Eq. (3.44) gives reason of the results by Marsden and Barratt shown in Fig. 1.16b.

The expectation of $f(t)$, is therefore the time we typically have to wait before observing an event. We have

$$\langle t \rangle = \int_0^\infty t \alpha e^{-\alpha t} dt = -\alpha \frac{d}{d\alpha} \int_0^\infty e^{-\alpha t} dt = \frac{1}{\alpha} = \tau, \tag{3.45}$$

i.e., the time we typically have to wait coincides with the *average* time between two events. The start time has not entered the calculation, thus the waiting time $\langle t \rangle$ remains the same regardless of the time that passed since the last event. This is because, for independent events, the conditional probability of observing an event at time t, knowing that another event occurred at time $t' < t$, is still equal to the simple probability of observing an event at time $t$. If instead of radioactive decays we think of the time we have to wait before a number we bet on in a lottery is extracted, this also means that any betting system or strategy is useless.

Nevertheless, a doubt may still torment your soul. What is wrong with saying that, since we start watching for an event at a random instant within the time interval between two of them, the typical waiting time should be *less* than $1/\alpha$, maybe, say, $1/(2\alpha)$? Indeed. If we start biding during a time interval of *generic* duration $t$, the waiting time $t_a$ must be a variable uniformly distributed between 0 and $t$, with an expectation $t/2$. For what we have just said, however, the time $t$ between two events is not always the same, but is rather distributed according to (3.45). The expectation for $t_a$ depends then on how we *sample* this distribution. The fact is, if we start watching for an event at a "random" moment we do not sample the distribution uniformly, but rather select the *longer* intervals.

Let's see why. Consider $N$ intervals distributed according to $f(t) = \alpha e^{-\alpha t}$, and therefore with an expectation for the duration $\langle t \rangle = 1/\alpha$. The intervals with a duration between $t$ and $t + dt$ take up a fraction of the total time, $T = N/\alpha$, equal to

$$\frac{Ntf(t)dt}{N\,\langle\tau\rangle} = \alpha t f(t)dt = \alpha^2 t \exp(-\alpha)dt.$$

If $N$ is very large, the probability of start waiting within an interval of length $(t, t + dt)$ will be almost equal to the fraction of the total time occupied by this type of intervals, which is *different* from the probability $f(t)$ that an interval has a duration between $t$ e $t + dt$. The expectation of the waiting time is then correctly given by

$$\langle t_a \rangle = \int_0^\infty \frac{t}{2}\alpha t p(t)dt = \frac{\alpha^2}{2}\int_0^\infty t^2 \exp(-\alpha t)dt = \frac{2}{\alpha^2}.$$

Therefore we have, once again and without any hope for the system bettor,

$$\langle t_a \rangle = \frac{1}{\alpha} = \tau.$$

**Example 3.19** (*The exponential and Laplace distributions*) The distribution of waiting times we just encountered is an example of an *exponential probability density*, which is the continuum version of the geometric distribution we discussed in Example 3.2. As such, it also describes for instance the probability of failure of a component or system for a given *time-independent* failure rate. The exponential decay may of course take place in space instead of time, so in general we write the exponential p.d.f. as

$$f(x) = \begin{cases} \dfrac{1}{x_0}\exp\left(-\dfrac{x}{x_0}\right) & x \geq 0 \\ 0 & x < 0 \end{cases} \qquad (3.46)$$

where $x_0$ has the same units of $X$ and is for example a characteristic decay time or length. The corresponding cumulative distribution, giving for instance the probability that a failure takes place before a given time, is

$$F(x) = \begin{cases} 1 - \exp\left(-\dfrac{x}{x_0}\right) & x \geq 0 \\ \\ 0 & x < 0 \end{cases} \tag{3.47}$$

From the definition and properties of the gamma function given in the appendix, the moments of the exponential distribution can be easily found to be $\langle x^r \rangle = r!\, x_0^r$, hence

$$\langle x \rangle = x_0 \;\; ; \;\; \sigma_x^2 = x_0^2 \tag{3.48}$$

The mode of the distribution is of course $x_{max} = 0$, whereas the median $x_m$ satisfies

$$1 - e^{-x_m/x_0} = \frac{1}{2} \implies x_m = x_0 \ln 2.$$

All we have said about waiting times, nuclear decays, and lotteries is just a consequence of a general feature of the exponential distribution: it is *memoryless*. Indeed, suppose that the probability density for an event to occur at time $t$ is given by (3.44). We wonder whether the knowledge that an event has *not* occurred up to a time $t^*$ changes or not the probability that the same event will occur at a later time $t + t^*$. In other words, we look for the conditional probability

$$P\left(T > t + t^* \mid T > t^*\right) = \frac{P\left[(T > t + t^*) \cap (T > t^*)\right]}{P\left(T > t^*\right)} = \frac{P\left(T > t + t^*\right)}{P\left(T > t^*\right)}$$

Now, the probability $P(T > t)$ that the event occurs after a generic lag time $t$ is just the complement $1 - F(t)$ to the cumulative distribution (3.46) with $x_0 = 1/\alpha$, so we have

$$P(T > t + t^* \mid T > t^*) = \frac{1 - F(t + t^*)}{1 - F(t^*)} = \frac{e^{-\alpha(t+t^*)}}{e^{-\alpha t^*}} = e^{-\alpha t} = P(T > t).$$

As you can see, the future exponential distribution has no memory of the past.

The domain of the exponential p.d.f. is restricted to positive values of $x$, or, as we will say, the distribution has a *positive support*. In some applications, however, it would be advantageous to utilize a distribution defined on the entire real axis that exhibits exponential tails on both sides. A p.d.f. with these features, originally introduced by Laplace in a first attempt to obtain the distribution of experimental errors, is

$$f(x) = \frac{1}{\sigma\sqrt{2}} e^{-\sqrt{2}|x|/\sigma} \tag{3.49}$$

Because of the symmetry about the origin,[23] $\langle x \rangle = x_{max} = 0$, while with the same method used for the exponential distribution one finds $\sigma_x^2 = \sigma^2$.

**Example 3.20** (*Waiting time for the following events*)  Going a little further in the reasoning made in the previous example, we can evaluate the probability that the $k$-th event occurs in the interval $(t, t + dt)$, i.e., the probability distribution of waiting times to have $k$ events. As before, this will be given by the probability of having observed exactly $k - 1$ events at time $t$ times the probability of observing the $k$-th event in the time interval $dt$, i.e.

$$P_k(t, t + dt) = \frac{(\alpha t)^{k-1} e^{-\alpha t}}{(k-1)!} \alpha dt.$$

It is useful to introduce the variable $t^* = \alpha t = t/\tau$, thus measuring time in units of the average time between two events. Since $dt^* = dt/\alpha$, we have

$$P_k(t^*, t^* + dt^*) = \frac{(t^*)^{k-1} e^{-t^*}}{(k-1)!} dt^*.$$

This means that the probability density of observing exactly $k$ events in a time $t$ is

$$\boxed{f_k(t^*) = \frac{(t^*)^{k-1} e^{-t^*}}{(k-1)!}} \tag{3.50}$$

We emphasize again that while the number of events in a fixed interval is given by the discrete Poisson distribution, the waiting time before the $k$-th event is a continuous variable. For $k = 1$ we obviously obtain the exponential behavior we have just studied, while for $k > 1$ the distribution shows a peak for a value of $\tau$ that, as shown in Fig. 3.10, increases with $k$. Since integrating repeatedly by parts gives $\int_0^\infty x^n \exp(-x) dx = n!$, the distribution is normalized. Moreover, $\langle t^* \rangle = k$, i.e., $\langle t \rangle = k/\alpha$, while the maximum of the distribution is obtained instead for:

$$\frac{d}{dt^*} \left[ \frac{(t^*)^{k-1} e^{-t^*}}{(k-1)!} \right] = 0 \Rightarrow t^* = k - 1 \Rightarrow t = \frac{k-1}{\alpha}.$$

## 3.6  The Normal (or Gaussian) Distribution

We are now ready to perform to study the limiting behavior of the binomial distribution for $n \to \infty$. This is actually the content of the DeMoivre–Laplace theorem.

---

[23] To center the p.d.f. around any real value $\mu \neq 0$, one can just displace $x \to x + \mu$.

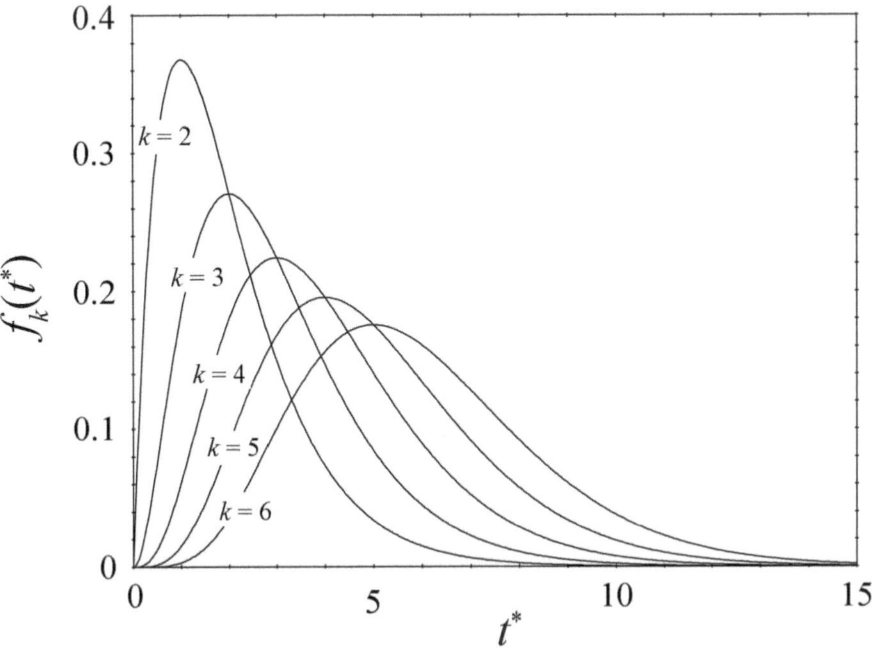

**Fig. 3.10**  Distribution of the scaled waiting time $t^* = \alpha t$ for observing $k$ events

### The DeMoivre–Laplace theorem

For large $n$, all values of $k$ with a significant probability in (3.23) will be large too. Thus, we can use Stirling's approximation for all the factorials in the binomial. With some simple algebraic steps, we find

$$B(k; n, p) \xrightarrow[n \to \infty]{} \sqrt{\frac{n}{2\pi k(n-k)}} \frac{n^n}{k^k (n-k)^{n-k}} \, p^k (1-p)^{n-k}, \tag{3.51}$$

which, in particular, in correspondence to the expectation $\langle k \rangle = np$ gives

$$B(np; n, p) \xrightarrow[n \to \infty]{} \frac{1}{\sqrt{2\pi}} \frac{1}{\sqrt{np(1-p)}} = \frac{1}{\sigma_k \sqrt{2\pi}}. \tag{3.52}$$

We wish to find an approximation of Eq. (3.51) around $\langle k \rangle$ that, for $n \to \infty$, we will show to be its maximum too. This requires to expand in series $B(k; n, p)$, taking onto account only the first terms of the expansion. As $n$ gets large, however, the relative standard deviation $\sigma_k / \langle k \rangle \sim \langle k \rangle^{-1/2}$ becomes smaller and smaller, thus the binomial takes on the appearance of a strongly peaked function. A series expansion to the low order of $B(k; n, p)$ will be a good approximation only in a very limited range around its maximum. However, the *logarithm* of a rapidly–varying function $f(x)$ has a much "milder" behavior.[24] Besides, the logarithm is a *monotonic* function, thus it has the same maxima and minima of its argument. An expansion of $\ln[B(k; n, p)]$ works then in a much wider region around the maximum. We have

$$\frac{d}{dk} \ln[B(k; n, p)] = -\frac{d}{dk} \ln(k!) - \frac{d}{dk} \ln(n - k)! + \ln p - \ln(1 - p).$$

Observing that, for large $r$,

$$\frac{d}{dr} \ln(r!) \simeq \frac{d}{dr} \left[ \left( r + \frac{1}{2} \right) \ln r - r - \frac{1}{2} \ln(2\pi) \right] = \ln r + \frac{1}{2r} \xrightarrow[r \to \infty]{} \ln r,$$

we find

$$\frac{d}{dk} \ln B(k; n, p) \simeq -\ln k + \ln(n - k) + \ln p - \ln(1 - p),$$

which vanishes when

$$\ln \frac{p(n - k)}{k(1 - p)} = 0 \implies \frac{p(n - k)}{k(1 - p)} = 1,$$

i.e., for $k = np$. Since the second derivative in $k = np$

$$\left[ \frac{d^2}{dk^2} \ln B(k; n, p) \right]_{k=np} \simeq \left[ -\frac{1}{k} - \frac{1}{nk} \right]_{k=np} = -\frac{1}{np(1 - p)} = -\frac{1}{\sigma_k^2} \qquad (3.53)$$

is negative, $k = np$ is a maximum, as anticipated.

Expanding now $\ln B(k; n, p)$ to second order around $k = np$ and using (3.52), (3.53) we have

$$\ln B(k; n, p) \simeq \ln \left( \frac{1}{\sigma_k \sqrt{2\pi}} \right) - \frac{(k - np)^2}{2\sigma_k^2},$$

so that

$$B(k; n, p) \xrightarrow[n \to \infty]{} \frac{1}{\sigma_k \sqrt{2\pi}} \exp \left[ -\frac{(k - \langle k \rangle)^2}{2\sigma_k^2} \right], \qquad (3.54)$$

which is the DeMoivre–Laplace theorem.

Using a very similar method, it is easy to obtain an analogous result for the Poisson distribution $P(k; a)$. Indeed, for $a \to \infty$ one has $P(a; a) \simeq 1/\sqrt{2\pi a}$. Besides, in this case too, for $k \to \infty$ the maximum of the distribution approaches $\langle k \rangle = a$, while the counterpart of (3.53) is

---

[24] For example, if a function decreases as fast as $\exp(-a(x - x_0)^2)$, its logarithm decreases only as $-a(x - x_0)^2$.

$$\left[ \frac{d^2}{dk^2} \ln P(k, a) \right]_{k=a} \simeq -\frac{1}{a}.$$

Expanding to second order the logarithm, we finally obtain

$$P(k; a) \xrightarrow[a \to \infty]{} \frac{1}{\sqrt{2\pi a}} \exp\left[ -\frac{(k-a)^2}{2a} \right]. \tag{3.55}$$

Since $a$ is both the expectation and the variance of the Poisson, we realize that the limiting distribution is entirely analogous to that obtained from the binomial.

## The Gaussian: Shape, Expectation, and Variance

The previous results lead us to consider the curve that interpolates both the limiting distributions we found, i.e., the continuous envelope of the two discrete distributions. This is the world–famous *normal distribution*, in physics commonly known as a *Gaussian*, with a somewhat arbitrary tribute to Gauss.[25] The Gaussian distribution, which is of course a probability *density*, has then the functional form

$$\boxed{g(x; \mu, \sigma) = \frac{1}{\sigma\sqrt{2\pi}} \exp\left[ -\frac{(x-\mu)^2}{2\sigma^2} \right]} \tag{3.56}$$

Quantitatively, a value of $n \sim 10 - 20$, or respectively a $a \sim 5 - 10$ is generally sufficient for the binomial and Poisson to be approximated quite well by their limiting expressions. However, we must introduce a note of caution. Both the binomial and the Poisson converge rapidly to the Gaussian in the "central" region, i.e., for values close to the expectation, but much more slowly as we move away towards the tails of the distribution. In other words, the convergence is not uniform.

Figure 3.11 shows that the normal probability density is mainly concentrated in an interval of one or two $\sigma$ around the value $x = \mu$, while it almost vanishes for $|x - \mu| > 3\sigma$. Because of the way we obtained it, we expect that the Gaussian is correctly normalized, and that its expectation, variance, and skewness are given by

$$\boxed{\langle x \rangle = \mu; \quad \sigma_x^2 = \sigma^2; \quad \gamma = 0.} \tag{3.57}$$

It is nevertheless instructive to show this directly by evaluating some basic integrals related of the form

$$\int_{-\infty}^{\infty} x^n e^{-ax^2} dx,$$

---

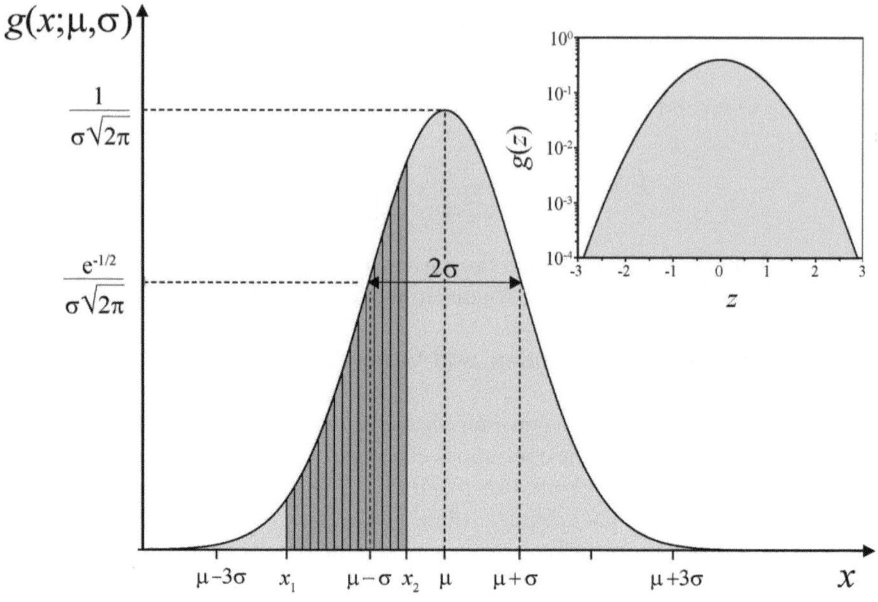

**Fig. 3.11** The normal (Gaussian) distribution, shown in the semi-log inset as a function of the scaled variable $z = (x - \mu)/\sigma$

where $a > 0$ is then related to the standard deviation $\sigma$ of a Gaussian distribution centered on the origin ($\mu = 0$ by $a = (2\sigma^2)^{-1}$. In the appendix we show that, if $r$ is a positive integer,

$$1. \int_{-\infty}^{+\infty} e^{-ax^2}\,dx = \sqrt{\frac{\pi}{a}} = \sqrt{2\pi}\,\sigma \tag{3.58a}$$

$$2. \int_{-\infty}^{+\infty} x^2 e^{-ax^2}\,dx = \frac{1}{2a}\sqrt{\frac{\pi}{a}} = \sqrt{2\pi}\,\sigma^3 \tag{3.58b}$$

$$3. \int_{-\infty}^{+\infty} x^{2r-1} e^{-ax^2}\,dx = 0. \tag{3.58c}$$

The appendix also shows that *all* even moments can be expressed in terms of the gamma function as

$$\int_{-\infty}^{+\infty} x^2 e^{-ax^2}\,dx = \frac{\Gamma(n + 1/2)}{a^{n+1/2}} \tag{3.58d}$$

By using (3.58a) and putting $y = x - \mu$, we have

$$\frac{1}{\sigma\sqrt{2\pi}} \int_{-\infty}^{\infty} \exp\left[-\frac{(x-\mu)^2}{2\sigma^2}\right] dx = \frac{1}{\sigma\sqrt{2\pi}} \int_{-\infty}^{\infty} \exp\left[-\frac{y^2}{2\sigma^2}\right] dx = \frac{1}{\sigma\sqrt{2\pi}}\sqrt{2\pi\sigma^2} = 1,$$

thus the Gaussian is correctly normalized. For the expectation, we can write

$$\langle x \rangle = \frac{1}{\sigma\sqrt{2\pi}} \int_{-\infty}^{\infty} x \exp\left[-\frac{(x-\mu)^2}{2\sigma^2}\right] dx =$$

$$= \frac{1}{\sigma\sqrt{2\pi}} \int_{-\infty}^{\infty} (x-\mu) \exp\left[-\frac{(x-\mu)^2}{2\sigma^2}\right] dx + \frac{\mu}{\sigma\sqrt{2\pi}} \int_{-\infty}^{\infty} \exp\left[-\frac{(x-\mu)^2}{2\sigma^2}\right] dx.$$

Putting again $y = x - \mu$ the first integral vanishes, so using (3.58c)

$$\langle x \rangle = \frac{\mu}{\sigma\sqrt{2\pi}} \int_{-\infty}^{\infty} \exp\left[-\frac{(x-\mu)^2}{2\sigma^2}\right] dx = \mu.$$

Finally we have

$$\sigma_x^2 = \frac{1}{\sigma\sqrt{2\pi}} \int_{-\infty}^{\infty} y^2 \exp\left[-\frac{y^2}{2\sigma^2}\right] dx = \frac{1}{\sigma\sqrt{2\pi}} \frac{2\sigma^2}{2} \sqrt{2\pi\sigma^2} = \sigma^2.$$

**Example 3.21** (*From the binomial to the Gaussian*) Let us gather the decimals of $\pi$ into groups of 25 digits, and evaluate the number of odd digits within each group. Assuming that $\pi$ is a normal number, the distribution of odd digits will approach, as the number of considered groups increases, the binomial distribution $B(k; 25, 0.5)$, which we expect to be approximated by a Gaussian $g(x; 12.5, 2.5)$. Figure 3.12, where the frequency distribution obtained considering 400 groups of 25 decimals each and the continuous Gaussian limit are compared, show that this is indeed the case.

**Example 3.22** (*A long (random) walk*) In Example 3.8 we have seen that the distribution of the number of steps in a one-dimensional random walk is a binomial that allows the distribution of the final position to be obtained. As the number of steps increases, the distribution is better and better approximated by a Gaussian with $\mu = 0$ and $\sigma = L\sqrt{N}$, which makes us understand the origin of the bell curve found in the simulation presented in Chap. 1. Let us now analyze a RW in time, by introducing the time $\tau$ it takes the walker for a step. The number of steps that take place in a time $t$ can then be written as $N = t/\tau$, and the variance of the Gaussian distribution as $\sigma^2 = 2Dt$, where

$$D = \frac{L^2}{2\tau} = \frac{\langle x^2 \rangle}{2t}. \tag{3.59}$$

Since $\langle x^2 \rangle$ grows linearly in time, the coefficient $D$, indicating how quickly the distribution of positions widens and has the dimensions of a square of a length divided by a time, remains finite even for $t \to 0$. Therefore this important parameter, called the *diffusion coefficient*, does not depend on the choice of $\tau$. The distribution of positions at time $t$ is given by

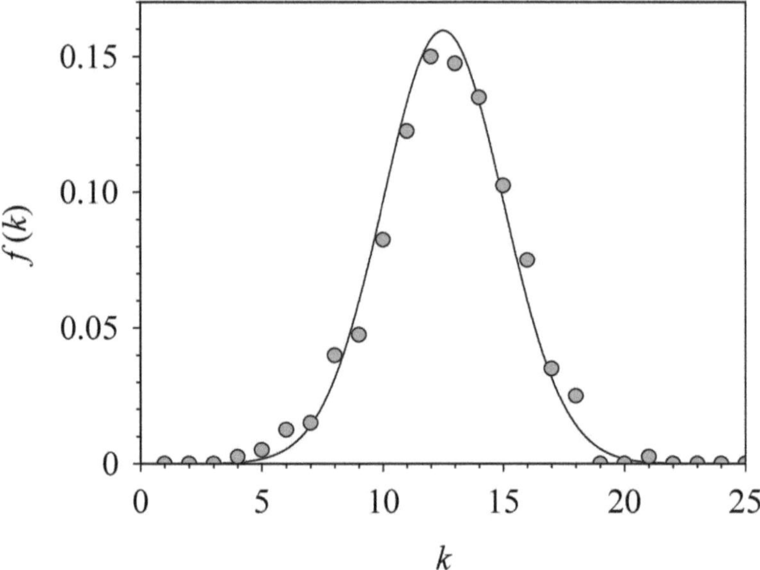

**Fig. 3.12** Distribution of the number of odd digits in groups of 25 decimals of $\pi$, compared with the normal distribution $g(x; 12.5, 2.5)$ (full line)

$$f(x, t) = g(x; 0, \sqrt{2Dt}) = \frac{1}{2\sqrt{\pi Dt}} \exp\left(-\frac{x^2}{4Dt}\right). \qquad (3.60)$$

This is the basic link between random walks, Brownian motion, and diffusion processes that we will extensively discuss in Chap. 5.

All the situations we have analyzed using the binomial or Poisson distribution can then be revisited in terms of the Gaussian distribution, when the expected value is high enough. Nevertheless, if the Gaussian distribution were only an approximation of the binomial or of the Poisson, its usefulness would be reduced to the limiting situations we have examined. However, the previous example shows that the distribution of the final position, which is the sum of many random steps, tends to become Gaussian, although a single step is a random variable that takes only the values $\pm 1$. Consider also the waiting time distributions discussed in Example 3.20 where we noticed that, as the number $k$ of events increases, $P_k(t)$ tends to become bell-shaped like a Gaussian. Since the waiting time for a single event has an exponential distribution, this means that summing many exponentially distributed random variables results in a random variable, the total waiting time, with a Gaussian distribution. These results are direct consequences of the Central Limit Theorem that we will address in the next chapter, which reveals the origin of the expression "normal distribution" and makes the significance of the Gaussian becomes truly disproportionate.

### 3.6.1 Cumulative Normal Distribution

The problem we will most often face is calculating the probability that a normally–distributed variable $x$ has a value between two limits $x_1$ and $x_2$, which means evaluating

$$\int_{x_1}^{x_2} g(x; \mu, \sigma)\mathrm{d}x = \frac{1}{\sigma\sqrt{2\pi}} \int_{x_1}^{x_2} \exp\left[-\frac{(x-\mu)^2}{2\sigma^2}\right]. \tag{3.61}$$

Unfortunately, there is no analytical expression for this integral. To simplify the matter, it is useful to introduce a "scaled" form of the distribution by writing it as a function $g(z)$ of the *dimensionless* variable

$$Z = \frac{(X-\mu)}{\sigma}. \tag{3.62}$$

As we will discuss in the next chapter, however, a change of variable requires the probability to be conserved. Namely, the probability that the variable $Z$ is in a neighborhood $\mathrm{d}z = \mathrm{d}x/\sigma$ of a value $z$ has to be equal to the probability that the variable $X$ is in a neighborhood $\mathrm{d}x$ around the inverse image of $z$,

$$g(z)\mathrm{d}z = g(x; \mu, \sigma)\mathrm{d}x,$$

where $x = g^{-1}(z)$. This yields the normal distribution in *standard form*

$$g(z) = \frac{1}{\sqrt{2\pi}} \exp\left(-\frac{z^2}{2}\right). \tag{3.63}$$

The inset of Fig. 3.12 shows $g(z)$ on a semi-logarithmic scale. If we compare this graph with the one in the inset of Fig. 3.9, we can notice that the tails of the Gaussian decrease much more rapidly than those of the Cauchy distribution.

The area under the distribution is then given by

$$\int_{x_1}^{x_2} g(x; \mu, \sigma)\mathrm{d}x = \frac{1}{\sqrt{2\pi}} \int_{z_1=(x_1-\mu)/\sigma}^{z_2=(x_2-\mu)/\sigma} \exp\left(-\frac{z^2}{2}\right) \mathrm{d}z.$$

Thus, by writing the cumulative distribution as

$$G(z) = \frac{1}{\sqrt{2\pi}} \int_{-\infty}^{z} \exp\left(-\frac{t^2}{2}\right) \mathrm{d}t, \tag{3.64}$$

we have

$$P(x_1 < x < x_2) = G(z_2) - G(z_1). \tag{3.65}$$

A table of $G(z)$ for $0 \leq z \leq 3.5$ is given in the appendix.[26] By introducing the *error function*

$$\mathrm{erf}(t) = \frac{2}{\sqrt{\pi}} \int_0^t \mathrm{e}^{-y^2} \mathrm{d}y,$$

a function with a sigmoid shape varying between -1 and +1 that occurs in several problems of mathematical physics, one can also write

$$P(x_1 < x < x_2) = \frac{1}{2} \left[ \mathrm{erf}\left(\frac{z_2}{2}\right) - \mathrm{erf}\left(\frac{z_1}{2}\right) \right].$$

**Asymptotic Behavior and Approximations for G(z).**

An asymptotic behavior of $G(z)$ for large $z$, useful for estimating the cumulative probability of very rare events, can be found as follows. We first observe that, for every $t$,

$$\left(1 - \frac{3}{t^4}\right) g(t) < g(t) < \left(1 + \frac{1}{t^2}\right) g(t),$$

since the quantities we subtract on the left and add on the right are certainly positive. Noticing that $\mathrm{d}g(t)/\mathrm{d}t = -tg(t)$, this expression can be written as

$$-\frac{\mathrm{d}}{\mathrm{d}t}\left[\left(t^{-1} - t^{-3}\right) g(t)\right] < g(t) < -\frac{\mathrm{d}}{\mathrm{d}t}\left(t^{-1}g(t)\right),$$

wherefrom, by integrating on $t$ from $z$ to $+\infty$, we obtain

$$\left(z^{-1} - z^{-3}\right) g(z) < 1 - G(z) < z^{-1}g(z).$$

However, for $z \to \infty$, $z^{-3} \ll z^{-1}$, so the upper and lower bounds become the same and we have

$$1 - G(z) \simeq \frac{g(z)}{z} = \frac{\exp(-z^2/2)}{z\sqrt{2\pi}}. \tag{3.66}$$

For a generic value of $z$ it is finally possible to give a simple approximate expression,[27] sufficiently accurate for our purposes, which overestimates by less than 1% the integral of the Gaussian between 0 and $z$,

---

[26] As an ageing physicist, I still insist on providing tables. Today, you can of course take advantage of great free online resources such as Wolfram|Alpha. But I still think that having a table on hand for a quick look does not hurt....

[27] This expression is due to J. D. Williams, *Ann. Math. Stat.* **17**, 373 (1946).

$$f(z) = \frac{1}{\sqrt{2\pi}} \int_0^z \exp\left(-\frac{t^2}{2}\right) dt \simeq \frac{\sqrt{1 - \exp(-2z^2/\pi)}}{2}. \tag{3.67}$$

Then, $G(z) = 1/2 \pm f(z)$ when $z \gtrless 0$, respectively.

Therefore, the strategy we have to follow to evaluate the probability that a continuous variable $X$, distributed according to a Gaussian with expectation $\mu$ and standard deviation $\sigma$, has a value between $x_1$ and $x_2$ is ultimately the following:

1.  Evaluate $z_1 = (x_1 - \langle x \rangle)/\sigma$ e $z_2 = (x_2 - \langle x \rangle)/\sigma$.
2.  Find $G(z_1)$ e $G(z_2)$ from the table or from (3.67).
3.  Calculate $P(x_1 < x < x_2) = G(z_2) - G(z_1)$.

It is finally useful to have an estimate of the chance that a Gaussian variable differs from the expectation by less than one, two, or three standard deviations. One finds

$$\begin{cases} P(\mu - \ \sigma < x < \mu + \ \sigma) = 0.683 \\ P(\mu - 2\sigma < x < \mu + 2\sigma) = 0.955 \\ P(\mu - 3\sigma < x < \mu + 3\sigma) = 0.997 \end{cases} \tag{3.68}$$

Therefore, when measuring a normally–distributed variable, we expect that about 2/3 of the results should fall within an interval of width $\sigma$ around $\sigma$ while almost all data will fall within $3\sigma$ from $\mu$.

**Example 3.23** (*The way height is distributed*) Let's go back to Example 1.4, where we saw that the height distribution of the 1900 draft class has a bell shape that closely resembles a Gaussian. In the next chapter, we will see that this has a clear theoretical justification, but to make a more quantitative comparison it is useful to review the data. In their article, A'Hearn *et al.* point out that the reported values can be altered by a series of spurious factors, above all by the heterogeneity in age of the examined subjects, which may make the less reliable. The 1900 draft class had in fact to deal with the final stages of World War I, thus all fit individuals, including those under 18 years old who have not fully grown yet, were called to arms.

The authors have then corrected the data by means of a careful statistical analysis on a standardized sample, obtaining the distribution shown by the full dots in Fig. 3.13, which shows a slightly higher mean ($\overline{h} \simeq 164$ cm), a smaller variance (($\sigma_h \simeq 6.3$ cm), and almost no skewness.

The figure shows that the data are very well fitted by a Gaussian with expectation $\langle h \rangle = 164$ cm and $\sigma_h = 6.3$ cm. Therefore, to estimate the probability of finding, at the beginning of the XX century, an Italian man taller than the author (who is 182 cm tall), we first find the value of the normalized variable as $z = (182 - 164)/6.3 \simeq 2.86$, and then use the table in the appendix (or Eq. (3.67), which gives almost same result) to obtain

$$P(h > 182) = 1 - P(h < 182) = 1 - G(2.86) = 1 - 0.9979 \simeq 0.002.$$

Back then, I would have been a real giant!

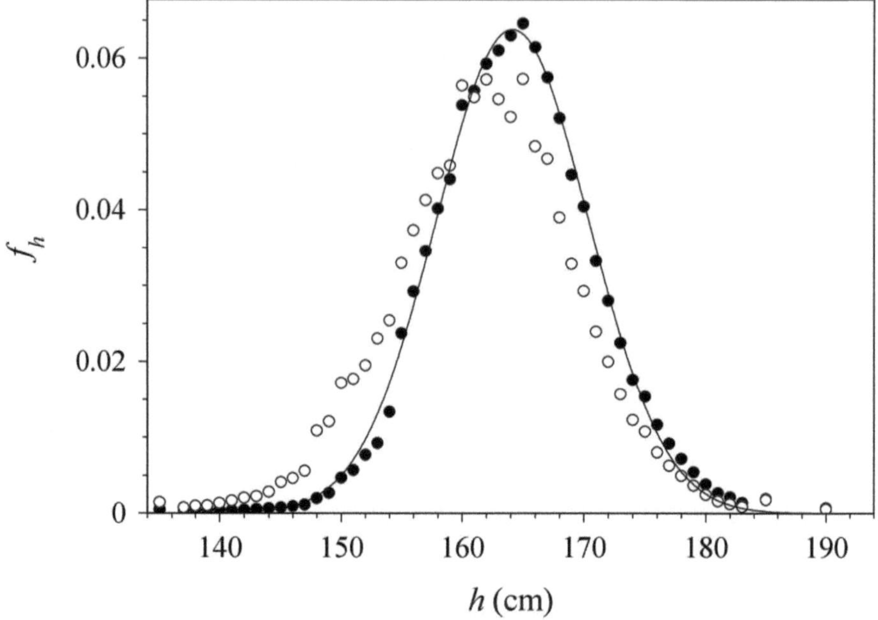

**Fig. 3.13** Comparison between the height distribution for the 1900 draft class (●), corrected with respect to the raw data (○), and the normal distribution $g(h; 164, 6.3)$ (solid line)

When we use the Gaussian as an approximation of a binomial or of a Poisson, we must pay some attention to the choice of the value of $z$, as in the example we now consider.

**Example 3.24** (*The right choice of* **z**) A die is rolled 120 times. We want to calculate the probability that the face "4" shows up:

(a)  Less than 18 times.
(b)  More than 24 times
(c)  Between 15 and 25 times

(a)  The binomial distribution for the outcomes is approximated by a Gaussian of expectation and variance

$$\mu = 120/6 = 20 \; ; \; \sigma^2 = 20 \cdot \left(\frac{1}{6}\right) \cdot \left(\frac{5}{6}\right) \simeq 16.7.$$

Now we have to choose a value for $z$. But what do we choose as $x$? The binomial actually collects in the single point $k = 18$ what in the Gaussian is distributed in a continuous interval of width $\Delta x = 1$ around this value. Then it is better to understand the expression "less than 18 times" as $x < 17.5$, not $x < 18$), and therefore take

$$z = (17.5 - 20)/4.1 \simeq -0.61.$$

From the table we have $G(0.61) \simeq 72.9\%$, hence $G(-0.61) \simeq 27.1\%$.

(b)  In this case, we should look for the probability $P(x > 24.5)$, i.e., $1 - P(x < 24.5)$. Since $z = (24.5 - 20)/4.1 \simeq 1.10$, we have

$$P(x > 24.5) = 1 - G(1.10) = 13.56\%.$$

(c)  Taking the values of $z$ corresponding to $x_1 = 14.5$ and $x_2 = 25.5$ we have $z_1 = -1.34$, $z_2 = +1.34$, hence

$$P(14.5 < x < 25.5) = G(1.34) - (1 - G(1.34)) = 2G(1.34) - 1 \simeq 82\%$$

## 3.7  A Selection of Useful Distributions

Besides the basic models we discussed, books on probability often present a variety of distributions for all tastes that are certainly useful for statisticians, but unlikely to pique the interest of a budding physicist. Thus, I will confine myself to a few examples that I have personally found particularly useful, trying to highlight the probabilistic model they originate from as much as possible.

### 3.7.1  The Negative Binomial Distribution

The binomial distribution gives the probability of obtaining $k$ successes in $n$ Bernoulli trials, where the probability of success in a single trial is $p$. However, we can look at this question the other way around, asking ourselves how many trials $n$ we expect to make before we obtain $r$ successes. Here it is $r$, and not $n$, to be regarded as a fixed parameter. Since the last trial is by definition a success, we have to make $r - 1$ successes in $n - 1$ trials. Therefore

$$P(n; r, p) = \binom{n-1}{r-1} p^r (1 - p)^{n-r} \quad (n \geq r), \tag{3.69}$$

which is called a *negative binomial distribution*. Note that $P(n; r, p)$ is related to the binomial $B(r; n, p)$ by

$$P(n; r, p) = \frac{(n-1)!}{(r-1)!(n-r)!} p^r (1 - p)^{n-r} = \frac{r}{n} B(r; n, p),$$

while the expected number of trials to make $r$ successes is evidently $\langle n \rangle = r/p$. By writing $n = r + k$, where $k$ is the number of *failures*, we can write the negative binomial in a different and often more useful form,

$$P(k; r, p) = \binom{r + k - 1}{k} p^r (1 - p)^k = \frac{(r + k - 1)!}{k!(r - 1)!} p^r (1 - p)^k \qquad (3.70)$$

where now $k$ can take all positive integer values. Thus, $P(k, r, p)$ responds to the question of *how many failures are expected before achieving a desired number of successes*, an extension of (3.4) that is surely of practical interest too. To show that (3.70) is normalized, put $q = 1 - p$ and note that, using Eqs. (2.25) and (2.26),

$$\sum_{k=0}^{\infty} P(k; r, p) = p^r \sum_{k=0}^{\infty} \binom{r + k - 1}{k} q^k = p^r \sum_{k=0}^{\infty} \binom{-r}{k} (-q)^k = p^r (1 - q)^{-r} = 1$$

The relation of $P(k; r, p)$ with Newton's binomial theorem for a negative power is, by the way, the origin of the specification "negative" for this distribution. The expectation of $P(k; r, p)$ can be intuitively found by noticing that, if the expected number of trials for $r$ successes is $\langle n \rangle = r/p$, then

$$\langle k \rangle = \langle n \rangle - r = \frac{1 - p}{p} r.$$

More rigorously, we can write the total number of failures before obtaining $r$ successes as $k = \sum_{s=i}^{r} k_s$, where $k_s$ is the number of failures between success $s - 1$ and success $s$. Thus, $k$ is the sum of $r$ independent variables with the geometrical distribution (3.4), and therefore

$$\langle k \rangle = \sum_{s=i}^{r} \langle k_s \rangle = \frac{1 - p}{p} r \ , \ \ \sigma_k^2 = \sum_{s=i}^{r} \sigma_s^2 = \frac{1 - p}{p^2} r. \qquad (3.71)$$

Using as a variable the number of failures instead of the number of expected trials has several advantages, first because $k$ can take any value (we do not have the condition $n \geq r$), and second because (3.70) only contains factorials of numbers that, if $r$ is small, are not too large.

You should not find too difficult to show that, in the limit $r \to \infty$ with constant $\mu$, the negative binomial (3.70) converges to the Poisson $P(k; \mu)$, which can easily be understood. Indeed, since $r \to \infty$ implies $p = r/(r + \mu) \to 1$, this simply means that, when successes are almost certain, failures are rare independent events that follow the Poisson distribution. However, compared to the latter, the negative binomial has a variance that is always *larger* than its expectation,

$$\sigma_k^2 = \frac{\langle k \rangle}{p} = \langle k \rangle + \frac{\langle k \rangle^2}{r} > \langle k \rangle \ ,$$

where $1/r$ quantifies how much the variance exceeds the mean, a feature usually dubbed *overdispersion*. This makes the negative binomial a suitable replacement for fitting frequency distributions that, although qualitatively resembling the shape of a Poisson distribution, display a consistent overdispersion, which arises naturally in biodiversity studies, for example because individuals of a populations tend to cluster or aggregate within a preferred habitat.[28] Similar clustering effects, however, also occur in physical situations ranging from the multiplicity of charged particles produced in high energy collisions to the distribution of galaxies.[29] The efficacy of the negative binomial as a fitting function of large-scale galaxy clustering can be understood using a model of spatial distribution of points where the probability that a point belongs to a particular cell is proportional to the number of points already occupying that cell.[30] It is possible that a similar model can also be used to explain excess clustering with respect to Poisson in other fields too. However, as we shall shortly see, another possible explanation of the widespread occurrence of data that can be fitted with a negative binomial comes from seeing the latter as a superposition of Poisson distributions with a precise distribution of the expectation values.

The power of the negative binomial as a model distribution to fit overdispersed data can be enhanced if we disregard its combinatorial origin, consider $r$ as a real, non necessarily integer number, and re-parameterize the distribution by putting $\mu = \langle k \rangle$ and $p = r/(r + \mu)$. This yields

$$P(k; r, p) = \frac{\Gamma(r + k)}{k!\,\Gamma(r)} \left( \frac{r}{r + \mu} \right)^{r} \left( \frac{\mu}{r + \mu} \right)^{k} \tag{3.72}$$

### 3.7.2  *The Hypergeometric Distribution*

We have seen that the binomial can be seen as the distribution for the number of "successes" in a sequence of independent samplings from a "dichotomous" population, that is, a population that consists of just two kinds of objects. An example is the number of draws with replacement from a two-color urn which contains $k$ balls of a selected color on a total of $n$. If we conveniently write a sequence of $m$ Bernoulli trials as an array $(x_1, x_2, \ldots, x_m)$ , where

$$x_i = \begin{cases} 1 \text{ for a success} \\ 0 \text{ for a failure,} \end{cases}$$


[28] J. Stoklosa, R. V. Blakey, and F. K. C. Hui, *Diversity* **14**, 320 (2022).

[29] B. J. T. Jones et al., *Rev. Mod. Phys.* **76**, 1211 (2004).

[30] E. Elizalde and E. Gaztanaga, Mon. Not. R. Ast. Soc. **254**, 247 (1992).

then the total number of successes is simply given by $k = \sum_{i=1}^{m} x_i$ and, for the case we are considering, the probability $p = k/n$ of a success (drawing a ball of the selected color) does not depend on $i$.

But what about sampling *without* replacement? Suppose for instance that we draw a sample of $n$ balls from a urn containing $r$ red and $N - r$ blue balls, without putting them back in the urn. In this case, the draws are *no more* independent, because the probability of a drawing a red ball at trial $i$ does not depend only on $r/N$, but also on all the values $x_j$ for $j = 1, \ldots i - 1$. So, to find the probability of having $k$ successes, we do not have anymore a fixed parameter $p$ like in the binomial. Nevertheless, the probability of having $k$ successes in can still be calculated as a function using $r$ and $N$ as additional free parameters.

We have indeed to select $k$ red balls out of $r$, which can be done in $\binom{r}{k}$ ways, and $n - k$ blue balls out of $N - r$, which can be done in $\binom{N-r}{n-k}$ ways.[31] The probability we are looking for, is then the ratio between the product of these two binomial coefficients and the total number of ways, $\binom{N}{n}$, of selecting $n$ balls out of $N$. Thus we have the so-called *hypergeometric distribution*,[32]

$$
P(k; n, r, N) = \frac{\binom{r}{k}\binom{N-r}{n-k}}{\binom{N}{n}} = \binom{n}{k}\frac{(r)_k(N - r)_{n-k}}{(N)_n}
\tag{3.73}
$$

where the second expression comes from the definition (2.14) of the falling factorials (the $k$-permutations) in Chap. 2, and we take by convection $(n)_m = 0$ for $m < n$ (so allowing all the values $0 \le k \le n$). You can easily check that the distribution is normalized using Vandermonde's identity (2.29).

A very interesting feature of the hypergeometric distribution is the following. While the probability that the outcome of a single trial is 0 or 1 depends on the outcomes of the previous ones (as we said, the trials are not independent), the total number of successes does not depend on the *order* of the outcomes. For example, the sequences $(x_1, \ldots, x_i, \ldots, x_j, \ldots x_n)$ and $(x_1, \ldots, x_j, \ldots, x_i, \ldots x_n)$ yield the same value of $k$. In other words the variables $X_i$ are *exchangeable*, an important concept that we will further investigate in Chap. 6.

Exchangeability has some important consequence. For instance, the expectation of $X_i$ should not depend on $i$, and must therefore coincide with the fraction of red balls $f_N = r/N$. Because of linearity, we must therefore have

$$
\langle k \rangle = \sum_{i=1}^{n} \langle x_i \rangle = \frac{nr}{N},
$$

---

[31] Which is what we found in Example 2.19.

[32] This rather awkward name comes from an important connection between this distribution and the hypergeometric function, which plays an important role in mathematical physics.

a result that is not so readily obtained from the definition of $\langle k \rangle$ (try and see). Similarly, the variance of each Bernoulli variable will be $\sigma_i^2 = f_N(1 - f_N)$. However, this does *not* mean that the $\sigma_k^2 = n\sigma_i^2$, again because of the lack of independence and the occurrence of *correlations* between the $X_i$ that we will calculate in the next chapter. Let me just anticipate that one obtains

$$\sigma_k^2 = \left(1 - \frac{n-1}{N-1}\right) n\sigma_i^2.$$

Note that, for $n > 1$, $\sigma_k^2$ is *smaller* than $n\sigma_i^2$, which is the variance of a binomial distribution with $p = f_N$.

With a bit of algebra (understatement), you can show that

$$P(k+1; n, r, N) = \frac{(r-k)(n-k)}{(k+1)(N-r-n+k+1)} \, P(k; n, r, N),$$

so that

$$\frac{P(k+1; n, r, N)}{P(k; n, r, N)} \gtrless 1 \quad \Leftrightarrow \quad k \gtrless \frac{(n+1)(r+1)}{N+2} - 1$$

Thus $P(k; n, r, N)$ increases with $k$, reaching a maximum at the greatest integer that does not exceed $c = (n+1)(r+1)/(N+2)$, (the mode of the distribution that, when $N$ and $r$ are large, is very close to $\langle k \rangle$) and then decreases.

Looking at the previous results, you have possibly noticed the similarity between the role of the fraction $f_N$ of red balls in the urn and that of the probability $p$ of a success in a binomial distribution. Besides, writing $r = pN$, for $N \gg n$ the expectation and the variance of $P(k; n, r, N)$ respectively become $Np$ and $Np(1 - p)$. You may then wonder whether, when the total number $N$ of balls in the urn is very large, the hypergeometric distribution approaches the binomial. This is indeed the case. Indeed, we have from (3.73)

$$P(k) = \binom{n}{k} \frac{pN \times \ldots \times (pN - k + 1)[(1-p)N] \times \ldots \times [(1-p)N - n - k + 1]}{N \times \ldots \times (N - n + 1)},$$

so that, for fixed $k = 0, 1, \ldots n$ and $N \to \infty$,

$$P(k) \longrightarrow \binom{n}{k} \frac{(pN)^k[(1-p)N]^{n-k}}{N^n} = \binom{n}{k} p^k (1-p)^{(n-k)}.$$

The hypergeometric distribution is very useful in many problems of statistical estimation, in particular when trying to assess the value of some parameter describing a population from a limited sample. For example, imagine that the election in a constituency ends with the red candidate (R) receiving $r$ votes and the blue candidate (B) receiving $N - r$ votes, and that we want to check if the votes have been accurately counted (i.e., to perform an election audit). Checking all the ballots would be impractical and time-consuming, so we test only $n$ of them, finding $k$ votes for R and $n - k$ for B. Then, the hypergeometric distribution allows us to calculate the probability of obtaining $k$ votes for R if the official result is correct. A related but in some sense inverse problem is estimating the unknown number $r$ of defective components in a large batch of $N$ manufactured items, by testing only $n \ll N$ of them, an important issue in industrial quality control.

**Example 3.25** (*A gas of repulsive particles*) Applications of the hypergeometric distribution in physics are not many, but one of them is simple and useful.[33] Suppose that have an ideal gas of $N$ atoms in a volume $V$, and consider a sub-volume $v$. Each atom has a probability $p = v/V$ to be in $v$ that, if the atoms are point-like, does not depend on the presence of other atoms in $v$. Then, the probability of having $n$ paricles in $v$ and the remaining in $V - v$ is simply given by the binomial distribution

$$B\left(n; N, p = \frac{v}{V}\right) = \binom{N}{n}\left(\frac{v}{V}\right)^n\left(1 - \frac{v}{V}\right)^{N-n},$$

which in the limit $v \ll V$ becomes, as we have already seen, the Poisson distribution

$$P\left(n; \frac{Nv}{V}\right) = \frac{1}{n!}\left(\frac{Nv}{V}\right)^n e^{-Nv/V}$$

However, if the atoms have a finite size, they reduce the volume available to the other particles. Let us say that each atom excludes a volume $b$, which is actually the *excluded volume* parameter entering in the van der Waals equation of state for a real fluid.[34] In this case, the probability of a particle to be in $v$ depends on the presence of other atoms.

What are the effects of the excluded volume on the probability distribution? A simple reflection shows that there is a full correspondence between this problem and that of extraction without replacement from a dichotomous urn. Indeed, let us divide the volume $V$ in $C = V/b$ elementary cells of volume $b$, of which $N$ are occupied by an atom (let's call them "filled balls", and $C - N$ empty ("hollow balls"). Then, denote by $c = v/b$ the number of cells in the sub-volume $v$. Then, you should easily convince yourself that the probability of finding $n$ atoms in $v$ is equal to the probability of drawing, without replacement, a random sample of $c$ balls, of which $n$ filled, from an urn containing $N$ filled and $C - N$ hollow balls, that is, the hypergeometric distribution

$$P(n; N, V, v) = \frac{\binom{N}{n}\binom{C-N}{c-n}}{\binom{N}{n}} = \frac{\binom{N}{n}\binom{V/b-N}{v/b-n}}{\binom{N}{n}}.$$

---

[33] See, for instance, J. Güémez and S. Velasco, Am. J. Phys. **55**,154 (1987).

[34] If the atoms behave as hard-spheres ("billiard balls"), then $b$ is simply four times the volume of a sphere, but an excluded volume parameter can be calculated for any kind of short-ranged repulsive potential between the atoms.

The average number of particles in $v$ is therefore $\langle n \rangle = Nv/V$ and the variance

$$\sigma_n^2 = c\left(1 - \frac{c-1}{C-1}\right)\frac{N}{C}\left(1 - \frac{N}{C}\right) = \frac{V-v}{V-b}\frac{v}{b}\,\phi(1-\phi),$$

where $\phi = N/C$ is the particle volume fraction (the fraction of the total volume occupied by the particles). Note that $\sigma_n^2$ is larger than the variance of a binomial with $p = N/C$. As qualitatively discussed in Sect. 3.4, we indeed expect repulsive interaction to reduce fluctuations.

A related problem is that of the distribution statistics of gases in the cavities of zeolites, a family of microporous, crystalline materials commonly used as adsorbents and catalysts. Zeolites provide ordered matrices that can trap clusters of atoms and molecules in their cavities. By increasing the loading pressure of a noble gas like xenon, the distribution of the atoms in the cavities changes from binomial to hypergeometric because of exclude volume effects.[35]


### 3.7.3  The Beta Distribution


It will be a long way before we discuss in detail what "testing an hypothesis" means. Nevertheless, it is worth doing some warm-up exercises. Suppose you flip a coin ten times, and you get 7 heads and 3 tails. You may wonder whether this is really a fair coin, which means finding the probability of the previous outcome using the binomial distribution

$$B(7;\, 10,\, 1/2) = \binom{10}{7}\left(\frac{1}{2}\right)^{10} \simeq 0.117.$$

This is indeed a rather low value, thus you may begin to suspect that the coin is actually a scam, with a preference for heads. So you expect that $p > 0.5$, but how much? A natural choice would be $p = 7/10$, giving

$$B(7;\, 10,\, 0.7) = \binom{10}{7}(0.7)^7(0.3)^3 \simeq 0.267,$$

which is surely a better figure. But is it really the best? If you try the mean value $p = 0.6$, you will find $B(7;\, 10,\, 0.6) \simeq 0.215$, which is not that worse!

Going on with trials and errors may not be a smart choice. We better look for a continuous variable $x$ with support in $[0, 1]$, giving us the entire distribution for the probability of heads (namely, a probability distribution for a probability value). With $k$ heads in $n$ trials, it is likely that $f(x)$ will be proportional to $x^k(1-x)^{n-k}$. But it is tempting to discuss a more general case in which the number of successes and the

---

[35] B. F. Chmelka et al., Phys. Rev. Lett. **66**, 580 (1991).

number of failures are not constrained to sum up to a fixed number like in binomial, but are rather two independent parameters. So, putting $\alpha = k + 1$ and $\beta = n - k + 1$ we look for a two-parameter probability density in [0, 1] of the form

$$f(x; \alpha, \beta) = \frac{1}{Z} x^{\alpha-1}(1 - x)^{\beta-1},$$

where the constant $Z$ is fixed by the normalization condition[36] $\int_0^1 f(x; \alpha, \beta)dx = 1$. Using Eq. (A.5) in the appendix, one has

$$Z = \frac{1}{B(\alpha, \beta)} = \frac{\Gamma(\alpha + \beta)}{\Gamma(\alpha)\Gamma(\beta)},$$

where $B(\alpha, \beta)$ is the beta function. This allows to define, the *beta distribution*

$$\boxed{f(x; \alpha, \beta) = \frac{1}{B(\alpha, \beta)} x^{\alpha-1}(1 - x)^{\beta-1} \quad (0 \le x \le 1, \alpha, \beta > 0)} \tag{3.74}$$

Note that (3.74) is symmetric upon reflection with parameter swap,

$$f(x, \alpha, \beta) = f(1 - x, \beta, \alpha).$$

In particular, when $\alpha = n + 1$ and $\beta = m + 1$ , with $n$ and $m$ integers, we have

$$f(x; n, m) = \frac{(n + m + 1)!}{n!\, m!} x^n (1 - x)^m. \tag{3.75}$$

The expectation of the beta distribution is

$$\langle x \rangle = \frac{1}{B(\alpha, \beta)} \int_0^1 x^\alpha (1 - x)^{\beta-1} dx = \frac{B(\alpha + 1, \beta)}{B(\alpha, \beta)} = \frac{\Gamma(\alpha + 1)\Gamma(\alpha + \beta)}{\Gamma(\alpha)\Gamma(\alpha + \beta + 1)},$$

and its second moment is

$$\langle x^2 \rangle = \frac{1}{B(\alpha, \beta)} \int_0^1 x^{\alpha+1}(1 - x)^{\beta-1} dx = \frac{B(\alpha + 2, \beta)}{B(\alpha, \beta)} = \frac{\Gamma(\alpha + 2)\Gamma(\alpha + \beta)}{\Gamma(\alpha)\Gamma(\alpha + \beta + 2)},$$

so that. using $\Gamma(x + 1) = x\Gamma(x)$, we obtain

$$\boxed{\langle x \rangle = \frac{\alpha}{\alpha + \beta} \; ; \; \sigma_x^2 = \frac{\alpha\beta}{(\alpha + \beta)^2(\alpha + \beta + 1)}} \tag{3.76}$$

---

[36] You may have noticed that the symbol $Z$ is not accidental, for this is the way a partition function is defined in statistical mechanics.

Note that, swapping $\alpha \rightleftarrows \beta$, $\langle x \rangle$ becomes $1 - \langle x \rangle$, while $\sigma_x$ does not change. By imposing $\mathrm{d}f(x)/\mathrm{d}x = 0$, one finds a single extremum,

$$\bar{x} = \frac{\alpha - 1}{\alpha + \beta - 2}$$

which, for $\alpha$ and $\beta > 1$, is the most probable value (the mode $x_{max}$) of the distribution, while it is a minimum for $\alpha$ and $\beta < 1$ The beta density function can take a variety of shapes depending on the values of $\alpha$ and $\beta$, which is the main reason of its extensive applications for modeling the distribution of any parameter that, like a probability, takes a value in the range $[0, 1]$:

$\boxed{\alpha = \beta}$  When the two parameters are equal, the distribution is *symmetric* around $x = 1/2$. We can distinguish three cases:

$\boxed{\alpha = \beta = 1}$  $f(x; \alpha, \beta)$ reduces to the *uniform density* $f(x) \equiv 1$ in $[0, 1]$.

$\boxed{\alpha = \beta < 1}$  The distribution is "U-shaped" with a minimum in $x = 1/2$ and, for $\alpha, \beta \to 0$, becomes more and more concentrated around the extremes $x = 0, 1$. The limiting distribution is $f(x; 0, 0) = [\delta(x) + \delta(x - 1)]/2$, which can be seen as a discrete distribution for a single Bernoulli trial.

$\boxed{\alpha = \beta > 1}$  The distribution has a maximum in $x = 1/2$. For $\alpha = \beta = 2$ it has a parabolic shape, while it becomes progressively more bell-shaped when $\alpha = \beta > 2$.

$\boxed{\alpha \neq \beta}$  The distribution is skewed, with a tail to the right or the left if $\alpha < \beta$ or $\alpha > \beta$, respectively. A special case, however, is $\alpha = 1, \beta = 2$, when the distribution reduces to a straight line with slope -2.

Several of these cases are illustrated in Fig. 3.14. In particular, the distributions labelled with $(6, 6)$ and $(8, 6)$ would be appropriate for the probability $p$ of heads for a coin that has been flipped 10 times, obtaining 5 or 7 heads respectively. The expectation, mode, and standard deviations of these distributions are

$$\begin{cases} f(x; 6, 6) : \langle x \rangle = 0.5, \ x_{max} = 0.5, \ \sigma_x \simeq 0.139 \\ f(x; 8, 4) : \langle x \rangle = 2/3, \ x_{max} = 0.7, \ \sigma_x \simeq 0.067 \end{cases}$$

Note that, for the second distribution, we have $f(0.5; 8, 6) \simeq 0.43 f(0.5; 8, 6)$, namely, that the probability density around the value $p = 1/2$ for a fair coin is less than half the maximum value. Whether this is sufficient to assert that the coin is a scam is something we will discuss later.

**Remark 3.25.1** (*Von Neumann's fair coin*) Given the previous considerations, you may have become doubtful of the possibility of finding a coin that is perfectly fair. Don't worry: if it does not exist, we can *make* it, thanks to the suggestion of that

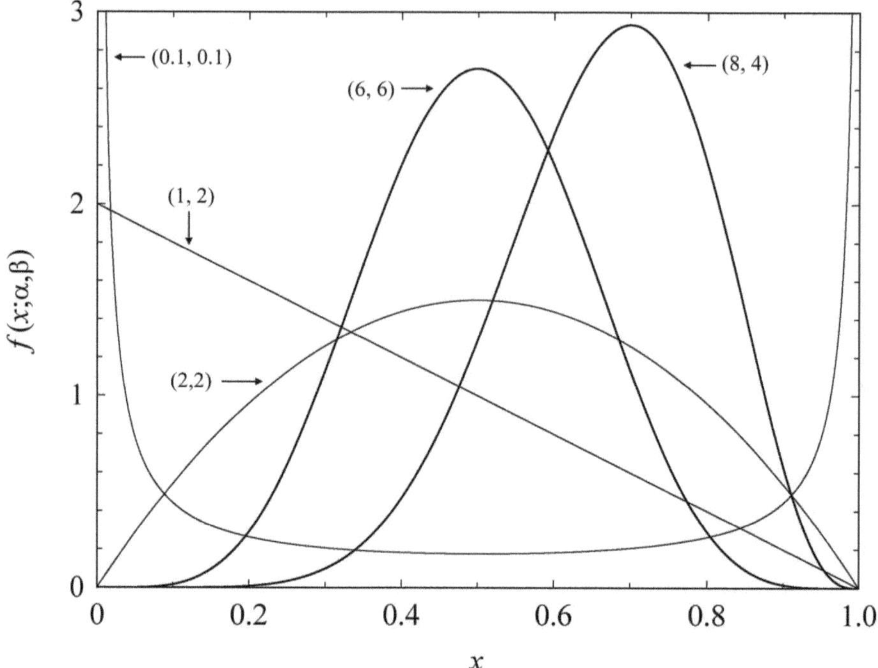


**Fig. 3.14** Beta distribution for several pair of values $(\alpha, \beta)$ indicated near to each curve

polymath genius of John von Neumann.[37] The procedure is simple. Flip a coin twice, no matter if biased. If you get two heads or two tails, throw the data away. Otherwise, set HT = 1 and TH = 0, or viceversa. Since $P(\text{HT}) = p(1 - p) = P(\text{TH})$, no matter what is the probability $p$ of a tail for the original coin, by repeating many times the experiment you get a Bernoulli sequence where the probabilities of 0 and 1 are exactly 1/2. This simple idea by von Neumann is at the origin of a class of "tree algorithms" of interest in computer science.[38] After all, we are speaking of the principal deviser of what we now call "computer"!

### 3.7.4 The Log-Normal Distribution

The normal distribution has a pivotal role in statistics, mainly for the reasons we will explore in the next chapter. However, several statistical variables can only take positive values and their distribution is not symmetric, two features that the Gaussian cannot definitely account for. A very flexible distribution with a positive support

---

[37] J. von Neumann, J. Res. Nat. Bur. Stand. Appl. Math. Series **3**, 36 (1951).
[38] Q.F. Stout and B. Warren, Ann. Prob. **12**, 212 (1984).

that displays a variable skewness is the *log-normal distribution*. A random variable is log-normally distributed if its logarithm has a normal distribution. Specifically, if $Z$ has the standard normal distribution (3.63), the probability density for $X = x_0 \exp(\sigma Z)$ is[39]

$$f(x) = \frac{1}{x\sigma\sqrt{2\pi}} \exp\left[ -\frac{1}{2}\left(\frac{\ln(x/x_0)}{\sigma}\right)^2 \right],$$

or, putting $\mu = \ln(x_0)$,

$$f(x; \mu, \sigma) = \frac{1}{x\sigma\sqrt{2\pi}} \exp\left[ -\frac{(\ln(x) - \mu)^2}{2\sigma^2} \right] \tag{3.77}$$

Note that $f(x)$ is defined only for $x > 0$. Moreover, the argument of $\ln(x_0)$ is *not* dimensionless, and therefore $\mu$ cannot be given any physical meaning, while $\sigma$ must be a pure number.[40] Thus, in contrast with the parameters of the Gaussian with the same name, $\mu$ and $\sigma$ cannot be the expectation and standard deviation of the log-normal distribution. Indeed, we will show (see Sect. 4.5.3) that the $r$-th moment of the log-normal distribution is non-trivially related to the parameters $\mu$ and $\sigma$ by

$$\langle x^r \rangle = \exp(r\mu) \exp\left(\frac{r^2\sigma^2}{2}\right), \tag{3.78}$$

thus the expectation and variance can be easily found to be

$$\langle x \rangle = \exp\left(\mu + \tfrac{\sigma^2}{2}\right); \quad \sigma_x^2 = \langle x \rangle^2 \left(e^{\sigma^2} - 1\right). \tag{3.79}$$

Thus, the relative standard deviation (the "relative width")

$$\sigma_r = \frac{\sigma_x}{\langle x \rangle} = \sqrt{\exp(\sigma^2) - 1}$$

does *not* depend on $\mu$, and we have $\lim_{\sigma \to 0} \sigma_r = \sigma$. One can also show that the median $x_m$ and the mode (the maximum) $x_{max}$ of the distribution are given by

$$x_m = e^{\mu}, \quad x_{max} = e^{\mu - \sigma^2},$$

[39] This result will be clear when we discuss the distribution of a function $Y(X)$ of a random variable $X$.

[40] One can alternatively parameterize the distribution using $\mu^* = \exp(\mu)$, which has the dimensions of $X$, and $\sigma^* = \sqrt{\exp(\sigma^2)}$.

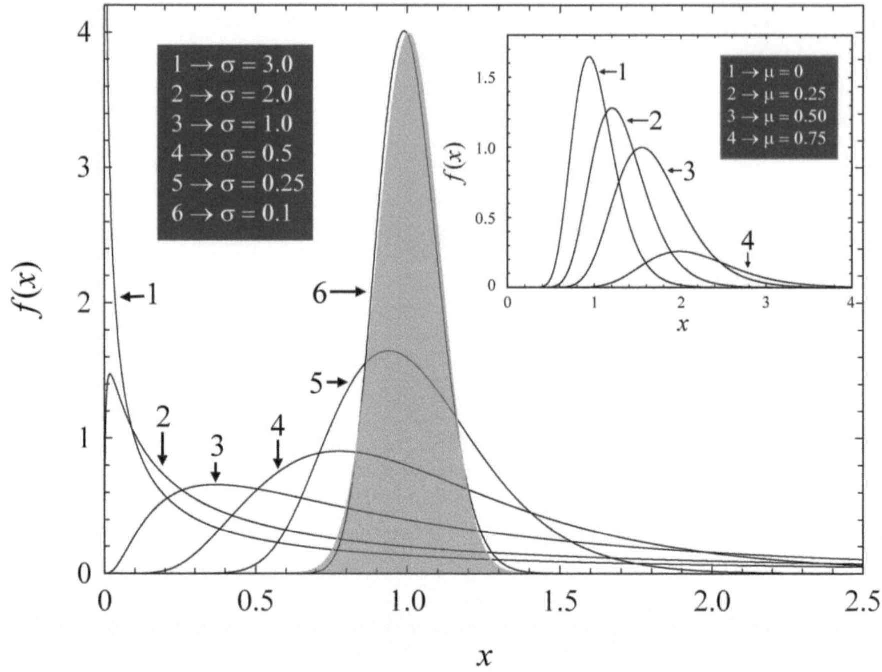

**Fig. 3.15** Body: Log-normal distribution for $\mu = 0$ and several values of $\sigma$. The distribution for $\sigma = 0.1$ is almost identical to a Gaussian with the same expectation and variance (shown in grey). Inset: the same for some values of $\mu$ and fixed $\sigma = 0.25$

and (with some effort!) that the skewness parameter is

$$\gamma_x = [\exp(\sigma^2) + 2]\sigma_r, \tag{3.80}$$

thus, for small $\sigma$, $\gamma_x \simeq 3\sigma$. Note that, like $\sigma_r$, $\gamma_x$ does not depend on $\mu$ either.

By changing $\mu$, therefore, the shape characteristics $\sigma_r$ and $\gamma$ of the distribution do not change. It is then worth visualizing the shape of the log-normal distribution for $\mu = 0$ as a function of $\sigma$. The body of Fig. 3.15 shows that, by decreasing $\sigma$ from 3 to 0.1, the shape of the distribution changes drastically, from a monotonically and rapidly decreasing function to an almost symmetric distribution (for $\sigma = 0.1$, $\gamma \simeq 0.3$) that resembles very much a Gaussian with the same expectation and variance. The figure inset seems to suggest that, by increasing $\mu$ at constant $\sigma = 0.25$, the shape of the distribution changes significantly. Yet, this is solely an illusion as all the curves have identical values for $\sigma_r \simeq 0.25$ and $\gamma \simeq 0.31$. So much for our ability to assess the shape of a distribution by eye!

It is also useful to point out that, when $\sigma$ is large, the log-normal can behave in a rather extended region as a power law. Indeed, taking the logarithm of $f(x)$, with simple steps we obtain

$$\ln f(x) = -\frac{(\ln x)^2}{2\sigma^2} + \left(\frac{\mu}{\sigma^2} - 1\right)\ln x - \ln\sqrt{2\pi}\sigma - \frac{\mu^2}{2\sigma^2}.$$

Thus, provided that $\sigma \gg 1$ and $\sigma \gg \mu$), $\ln f(x)$ is approximately linear in $\ln x$ over a rather large range of $x$, with a slope that is very close to -1. This is quite interesting, because there are several physical quantities whose probability density scales as the inverse of the variable in a wide range.

By playing with the two parameters $\mu$ and $\sigma$ we can then generate a wide spectrum of positive–definite distribution with prescribed characteristics. More specifically, using (3.79) we see that, to obtain a log-normal distribution of selected expectation $\langle x \rangle$ and variance $\sigma_x^2$, we have to choose

$$\mu = \ln\left[\frac{\langle x \rangle^2}{\sqrt{\langle x \rangle^2 + \sigma_x^2}}\right]; \quad \sigma^2 = \ln\left[1 + \frac{\sigma_x^2}{\langle x \rangle^2}\right]. \tag{3.81}$$

The versatility and effectiveness log-normal in dealing with skewed distributions with a positive support has led to a large number of applications ranging from economy to medicine, from linguistics to food technology[41] where it can often outperform the Gaussian. In some instances, the log-normal distribution is just used empirically to fit a data set, but some of the most interesting applications arise from an important connection between this and the normal distribution.

To bring out this link, we have to anticipate the main message of the Central Limit Theorem (CLT), which is the following: the probability density of the sum of a sufficiently large number of independent random variables is normal, provided that these variables satisfy some rather loose conditions (essentially, they have a finite variance and none of them has a "disproportionate weight"). If now we consider a variable $X$ which is the *product* on $n$ variables $X_i$, we have

$$X = \prod_{1=1}^{n} X_i \implies \ln X = \sum_{1=1}^{n} \ln X_i,$$

thus, if the RVs $\ln X_i$ satisfy the requirements of the CLT then, as $n$ increases, the distribution of $\ln X$ approaches better and better a Gaussian, and therefore $X$ is approximately a log-normal variable.

As a consequence, a large number of mechanisms can lead to the log-normal distribution. Indeed, consider any process that can be broken down in $n$ independent *serial* steps,[42] each one with a probability $p_i$ to be accomplished, so that the

---

[41] For a review, see E. Limpert, W. A. Stahel, and M. Abbt, BioScience **51**, 341 (2001), where an ingenious mechanical model based on a modified Galton board that generates a log-normal distribution is also presented.

[42] In fact, events that lead to log-normal distributions are analogous to series networks of AND gates, while events that lead to a Gaussian distribution are the analogs of parallel networks of OR gates.

probability of success for whole process is $P = p_1 p_2 \ldots p_n$. This idea was exploited by Kolmogorov[43] to find the size distribution of crushed ore, and presented in general form by William Shockley to discuss individual variations of productivity in research labs.[44] A very similar "multiplicative model", the so-called Gibrat's law of proportionate effect, is employed to explain the occurrence of log-normal distribution in the growth of organisms, the differentiation of biological species, and even the size of firms. More generally, the law of proportionate effects describes a stepwise process where at each step the change of a random variable is a random proportion of its current value. In the continuum limit, this leads to the differential equation $\dot{x}(t) = kx(x)$, where the first-order rate constant $k$ fluctuates in time according to an arbitrary probability distribution.[45] This is for instance the reason why the distribution of polymer molecular weight or nanoparticle size obtained by radical polymerization is often well described by a log-normal probability density.

### 3.7.5   *The Weibull and Pareto Distributions*

We have seen that, provided that the probability density for the failure of a system or an item to take place at time $t$ is given by the exponential distribution (3.46). However, this is true only if the decay rate $\alpha$ is a constant, but this is seldom the case. Quite often, a component wears out with time, so its failure becomes more likely as time goes on. On the other hand, observing something is working at the beginning may give us confidence that it will continue operating in the long run. In other words, we expect that if that system is going to fail, it will more likely fail at the start.

It is then worth trying to find a model that also accounts for these opposite behaviors. Generalizing (3.44), we write

$$P(t; \mathrm{d}t) = f(t)\mathrm{d}t = S(t)\alpha(t)\mathrm{d}t, \tag{3.82}$$

where $\alpha(t)$ is a time-dependent failure rate (often called the *hazard function*) and the *survival function* $S(t)$ is the probability that no failure takes place up to time $t$, which is relate to the cumulative probability distribution by

$$S(t) = 1 - F(t) = 1 - \int_{-\infty}^{t} f(t')\mathrm{d}t'.$$

---

[43] A. N. Kolmogorov, C. R. Acad. Sci. USSR **31**, 99 (1941) (for an English translation, see NASA Technical Translation F-12, 287, 1969).

[44] W. Shockley, Proc. IRE **45**, 279 (1957).

[45] See for instance D. B. Siano, J. Chem. Education **49**, 755 (1972).

Since $f(t) = \mathrm{d}F(t)/\mathrm{d}t$, the hazard and survival functions are related by

$$\alpha(t) = \frac{f(t)}{S(t)} = -\frac{1}{S(t)}\frac{\mathrm{d}S(t)}{\mathrm{d}t} = -\frac{\mathrm{d}\ln[S(t)]}{\mathrm{d}t}.$$

From Eq. (3.47), we see that for a constant failure rate $S(t) = \exp(-t/\tau)$, where $\tau = 1/\alpha$. To account for the other situations, we seek a function that decays respectively, slower or faster than a simple exponential. A natural choice is

$$S(t) = \exp\left(-\frac{t}{\tau}\right)^{\beta}. \tag{3.83}$$

For $\beta < 1$ this is called a *stretched exponential*, a function often used as a phenomenological description of relaxation in disordered systems like glasses, dielectric spectra of macromolecules (where its counterpart in the frequency domain is known as the Kohlrausch-Williams-Watts relaxation), and a multitude of other phenomena—to be honest, probably too many. The opposite case, $\beta > 1$, known as a *compressed* exponential, has been recently applied to study the glass relaxation in soft materials[46] (of course, for $\beta = 2$ it also describes the tails of a Gaussian). With this choice we have

$$\alpha(t) = \frac{\beta}{\tau}\left(\frac{t}{\tau}\right)^{\beta-1},$$

which correctly yields a constant failure rate for $\beta = 1$, and

$$f(t;\beta,\tau) = \begin{cases} \dfrac{\beta}{\tau}\left(\dfrac{t}{\tau}\right)^{\beta-1}\exp\left(-\dfrac{t}{\tau}\right)^{\beta} & t \geq 0 \\[2mm] 0 & t < 0 \end{cases} \tag{3.84}$$

which is known as the *Weibull distribution*.[47] All moments of the distribution can be related to the Gamma function by observing that[48]

$$\int_0^{\infty} t^{n-1}\mathrm{e}^{-(t/\tau)^{\beta}} = \frac{\tau^n}{\beta}\Gamma\left(\frac{n}{\beta}\right), \tag{3.85}$$

which first of all shows that the distribution is correctly normalized, and then allows us to obtain

---

[46] see, for instance, A. M. Philippe et al., Phys. Rev. E **97**, 040601 (2018).

[47] W. Weibull, J. Appl. Mech. **18**, 293 (1951).

[48] Equation (3.85) implies that the moments of the *positive half* of the Gaussian ($\beta = 2$) are given by

$$\int_0^{+\infty} x^k \mathrm{e}^{-ax^2} = \frac{\Gamma\left(\frac{k+1}{2}\right)}{2a^{\frac{k+1}{2}}}.$$

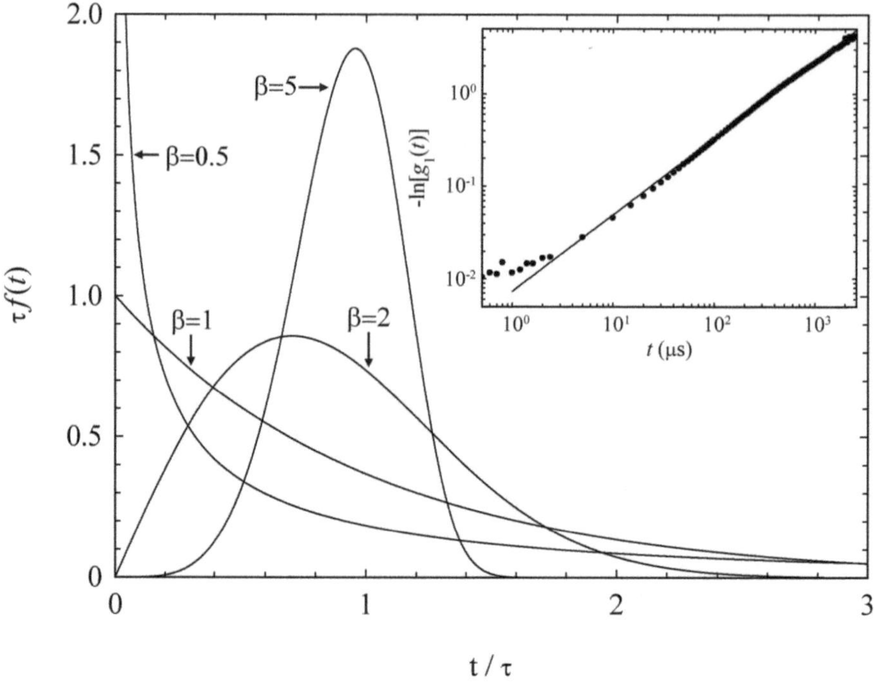



**Fig. 3.16** Body: Weibull distribution for several values of $\beta$. Inset: Example of stretched exponential relaxation, with $\beta \simeq 0.8$, for the Dynamic Light Scattering correlation function $g_1(t)$ from a soft polymer gel, plotted as discussed in the text

$$\langle t \rangle = \tau \Gamma(1 + 1/\beta); \quad \sigma_t^2 = \tau^2 \left[ \Gamma\left(1 + \frac{2}{\beta}\right) - \Gamma^2\left(1 + \frac{1}{\beta}\right) \right]. \tag{3.86}$$

The body of Fig. 3.16 shows that the shape of the Weibull distribution remarkably depends on $\beta$, changing from a monotonically decreasing function for $\beta \leq 1$ to a distribution with a single maximum in $t_{max} = \tau(1 - 1/\beta)$ for $\beta > 1$, which becomes more and more symmetric by increasing $\beta$.

By using Eq. (3.85), we can also grasp the meaning of the parameter $\tau$ better. The average relaxation time $\langle \tau \rangle$ of a generic relaxation is often taken as the area under $S(t)$, which for a simple exponential $\exp(-t/\tau)$ coincides in fact with $\tau$. For a stretched exponential, however, we have

$$\langle \tau \rangle = \int_0^{\infty} S(t)\mathrm{d}t = \frac{1}{\beta}\Gamma\left(\frac{1}{\beta}\right)\tau, \tag{3.87}$$

which, since $\beta < 1$ is generally *longer* than $\tau$. For example, when $\beta = 1/n$, with $n$ integer, $\langle \tau \rangle = n\Gamma(n)\tau = n(n - 1)!\tau$. Note that the logarithm of $f(t) = \exp[-(t/\tau)^{\beta}$ is a power-law in $t$,

$$\ln f(t)) = -\left(\frac{t}{\tau}\right)^{\beta}.$$

Thus, a double-log plot of $-\ln f(t)$ is a straight line with slope $\beta$, which is a rather standard way to check whether and on which range of $t$ a relaxation function is a stretched (or compressed) exponential. The inset in Fig. 3.16 shows for instance that the decay of the correlation function of the concentration fluctuations obtained by Dynamic Light Scattering in my lab on a soft polymer gel is fitted by a stretched exponential over three decades in time.

The Weibull distribution is extensively exploited to analyze survival statistics such as life expectancy after a cancer diagnosis or the probability of fracture of brittle materials like glass. As we shall later discuss, it is also useful to describe (not so much to *predict*) extreme phenomena like floods, windstorm and torrential rains. I must stress, however, that in most cases these are just empirical models that do not address the physical processes leading to this peculiar distribution.

However, there are also situations where the survival function does not show a characteristic decay time like a simple or even a stretched exponential. In these cases, one may think of assuming for $S(t)$ a scale-free form like a power law. An example is the so-called "Lindy's Law", according to which items that have been around for a long time, like successful inventions, books, songs, shows, are also prone to staying around for a longer periods of time.[49] Indeed, suppose that the expectation of the remaining lifetime of something that has survived up to time $t_0$ is proportional to $t_0$, $\langle t - t_0 \rangle \propto p t_0$: then,[50] the survival function turns out to be proportional to $t^{-(1+p)/p}$.

Yet, no probability distribution can be a power-law over the whole positive real axis, because $\int_a^b t^{\beta} dt$ diverges for $b \to \infty$ when $\beta \leq 1$ and for $a \to 0$ when $\beta \geq 1$. So, we may rather take

$$S(t) = \left(\frac{t_0}{t}\right)^{\beta}, \text{ with } t \in [t_0, \infty], \beta > 0.$$

For the hazard function we then have $\alpha(t) = -\beta/t$, and for the probability density

$$f(t; \beta, t_0) = \begin{cases} \dfrac{\beta}{t_0}\left(\dfrac{t_0}{t}\right)^{\beta+1} & t \geq t_0 \\ 0 & \text{otherwise} \end{cases} \tag{3.88}$$

which is known among the economists as the *Pareto distribution* from the name of the Italian engineer, economist, and sociologist Vilfredo Pareto, who originally

---

[49] See, for instance, I. Eliazar, Physica A **486**,797 (2017).

[50] This is actually a *conditional expectation*, a concept we will develop in the next chapter.

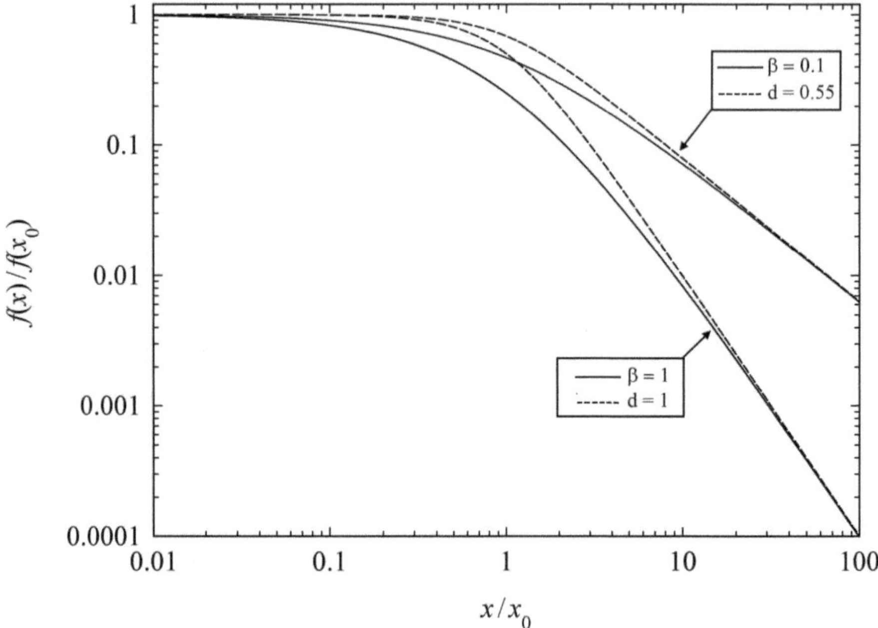



**Fig. 3.17** Lomax distribution for $\beta = 0.1$ and $\beta = 1$ (full lines) compared to the Fisher-Burford function, Eq. (1.19), for the corresponding values $d = 0.55$ and $d = 1$ (broken lines)

applied it to describe the distribution of wealth.[51] You can check that the distribution is correctly normalized. However, the observations we made about the integral of a power-law should convince you that the expectation of the distribution exists only for $\beta > 1$, and its variance only for $\beta > 2$, in which case we have

$$\langle t \rangle = \frac{\beta}{\beta - 1}\, t_0 \;\; ; \; \sigma_t^2 = \frac{\beta}{(\beta - 1)^2(\beta - 2)}\, t_0^2 \tag{3.89}$$

while the mode is (of course) $t_0$ and the median $2^{1/\beta} t_0$.

In Chap. 1 we have seen that many experimental frequency distribution show indeed an extended power law tail. A limitation of the Pareto distribution, however, is that it cannot describe the *entire* probability distribution of a positive-definite R.V., but only *part* of it. An interesting probability density which smoothly joins a constant plateau for small $x$ to a power law decay for large $x$ is the *Lomax distribution*,

---

[51] In this case, the random variable is not of course time but, for instance, the individual's fortune. Actually, by analyzing the distribution of wealth in Italy at the beginning of the 20th century (20% of the population holding approximately 80% of the resources) Pareto introduced the general "80-20 rule" stating that 80% of the outcomes is due to 20% of the causes, just an often useful rule-of-thumb rule that, however, sociologists consider sometimes as a Fundamental Law of Nature. The 80-20 rule applies of course only when $\beta = \ln 5 / \ln 4 \simeq 1.16$.

$$f(x; \beta, x_0) = \frac{\beta}{x_0} \left( \frac{1}{1 + \frac{x}{x_0}} \right)^{\beta+1} \quad \text{(with } \beta > 0\text{)}, \tag{3.90}$$

also called Pareto Type II, which is shown in Fig. 3.17 for two values of $\beta$. The Lomax distribution is remarkably similar to the Fisher-Burford function (1.19) that we used in Chap. 1 to fit the size distribution of the Italian municipalities and the gross income of the US residents. The latter however shows a much sharper transition to the power law regime than the Lomax distribution.

### 3.7.6 A Multifaceted Distribution: The Generalized Gamma

The main limit of the distributions we have discussed so far is the lack of a physical model supporting them. A general approach leading to a very general probability distribution with many applications in risk and survival analysis has been presented by Lienhard and Meyer.[52] I will just sketch their model, which is based on standard methods in statistical mechanics and has a lot in common with the concept of statistical entropy we shall later discuss.

Consider a series of events occurring in time that may lead to the failure of a system due to some kind of stress it has been subjected to. We call $N_k$ the number of events occurring in the time interval $[t_k - \Delta t, t_k]$ and assume that:

1. The total number of events $\sum_{k=1}^{\infty} N_k = N$ is fixed and finite.
2. Each of the $N_k$ events can take place in $g_k(t)$ distinguishable ways that, to account for the system aging or rejuvenation, can depend on time. Specifically, we assume $g_k \propto t_k^{\alpha-1}$ with $\alpha > 0$.
3. There is an integer $\beta$ such that the moment of order $\beta$ of the distribution of events in time is finite, $\sum_k (N_k/N)t_k^{\beta} = C$.

The total number of ways an event can happen for a fixed distribution $\{N_1, N_2, \ldots, N_k, \ldots\}$ of events in $\{t_1, t_2 \ldots, t_k, \ldots\}$ is then $\prod_{k=1}^{\infty} g_k^{N_k}$ , which, multiplied by the number of distinguishable permutations of the $N_k$'s, the multinomial coefficient $M_N^{\infty} = N!/\prod_{k=1}^{\infty} N_k!$, yields a total number of ways an event can occur at any moment in time,

$$W = M_N \prod_{k=1}^{\infty} g_k^{N_k} = N! \prod_{k=1}^{\infty} \left( \frac{g_k^{N_k}}{N_k!} \right).$$

The idea is finding the distribution $\{\bar{N}_k\}$ that *maximizes* $W$ (or better, its logarithm) with the constraints $\sum_{k=1}^{\infty} N_k = N$ and $\sum_k (N_k/N)t_k^{\beta} = C$ , because for the distribution $\{\bar{N}_k\}$ the maximum of $W$ will be (exponentially) larger than for any other

---

[52] J. H. Lienhard and P.L. Meyer, *A physical basis for the generalized gamma distribution*, Q. Appl. Math. **25**, 330 (1967).

sequence $\{N_k\}$ that significantly differs from it.[53] Putting then $\bar{N}_k/N = \int_{t_k-\Delta t}^{t_k} f(t)\mathrm{d}t$ and passing to the limit for $\Delta t \to 0$, one obtains

$$f(t; \alpha, \beta, \tau) = \frac{\beta}{\tau^\alpha \Gamma(\alpha/\beta)}\, t^{\alpha-1} \mathrm{e}^{-(t/\tau)^\beta} \tag{3.91}$$

where $\tau = (\beta C/\alpha)^{1/\beta}$. This probability density, known as the *generalized gamma distribution*, is really a multi-faceted distribution. Indeed, different choices of the three parameters $\alpha, \beta, \tau$ generate many other useful distributions. Moreover, although in the model Lienhard and Meyer the variable is time, or possibly the amount of stress a system is subjected to, the generalized gamma density is effective for problems that have little to do with risk analysis. We now introduce the most successful "offsprings' of the generalized gamma distribution, starting with the distribution that inspired its name.

### 3.7.6.1   The Gamma Distribution ($\beta = 1$)

For $\beta = 1$ we obtain the standard *gamma distribution* a versatile two-parameter distribution that, for a generic nonnegative variable $X$, can be written

$$f(x; \alpha, x_0) = \frac{1}{\Gamma(\alpha)x_0^\alpha}\, x^{\alpha-1} \mathrm{e}^{-x/x_0} \quad (\alpha > 0) \tag{3.92}$$

where $\alpha$ fixes the *shape* and $x_0$ the *scale* of the distribution. Using the definition of the gamma function in the appendix, it is easy to show that

$$\langle x^n \rangle = \frac{1}{\Gamma(\alpha)x_0^\alpha} \int_0^\infty x^{n+\alpha-1}\mathrm{e}^{-x/x_0}\mathrm{d}x = \frac{\Gamma(\alpha+n)}{\Gamma(\alpha)}\, x_0^n, \tag{3.93}$$

wherefrom we obtain

$$\langle x \rangle = \alpha x_0 \ ; \ \sigma_x^2 = \alpha x_0^2 \ ; \ \gamma = \frac{2}{\sqrt{\alpha}} \tag{3.94}$$

The expectation of $X$ is then the product of the shape and the scale parameters, while the asymmetry of the distribution is solely determined by $\alpha$. Setting to zero the derivative of (3.92) we see that the mode of the distribution is

---

[53] This approach fully parallels the strategy used to link statistical mechanics to thermodynamics by stating that macroscopic equilibrium is given by the distribution of microstates with the highest probability.

$$x_{max} = \begin{cases} (\alpha - 1)x_0 & \text{if } \alpha \geq 1 \\ 0 & \text{otherwise} \end{cases} \tag{3.95}$$

For $\alpha = 1$ the gamma probability density reduces to the exponential distribution, while for a dimensionless variable $x = \chi^2$ with $x_0 = 2$ and $\alpha = \nu/2$, where $\nu$ is an integer, is the distribution of a variable called $\chi^2$, an important tool in data analysis and hypothesis testing that we will meet in Chap. 8. Even more interestingly, when $\alpha$ is a positive integer $k$, putting $t^* = x/x_0$ and taking into account that $\Gamma(k) = (k-1)!$, $f(t^*)$ coincides with the distribution (3.50) of the waiting time for $k$ events, shown for some values of $t^*$ in Fig. 3.10. Therefore, in this case the gamma density (also known in this case as the *Erlang distribution*[54]) also counts *the amount of time until the occurrence of a fixed number of Poisson events*.

The gamma distribution has a large number of applications ranging from shelf-life testing to cell cycle time distribution.[55] Here, however, I will only focus on a recent and highly intriguing application in physics that concerns the random packing of granular systems, which include materials that are common in our daily lives, such as sand, powders, gravel, whose fundamental description has proved to be exceedingly challenging. Indeed, in the absence of an external driving force granular materials do not explore different equilibrium configurations, both because thermal fluctuations are negligible and because friction quickly dissipate the grain kinetic energy. In fact, granular systems are usually trapped in *jammed states* where particle motion is prevented due to the confinement by the neighboring grains, and the system dynamics basically consists in jumps from a jammed state to another one. A fundamental step towards the understanding of granular materials has been made by Sam Edwards, who conceived that the static properties of granular matter could be obtained as an average over these jammed configurations, regarded as equiprobable, with the volume occupied by the grain taking the role of the energy for an isolated system in equilibrium statistical mechanics.[56]

There is extensive experimental evidence that the statistical properties of the jammed configurations of spherical grains show an interesting universality, which does not depend on the protocol used to obtain them. Specifically, suppose we subdivide the packing into a set of $k$ local elementary cells. Then, the volume fluctuations over these cells can be described by a gamma distribution with shape parameter $\alpha = k$ and a scale parameter $x_0 = \langle V \rangle /k - v_{min}$, where $v_{min}$ is the minimum volume a cell can attain. Following Edwards' ideas, Aste and Di Matteo[57] have recently

---

[54] From the name of Agner Erlang, who introduce it in 1909 to evaluate how many telephone operators were needed to handle a given volume of calls.

[55] A particularly interesting application of the gamma distribution concerns the synthesis of polymers. A polymerization reaction terminated by the recombination of free radicals is indeed expected to generate a molecular mass distribution given by a gamma distribution with scale and shape factors related by $x_0 = 1/\alpha$, known in the field of polymer science as a Schulz–Zimm distribution.

[56] For a review, see A. Baule et al.., Rev. Mod. Phys. **90**, 015006 (2018).

[57] T. Aste and T. Di Matteo, Phys. Rev. E **77**, 021309 (2008).

shown that this results follow from maximizing the Gibbs entropy $S(V)$ of the configuration, and that the scale parameter $x_0$ coincides with the "granular temperature" $\chi = [\partial S(V)/\partial V]^{-1}$ defined by Edwards in analogy with the thermodynamic temperature $T = [\partial S(E)/\partial E]^{-1}$.

As we anticipated, the gamma distribution provides a very interesting connection between the Poisson and the negative binomial distribution. Suppose indeed that we consider a random variable which is distributed as a Poisson $P(k; \lambda)$, but with an expectation $\lambda$ which is itself a (continuous) random variable distributed as a gamma, $f(\lambda; \alpha, \lambda_0)$. The resulting probability distribution is then

$$P(k; \alpha, x_0) = \frac{1}{\Gamma(\alpha)\lambda_0^\alpha} \int_0^{+\infty} \frac{e^{-\lambda}\lambda^k}{k!} \lambda^{\alpha-1} e^{-\lambda/\lambda_0} =$$

$$= \frac{1}{k!\Gamma(\alpha)\lambda_0^\alpha} \int_0^{+\infty} \lambda^{\alpha+k-1} e^{-\lambda(1+1/\lambda_0)},$$

which, using

$$\int_0^{+\infty} t^{s-1} e^{-t/t_0} = \Gamma(s) t_0^s,$$

yields

$$P(k; \alpha, x_0) = \frac{\Gamma(\alpha+k)}{k!\Gamma(\alpha)\lambda_0^\alpha} \left(\frac{\lambda_0}{1+\lambda_0}\right)^{\alpha+k} = \frac{\Gamma(\alpha+k)}{k!\Gamma(\alpha)} \left(\frac{1}{1+\lambda_0}\right)^\alpha \left(1 - \frac{1}{1+\lambda_0}\right)^k,$$

which is a negative binomial distribution with $r = \alpha$ and $p = 1/(1 + \lambda_0)$.

### 3.7.6.2  The Rayleigh Distribution ($\alpha = \beta = 2$)

When $\beta = \alpha$ the generalized gamma becomes the Weibull distribution (3.84) that we have already met. We have seen that for $\alpha = 1$ the latter reduces again to the exponential distribution (3.46). For $\alpha = 2$ we obtain instead a probability density of relevant interest in physics, the *Rayleigh distribution*

$$\boxed{f(r) = \frac{r}{\sigma^2} \exp\left(-\frac{r^2}{2\sigma^2}\right) \quad (x > 0)} \tag{3.96}$$

where, for reasons that will soon be evident, we have renamed $\{t, \tau\} \to \{r, \sqrt{2}\sigma\}$. The expectation and variance of the distribution can be obtained from (3.85),

$$\boxed{\langle r \rangle = \sqrt{\frac{\pi}{2}}\, \sigma \; ; \; \sigma_r^2 = \left(2 - \frac{\pi}{2}\right)\sigma^2.} \tag{3.97}$$

This distribution was originally derived in 1880 by Rayleigh (John William Strutt) to account for the distribution of the amplitude of the resultant of a large number of vibration with random phases.[58] In fact, we will see that Rayleigh's investigation bears many points in common with the problem of finding the intensity distribution of laser speckles. A quarter of century later, however, Karl Pearson, arguably the father of mathematical statistics, posed a very different question, at least apparently:

> A man starts from a point O and walks $l$ yards in a straight line; he then turns through any angle whatever and walks another $l$ yards in a second straight line. He repeats this process $n$ times. I require the probability that after these $n$ stretches he is at a distance between $r$ and $r + \delta r$ from the from his starting point, O.[59]

You may have already realized that Pearson's query is nothing but the problem of a random walk in two dimensions. The brilliant answer came in the same issue of Nature by Rayleigh himself, who replied that:

> This problem, proposed by Prof. Karl Pearson in the current number of NATURE, is the same as that of the composition of $n$ iso-periodic vibrations of unit amplitude and of phases distributed at random [...]. If $n$ be very great, the probability sought is[60]

$$\frac{2}{n} e^{-r^2/n} r \, dr.$$

In the next chapter, we will indeed find that the Rayleigh distribution is the exact solution for the r.m.s. displacement $r$ in a bi-dimensional RW when the number of steps $n \to \infty$, a result confirmed in Fig. 3.18 where the frequency distribution for $r$ obtained from the numerical simulations in Fig 1.11b are fitted by a Rayleigh probability density with $\sigma \simeq 21$. Note that for this distribution $\langle r \rangle \simeq 26$, which is very close to the square root of the number of steps in the simulations.

### 3.7.6.3  The Maxwell Molecular Speed Distribution ($\alpha = 3, \beta = 2$)

Still for $\beta = 2$, but $\alpha = 3$, and renaming this time $\{t, \tau\} \to \{v, \sqrt{2}\sigma\}$ we find instead

$$f(v) = \sqrt{\frac{2}{\pi} \frac{v^2}{\sigma^3}} \exp\left(-\frac{v^2}{2\sigma^2}\right) \quad (x > 0) \tag{3.98}$$

---

[58] Lord Rayleigh, F.R.S., Phil. Mag. Series 5, 10, 73 (1880).

[59] K. Pearson, Nature **72**, 294 (1905). Note that at that time Einstein had already published his first paper on Brownian motion: evidently, the connection between this effect and RW had not been appreciated yet.

[60] So, here $\sigma^2 = n/2$.

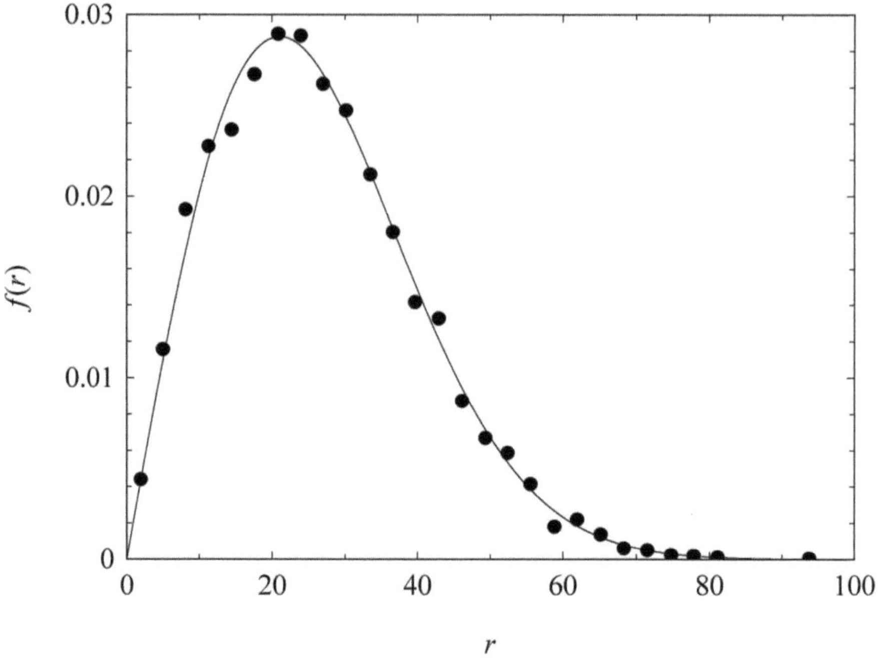

**Fig. 3.18** Comparison between the frequency distribution of the r.m.s. displacement in 2500 bi-dimensional random walks, shown in Fig. 1.11b, and a Rayleigh distribution with $\sigma \simeq 21$

which, taking as scale parameter $\sigma = \sqrt{k_B T/m}$, where $m$ is the molecular mass, coincides with the probability distribution for the molecular speed obtained by Maxwell in his kinetic theory of ideal gases.[61] Using again Eq. (3.85), the first and second moment of the Maxwell distribution are found to be

$$\langle v \rangle = 2\sigma \sqrt{\frac{2}{\pi}} = \sqrt{\frac{8k_B T}{\pi m}} \; ; \; \langle v^2 \rangle = 3\sigma^2 = \frac{3k_B T}{m} \tag{3.99}$$

In full analogy with what we said for the Rayleigh distribution, however, the Maxwell probability density also yields the distribution of the r.m.s. displacement in a *three* dimensional random walk (say, the spacewalk of a drunken astronaut). So, the results of Maxwell's kinetic theory, based on some heuristic assumptions about the distribution of molecular velocities at equilibrium, coincides with those for a random walk in the velocity space (which is not at all obvious).

---

[61] More generally, this is the probability density for the magnitude of the velocity of a system of classical, independent, non-relativistic particles.

### 3.7.7   *Extreme Values*

Many of us may, unfortunately, have witnessed events that are rare but, when they occur, have devastating consequences: hurricanes, earthquakes, heat waves, blizzards, wildfires, flooding, drought, pathogen spillovers are just some examples in the wild, sometimes triggered or exacerbated by anthropogenic factors. But also stock market crashes, social collapses, chemical or nuclear disasters are extreme and apparently unpredictable events that fiercely affect the human society. Forecasting based on previous observations whether random variables such as wind speed, sea level, hotness, precipitations, financial indices, leakages will exceed or fall below an alarming level is the subject of Extreme Value Analysis (EVA).

The dramatic impact of extreme events in fields like civil engineering is palpable. For instance, the floodgates of MOSE, the system intended to protect Venice and the surrounding lagoon from high tides, are designed to be raised when the tide exceeds 130 cm, which was estimated to happen 4–5 times a year: since MOSE entered into operation in October 2020, they were activated more than 80 times…telling a lot about the need for an accurate EVA. In recent years, however, EVA has acquired an equally important role in various physical contexts, ranging from fracture, to disordered systems, nonlinear optics, diffusion in random media.

Providing an adequate survey of this rich field of investigation in probability and statistics goes well beyond the scope of this book. I will just sketch the basics of what is usually called "classical" EVA, which starts from considering a series of *independent* measurements $x_1, x_2, \ldots x_n$ of a random variable $X$ with a p.d.f. $f(x)$. Because the variables $x_i$ are independent, the probability that *all* these values do not exceed a given value $x$ is simply

$$\Phi_N(x) = P[(x_1 \le x) \cap \ldots \cap (x_N \le x)] = \left[ \int_{-\infty}^{x} f(x)\mathrm{d}x' \right]^N = [F(x)]^N,$$

(3.100)

where $F(x)$ is the cumulative distribution.[62] We may then wonder whether, for $N$ sufficiently large, $\Phi_N(x)$ approaches a limiting distribution $\Phi(x)$.

The answer is the content of an important and rather surprising result obtained in its most general form by Boris Gnedenko, a student of Kolmogorov. Let us first observe that, since $F(x) \le 1$ monotonically increases with $x$, the dominant contribution to (3.100) comes for values of $x$ for which $F(x) \simeq 1$. Then, Gnedenko first showed that

---

[62] The distribution for the minimum value is also simply related to $F(x)$.

$$P[(x_1 \ge x) \cap \ldots \cap (x_N \ge x)] = \left[ \int_{x}^{\infty} f(x)\mathrm{d}x' \right]^N = [1 - F(x)]^N.$$

we can always find two sequences of real numbers $s_N$ and $\mu_N$ such that introducing a rescaled variable[63]

$$Z = \frac{X - \mu_N}{s_N},$$

makes the series $[F(z)]^N = [F(s_N x + \mu_N)]^N$ converge to a limit function, i.e.,

$$\lim[F(s_N x + \mu_N)]^N \xrightarrow[N\to\infty]{} \Phi(z).$$

The p.d.f. of the maximum value is obtained from the cumulative distribution $\Phi(z)$ as $\phi(z) = d\Phi(z)/dz$, substituting then back $x = s_n z + \mu_N$ to obtain $\phi(x)$. There is no general rule to find the sequences $\mu_N$ and $s_N$. They can be taken as the expectation and standard deviation of $\Phi_N(x)$ in the limit of large $N$, when these values can easily be calculated. In general, however, it is easier to consider the asymptotic expansion of $F(x)$ for large $x$, choosing the coefficients so that $z$ becomes independent of $N$.

The most interesting consequence that can be drawn from Gnedenko's theorem, however, is that $\Phi(u)$ only depends on some very general features of $f(x)$, which leads to only three classes of limiting probability densities. By identifying these limiting distributions with some useful examples, we can also clarify the procedure we have just outlined.

### (A) The Weibull Limit

Let us start from a simple example. Consider the uniform p.d.f. in [0,1], $f(x) \equiv 1$, whose cumulative distribution is $Fx) = x$. For fixed $N$, the p.d.f. of the maximum is then

$$\phi_N(x) = \frac{d}{dx}[F(x)]^N = Nx^{N-1},$$

whose expectation and variance are easily found to be

$$\langle x \rangle_N = N \int_0^1 x^N dx = \frac{N}{N+1} \xrightarrow[N\gg1]{} 1$$

$$\sigma_N^2 = \langle x^2 \rangle_N - \langle x \rangle_N^2 = \frac{N}{(N+1)^2(N+2)} \xrightarrow[N\gg1]{} \frac{1}{N^2}$$

We can then choose $\mu_N = 1$ and $s_N = 1/N$, so that

$$\Phi_N(s_N z + \mu_N) = \left(1 + \frac{z}{N}\right)^N \xrightarrow[N\to\infty]{} \Phi(z) = \begin{cases} e^{-(-z)} & z \leq 0 \\ 0 & z > 0 \end{cases}$$

---

[63] Note that the dimensionless variable $Z$ is obtained from $X$ by subtracting out a "location factor" $\mu$ and dividing by a "scale factor" $s_N$, similarly to what we did when we defined a scaled form for the Gaussian.

Note that, when $x$ varies from 0 to 1, $z = (x - 1)/N$ varies from $-\infty$ to 0. To have a RV with positive support, it may then be convenient defining $t = -z$, so to obtain the limiting distribution

$$\phi(t) = -\frac{d\Phi(t)}{dt} = \begin{cases} e^{-t} & t \geq 0 \\ 0 & t < 0 \end{cases} \tag{3.101}$$

For a more general result, consider the p.d.f.

$$f(x) = \begin{cases} \beta(1 - x)^{\beta-1} & 0 \leq x \leq 1 \\ 0 & \text{otherwise} \end{cases} \tag{3.102}$$

with $\beta > 0$ (which reduces for $\beta = 1$ to the uniform distribution we just considered), whose cumulative distribution is

$$F(x) = \begin{cases} 1 - (1 - x)^\beta & 0 \leq x \leq 1 \\ 0 & \text{otherwise} \end{cases} \tag{3.103}$$

Although there is no simple analytical expression for the expectation and variance of $\phi_N(x)$, we can try to retain $\mu_N = 1$ and assume that $s_N$ scales with $N^{-1/\beta}$, so that $z = N^{1/\beta}(x - 1)$, again with $z \in (-\infty, 0]$. Thus we have

$$[F(z)]^N = \left[1 - \frac{(-z)^\beta}{N}\right]^N \xrightarrow[N \to \infty]{} \Phi(z) = \begin{cases} \exp\left[-(-z)^\beta\right] & z \leq 0 \\ 0 & z > 0 \end{cases}$$

so that, putting again $t = -z$

$$\phi(t) = -\frac{d\Phi(t)}{dt} = \begin{cases} \beta t^{\beta-1} e^{-t^\beta} & t \geq 0 \\ 0 & t < 0 \end{cases} \tag{3.104}$$

which is a *Weibull distribution*, Eq. (3.84), with $\tau = 1$ (note that the exponential p.d.f. (3.101) can be regarded as a Weibull distribution too with $\beta = 1$). We can express (3.101) in terms of the original variable $x$ by substituting $t = N^{1/\beta}(1 - x)$, so that $\Phi(x) = \exp\left[-N(1 - x)^\beta\right]$, obtaining

$$\phi(x) = \begin{cases} \beta N(1 - x)^{\beta-1} e^{-N(1-x)^\beta} & 0 \leq x \leq 1 \\ 0 & \text{otherwise} \end{cases} \tag{3.105}$$

**(B) The Fréchet limit**
Suppose now that $f(x)$ decays for $x \to \infty$ as a power law, $f(x) \sim k\beta x^{-(1+\beta)}$, where $\beta > 0$ and $k$ is a positive constant, so that, for large $x$, its cumulative density behaves as $F(x) \xrightarrow[x \to \infty]{} 1 - kx^{-\beta}$. If we introduce the variable $Z = X(kN)^{-1/\beta}$ (corresponding to $\mu_N = 0$ and $s_N = (kN)^{-1/\beta}$), the cumulative distribution for the maximum becomes

$$\Phi_N(x) = \left[1 - kx^{-\beta}\right]^N = \left[1 - k\left(k^{1/\beta}N^{1/\beta}z\right)^{-\beta}\right]^N = \left[1 - \frac{z}{N}\right]^N,$$

which, in the limit $N \to \infty$, yields

$$\Phi(z) = \begin{cases} \exp\left(-z^{-\beta}\right) & z > 0 \\ 0 & z \leq 0 \end{cases}$$

to which correspond the (standard) *Fréchet probability density*[64]

$$\phi(z) = \begin{cases} \dfrac{\beta}{z^{1+\beta}} \exp\left(-z^{-\beta}\right) & z > 0 \\ 0 & z \leq 0 \end{cases} \tag{3.106}$$

The moments $\langle z^r \rangle$ of the Frechét distribution are defined only for $r < \beta$, in which case they are given by[65]

$$\langle z^r \rangle = \Gamma\left(1 - \frac{r}{\beta}\right) \tag{3.107}$$

Therefore, the Frechét distribution has a finite expectation $\langle r \rangle = \Gamma(1 - 1/\beta)$ only for $\beta > 1$, and a finite variance $\sigma_r = \Gamma(1 - 2/\beta) - \Gamma^2(1 - 1/\beta)$ only for $\beta > 2$. By writing $Z = (x - \mu)/s$, the standard Fréchet distribution can be generalized to a p.d.f $f(x)$ that includes both a location and a scale parameter,

$$f(x) = \frac{\beta}{s}\left(\frac{x - \mu}{s}\right)^{-(1+\beta)} e^{-\left(\frac{x-\mu}{s}\right)^{-\beta}} \tag{3.108}$$

Note that $\mu$ and $s$ do not coincide with the expectation and the standard deviation of the distribution ($\mu$ is however the minimum, $f(x) = \beta/s$.) As we will see in the next chapter (see 4.1), the Fréchet and Weibull distribution are closely related.

### (G) The Gumbel Limit

The last and arguably more interesting class of limiting distribution primarily concerns those probability densities that decays *faster* then a power law. Let us start with two important examples.

---

[64] You may want to check that, in terms of the original variable,

$$\phi(x) = \begin{cases} \dfrac{N\beta}{x^{1+\beta}} \exp\left(-Nx^{-\beta}\right) & x > 0 \\ 0 & x \leq 0 \end{cases}$$

[65] To obtain (3.107), substitute $t = z^{-\beta}$ in the integral for $\langle z^r \rangle$ and use the definition of the gamma function.

EXPONENTIAL DENSITY. For the exponential p.d.f. (3.46), with cumulative distribution (3.47), we have

$$\Phi_N(x) = \left[1 - e^{-x/x_0}\right]^N \quad (x \geq 0).$$

Guided from our previous experience, we can try to choose $s_n = x_0$ and $\mu_N = x_0 \ln N$, so that $x = x_0(z + \ln N)$ (with $z$ non necessarily positive) obtaining

$$\Phi_N(z) = \left[1 - \frac{e^{-z}}{N}\right]^N,$$

which in the limit $N \to \infty$ becomes

$$\Phi(z) = \exp\left(-e^{-z}\right) \tag{3.109}$$

The limiting probability density is then given by the (standard) *Gumbel distribution*

$$\boxed{\phi(z) = \exp\left[-(z + e^{-z})\right] \quad (z \in \mathbb{R})} \tag{3.110}$$

GAUSSIAN DENSITY. Consider now a standard gaussian,

$$g(x) = \frac{1}{2\pi} \exp\left(-\frac{x^2}{2}\right).$$

In Sect. 3.6.1 we have seen that its cumulative distribution has, for large $x$, the asymptotic form (3.66), so we can write

$$\Phi_N(x) \simeq \left[1 - \frac{\exp(-x^2/2)}{x\sqrt{2\pi}}\right]^N = \exp\left[N \ln\left(1 - \frac{e^{-x^2/2}}{x\sqrt{2\pi}}\right)\right].$$

For large $x$ the second term in the logarithm is very small, so, approximating $\ln(1 - \epsilon) \simeq -\epsilon$, we have

$$\Phi_N(x) \simeq \exp\left[-N\frac{e^{-x^2/2}}{x\sqrt{2\pi}}\right]$$

Substituting $x = s_N z + \mu_N$ we obtain

$$\Phi_N(z) \simeq \exp\left[-N\frac{e^{-(s_N^2 z^2 + 2s_N \mu_N z + \mu_N^2)/2}}{\sqrt{2\pi}(s_N z + \mu_N)}\right],$$

which looks like a rather complicated expression. However, we can anticipate (and we will shortly prove) that, for large $N$, $s_N \ll \mu_N$, so that we can drop the quadratic term in $s_N$ and approximate $s_N z + \mu_N \simeq \mu_N$, obtaining

$$\Phi_N(z) \underset{N \gg 1}{\rightarrow} \exp\left[ -\frac{N e^{-\mu_N^2/2}}{\mu_N \sqrt{2\pi}} \exp(-s_N \mu_N z) \right],$$

Therefore, if we chose $\mu_N$ and $s_N$ such that[66]

$$\frac{N e^{-\mu_N^2/2}}{\mu_N \sqrt{2\pi}} = 1, \text{ and } s_N \mu_N = 1$$

we obtain *again* the Gumbel cumulative distribution

$$\Phi(z) = \exp\left(-e^{-z}\right).$$

These two examples suggest that the Gumbel distribution is the limiting distribution for the extremes for all those probability densities that decay exponentially. Actually, a sufficient condition for the Gumbel distribution to be the extreme distribution for a p.d.f. $f(x)$ is that[67]

$$\lim_{x \to \infty} \frac{d}{dx}\left[ \frac{1 - F(x)}{f(x)} \right] = 0,$$

which is satisfied by a large class of rapidly decaying densities.

It is then worth giving a look to this important distribution. By substituting $t = \exp(-z)$ in Eq. (3.110), is immediate to show that the Gumbel p.d.f. is correctly, normalized,

$$\int_{-\infty}^{+\infty} e^{-z} \exp\left(-e^{-z}\right) dz = \int_0^\infty e^{-t} dt = 1.$$

With same change of variable, we have

$$\langle z \rangle = \int_{-\infty}^{+\infty} z e^{-z} \exp\left(-e^{-z}\right) dz = -\int_0^\infty \ln(t)\, e^{-t} dt = \gamma. \tag{3.111}$$

---

[66] Note that, at leading order in $N$, $\mu_N \approx \sqrt{2 \ln N}$. A posteriori, this confirms that $s_N \approx 1/\sqrt{2 \ln N} \ll \mu_N$.

[67] See, for instance, A. Hansen, Front. Phys. **8**, 604053 (2020).

where $\gamma = 0.57721\ldots$ is the Euler-Mascheroni constant.[68] With some effort, one can also find that the variance of the Gumbel is given by

$$\sigma_z^2 = \int_{-\infty}^{+\infty} (z - \gamma)^2 e^{-z} \exp\left(-e^{-z}\right) dz = \frac{\pi^2}{6} \tag{3.112}$$

It is also easy to check (see footnote 62) that, when the r.v. $z$ has a Gumbel distribution for the maximum, the variable $-z$ has a Gumbel distribution for the minimum.

The Gumbel distribution can be generalized by introducing a location factor $\mu$ and a scale factor $s$ and substituting $z = (x - \mu)/s$,

$$\boxed{f(x; \mu, s) = \frac{1}{s} e^{-\left(\frac{x-\mu}{s} + e^{-(x-\mu)/s}\right)} \;;\; F(x; \mu, s) = e^{-e^{-(x-\mu)/s}}} \tag{3.113}$$

The mode of the distribution is always $x = \mu$, while, from (3.111), (3.112), the expectation and standard deviation are

$$\boxed{\langle x \rangle = \gamma s + \mu \;;\; \sigma_x = \frac{\pi}{\sqrt{6}} s} \tag{3.114}$$

The Gumbel p.d.f. is plotted in Fig. 3.19 for $\mu = 0$ and three values of $s$. While on a linear scale, the distribution for $s = 1$ looks like a slightly skewed Gaussian, the logarithmic plot in the inset shows that the tails of two distributions are actually very different.

The following example presents an application of the EVA to the investigation of spontaneous restructuring effects taking place in soft solids.

**Example 3.26** (*Quakes in gels*) Relaxation of internal stresses through a cascade of microscopic restructuring events is the hallmark of many materials, ranging from amorphous solids like glasses and gels to geological structures subjected to a persistent external load. For instance the time evolution of the strain in materials subjected to a constant load can be "jagged", i.e., these systems do not deform smoothly but rather through a series of abrupt rearrangements, usually dubbed as "plastic avalanches" or "quakes".

Spontaneous restructuring effects of this kind are shown by colloidal gels generated by nanoparticle aggregation. These tenuous networks are suitable to be investigated by optical spectroscopy techniques, which, in essence, are based on monitoring the degree of correlation $c(\tau)$ between the intensity of the light scattered by the gel at two different times, $t$ and $t + \tau$.

---

[68] You can also check that $\gamma = \Gamma'(1)$, where $\Gamma'(1)$ is the derivative of the gamma function.

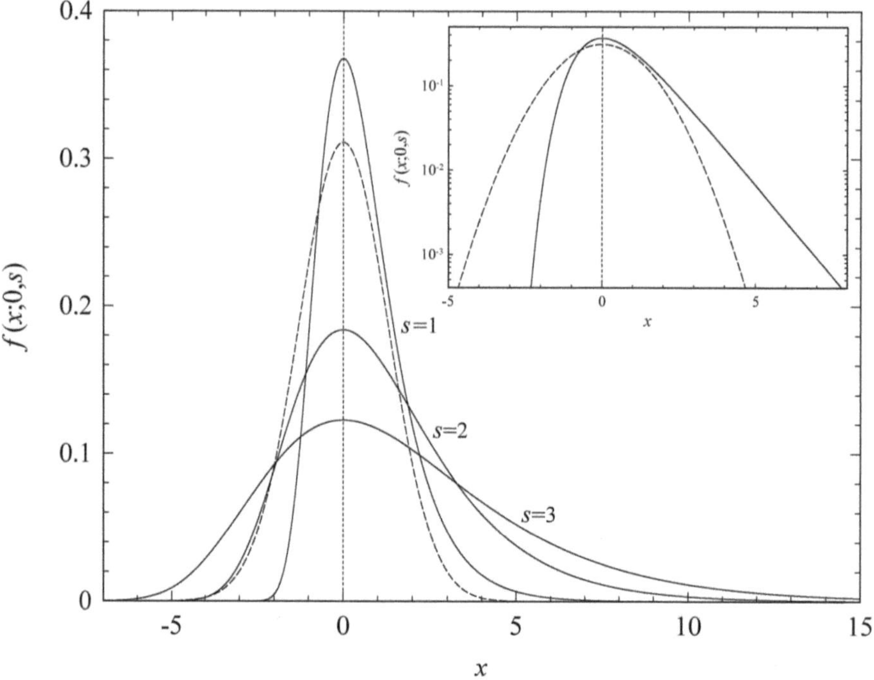

**Fig. 3.19** Gumbel distribution for three values of $s$. The p.d.f. with $s = 1$ is compared in the body and in the inset with a Gaussian centered at the origin with the same standard deviation $\sigma_x$

In a series of experiments recently performed by my group we have followed for about two weeks the time evolution of a colloidal gel obtained by a temperature-induced aggregation of spherical nanoparticles with a size of 50 nm. After an initial settling stage of a few hours, during which the newly formed gel compresses under its own weight, the network undergoes an extremely long restructuring stage consisting in a sequence of thousands of structural "micro-quakes" evidenced by a sudden decreases of $c(\tau)$, which is shown in the inset of Fig. 3.20. From the magnitude of these "decorrelation spikes", one can estimate how large is the average spatial rearrangement $\delta$ of the gel network resulting from a restructuring effect, which is typically in the tens of nanometers range.

If the delay time $\tau$ used to gauge the degree of correlation is much shorter than the average time between two micro-quakes, the latter can be regarded as rare events, whose largest magnitude $\delta_m$ should be distributed according to one of the extreme value probability densities we have discussed. Indeed, the body of Fig. 3.19 shows that the distribution of $\delta_m$, obtained by splitting the sequence of values for $\delta$ in blocks

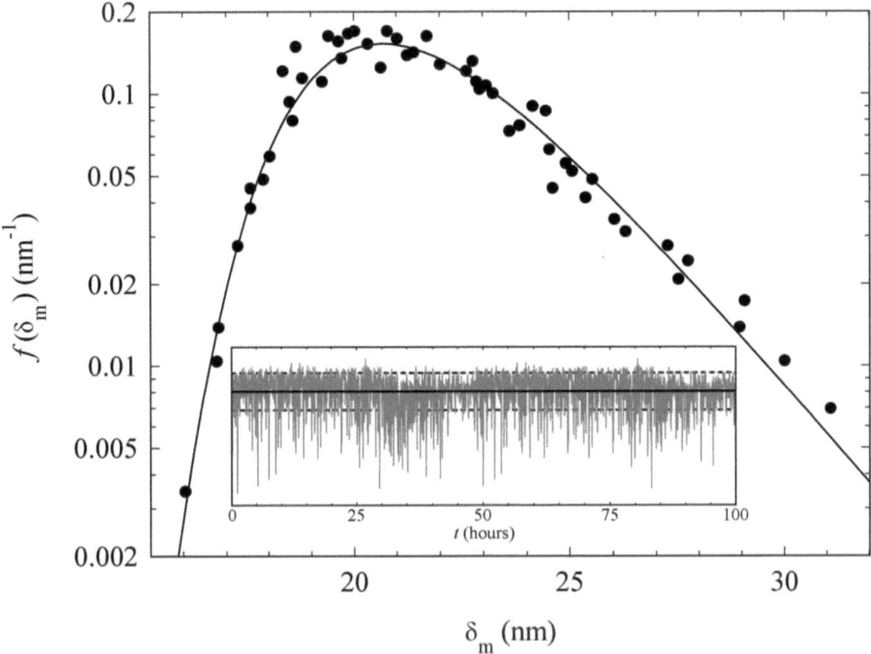



**Fig. 3.20** Inset: Time evolution over 100 h of the degree of correlation $c(\tau)$, with $\tau = 60$ s. The broken lines indicate an interval of plus/minus one standard deviation around $\overline{c(\tau)}$, shown by the full line. Sudden large-magnitude decorrelation events evidence the occurrence of "micro-quakes" in the gel. Body: Distribution of the maximal structural rearrangements $\delta_m$, fitted with a Gumbel distribution with parameters $\langle \delta_m \rangle \simeq 20$ nm, $\sigma(\delta_m) = 3$ nm

corresponding to an observation time of 100 min, is nicely fitted by a Gumbel distribution. In fact, a more detailed analysis shows that the distribution of the scattered intensity fluctuations is highly non-Gaussian, with exponential tails that suggest a Gumbel distribution of the extremes.[69]

---

[69] Z. Filiberti, R. Piazza, S. Buzzaccaro, Phys. Rev. E **100**, 042607 (2019).

# Chapter 4
# Probability: Accessories for Use

*When the going gets tough*
*the tough gets going*

*J. Belushi*

This chapter is surely more technical than the previous ones, but the concepts we will develop are crucial for grasping the power of probability methods. So, try to follow me, although this may require some effort. In a nutshell, the most significant questions we will address in this chapter are:

1. What is the probability distribution for a variable $Y$ obtained *as a function* $y = \phi(x)$ of another random variable $X$?
2. Given *two* random variables $X$ ed $Y$, can we find the probability

$$P(x < X < x + \mathrm{d}x \, ; \, y < Y < y + \mathrm{d}y)$$

   that the variable $X$ takes a value between $x$ and $x + \mathrm{d}x$ and, concurrently, the variable $Y$ a value between $y_0$ e $y + \mathrm{d}y$? In other words, can we define a *joint* probability distribution for two or more RVs?
3. What is the probability distribution for a *function* of two random variables $X$ and $Y$, and in particular of a linear combination of them? Moreover, can we say something special about the sum of *many* independent random variables?
4. Can we somehow quantify the "amount of information" that a probability distribution conveys?

The third question, in particular, will introduce us to concepts like characteristic functions and cumulants providing a thorough and exhaustive description of a probability

© The Author(s), under exclusive license to Springer Nature Switzerland AG 2025  181
R. Piazza, *An Invitation to Probability and Data Analysis for Physicists*, UNITEXT for Physics, https://doi.org/10.1007/978-3-031-83856-9_4

distribution, and will lead us to discover the most important result of probability theory, the Central Limit Theorem. The last question will finally bring us to the frontier between probability and statistical physics.

## 4.1  Functions of a Random Variable

We have already dealt with the expectation of a function of a random variable. Now we tackle a more general problem. Namely, given a random variable $Y$, which is a continuous and differentiable function $y = \phi(x)$ of another random variable $X$ with probability density $f_x(x)$, can we find a probability density $f_y(y)$ for $Y$?[1] The simplest case is that of a monotonic function, like the strictly increasing one shown in panel A of Fig. 4.1. From the figure, you can notice that $Y$ is in a neighborhood $dy$ of $y$ *if and only* if $X$ is in a neighborhood $dx$ of the inverse image $x = \phi^{-1}(y)$, which is *unique* because $f$ is monotonic. Then, we must have $P(y; dy) = P(x; dx)$, or, introducing probability densities,

$$f_y(y)\,|dy| = f_x(x)\,|dx|$$

Note that, to ensure that both sides of this equation are positive (they are probabilities!), we must use the *absolute values* of $dx$ and $dy$, because we are actually

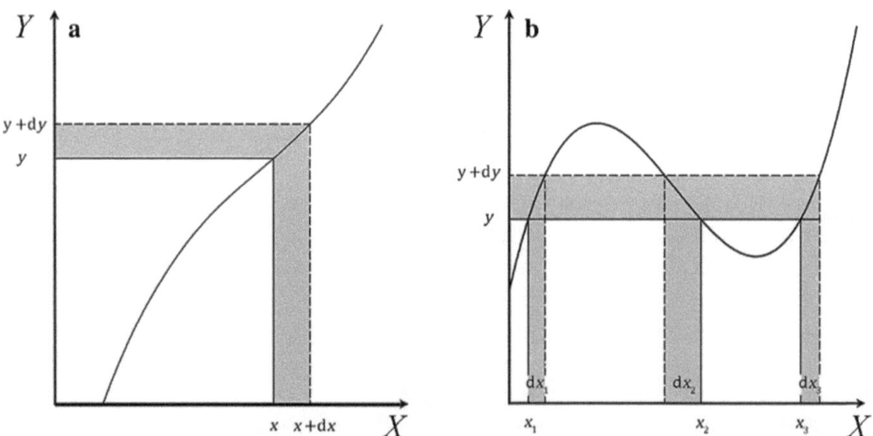


**Fig. 4.1**  Graphical construction of $f_y(y)$ for monotonic function (panel A) and for a generic function $y(x)$ (panel B) of a RV $x$

---

[1] The two distributions are labeled with different subscripts because they are in general *different functions* (the values in brackets indicate instead the *arguments* of these functions).

referring to the *length* of these intervals.[2] The former relation can be written using the inverse derivative

$$\boxed{f_y(y) = \left|\frac{dx(y)}{dy}\right| f_x[x(y)]}$$

(4.1)

where, at the right–hand side, it is understood that $x$ is expressed as $\phi^{-1}(y)$. Let us make a couple of simple examples.

$\boxed{y = ax + b}$   The function is monotonic with inverse $x = (y - b)/a$, hence

$$\left|\frac{dx}{dy}\right| = \frac{1}{a}.$$

Therefore

$$f_y(y) = \frac{1}{a} f_x\left(\frac{y - b}{a}\right).$$

(4.2)

For example, if $f_x(x)$ is a Gaussian with unit variance centered on the origin,

$$f_x(x) = \frac{1}{\sqrt{2\pi}} \exp\left(-\frac{x^2}{2}\right),$$

we obtain

$$f_y(y) = \frac{1}{\sqrt{2\pi}} \exp\left[-\frac{(y - b)^2}{2a^2}\right],$$

which is still a Gaussian with $\langle y \rangle = b$ and $\sigma_y = a$.

$\boxed{y = 1/x}$   The function is again monotonic with inverse $x = 1/y$. Thus

$$f_y(y) = \frac{1}{y^2} f_x\left(\frac{1}{y}\right).$$

(4.3)

Therefore, if

(a)   $f_x(x)$ is a *uniform* distribution:

$$f_x(x) = \frac{1}{|b - a|} \implies f_y(y) = \frac{|b - a|}{y^2},$$

which is *not* uniform, ma rather displays, between $y = 1/b$ and $y = 1/a$, a power-law decay with exponent $-2$.

(b)   $f_x(x)$ is *Gaussian* distribution:

---

[2] In fact, when $f$ is a monotonically *decreasing* function, the differentials $dy$ and $dx$ have instead opposite sign.

$$p_x(x) = \frac{1}{\sqrt{2\pi}} \exp\left(-\frac{x^2}{2}\right) \implies p_y(y) = \frac{1}{y^2\sqrt{2\pi}} \exp\left(-\frac{1}{2y^2}\right),$$

which is *not* a normal distribution.

(c)    $f_x(x)$ is a *Cauchy* distribution:

$$f_x(x) = \frac{\alpha}{\pi(x^2 + \alpha^2)} \implies f_y(y) = \frac{1/\alpha}{\pi(y^2 + 1/\alpha^2)},$$

which is *still* a Cauchy distribution of "width" $1/\alpha$.

(d)    $f_x(x)$ is a *Fréchet* distribution (3.108) with $\mu = 0$:

$$f(x) = \frac{\beta}{s}\left(\frac{x}{s}\right)^{-(1+\beta)} e^{-\left(\frac{x}{s}\right)^{-\beta}} \implies f_y(y) = \beta s (sy)^{\beta-1} e^{-y^\beta},$$

which is a Weibull distribution (3.84) with the same $\beta$ and $\tau = 1/s$. This is why the Fréchet distribution is also known as the *inverse Weibull* distribution.

The "golden rule" (4.1) for changing the variables of a probability density can be easily generalized to a generic function $\phi(x)$ where there are several points $x_i$ that are mapped onto the same value $y$ of $Y$. For example, in the graph shown in panel B of Fig. 4.1, $y$ is at the same time the image of $x_1$, $x_2$, and $x_3$. Hence, we must have

$$P(y; dy) = P(x_1; dx_1) + P(x_2; dx_2) + P(x_3; dx_3),$$

In general, therefore, we must find all the counter–images $x_i = \phi^{-1}(y)$ of $y$ and evaluate

$$\boxed{f_y(y) = \sum_i \left|\frac{dx}{dy}\right|_{x_i} f_x(x_i)} \tag{4.4}$$

which means splitting the domain of $\phi(x)$ in intervals where the function is monotonic, applying Eq. (4.1) to each one of these sub-domains, and finally summing the contributions from all the $x_i = \phi^{-1}(y)$.

An important case is that of a quadratic relation $y = x^2$. If $y < 0$, there is *no x* such that $y = x^2$, hence $f_y(y)$ must vanish. For $y > 0$, we have instead *two* values, $x_{1,2} = \pm\sqrt{y}$, which satisfy $y = x^2$. In both cases $|dx/dy| = (2\sqrt{y})^{-1}$, hence

$$f_y(y) = \begin{cases} 0 & (y < 0) \\ \dfrac{1}{2\sqrt{y}}\left[f_x(\sqrt{y}) + f_x(-\sqrt{y})\right] & (y \geq 0). \end{cases} \tag{4.5}$$

Let us apply this result to a Gaussian distribution centered on the origin and with unit variance, $f_x(x) = (2\pi)^{-1/2}\exp(-x^2/2)$. We easily obtain

$$f_y(y) = \frac{1}{\sqrt{2\pi y}} \exp\left(-\frac{y}{2}\right).$$

Hence, for large $y$, the probability distribution for the *square* of a Gaussian variable decreases *exponentially*.[3] This result is very useful if we wish to compare the probability distribution for the *intensity* of a random quantity with the probability distribution for its *amplitude*.

The following examples are particularly relevant for physics.

**Example 4.1** (*Harmonic oscillator*) A point moves in uniform circular motion along a circle of unit radius centered on the origin of a system of axes. At random instants, we record the $x$-coordinate of the point, that is, its projection on the $x$-axis. What is the probability distribution for $x$?

Since the point moves uniformly, the angle $\vartheta$ that the position vector forms with the $x$-axis will be a random variable distributed uniformly between 0 and $2\pi$, namely, $f(\vartheta) = 1/2\pi$. We then have $x = \cos(\vartheta)$ and therefore $\vartheta = \arccos(x)$. Thus,

$$\left|\frac{\mathrm{d}\vartheta}{\mathrm{d}x}\right| = \frac{1}{\sqrt{1 - x^2}}.$$

We must again be careful, because in the interval we consider there are two inverse images that, as for a quadratic relation, give an equal contribution to the probability density for $x$. Therefore, we have

$$f_x(x) = \frac{1}{\pi\sqrt{1 - x^2}},$$

which is a rather strange probability distribution, since it has the maximum value (in fact, it diverges) at the ends of the range of $x$.

If we remember that the projection of a point in uniform circular motion moves harmonically, we can observe that this is also the probability distribution for the position of a harmonic oscillator that oscillates with unit amplitude and is observed at random instants. Physically, having found that the probability density is maximum at the ends of the oscillation simply corresponds to the fact that the oscillator spends most of the time near these points, where its speed is minimum.

**Example 4.2** (*Cauchy's illumination*) A light bulb, placed at a distance $d$ from a vertical wall, can be considered in a first approximation as an isotropic source that emits light uniformly. Then, let us consider a horizontal plane where the $x$-axis is perpendicular from the light bulb to the wall and the origin is where the perpendicular begins. We want to find the intensity distribution of the light along the $Y$-axis (which is therefore a horizontal axis on the wall plane).

---

[3] Actually, $p_y(y)$ diverges at the origin, but this is not a problem, because what matters is the *integral* of the probability density, which is convergent.

For convenience, it is useful to think of the light emitted by the light bulb as a stream of photons, so that the light intensity in a certain position $y$ will simply be proportional to the number of photons that fall per unit of time and surface in a neighbour of $y$. The angle $\vartheta$ between the $x$-axis and the direction in which is a emitted a photon that propagates in the $xy$ plane and *hits* the wall will therefore have a uniform distribution in the range $\vartheta \in [-\pi/2, \pi/2]$, hence $f_\vartheta(\vartheta) = 1/\pi$. Moreover, the point where the photon reaches the wall is $y = d\tan(\vartheta)$, namely, $\vartheta = \arctan(y/d)$. The intensity distribution along $y$ will therefore be proportional to the probability density of the arrival points, given by:

$$p_y(y) = \frac{d}{\pi(d^2 + y^2)},$$

which is a Cauchy distribution with parameter $\alpha = d$.

## 4.2  Bivariate Distributions

In this section, we will extend several of the ideas and results obtained in the previous chapter to a multivariate distribution. i.e., a probability distribution involving more than one variable. We will mostly focus on bivariate distributions, although several of the concepts we introduce apply can be easily extended to distributions of more than two variables.

### 4.2.1  Joint and Marginal Distributions

Let us first consider two *discrete* random variables $X$ and $Y$, where $X$ takes only $n$ values $x_i$ and $Y$ only $m$ values $y_j$. As we have done for a single RV, we can then call *joint probability distribution* of $X$ and $Y$ the function $P(x, y)$ that associates to each pair of values $\{x_i, y_j\}$ the probability $P(x_i, y_j)$ that, simultaneously, $X$ is equal to $x_i$ and $Y$ to $y_j$. The joint distribution can then be specified by a $n \times m$ matrix $P(x_i, y_j)$ with the following features:

(a) Summing all the $n \times m$ values $P(x_i, y_j)$ we get the probability that $X$ and $Y$ take any possible value, thus we have again a normalization condition

$$\sum_{i=1}^{n} \sum_{j=1}^{m} P(x_i, y_j) = 1. \tag{4.6}$$

(b) If we just sum $P(x_i, y_j)$ over all values $y_j$ for a *fixed* value $X = x_i$, we instead get the probability $P_x(x_i)$ that $X$ is equal to $x_i$ regardless of the value of $y$, which corresponds to what we simply called $P_x(x_i)$ in the previous chapter,

$$P_x(x_i) = \sum_{j=1}^{m} P(x_i, y_j).$$ (4.7)

When dealing with probability distributions of more than one variable, $P_x$ is usually called *marginal distribution* for $X$. Of course, we can instead sum $P(x_i, y_j)$ over all values of $X$, obtaining in this case the marginal probability distribution $P(y)$ for $Y$.

**Example 4.3** (*A tricolor die*) Imagine we roll a die that has faces 1 and 3 colored in red (R), face 4 in blue (B), and the remaining ones in green (G). If we take as variables $X$ and $Y$ the value and color of a face, the table for $P(x, y)$ is then the following

|   | 1   | 2   | 3   | 4   | 5   | 6   |
|---|-----|-----|-----|-----|-----|-----|
| R | 1/6 | 0   | 1/6 | 0   | 0   | 0   |
| B | 0   | 0   | 0   | 1/6 | 0   | 0   |
| G | 0   | 1/6 | 0   | 0   | 1/6 | 1/6 |

### 4.2.2  Independent Variables

Therefore, extending the concept of probability distribution to more than one variable is easy, but we to have to be sure that two values $x_i$ and $y_j$ can really be obtained "simultaneously".[4] We have seen in Chap. 2 that the probability that two events $A$ e $B$ jointly take place is equal to the product $P(A)P(B)$ only if $A$ and $B$ are *independent* events. If now we identify $A$ with "$X$ is equal to $x_i$", and $B$ with "$Y$ is equal to $y_j$", it is generally *not* true that $P(x_i, y_j) = P_x(x_i)P_y(y_j)$. In the previous example, for instance, the probability to roll a 2 regardless of the color of the face is 1/6, while the probability that a face is red is 1/3. But the probability to roll at the same time a score of 2 and a red face is zero, because the face with a 2 is green. Another example may shed some more light on the question.

**Example 4.4** (*A three-state system*) According to the classical Maxwell-Boltzmann statistics, if $n$ distinguishable particles are distributed over three states with the same energy the probability that $n_1$ are in the first state and $n_2$ in the second one is

$$P(n_1, n_2) = \frac{1}{3^n} \frac{n!}{n_1! n_2! (n - n_1 - n_2)!}.$$

We can regard $n_1$ and $n_2$ as the values of two random variables, the occupation numbers $N_1$ ed $N_2$ of the first two states, of which $P(N_1, N_2)$ is the joint probability

---

[4] In this context, "simultaneously" does not necessarily mean "at the same time": it could also indicate "in the same place", "in the same experiment", or any other form of coincidence.

distribution. The last term at the denominator, however, does not allow us to write $P(N_1, N_2)$ as the product of two function of $N_1$ and $N_2$ alone, hence $N_1$ and $N_2$ are *not* independent.

To obtain the marginal probability distribution for $N_1$ we have to sum over all possible values of $N_2$ that, for fixed $N_1 = n_1$, are all values $n_2$ from 0 to $n - n_1$,

$$P_{N_1}(n_1) = \frac{1}{3^n}\binom{n}{n_1}\sum_{n_2=0}^{n-n_1}\frac{(n-n_1)!}{n_2!\,(n-n_1-n_2)!},$$

where we have multiplied and divided by $(n - n_1)!$. Because of the binomial theorem, this summation is simply equal to $2^{n-n_2}$. Therefore

$$P_{N_1}(n_1) = \frac{1}{3^n}\binom{n}{n_1}2^{n-n_2} = \binom{n}{n_1}\left(\frac{1}{3}\right)^{n_1}\left(\frac{2}{3}\right)^{n-n_1},$$

which, as we could expect,is a binomial distribution corresponding to get $n_1$ "successes" (a particle in the 1$^{\text{st}}$ state) in $n$ "trials". Of course, the same is for $N_2$. Thus, once again $P_{N_1}(n_1)P_{N_2}(n_2) \neq P(n_1, n_2)$.

The observation we made suggests to state that two random variables are independent if and only if for *all* pairs $(x_i, y_j)$ we have $P(x_i, y_j) = P_x(x_i)P_y(y_j)$, i.e., *when their joint probability distribution factorizes in the product of the marginal probability distributions,*

$$\boxed{P(x, y) = P_x(x)P_y(y)} \tag{4.8}$$

In order to ensure that two variables $X$ and $Y$ are independent, we must verify that their joint distribution can be written as the sum of two functions with only $X$ and only $Y$, respectively. As a matter of fact, however, while it is often easy to guess the marginal probability distributions for $X$ and $Y$, it is much harder to say something about their joint distribution. To put it differently, the most challenging issue is determining whether or not two RVs are independent.

### 4.2.3  Conditional Distributions

We have introduced the marginal distribution for $X$ as the probability distribution that is obtained by summing over all values of $Y$. In contrast, what is the probability distribution of $X$ for a *certain* value of $Y$, that is, when the value $y_j$ of $Y$ is pre-assigned? What we are looking for is nothing but the function giving the conditional probability $P(x_i|Y = y_j)$ to get $x_i$ once $Y = y_j$. If we fix the value of $Y$ in the joint probability distribution we do find a function of $X$ alone, $P(x, y_j)$, which however depends on the probability of obtaining the value $Y = y_j$. To get rid of this feature, we define the probability distribution of $X$ *conditional* on $Y = y_j$ as

$$P(x \mid Y = y_j) = \frac{P(x, y_j)}{P_y(y_j)} \tag{4.9}$$

It is clear that an equivalent definition can be made for the distribution of $Y$ conditional on $X = x_i$. You can easily check that, when $X$ ed $Y$ are independent RVs, we simply have $P(x|y_j) = P_x(x)$ and $P(y|x_i) = P_y(y)$. However, in general this is not true, i.e., the conditional distribution obtained by fixing a value for $Y$ is *different* from the marginal distribution of $X$.

### 4.2.4  Expectation Values and Correlation Coefficients

The concept of expectation value can be simply extended to a function $\phi(x, y)$ of two random variables $X$ and $Y$ by defining

$$\langle \phi(x, y) \rangle = \sum_{i=1}^{n} \sum_{j=1}^{m} \phi(x_i, y_j) P(x_i, y_j). \tag{4.10}$$

The simplest situation is when the function matches one of the two variables, that is, $\phi(x, y) = x$ or $\phi(x, y) = y$. Taking into account Eq. (4.7), or the equivalent definition for $Y$, we obtain

$$\langle x \rangle = \sum_{i=1}^{n} x_i \sum_{j=1}^{m} P(x_i, y_j) = \sum_{i=1}^{n} x_i P_x(x_i) = \langle x \rangle_x$$

$$\langle y \rangle = \sum_{j=1}^{m} y_j \sum_{i=1}^{n} P(x_i, y_j) = \sum_{j=1}^{m} y_j P_y(y_j) = \langle y \rangle_y \, ,$$

that is, the expectation values of $X$ ed $Y$ coincide with the values $\langle x \rangle_x$, $\langle y \rangle_y$ found by using the marginal distributions.

If we now consider the sum of two variables, $\phi(x, y) = x + y$, we formally recover a result that we already used in the previous chapter,

$$\langle x + y \rangle = \sum_{i=1}^{n} \sum_{j=1}^{m} (x_i + y_j) P(x_i, y_j) = \sum_{i=1}^{n} x_i P_x(x_i) + \sum_{j=1}^{m} y_j P_y(y_j) = \langle x \rangle + \langle y \rangle ,$$

$$\tag{4.11}$$

namely, *the expectation of the sum of two RVs is the sum of the expectations* (which means that $\langle \cdot \rangle$ is a *linear* operator). Note that, in general, for the *product* of $X$ and $y$ we have instead

$$\langle xy \rangle = \sum_{i=1}^{n} \sum_{j=1}^{m} x_i y_j P(x_i, y_j) \neq \langle x \rangle \langle y \rangle . \tag{4.12}$$

In Chap. 1 we have seen that, when the experimental mean of the product of two fluctuating quantities differs from the product of their means, two variables can be said to be correlated. We can extend this observation for a statistical sample to the population wherefrom the sample is extracted by stating that *two random variables X and Y are uncorrelated if* $\langle xy \rangle = \langle x \rangle \langle y \rangle$.

Moreover, mirroring what we did for experimental correlations, we can define a *correlation coefficient* between $X$ and $Y$,

$$\rho_{xy} = \frac{\langle xy \rangle - \langle x \rangle \langle y \rangle}{\sigma_x \sigma_y}. \tag{4.13}$$

The quantity $\sigma_{xy} = \langle xy \rangle - \langle x \rangle \langle y \rangle$, which is the theoretical analogous of the experimental mutual standard deviation, is also called *covariance* of $X$ ed $Y$. Note that, like for the variance of a single variable, we can also write

$$\sigma_{xy} = \langle (x - \langle x \rangle)(y - \langle y \rangle) \rangle . \tag{4.14}$$

The correlation coefficient of two variables $X$ and $Y$ always satisfies $|\rho_{xy}| \leq 1$. Indeed we have,

$$\langle (x + ay)^2 \rangle = \langle x^2 \rangle + 2a \langle xy \rangle + a^2 \langle y^2 \rangle \geq 0.$$

Since this is true for any for any $\alpha \in \mathbb{R}$, it holds also for the value that minimizes the term at the l.h.s., which is $a = - \langle xy \rangle / \langle y^2 \rangle$, so easily obtain

$$\langle xy \rangle^2 \leq \langle x^2 \rangle \langle y^2 \rangle \tag{4.15}$$

Replacing $X$ with $X - \langle y \rangle$ and $Y$ with $Y - \langle y \rangle$ we have

$$\boxed{|\rho_{xy}| = \frac{\sigma_{xy}}{\sigma_x \sigma_y} \leq 1} \tag{4.16}$$

Let me emphasize two crucial points:

- It is easy to see that *two independent variables are also necessarily uncorrelated,*

$$\langle xy \rangle = \sum_{i=1}^{n} \sum_{j=1}^{m} x_i y_j P(x_i, y_j) = \sum_{i=1}^{n} x_i P_x(x_i) \sum_{j=1}^{m} y_j P_y(y_j) = \langle x \rangle \langle y \rangle .$$

  The opposite, however, is not necessarily true, that is, the condition of independence is *stronger* than the lack of correlation. Indeed, the latter just implies that $\langle xy \rangle$ factorizes, while independence requires that the *whole* joint distribution factorizes into the marginal distributions.
- It is important to keep in mind that $\rho_{xy}$ is a *linear* correlation coefficient. In fact, two RVs can be uncorrelated even if they are deterministically connected by a

nonlinear function. Consider for example a quadratic relation $Y = X^2$, where $X$ can only take the two values $x = \pm 1$, both with probability $p_x = 1/2$. Then, $y = 1$ with probability $p_y = 1$, so that $\langle x \rangle$, $\langle xy \rangle$ and $\rho_{xy}$ vanish.

Summing up, independence implies lack of correlation, but lack of correlation does not imply independence.

Finally, using the definition of conditional distribution (4.9) we can introduce the concept of *conditional expectation* as the expected value of a variable $X$ evaluated with respect to $P(x \mid Y = y_j)$,

$$\langle x \mid Y = y_j \rangle = \sum_{i=1}^{n} x_i P(x \mid Y = y_j) = \sum_{i=1}^{n} \frac{P(x, y_j)}{P_y(y_j)}. \tag{4.17}$$

Of course, when $X$ is independent of $Y$, $\langle x \mid Y = y_j \rangle = \langle x \rangle_x$.

**Example 4.5** (*Random walk in more than one dimension*)  Assuming that the displacement in one direction is independent from that in another, the simple result found in Example 3.8 is readily extended to a random walk in two or more dimensions. For example, suppose that in a planar RW a step is chosen to be made along $x$ or $y$ by flipping a coin, so that, on a total of $N$ steps, half is typically made along $x$ and the other along $y$. Then, if the probability of making a step towards the positive or negative direction on each axis is $p = 1/2$, after $N$ steps we will have $\langle x^2 \rangle = \langle y^2 \rangle = (N/2)L^2$. Since $X$ and $Y$ are independent variables the r.m.s. displacement will still be

$$\langle r^2 \rangle^{1/2} = \left( \langle x^2 \rangle + \langle y^2 \rangle \right)^{1/2} = L\sqrt{N}.$$

### 4.2.5  Continuous Variables

We can extend the previous definitions and considerations to a pair of continuous RVs $X$ and $Y$ by writing the joint probability that $X$ is in a neighbor $dx$ of $x$ and $Y$ in a neighbor $dy$ of $y$ as

$$P(x < X < x + dx, y < Y < y + dy) = f(x, y)dxdy \tag{4.18}$$

which defines the *joint probability density* $f(x, y)$ for $X$ and $Y$ that must satisfy the normalization condition

$$\int_{-\infty}^{+\infty} \int_{-\infty}^{+\infty} f(x, y)dxdy = 1$$

By integrating $f(x, y)$ over all values for one of the two variables, we can also define the *marginal probability densities* for $X$ and $Y$

$$f_x(x) = \int_{-\infty}^{+\infty} f(x, y)\mathrm{d}y \tag{4.19a}$$

$$f_y(y) = \int_{-\infty}^{+\infty} f(x, y)\mathrm{d}x, \tag{4.19b}$$

and we can regard two continuous variables as independent when their joint probability density factorizes into the marginal densities,

$$f(x, y) = f_x(x)f_y(y). \tag{4.20}$$

Drawing a parallel with (4.10), the expectation of a function of $X$ and $Y$ will be given by an integral over both variables,

$$\langle \phi(x, y) \rangle = \int_{-\infty}^{+\infty} \int_{-\infty}^{+\infty} \phi(x, y)f(x, y)\mathrm{d}x\mathrm{d}y, \tag{4.21}$$

so that

$$\langle x \rangle = \int_{-\infty}^{+\infty} \mathrm{d}x\, x \int_{-\infty}^{+\infty} \mathrm{d}y\, f(x, y) = \int_{-\infty}^{+\infty} xf_x(x)\mathrm{d}x = \langle x \rangle_x$$

$$\langle y \rangle = \int_{-\infty}^{+\infty} \mathrm{d}y\, y \int_{-\infty}^{+\infty} \mathrm{d}x\, f(x, y) = \int_{-\infty}^{+\infty} yf_y(y)\mathrm{d}y = \langle y \rangle_y$$

coincide with the expectations evaluated on the marginal probability densities. We can finally define the conditional density for $X$ given that $Y = y$ as

$$f(x \mid Y = y) = \frac{f(x, y)}{f_y(y)} \tag{4.22}$$

and the corresponding conditional expectation for $X$ as

$$\langle x \mid Y = y) \rangle = \int_{-\infty}^{+\infty} xf(x \mid Y = y)\,\mathrm{d}x = \frac{1}{f_y(y)} \int_{-\infty}^{+\infty} xf(x, y)\,\mathrm{d}x \tag{4.23}$$

### 4.2.6   *Bivariate Normal Distribution*

We aim to introduce a distribution for two (or possibly more) variables that is akin to the normal distribution for a single RV. For two independent variables $\tilde{X}$ and $\tilde{Y}$, both distributed according to a standard Gaussian (3.63), we can simply put, using (4.20),

$$g_{ind}(\tilde{x}, \tilde{y}) = g_{\tilde{x}}(\tilde{x})g_{\tilde{y}}(\tilde{y}) = \frac{1}{2\pi} \exp\left[ -\frac{1}{2}(\tilde{x}^2 + \tilde{y}^2) \right]. \tag{4.24}$$

But what can we do if $\widetilde{X}$ and $\widetilde{Y}$ are *not* independent, so that thor joint distribution does not factorize? Tentatively, we may look for an extension of (4.24) where he argument of exponential is a generic quadratic form of the two variables,

$$g(\tilde{x}, \tilde{y}) = g_0 e^{-(a\tilde{x}^2 + b\tilde{x}\tilde{y} + c\tilde{y}^2)}.$$

However, $g(\tilde{x}, \tilde{y})$ has to satisfy some consistency requirements. Specifically, we ask the distribution to be normalized and that it reduces to $g_{ind}(\tilde{x}, \tilde{y})$ for independent variables. Besides, the marginal distributions for $\widetilde{X}$ and $\widetilde{Y}$ must still be standard Gaussians. With some effort, one finds that these requirements are satisfied if and only if the constants $g_0, a, b, c$ are related to the correlation coefficient $\rho(\tilde{x}, \tilde{y})$ by

$$a = c = \frac{1}{2(1 - \rho^2)}; \qquad b = -\frac{\rho}{1 - \rho^2}; \qquad g_0 = 2\pi\sqrt{1 - \rho^2}.$$

Then, a consistent joint normal distribution of $\widetilde{X}$ and $\widetilde{Y}$ is

$$g(\tilde{x}, \tilde{y}) = \frac{1}{2\pi\sqrt{1 - \rho^2}} \exp\left(-\frac{\tilde{x}^2 + \tilde{y}^2 - 2\rho\tilde{x}\tilde{y}}{2(1 - \rho^2)}\right) \tag{4.25}$$

which is therefore a *standard bivariate normal distribution*. Indeed, by adding and subtracting $\rho^2\tilde{x}^2$ to the exponent and putting $t = (\tilde{y} - \rho\tilde{x})/\sqrt{1 - \rho^2}$, the marginal distribution for $\widetilde{X}$ is found to be

$$g_{\tilde{x}}(\tilde{x}) = \frac{e^{-\tilde{x}^2/2}}{2\pi\sqrt{1 - \rho^2}} \int_{-\infty}^{\infty} \exp\left[-\frac{(\tilde{y} - \rho\tilde{x})^2}{2(1 - \rho^2)}\right] d\tilde{y} = \frac{e^{-\tilde{x}^2/2}}{2\pi},$$

and the marginal distribution for $\widetilde{Y}$ is similarly obtained. Moreover, if the two variables are uncorrelated ($\rho = 0$) we have

$$g(\tilde{x}, \tilde{y}) = \frac{1}{2\pi} e^{-(\tilde{x}^2 + \tilde{y}^2)/2} = \left[\frac{1}{\sqrt{2\pi}} e^{-\tilde{x}^2/2}\right]\left[\frac{1}{\sqrt{2\pi}} e^{-\tilde{y}^2/2}\right], \tag{4.26}$$

that is, the joint probability distribution factorizes into two standard Gaussians. Therefore, two uncorrelated variables with a joint normal distribution are also independent. Remember however that, in general, this is not the case.

To find the joint distribution of two Gaussian variables $X$ and $Y$ with generic expectation and variance, we can simply substitute in Eq. (4.25)

$$\tilde{x} = \frac{x - \langle x \rangle}{\sigma_x}; \quad \tilde{y} = \frac{y - \langle y \rangle}{\sigma_y},$$

so that, taking into account that $|\mathrm{d}\tilde{x}/\mathrm{d}x| = 1/\sigma_x$ and $|\mathrm{d}\bar{y}/\mathrm{d}y| = 1/\sigma_y$,

$$
g(x, y) = \frac{1}{2\pi\sigma_x\sigma_y\sqrt{1-\rho^2}}\, e^{-\frac{1}{2(1-\rho^2)}\left[\left(\frac{x-\langle x\rangle}{\sigma_x}\right)^2+\left(\frac{y-\langle y\rangle}{\sigma_y}\right)^2\right]}
\tag{4.27}
$$

## 4.3  Probability Distribution for a Function of Two RVs

We would like to extend the results of Sect. 4.1 to find the probability distribution of
a function $\phi(x, y)$ of two continuous RVs $X$ and $Y$ whose joint probability density
$f(x, y)$ is known. This task is not trivial, but it becomes easier when we consider an
apparently more complicated problem. Suppose we want to change variables from
$\{X, Y\}$ to $\{Z, T\}$, where $z = z(x, y)$ e $t = t(x, y)$ are known monotonic functions
of the original variables. Because they are monotonic, we can invert them to obtain
$X$ and $Y$ as a function of $Z$ and $T$:

$$
x = x(z, t) \,; \ \ y = x(z, t).
$$

Following the same approach we used for a function of a single variable, we can
state that the joint probability that $Z$ and $T$ fall within $\mathrm{d}z, \mathrm{d}t$ around those values $x$
ed $y$ such that $z = z(x, y)$ and $t = t(x, y)$ is

$$
f_{zt}(z_0, t_0)\mathrm{d}z\mathrm{d}t = f_{xy}(x_0, y_0)\mathrm{d}x\mathrm{d}y,
\tag{4.28}
$$

The differential $\mathrm{d}x$ and $\mathrm{d}y$ can be expressed as a function of $\mathrm{d}z$ and $\mathrm{d}t$ as $\mathrm{d}x\mathrm{d}y = |J|\mathrm{d}z\mathrm{d}t$, where $|J|$ is the Jacobian determinant

$$
|J| = \begin{vmatrix} \partial x/\partial z, & \partial x/\partial t \\ \partial y/\partial z, & \partial y/\partial t \end{vmatrix} = \frac{\partial x}{\partial z}\frac{\partial y}{\partial t} - \frac{\partial x}{\partial t}\frac{\partial y}{\partial z}.
\tag{4.29}
$$

For e joint distribution of $Z$ and $T$, Eq (4.1) is then generalized by

$$
f_{zt}(z_0, t_0) = |J|f_{xy}(x_0, y_0).
\tag{4.30}
$$

But what use do we have for this result? What we actually need is to obtain the
probability density for a *single* function of $X$ and $Y$: where do we find the second
variable? The answer is: it's up to us choosing it!

Let's try to understand what we have to do in one of the most interesting cases,
which is that of a quantity obtained as the sum of two others. We have already seen that
the probability distribution for the sum of two discrete uniformly distributed variables
is not uniform, but takes on a triangular shape. Now we want to ask, more generally,
how to calculate the probability distribution of a linear combination $Z = aX + bY$
when $f_x(x)$ and $f_y(y)$ are known. We can use the method we have just outlined by

taking $Z$ as one of the two new variables, while we are free to arbitrarily choose the second: we then simply assume $T \sim Y$. The inverse relations are therefore:

$$\begin{cases} X = (Z - bT)/a \\ Y = T. \end{cases}$$

so the Jacobian determinant is

$$|J| = \begin{vmatrix} 1 & -b \\ 0 & 1 \end{vmatrix} = 1$$

and therefore $f_{zt}(z, t) = f_{xy}(x, y) = f_{xy}[(z - bt)/a, t]$.

However, we are not interested in the joint probability distribution of $Z$ with the "fictitious" variable $T$, but in the distribution of $Z$ *alone* independently of the value of $T$, namely, in its marginal distribution $f_z(z)$, which is obtained as

$$\boxed{f_z(z) = \int_{-\infty}^{\infty} f_{xy}\left(\frac{z - bt}{a}, t\right) \mathrm{d}t}$$

In particular, for the sum $Z = X + Y$ of two *independent* variables, we have

$$f_{xy}(z - t, t) = f_x(z - t) f_y(t),$$

and therefore

$$\boxed{f_z(z) = \int_{-\infty}^{\infty} f_x(z - t) f_y(t) \mathrm{d}t} \tag{4.31}$$

Given two functions $f_1$ and $f_2$, the function $f_1 * f_2$ defined as

$$[f_1 * f_2](x) \doteq \int f_1(x - x') f_2(x') \mathrm{d}x' \tag{4.32}$$

is called the *convolution* of $f_1$ and $f_2$, an operation between functions that comes up in several mathematics and physical problems. Equation (4.31) therefore tells us that *the probability distribution of the sum of two independent variables is the convolution of the probability distributions of the two variables.*

It is also useful to consider the *difference* $Z = Y - X$ of two independent variables, which is given by

$$f_z(z) = \int_{-\infty}^{\infty} f_x(t) f_y(z + t) \mathrm{d}t = \int_{-\infty}^{\infty} f_x(t' - z) f_y(t') \mathrm{d}t \tag{4.33}$$

This is another important operation, known as the *cross-correlation* between two RVs, which the r.h.s. shows to be quite similar to a convolution, but with an inverted sign in the argument of $f_x$. In particular, when $X$ and $Y$ have the same probability density $f(t)$, the quantity

$$C(t) = \int_{-\infty}^{\infty} f(t')f(t+t')\mathrm{d}t' \tag{4.34}$$

is the *autocorrelation function* of $f(t)$, which we will extensively deal with in the next chapter.

**Example 4.6** (*Sum of two uniform continuous variables*) Let us find the probability density of $Z = X + Y$, where $X$ and $Y$ are two independent RVs uniformly distributed in $[0, a]$,

$$f_x(x) = f_y(y) = \begin{cases} 1/a & 0 \le x, y \le a \\ 0 & \text{otherwise.} \end{cases}$$

Equation (4.31) shows that, to obtain $f_z(z)$, we have to take the following steps:

1. Reflect $f_x(t)$ about the $y$-axis by changing $t \to -t$;
2. Shift $f_x(-t)$ by adding $z$ to its argument;
3. Multiply $f_x(z-t)$ by $f_y(t)$ and find the area under the product function.

This "graphical recipe" implies that $f_z(z)$ vanish for $z < 0$ (because we shift $f_x$ in the wrong direction, so $f_x$ does not overlap with $f_y$) and also for $z > 2a$ (because we shift too much). For $0 \le z \le a$ the product of the two functions is a rectangle of base $z$ and height $1/a^2$, while for $a < z \le 2a$ it is a rectangle of base $2a - z$ and height $1/a^2$ (see Fig. 4.2). So we obtain

$$f_z(z) = \begin{cases} z/a^2 & 0 \le z \le a \\ (2a - z)/a^2 & a < z \le 2a \\ 0 & \text{otherwise} \end{cases}$$

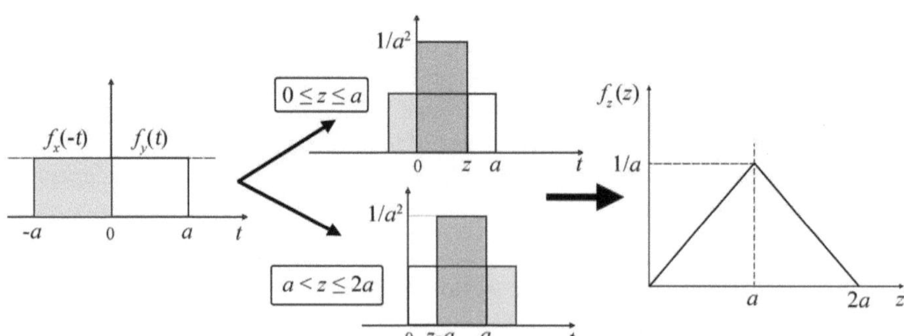


**Fig. 4.2**  Sum $Z = X + Y$ of two independent uniform variables

which has a triangular shape like the distribution of the sum of two discrete uniform variable in example 3.1.

**Example 4.7** (*Difference of two exponential probability densities*) Let us consider now two independent RVs with the same exponential density $f(x)$, given by Eq. (3.46), $f(x) = \lambda \exp(-\lambda x)$, where $\lambda = x_0^{-1}$. While $f(x)$ is defined only for $x \geq 0$, the probability density $f_z(z)$ for the difference $Y - X$ can be defined over the *whole* real axis. However, we have to be a bit careful, because ensuring that the argument of $f_y(z + t)$ in (4.33) is positive requires different integration limits for $z > 0$ and $z < 0$. Indeed, for:

$\boxed{z \geq 0}$ :   The argument of $f_y(z + t)$ is always positive, therefore

$$f_z(z) = \lambda^2 \int_0^{+\infty} e^{-\lambda(t)} e^{-\lambda(z+t)} dt = \lambda^2 e^{-\lambda z} \int_0^{+\infty} e^{-2\lambda(t)} dt = \frac{\lambda}{2} e^{-\lambda z}$$

$\boxed{z < 0}$ :   The argument of $f_y(z + t)$ is positive only when $t \geq -z$, so

$$f_z(z) = \lambda^2 \int_{-z}^{+\infty} e^{-\lambda(t)} e^{-\lambda(z+t)} dt = \lambda^2 e^{-\lambda z} \int_{-z}^{+\infty} e^{-2\lambda(t)} dt = \frac{\lambda}{2} e^{+\lambda z}$$

We can combine these two results obtaining, for all $z \in \mathbb{R}$, a *Laplace* distribution. Equation (3.49), with standard deviation $\sigma_x = \sqrt{2}/\lambda$,

$$\boxed{f_z(z) = \frac{\lambda}{2} e^{-\lambda|x|}} \tag{4.35}$$

## 4.4  Complex Random Variables and Distributions

So far, we have always being dealing with random variables defined on $\mathbb{R}$ or on a Borel subset of $\mathbb{R}$. In physics and engineering, however, it is quite often useful to introduce complex variables. So, for example, it is quite convenient to introduce a complex representation $u(t) = A \exp[-i(\omega t - \phi)]$ of an oscillating signal $u(t) = A \cos(\omega t - \phi)$, simply because it is much simpler dealing with exponentials than with trigonometric functions. A physical signal is of course a real quantity, so that a complex representation is just a useful description: at the end of a sequence of operations on the signal we must then consider only the *real* part of the result.[5]

---

[5] It is also important to stress that operating in the complex domain is generally useful when only *linear* transformations are involved. If you use a complex description to model a nonlinear optical effect, you'll really be in trouble!.

At first, the idea of a complex random variable $Z$ may seem senseless. Indeed, what could it mean finding the cumulative probability $P(Z \leq z)$? The complex field is not ordered, so stating that a complex number is less than another one is meaningless. Nevertheless, we can try to extend what we have done so far defining $Z$ as a pair of *real* random variables, its real part $X = \mathrm{Re}(Z)$ and its imaginary part $Y = \mathrm{Im}(Z)$, with a distribution given by the joint distribution of $X$ and $Y$. If we wish to consider it, at least formally, as a function of a complex variable, we have to be a bit careful. Indeed, while defining the expectation of $Z$ does not cause any problem,

$$\langle z \rangle = \langle x + \mathrm{i}y \rangle = \int_{-\infty}^{+\infty} \mathrm{d}x \int_{-\infty}^{+\infty} \mathrm{d}y (x + \mathrm{i}y) f(x, y) = \langle x \rangle + \mathrm{i} \langle y \rangle,$$

defining the *variance* of $Z$ as

$$\sigma_z^2 = \langle (z - \langle z \rangle)^2 \rangle = \langle (x - \langle x \rangle)^2 + \mathrm{i}(y - \langle y \rangle)^2 \rangle = \sigma_x^2 - \sigma_y^2 + 2\mathrm{i}\sigma_{xy}$$

does not make any sense, first because this is general a complex number, and we wish retain the meaning of $\sigma_z$ as a typical distance from $\langle z \rangle$, and second because, when $X$ and $Y$ are independent and have the same variance, $\sigma_z^2$ vanishes! Since the distance between two points $z, z'$ in the complex plane is $|z - z'|$, we better define[6]

$$\sigma_z^2 = \langle |z - \langle z \rangle|^2 \rangle = \frac{1}{2} \langle (z^* - \langle z^* \rangle)(z - \langle z \rangle) \rangle = \sigma_x^2 + \sigma_y^2, \qquad (4.36)$$

where $z^*$ is the complex conjugate of $z$.

The probability density of a complex variable whose real and imaginary part have both a normal distribution, is then given by (4.27). In particular if $X$ and $Y$ are independent, centered at the origin, and have the same standard deviation $\sigma$,

$$g(x, y) = \frac{1}{2\pi\sigma^2} \exp\left(-\frac{x^2 + y^2}{2\sigma^2}\right) \implies g(z) = \frac{1}{2\pi\sigma^2} \exp\left(-\frac{zz^*}{2\sigma^2}\right)$$

It is very interesting to express this result as function of the amplitude $A$ and phase $\theta$ of $z$, by substituting

$$\begin{cases} x = A\cos\theta \\ y = A\sin\theta \end{cases} \implies \begin{cases} A = \sqrt{x^2 + y^2} \\ \theta = \arctan\left(\frac{x}{y}\right) \end{cases}$$

so that, from the Jacobian

$$|J| = \begin{vmatrix} \cos\theta, & -A\sin\theta \\ \sin\theta, & A\cos\theta \end{vmatrix} = A$$

---

[6] Some texts include a factor of $1/2$ in (4.36), so that when $\sigma_x^2 = \sigma_y^2 = \sigma^2$, $\sigma_z^2 = \sigma^2$ too. This is fully legitimate, precisely because (4.36) is just a *definition*.

the joint probability density becomes

$$f_{A,\theta}(A,\theta) = \frac{A}{2\pi\sigma^2} \exp\left(-\frac{A^2}{2\sigma^2}\right) \quad (A \geq 0,\ \pi \leq \theta \leq \pi) \tag{4.37}$$

which actually does *not* depend on the phase $\theta$ and therefore is known as a *circularly symmetric Gaussian*. The marginal distribution for $A$ is then

$$\boxed{f_A(A) = \int_{-\pi}^{\pi} f_{A,\theta}(A,\theta) = \frac{A}{\sigma^2} \exp\left(-\frac{A^2}{2\sigma^2}\right) \quad (A \geq 0} \tag{4.38}$$

which is a *Rayleigh* distribution (3.96).


## 4.5 Characteristic Function

The calculation made in Sect. 4.3 to obtain the distribution of the sum of two independent random variables can be greatly simplified by exploiting the power of the Dirac $\delta$ as a "sampling tool". We can indeed think of obtaining the distribution for $Z = X + Y$ by summing over *all* the values of the joint density $f(x,y) = f_y(x)f_y(y)$, but sifting out those pairs $(x,y)$ that do not satisfy the constraint $x + y = z$ using $\delta(x + y - z)$,

$$f_z(z) = \int_{-\infty}^{\infty} dx \int_{-\infty}^{\infty} dy f_x(x) f_y(y) \delta(x + y - z). \tag{4.39}$$

this means that we can for example take $y$ as a free variable and eliminate the integral in $dx$ by substituting $x = z - y$,

$$f_z(z) = \int_{-\infty}^{\infty} f_x(z - y) f_y(y) dy,$$

which, apart from the different symbol for the integration variable, this expression is nothing but Eq. (4.31).

Using the properties of the Dirac $\delta$, we can reach an important conclusion about the convolution of two functions. Indeed, multiplying both sides of Eq. (4.39) by $\exp(ikz)$ and integrating over $z$,

$$\int_{-\infty}^{\infty} dz\, e^{ikz} f_z(z) = \int_{-\infty}^{\infty} dz\, \delta(x + y - z) e^{ikz} \int_{-\infty}^{\infty} dx \int_{-\infty}^{\infty} dy\, f_x(x) f_y(y),$$

and using again the sampling property of the $\delta$, we have

$$\int_{-\infty}^{\infty} e^{ikz} f_z(z) dz = \int_{-\infty}^{\infty} e^{ikx} f_x(x) dx \int_{-\infty}^{\infty} e^{iky} f_y(y) dy.$$

However, the three integrals that appear in the expression above are nothing but the expectation values of $\exp(ikz)$, $\exp(ikx)$, and $\exp(iky)$ on their respective distributions. Therefore

$$\langle e^{ik(x+y)} \rangle = \langle e^{ikx} \rangle \langle e^{iky} \rangle, \tag{4.40}$$

a much simpler relation than Eq.(4.31).

This result leads us to introduce an important quantity associated with a probability density $f(x)$ which we will call the *characteristic function* $\widetilde{f}(k)$ of the distribution,[7]

$$\boxed{\widetilde{f}(k) = \int_{-\infty}^{\infty} e^{ikx} f(x)\,\mathrm{d}x} \tag{4.41}$$

We have regarded $\widetilde{f}(k)$ as the expectation of $\exp(ikx)$, thought as a function of $k$. However, most of you will have already realized that Eq. (4.41) actually states that the characteristic function is the *Fourier transform*, with sign reversal (or "inverse" Fourier transform[8]) $\mathfrak{F}^{-1}[f]$ of $f(x)$. A probability density and its characteristic function are therefore a *Fourier transform pair*, so we can then also say that[9]

$$f(x) = \mathfrak{F}[\widetilde{f}(k)] = \frac{1}{2\pi} \int_{-\infty}^{\infty} e^{-ikx} \widetilde{f}(k)\,\mathrm{d}k. \tag{4.43}$$

where the pre-factor $1/2\pi$ ensures that $\mathfrak{F}^{-1}\mathfrak{F}[f] = f$. Equation (4.40) can then be written

$$\mathfrak{F}^{-1}[f * g] = \mathfrak{F}^{-1}[f]\,\mathfrak{F}^{-1}[g], \tag{4.44}$$

that is, the (inverse) transform of a convolution of two functions is simply the *product* of the their (inverse) transforms.

---

[7] The characteristic function is usually indicated as $\varphi(x)$. Here we adopt the notation $\widetilde{f}(x)$ to stress its Fourier transform relationship with the probability distribution.

[8] The "direct" Fourier transform is usually defined as

$$\mathfrak{F}[f(x)] = \widetilde{f}(k) = \int_{-\infty}^{+\infty} e^{-ikx} f(x)\,\mathrm{d}x,$$

but the inverse notation is not uncommon, in particular in physical optics. One should better say that $f(x)$ and $\widetilde{f}(k)$ are a Fourier transform pair. It is also useful to recall that, to have a Fourier transform, a function $f$ must satisfy some requirements. For example, all functions $f(x)$ that are rapidly decreasing, i.e., vanish with all their derivatives faster than any power of $1/|x|$ for $x \rightarrow \pm\infty$ have a Fourier transform.

[9] In the appendix it is shown that $\delta(x)$ can be formally related to the complex exponential $\exp(ikx)$ as

$$\delta(x) = \frac{1}{2\pi} \int_{-\infty}^{\infty} e^{-ikx}\,\mathrm{d}k, \tag{4.42}$$

which shows that $\delta(x)$ can formally be regarded as the Fourier transform of the constant unit function $\widetilde{f}(k) \equiv 1$.

Equation (4.40) can easily be generalized to the sum of $N$ independent RVs, $X = \sum_{i=1}^{N} X_i$. For example, when all the variables $X_i$ have the same probability density $f(x)$, which is the case that will mainly interest us, we can use the same method used to obtain (4.40) to find out that

$$\widetilde{f}_x(k) = \left[\widetilde{f}(k)\right]^N .$$

(4.45)

For what follows, it is useful to recall some basic properties of the Fourier transform. Indicating with $f(x) \overset{\mathfrak{F}}{\longleftrightarrow} \widetilde{f}(k)$ a Fourier transform pair, we have:

| | | | |
|---|---|---|---|
| Linearity | $ag(x) + bh(x)$ | $\overset{\mathfrak{F}}{\longleftrightarrow}$ | $a\widetilde{g}(k) + b\widetilde{h}(k)$ |
| Shifting | $g(x - x_0)$ | $\overset{\mathfrak{F}}{\longleftrightarrow}$ | $e^{-ikx_0}\widetilde{g}(k)$ |
| Scaling | $g(ax)$ | $\overset{\mathfrak{F}}{\longleftrightarrow}$ | $\dfrac{1}{|a|}\widetilde{g}\left(\dfrac{k}{a}\right)$ |
| Conjugation | $[g(x)]^*$ | $\overset{\mathfrak{F}}{\longleftrightarrow}$ | $[\widetilde{g}(-k)]^*$ |

The conjugation property, in particular, implies that the transform of a real function $g(x)$ is Hermitian, $\widetilde{g}(-k) = [\widetilde{g}(k)]^*$, and that the transform of a symmetric function is real, $[\widetilde{g}(k)]^* = \widetilde{g}(k)$. One of the most useful features of the Fourier transform is that it turns the derivative of a function into a simple multiplication by $-ik$. Indeed, if $f(x)$ is differentiable,

$$\frac{d}{dx} f(x) = \frac{1}{2\pi} \frac{d}{dx} \int_{-\infty}^{\infty} e^{-ikx} \widetilde{f}(k)dk = \frac{1}{2\pi} \int_{-\infty}^{\infty} e^{-ikx} [-ik\widetilde{f}(k)]dk.$$

Hence, comparing with Eq. (4.43),

$$\left[\frac{df(x)}{dx}\right] \overset{\mathfrak{F}}{\longleftrightarrow} -ik\widetilde{f}(k).$$

(4.46)

Similarly, using (4.41) to find $d\widetilde{f}(k)/dk$,

$$xf(x) \overset{\mathfrak{F}}{\longleftrightarrow} -i\left[\frac{d\widetilde{f}(k)}{dk}\right].$$

(4.47)

Finally, remember that Eq, (4.44) implies that the convolution of two functions and the product of their transforms are also a Fourier transform pair,

$$(f * g)(x) \overset{\mathfrak{F}}{\longleftrightarrow} \widetilde{f}(k)\widetilde{g}(k)$$

(4.48)

### 4.5.1  Some Properties of the Characteristic Function

First, note that the characteristic function is bounded,

$$\left| \widetilde{f}(k) \right| = \left| \int_{-\infty}^{+\infty} e^{ikx} f(x)dx \right| \le \int_{-\infty}^{+\infty} |f(x)|dx = 1.$$

Besides, since $f(x)$ is normalized and real,

$$\widetilde{f}(0) = 1 \; ; \; \widetilde{f}(-k) = [\widetilde{f}(k)]^*.$$

Moreover, when $f(x)$ is symmetric, $\widetilde{f}(k)$ is *real*, which can also be seen by observing that, in this case, the imaginary part of $\widetilde{f}(k)$ is the integral of an odd function,

$$\text{Im}\left[\widetilde{f}(k)\right] = \int_{-\infty}^{\infty} \sin(kx) f(x)dx$$

Another useful property is that, if we consider a variable $Y = aX + b$, where $a$ and $b$ are constants,

$$\widetilde{f}_y(k) = \left\langle e^{iky} \right\rangle = e^{ikb} \int_{-\infty}^{+\infty} e^{ikax} f_y(y)dy = e^{ikb}\, \widetilde{f}_x(ak), \tag{4.49}$$

which is easily found putting $f_y(y) = f_x(x)/|a|$. In particular, if $Y = -X$,

$$\widetilde{f}_y(k) = \widetilde{f}_x(-k) = [\widetilde{f}_x(k)]^*,$$

while, if $Y = X + b$,

$$\widetilde{f}_y(k) = e^{ibk}\, \widetilde{f}_x(k).$$

Thus, a translation of the variable $x$ results in a phase shift of the characteristic function.

### 4.5.2  Characteristic Functions of Noteworthy Distributions

We have introduced the characteristic function for continuous variables, but there is no problem in extending the definition to a distribution $P(j)$ of a discrete variable $J$, which takes the set of values $\{j\}$, simply replacing the integral with a summation,

$$\widetilde{P}(k) = \sum_{\{j\}} e^{ikj} P(j). \tag{4.50}$$

Let us then find the characteristic function of some notable probability distributions, both discrete and continuous, discussed in Chap. 3.

**Binomial**

The binomial distribution can be thought of as the sum of $n$ independent variables, the results of each single Bernoulli trial, which can only take the values $j = 1$ with probability $p$, and $j = 0$ with probability $1 - p$. All these variables have then the same characteristic function

$$\widetilde{P}_1(k) = e^{ik \cdot 1} p + e^{ik \cdot 0}(1 - p) = 1 + (e^{ik} - 1)p. \tag{4.51}$$

From (4.45), the characteristic function of the binomial distribution is then

$$\boxed{\widetilde{B}(k; n, p) = \left[1 + (e^{ik} - 1)p\right]^n} \tag{4.52}$$

**Poisson**

Substituting $a = np$ and taking the limit for $n \to \infty$, we have

$$\boxed{\widetilde{P}(k; a) = \lim_{n \to \infty} \left[1 + \frac{(e^{ik} - 1)a}{n}\right]^n = e^{a[\exp(ik) - 1]}} \tag{4.53}$$

**Uniform**

For a continuous and uniform variable $x$, defined for $a \le x \le b$, we obtain, with an elementary integration,

$$f(x) = \frac{1}{b - a} \implies \boxed{\widetilde{f}(k) = \frac{e^{ikb} - e^{ika}}{i(b - a)k}} \tag{4.54}$$

In particular, for a p.d.f. centered on the origin ($a = -b$) we have, using $\sin t = \frac{e^{it} - e^{-it}}{2i}$,

$$\widetilde{f}(k) = \frac{\sin(kb)}{kb}.$$

In Chap. 3, we have seen that, for a uniform variable strongly localized around $x_0$, $p(x) \to \delta(x - x_0)$. In this limit

$$\boxed{\widetilde{f}(k) = \int_{-\infty}^{\infty} e^{ikx} \delta(x - x_0) dx = e^{ikx_0}} \tag{4.55}$$

which reduces to $f(k) \equiv 1$ for a p.d.f. centered on the origin.[10]

---

[10] When $x$ is a time and therefore $k$ an angular frequency $\omega$, this is tantamount to saying that a unit pulse has a "white" spectrum.

**Exponential**

When $f(x) = \lambda e^{-\lambda x}$, with $\lambda = x_0^{-1} \geq 0$), we have

$$\widetilde{f}(k) = \lambda \int_0^\infty e^{(ik-\lambda)x} \mathrm{d}x.$$

which, if you are not familiar with complex integration, can be calculated by separating the real from the imaginary part and integrating both terms by parts. This gives

$$\boxed{\widetilde{f}(k) = \frac{\lambda}{\lambda - ik}} \tag{4.56}$$

**Cauchy**

Consider first a Laplace distribution, Eq. (3.49), with $\sigma_x = \sqrt{2}\lambda$ that, as shown in example 4.7, can be regarded as the p.d.f. of $x_1 - x_2$, where $x_1$ and $x_2$ are two independent RVs with the same exponential probability density $f_x(x) = \lambda \exp(-\lambda x)$. Then, since $\widetilde{f}(-k) = [\widetilde{f}(k)]^*$,

$$\widetilde{f}(k) = \left\langle e^{ik(x_1-x_2)} \right\rangle = \left\langle e^{ikx_1} \right\rangle \left\langle e^{ikx_2} \right\rangle^* = \left| \left\langle e^{ikx_1} \right\rangle \right|^2 = \frac{\lambda^2}{\lambda^2 + k^2}.$$

Therefore, the characteristic function is, within a factor $1/\pi$, a Cauchy distribution. Equation (3.42), with $\alpha = \lambda$. But then, because of the relation between a function and its Fourier transform, we also have

$$f(x) = \frac{\lambda^2}{\pi(\lambda^2 + x^2)} \implies \boxed{\widetilde{f}(k) = \mathfrak{F}^{-1}[f(x)] = e^{-\lambda|k|}} \tag{4.57}$$

**Gaussian**

The Gaussian has the very special property of "self-transforming", that is, the characteristic function of a Gaussian is still a Gaussian[11], This very interesting result can be easily obtained if you are familiar with the integration of complex functions, but, since this may not be the case, we will follow a different route that exploits the property of the Fourier transform of turning a derivative into a product. The derivative of a Gaussian distribution centered on the origin and with variance $\sigma^2$ is

---

[11] To be precise, within a normalization constant, because of the (standard) definition we gave of $\mathfrak{F}[f]$. Defining

$$\widetilde{f}(k) = \frac{1}{\sqrt{2\pi}} \int_{-\infty}^\infty e^{ikx} f(x) \mathrm{d}x.$$

would give an exact match.

$$\frac{d}{dx} g(x) = \frac{1}{\sigma\sqrt{2\pi}} \frac{d}{dx} \exp\left(-\frac{x^2}{2\sigma^2}\right) = -\frac{x}{\sigma^2} g(x).$$

Taking the Fourier transform of both sides and using (4.46) we have

$$ik\widetilde{g}(k) = -\frac{i}{\sigma^2} \frac{d\widetilde{g}(k)}{dk},$$

that is,

$$\frac{1}{\widetilde{g}(k)} \frac{d\widetilde{g}(k)}{dk} = -\sigma^2 k.$$

Integrating both sides between 0 and a generic $k$ we obtain

$$\ln[\widetilde{g}(k)] - \ln[\widetilde{g}(0)] = -\frac{\sigma^2 k^2}{2}.$$

Therefore, since $\widetilde{g}(0) = 1$,

$$\boxed{\widetilde{g}(k) = \exp\left(-\frac{\sigma^2 k^2}{2}\right)} \tag{4.58}$$

The characteristic function of a Gaussian with an expectation $\mu \neq 0$ is simply obtained using Eq. (4.49),

$$g(x) = \frac{1}{\sigma\sqrt{2\pi}} \exp\left[\frac{(x-\mu)^2}{2\sigma^2}\right] \implies \boxed{\widetilde{g}(k) = \exp\left(i\mu k - \frac{\sigma^2 k^2}{2}\right)} \tag{4.59}$$

**Example 4.8** (*Random walks and characteristic functions*) So far, we have discussed only a random walk with all the steps of the same length $L$. However, using characteristic functions we can investigate a much wider class of random walks. Suppose indeed that the steps do not have a fixed length, but are rather distributed according to a generic probability density $p(x)$, and consider a particle starting at $x = 0$. The probability density of the particle displacement after the first step then simply $P_1(x) = p(x)$. Since each step is independent from the previous ones, the probability of finding the particle at position $x$ after $N$ steps is the sum over all possible values $y$ of the position after $N - 1$ steps times the probability of a displacement $x - y$ in the last step,

$$P_N(x) = \int_{-\infty}^{+\infty} P_{N-1}(y) p(x - y) dy. \tag{4.60}$$

But this means that $\widetilde{P}_N(k) = \widetilde{P}_{N-1}(k)\tilde{p}(k)$, which is exactly the same recursive relation satisfied by a random walk with a fixed step. Therefore, the characteristic function after $N$ step is $\widetilde{P}_N(k) = [\tilde{p}(k)]^N$ and the probability distribution for the final position is

$$P_N(x) = \frac{1}{2\pi} \int_{-\infty}^{+\infty} [\tilde{p}(k)]^N e^{-ikx} dk \tag{4.61}$$

Since the steps are independent, the same result can of course be obtained by writing the position after $N$ step as $X = \sum_{i=1}^{N} x_i$ and using (4.45).

So, for example, if the probability density for a single step is a Gaussian centered at the origin and standard deviation $\sigma$, its characteristic function is given by (4.58). Therefore, $\widetilde{P}_N(k) = \exp(-N\sigma^2 k^2/2)$, whose transform is again a Gaussian with $\sigma_N = \sigma\sqrt{N}$,

$$P_N(x) = \frac{1}{\sigma_N\sqrt{2\pi}} \exp\left(-\frac{x^2}{2\sigma_N^2}\right).$$

### 4.5.3  A (Poor) Relative of $\widetilde{f}(k)$: The Moment Generating Function

The characteristic function is a valuable complement to a probability distribution, sometimes essential for obtaining basic results, with the disadvantage, however, of requiring complex variables. Nevertheless, we can easily conceive a function that *looks as* a close relative of $\widetilde{f}(k)$ by considering, instead of $\langle e^{ikx} \rangle$,

$$M_x(r) \equiv \langle e^{rx} \rangle = \int_{-\infty}^{+\infty} e^{rx} f(x) dx \tag{4.62}$$

where $r$ is a *real* number. *If* such a function exists and is finite, then, as we are going to show, it can be a very powerful tool to calculate all moments of $f(x)$. The problem is, while the characteristic function is well-defined for any "reasonable" $f(x)$, this is not the case for $M_x(r)$. Later, we will discuss the conditions for this to occur. For the moment, assume that $M_x(r)$ does exist, at least in a arbitrarily small interval around the origin. Then, its derivative with respect to $r$ is

$$\frac{dM_x(r)}{dr} = \frac{d}{dr} \int_{-\infty}^{\infty} e^{rx} f(x) dx = \int_{-\infty}^{\infty} x e^{rx} f(x),$$

so that, *if* the expectation $\langle x \rangle$ is finite,

$$\langle x \rangle = \frac{dM_x(r)}{dr}\bigg|_{r=0}.$$

By deriving a second time, it is easy to show that a similar relationship exists between the second moment of $f(x)$ and the second derivative of $M_x(r)$. We can generalize these results if we consider a probability density $f(x)$ that has finite moments $\langle x^n \rangle$ for *all* integers $n$. Since the series expansion of an exponential is $\exp(s) = \sum_{n=0}^{\infty}(s^n/n!)$, we can rewrite Eq. (4.41) as

$$M_x(r) = \int_{-\infty}^{\infty} e^{rx} f(x)\mathrm{d}x = \sum_{n=0}^{\infty} \frac{r^n}{n!} \int_{-\infty}^{\infty} x^n f(x)\mathrm{d}x = \sum_{n=0}^{\infty} \frac{\langle x^n \rangle}{n!} r^n. \qquad (4.63)$$

Therefore, the coefficients of the series expansion around $s = 0$ of $\widetilde{f}(s)$ are given by $\langle x^n \rangle /n!$, which means that

$$\boxed{\langle x^n \rangle = \frac{\mathrm{d}^n}{\mathrm{d}r^n} M_x(r) \bigg|_{r=0}} \qquad (4.64)$$

Therefore, *provided that it exists*, $M_x(r)$ allows all the moments of $f(x)$ to be recovered, and for this reason it is called the *moment generating function (MGF)*. We already said that a necessary condition for $M_x(r)$ to exist is that all moments of $f(x)$ exist and are finite. So, for example, the Cauchy distribution has *no* MGF, although it *does* have a characteristic function. But this is not enough, for it is also necessary that the series in (4.63) *converges*. This usually happens for probability densities that rapidly vanish for $x \to \pm\infty$, in which case one can simply substitute $ik \to r$ in the expression for the characteristic function. So, for example, the MGF of a normal distribution $g(x; \mu, \sigma)$ is simply $M_x(r) = \exp(\mu r) \exp(\sigma^2 r^2/2)$.

But consider now a *log*-normal distribution. We can easily show that its moments are finite, and are given by Eq. (3.78). Indeed if $X$ has a log-normal density, it can be written as $X = e^Y$, where $Y$ is normally distributed, so we have

$$\langle x^r \rangle = \langle e^{ry} \rangle = \exp(\mu r) \exp\left(\frac{\sigma^2 r^2}{2}\right)$$

because $\langle X^r \rangle$ is nothing but the MGF of a Gaussian. However, $M_x(r) = \infty$ for every $r > 0$, because, by expanding the exponential and using the previous expression for the moments,

$$M_x(r) = \langle e^{rx} \rangle = \left\langle \sum_{n^0}^{\infty} \frac{(rx)^n}{n!} \right\rangle = \sum_{n=0}^{\infty} \frac{\langle x^n \rangle}{n!} r^n = \sum_{n=0}^{\infty} \frac{e^{n\mu} e^{\frac{1}{2}n^2\sigma^2}}{n!} r^n,$$

which is clearly a divergent series.

The absence of a MGF has some subtle consequences. It is a rather widely accepted belief that the knowledge of all moments univocally determines a probability distribution. This is generally true, but not for those distribution that lack a MGF. Take for instance the log-normal distribution (3.77) with $\mu = 0$ and $\sigma = 1$,

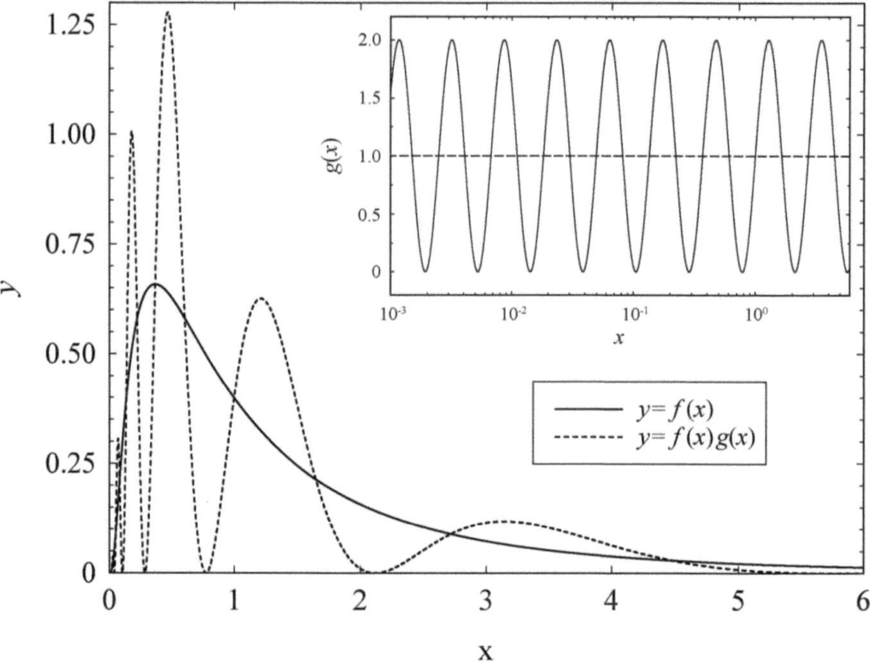

**Fig. 4.3** Comparison of the two distributions $f(x)$ and $f(x)g(x)$ with the same moments discussed in the text. The inset shows $g(x)$ on a logarithmic $x$-axis

$$f(x) = \frac{1}{x\sqrt{2\pi}} \exp\left[-\frac{1}{2}\ln^2(x)\right],$$

and consider the function $g(x) = 1 + \sin[2\pi \ln(x)]$: it can be proved that $f(x)g(x)$ is a normalized probability density too, and that *all* its moments coincide with those of $f(x)$.[12] And yet, the two distributions, shown in Fig. 4.3, look quite different! This is just an example of a very intriguing question in mathematical statistics, called the *problem of moments*, which involves delicate questions concerning convergence of probability distributions. It is hard to pinpoint any applications of probability in physics where this problem shows up. But we better be aware.

Setting aside the previous delicate question, I do not personally see any compelling reason to introduce the MGF. Indeed, by substituting $r \to ik$ in Eq. (4.63), we can formally write a series expansion for the characteristic function (which exists for all distributions that have a Fourier transform),

$$\tilde{f}(k) = 1 + \sum_{r=1}^{\infty} \frac{(-ik)^r}{r!} \langle x^r \rangle \tag{4.65}$$

---

[12] C. C. Heyde, J. R. Stat. Soc. Ser. B. **25**, 16 (1963).

where we have taken into account that $\widetilde{f}(0) = 1$. Therefore

$$\langle x^r \rangle = i^{-r} \left. \frac{d^r \widetilde{f}(k)}{dq^r} \right|_0 , \tag{4.66}$$

which is actually the most convenient way to find the moments of a distribution whose characteristic function is known. For example, since for a Poisson distribution $\widetilde{f}(s) = e^{a(\exp(s)-1)}$, we have

$$\widetilde{f}^{(1)}(s) = a e^s e^{a(\exp(s)-1)}$$
$$\widetilde{f}^{(2)}(s) = a e^s e^{a(\exp(s)-1)} + a^2 e^{2s} e^{a(\exp(s)-1)} ,$$

so its variance is

$$\sigma_k^2 = \widetilde{f}^{(2)}(0) - (\widetilde{f}^{(1)}(0))^2 = a.$$

Notice however that (4.65) entails that the moment of order $r$ exists if and only if $(d^r/dk^r)\widetilde{f}(k)$ is finite at $k = 0$. For example, $\exp(-\lambda|k|)$ has a cusp for $k = 0$, so its first derivative is non defined at the origin and the Cauchy distribution does not admit, as we have seen, a finite expectation.

Actually, the characteristic function generates the moments of a distribution about any point $x_0$, including then the central moments when $x_0 = \langle x \rangle$, by expanding

$$e^{ikx_0} \widetilde{f}(k) = \langle e^{-ik(x-x_0)} \rangle = \sum_{r=1}^{\infty} \frac{(-ik)^r}{r!} \langle (x - x_0)^r \rangle \tag{4.67}$$

### 4.5.4  Cumulants: Why the Gaussian Is So "Special"

Using moments about the origin to describe a distribution is not very practical because for a complete representation of $f(x)$ we have to provide *all* its moments, even if the distribution is characterized by just one or two parameters. In Chap. 3 we saw that the shape characteristics of a distribution, such as its width or its asymmetry, are better described by quantities like $\sigma_x$ or $\gamma_x$, related to the *central* moments. The purpose of this section is to show that it is generally possible to introduce more "efficient" descriptors, called *cumulants* and denoted with $\kappa_n$. The main reason why cumulants are more practical than moments about the origin is that, usually, just a few of them (those with the lowest values of the index $n$) are required for an overall description of $f(x)$. To introduce cumulants, let us consider the sum of two RVs $X$ and $Y$. If they are independent, their characteristic function satisfies

$$\langle e^{ik(x+y)} \rangle = \langle e^{ikx} \rangle \langle e^{iky} \rangle ,$$

which means that

$$\ln[\widetilde{f}_{x+y}(k)] = \ln[\widetilde{f}_x(k)] + \ln[\widetilde{f}_y(k)]. \tag{4.68}$$

This expression suggests to put a special emphasis on the *logarithm* of the characteristic function that, for independent RVs, is therefore *additive*. We define then the *cumulant generating function*,

$$C(k) = \ln[\widetilde{f}(k)], \tag{4.69}$$

which implicitly defines the cumulants $\kappa_n$ of the distribution through its series expansion,

$$C(k) = \sum_{r=1}^{\infty} \frac{(-ik)^r}{r!} \kappa_r. \tag{4.70}$$

Note that the expansion does not contain the $r = 0$ term since $C(0) = \ln[\widetilde{f}(0)] = 0$. To relate the cumulants to the moments, we have to impose $\widetilde{f}(k) = e^{C(k)}$. Differentiating both sides and writing for short $D^n[f(x)] = d^n[f(k)]/dk^n$, we have

$$D[\widetilde{f}(k)] = D[C(k)]\widetilde{f}(k),$$

so that, using the product rule,

$$D^n[\widetilde{f}(k)] = D^{n-1}\left[D[C(k)]\widetilde{f}(k)\right] = \sum_{m=0}^{n-1} \binom{n-1}{m} D^{n-m-1}\left[D[C(k)]\right]D^m[\widetilde{f}(k)]$$

that is,

$$D^n[\widetilde{f}(k)] = \sum_{m=0}^{n-1} \binom{n-1}{m} D^{n-m}[C(k)]D^m[\widetilde{f}(k)]. \tag{4.71}$$

Therefore, for $q = 0$, since $D^r[\widetilde{f}(0)] = \langle x^r \rangle$ and $D^r[C(0)] = \kappa_r$, we have

$$\langle x^n \rangle = \sum_{m=0}^{n-1} \binom{n-1}{m} \kappa_{n-m} \langle x^m \rangle,$$

Setting apart the $m = 0$ term and recalling that $\langle x^0 \rangle = \int_{-\infty}^{+\infty} f(x)\,dx = 1$, we find the very useful recursion relation

$$\boxed{\kappa_n = \langle x^n \rangle - \sum_{m=1}^{n-1} \binom{n-1}{m} \kappa_{n-m} \langle x^m \rangle} \tag{4.72}$$

For the first three cumulants we find

$$\boxed{\begin{aligned}
\kappa_1 &= \langle x \rangle \\
\kappa_2 &= \langle x^2 \rangle - \langle x \rangle^2 = \langle (x - \langle x \rangle)^2 \rangle = \sigma_x^2 \\
\kappa_3 &= \langle x^3 \rangle - 3 \langle x^2 \rangle \langle x \rangle + 2 \langle x \rangle^3 = \langle (x - \langle x \rangle)^3 \rangle = \sigma_x^3 \gamma.
\end{aligned}} \tag{4.73}$$

So the first cumulant is nothing but the expectation, the second one is the variance, and the third one is proportional to the skewness. Looking at (4.73), one is tempted to conclude that all cumulants of order $r > 1$ are nothing but the central moments of order $r$, but unfortunately this is not true. You can indeed show that, for example,

$$\kappa_4 = \langle (x - \langle x \rangle)^4 \rangle - 3\kappa_2^2.$$

Nevertheless, cumulants share with the central moments the property of *invariance by translation*. Indeed, using (4.49), a translation $x \to x + c$ yields

$$C_{x+c}(k) = \ln[\widetilde{f}_{x+c}(k)] = C_x(k) + ick. \tag{4.74}$$

which allows us to deduce that, in this transformation, all cumulants remain unchanged except the first one, which becomes $\kappa_1 + c$. If instead we scale $x \to ax$, Eq. (4.49) gives

$$C_{ax}(k) = C_x(as) = \sum_{n=1}^{\infty} \frac{\kappa_n a^n}{n!}(k)^n \implies \kappa_n(ax) = a^n \kappa_n. \tag{4.75}$$

One of the most interesting features of cumulants is they provide a unique definition and characterization of the normal distribution. Indeed, Eq. (4.59) gives, for the cumulant generating function,

$$C(k) = \mu k - \frac{\sigma^2}{2}k^2, \tag{4.76}$$

so that, for a Gaussian, $\kappa_1 = \mu$, $\kappa_2 = \sigma^2$ and, above all, $\kappa_n \equiv 0$ for all $n > 2$. Since $C(k) = f(ikx)$ uniquely determines $f(x)$, *the Gaussian is the only probability distribution that has all cumulants higher than the second equal to zero* and, conversely, *every probability distribution with this property is a Gaussian*. In Sect. 4.6 we will appreciate how relevant is this conclusion.

As we have seen and will understand why shortly, many probability densities become similar to a normal distribution in an appropriate limit, which makes the Gaussian a simple "model" distribution. Nevertheless, we might wonder if more refined models can be developed, capable of representing a wider class of limiting conditions. For example, we might ask if there is a distribution in which only the first *three* cumulants are non-zero. But this does not happen: it can be shown that the normal distribution is the only distribution having a finite number of non-zero cumulants.[13] In other words: either a probability distribution has only the first cumulant $\kappa_1 = \langle x \rangle$, and then it is a distribution for a "fully localized" RV, $f(x) = \delta(x - \langle x \rangle)$, or, if it is not a Gaussian, has *infinite* cumulants. For example, using (4.53) it is easy to show that all cumulants of a Poisson are equal to $a$. Nevertheless, when the cumulants are appropriately scaled, so to obtain dimensionless parameters like the relative variance and the skewness, the latter usually decrease rapidly with $\langle x \rangle$ (as $a^{-1/2}$ for the Poisson).

**Example 4.9** (*Cumulants and graph connectivity*) Graphs are an invaluable tool to model pair interactions in physical, biological, social and information systems.[14] The total number of possible (undirected) graphs connecting $n$ (labelled, i.e., *distinguishable*) nodes is simply given by $G_n = 2^{\binom{n}{2}}$, because each of the $n(n-1)/2$ pairs of nodes can be linked or not. Evaluating the number of possible *connected* graphs $C_n$, i.e., of graphs where every pair of nodes in is connected (directly or through other nodes) is a much harder problem. Evidently $C_2 = 1$, while you can easily convince yourself that $C_3 = 4$. But even for $n$ as small as 4, finding that $C_4 = 38$ is rather time-consuming (try, if you don't believe me!). In fact, the first $C_n$ given in *The On-Line Encyclopedia of Integer Sequences*[15] for $n \geq 1$ are

$$1, 1, 4, 38, 728, 26704, 1866256, 251548592, 66296291072, 34496488594816 \ldots$$

which looks like a rather unpredictable sequence[16]…Nevertheless there is a combinatorial strategy that provides a connection between the $C_n$'s and the $G_n$'s, and it has a lot to do with moments and cumulants.

Let us consider a graph with $n$ labelled nodes. We first pick up the a "pivot" node, for instance the node labelled with $n$. We then split the graph into two disjoint subgraphs, the first one fully connected and containing $n$, the second arbitrary. For the second one we can choose $m$ of the remaining $n-1$ nodes in $\binom{n-1}{m}$ ways, building with them an arbitrary graph in $G_k$ ways. With the remaining $n-1-m$ nodes, together with node $n$, we can build a connected graph in $C_{n-m}$ ways. Letting $m$ range from 1 to $n-1$ we obtain all *disconnected* graphs $D_n$, i.e., all the graphs where at least one node is isolated,

$$D_n = \sum_{m=1}^{n-1} \binom{n-1}{m} C_{n-m} G_m.$$

---

[13] J. Marcinkiewicz, Math. Z. **44**, 612 (1939).

[14] For instance, my daughter PhD in Mental Health Science has a lot to do with graphs. I'm not fully convinced, however, that the deep math required by graph theory is the best medicine for mental wellness….

[15] https://oeis.org.

[16] Remember our preliminary discussion of "randomness" in Chap. 1?.

Then, the number of fully *connected* graphs is

$$C_n = G_n - \sum_{m=1}^{n-1} \binom{n-1}{m} C_{n-m} G_m,\tag{4.77}$$

which, once we identify $C_n \leftrightarrow \kappa_n$ and $G_n \leftrightarrow \langle x^n \rangle$ is *identical* to Eq. (4.72). So, for example, using $G_n = 2^{\binom{n}{2}}$,

$$C_4 = 64 - 4 \times 8 \times 1 - 3 \times 4 + 12 \times 2 \times 1 - 6 \times 1 = 38$$

Of course, we also have the analogous of Eq. (4.5.4),

$$G_n = \sum_{m=0}^{n-1} \binom{n-1}{m} C_{n-m} G_m,$$

which suggest an interesting reinterpretation. Indeed, what we are actually doing is splitting $G_n$ in graphs containing a *cluster* of $n - m$ nodes, with $m$ ranging between 0 and $n - 1$. So, for example, total number of labelled graphs with four nodes can be split as

$$\langle x^4 \rangle = \left(\begin{smallmatrix}\bullet & \bullet \\ \bullet & \bullet\end{smallmatrix}\right) + 4\left(\begin{smallmatrix}\bullet \\ \bullet\bullet\end{smallmatrix}\right) + 3\left(\begin{smallmatrix}\bullet \\ \bullet\end{smallmatrix}\right)^2 + 6\left(\begin{smallmatrix}\bullet \\ \bullet\end{smallmatrix}\right) \times (\bullet)^2 + (\bullet)^4$$

where a group of $k$ points in brackets is the number of connected graphs with $k$ nodes. A decomposition of this kind is very useful when we consider a physical quantity whose value can be obtained by summing the contributions from, one, two, three, ...$k$ particles, if the importance of these contributions decreases with $k$, and is the starting point for computations based on a "diagrammatic expansion".

## 4.6  The Central Limit Theorem (CLT)

The distinctive feature of the normal distribution we just talked about is at the core of what is probably the most important result in probability theory, a result that also plays a crucial role in the analysis of experimental accuracy that we will later discuss. Let us consider again the sum $X = \sum_{i=1}^{N} X_i$ of $N$ independent RVs that have the same probability distribution $f(x)$, possessing all moments $\langle x_i^n \rangle$, equal for all the $x_i$. Therefore all cumulants $\kappa_n$ are defined. From Eq. (4.68) we have

$$C_x(s) = N C_{x_i}(s).$$

Hence, indicating with $\kappa_n(x)$ the cumulants of $X$, $\kappa_n(x) = N\kappa_n$ for all $n$. In particular, putting $\kappa_1 = \langle x_i \rangle = \mu$ and $\kappa_2 = \sigma^2$,

$$\kappa_1(x) = N\mu \, , \;\; \kappa_2(X) = N\sigma^2.$$

Then, for the variable $Z = (X - N\mu)/\sqrt{N}$, we have $\kappa_1(z) = \langle z \rangle = 0$, while, since a translation does not change $\kappa_n$ for $n > 1$, Eq. (4.75) gives

$$\kappa_n(z) = \kappa_n\left(\frac{x}{\sqrt{N}}\right) = N^{-n/2}\kappa_n(x) = N^{1-n/2}\kappa_n.$$

Therefore, we have $\kappa_2(z) = \sigma_Z^2 = \sigma^2$, while all cumulants with $n > 2$ *vanish* as $N$ increases. So, in the limit $N \to \infty$, $Z$ becomes a normal probability density with zero expectation and variance $\sigma^2$. But then $X = \sqrt{N}(Z + N\mu)$ will have a Gaussian distribution too,

$$f(x) = \frac{1}{\sqrt{2\pi N}\sigma} \exp\left[-\frac{(x - N\mu)^2}{2N\sigma^2}\right]. \tag{4.78}$$

What we have just demonstrated is nothing but the simplest form of the *Central Limit Theorem* (CLT), according to which the sum of a sufficiently large number of variables is Gaussian, despite the distributions of the individual variables being completely generic.[17]

Several of the simplifying assumptions we have made can be significantly weakened. First of all, it is not necessary for $f(x)$ to have all moments, but it is sufficient that only $\langle x \rangle$ and $\sigma$ are finite: if this is the case, the convergence to the Gaussian is only slower. But above all, it is not even necessary for the variables $x_i$ to have the *same* probability distribution. Actually, the CLT can be considerably extended, provided it is derived through much more advanced methods. More than "the" Central Limit Theorem, one should therefore speak of a *class* of theorems, which establish in an increasingly precise way the role of the Gaussian as a limiting distribution. Qualitatively, what happens is that by summing many random variables, the "fine details" of the individual distributions are progressively washed out, until a distribution fully characterized only by expectation and variance, i.e., a Gaussian, is obtained.

For our purposes, it is sufficient to state, in not very rigorous terms and without proving it, a form of the CLT that, although not the most general, allows us to grasp better the deep meaning of this result. Let us consider again $N$ independent variables $X_i$, each described by its own specific distribution with finite expectation, $\langle x_i \rangle = \mu_i$, and variance, $\sigma_i^2$. Then, as the number $N$ of considered variables becomes "sufficiently large", and provided that each variance $\sigma_i^2$ is "small" compared to the sum $\sigma^2 = \sum_{i=1}^{N} \sigma_i^2$ of the individual variances, namely, if

$$\max_{1 < i < n}\left(\frac{\sigma_i^2}{\sigma^2}\right) \xrightarrow[n\to\infty]{} 0, \tag{4.79}$$

---

[17] The DeMoivre-Laplace theorem (3.54) is just a particular case of the CLT.

the distribution of the sum, $X = \sum_{i=1}^{N} X_i$ "tends" to a Gaussian with

$$\begin{cases} \langle x \rangle = \sum_{i=1}^{N} \mu_i \\ \sigma_x^2 = \sigma^2 = \sum_{i=1}^{N} \sigma_i^2. \end{cases} \qquad (4.80)$$

What makes the CLT fundamental is precisely its generality. Yet, some caution is appropriate, which is why I put "sufficiently large" and "small" in quotes.

- The number of variables $x_i$ required to approximate a Gaussian to a given extent depends on how they are distributed. A few variables (typically 5–10) will be sufficient if their distributions are quite regular and symmetrical, whereas a sum of variables with strongly asymmetric distributions will converge to the normal distribution much more slowly.
- The convergence to the Gaussian of the sequence of partial sums is *not uniform*. That is, while in a neighborhood of $\langle x \rangle$ the distribution quickly takes on a Gaussian shape, the "tails" of the distribution converge more slowly.[18] Indeed the interval of convergence grows only as $N^{1/2}$.
- Condition (4.79) implies that, if the individual variables are summed with different weights, there must not be a "predominant" variable $x_i$. In other words, When 90% of $X$ is determined by a single variable and only 10% by the rest, its distribution will tend to reflect the characteristics of the leading variable distribution even for rather large values of $N$.

The main reason why the CLT assumes particular importance is that very often a random variable can be thought of as the final result of the effects of many concomitant variables that contribute to determine the value of the sum variable. For example, the stature of an individual is determined by many genetic, dietary, and environmental factors. If we approximately describe the fluctuation of the individuals stature with respect to the population average as due to a sum of a large number of contributions, than we expect it to have a normal distribution, a fact that, as we have seen, is well verified experimentally. Conversely, we have seen that the distribution of individuals' weights deviates significantly from a Gaussian and exhibits a marked positive asymmetry. This is probably due to the fact that, compared to other factors, dietary habits contribute predominantly to determining weight.

---

[18] A note in terminology. The English expression *Central Limit Theorem* should be sufficiently unambiguous, for the word "limit", playing the role of a *factual* adjectives (it *is* a "limit theorem"), comes after "central", expressing the firm *opinion* of George Pólya, who considered the CLT that he named a central theorem in probability theory. After all, the German expression he used, *Zentraler Grenzwertsatz*, leaves no room for doubt. Unfortunately, in Latin languages the terminology is sometimes confusing. For example, in French texts the expression *théorème de la limite centrale* is much more frequent than *théorème central limite* (considered an abominable Anglicism). Similarly, in Spanish one finds *teorema del límite central*, almost identical to the Italian frequently used expression *teorema del limite centrale*. But what could "theorem of the central limit" mean? Perhaps that convergence occurs more rapidly "at the center"? As far as I know, limits are not defenders of a soccer team….

What I have just presented is the customary reason why the normal distribution is often put on a pedestal, if not truly "sanctified".[19] There is, however, an hidden question: even though a quantity is determined by multiple factors, why should each factor contribute *additively*? What if the overall effect results for instance from the *product* of many factors? The CLT, because if $X = \prod_{i=1}^{N} X_i$, then $\ln X = \sum_{i=1}^{N} \ln X_i$. So, if $N$ is large, $\ln X$ will be a normally distributed variable, and therefore $X$ will have a *log*-normal distribution, as we already anticipated in Sect. 3.7.4 while discussing Kolmogorov's model of ore crushing and Gilbrait's law of proportionate effects.[20] This is the reason why the log-normal, despite lacking the simple structure of the Gaussian, is often competitive with the normal distribution, especially when fitting data with intrinsically positive values.

Nevertheless, a further reason why the normal distribution is capable of modeling data resulting from the combination, not necessarily additive, of many factors is probably related, as we discuss in the next section, to its "information content", which has to do with the versatility it offers in fitting data with only some basic statistical parameters known.

**Example 4.10** (*From random walks to speckle patterns*) We first review the one-dimensional random walk in the light of the CLT. Each step $x_i$ is a random variable that can only take the values $\pm L$ with probability $p = 0.5$, and therefore has an average value $\langle x_i \rangle = 0$ and variance $\sigma_i^2 = 0.5L^2 + 0.5L^2 = L^2$. If the number of steps $N$ is large, the final position $X$ of $N$ will therefore be distributed as a Gaussian of expectation $\langle x \rangle = 0$ and variance $\sigma_x^2 = NL^2$. But the CLT offers the chance to extend the previous results to a much wider class of random walks. Indeed, suppose that the steps do not have a fixed length, but are rather distributed according to a symmetric probability density $f(x_i)$ with an expectation $\langle x_i \rangle = 0$ and a finite variance $\sigma_i^2$. Then, the final position will *still* have a normal distribution with $\langle x \rangle = 0$ and $\sigma_x^2 = N\sigma_i^2$.

In Example 1.13 we have observed that an exponential frequency distribution is a suitable representation of the light intensity of a speckle pattern: what we have just said about random walks will help us understand why. Before that, let us first recall that speckles show up in two basic optical geometries. The first just consists in collecting on a multi-pixel sensor the light scattered off a rough surface (or scattered

---

[19] Someone, however, rather views the Gaussian as a "Great Intellectual Fraud", pointing out the key role played in many fields by the so called "black swans", highly improbable but devastating (for the CLT) events. Stressing the insidious nature of these circumstances is certainly creditable (and possibly beneficial to your career).

[20] This extension of the CLT has however an additional requirement, namely, that all the distributions of the $X_i$'s have a *strictly* positive support (no value $x_i$ where $f_i(x_i) = 0$).

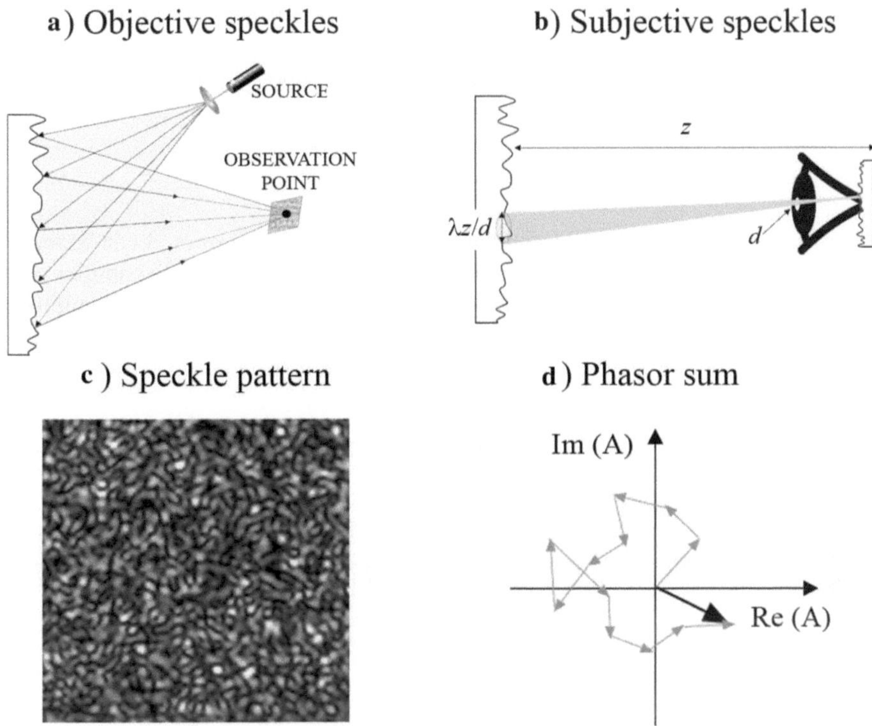



**Fig. 4.4** Speckle effects and analogy with a random walk in the complex plane

by a particle dispersion) illuminated by a spatially–coherent source[21] *without* a collection optics (see Fig. 4.4a), which produces on the detector a "speckled" pattern as in Fig. 4.4c.

A similar pattern also forms if we image the scattering surface on a plane with a lens system that has an entrance pupil of finite size $d$, since the optical signal at a point on the image plane actually results, because of diffraction, from a finite region of approximate size $\lambda z/d$ on the object plane. These speckles, actually generated by the imaging optics, are called "subjective", to distinguish them from the "objective" speckles obtained without imaging.

---

[21] Notice that, in principle, we don't need a laser, but just a light source of small size placed far away from the surface. Spatial coherence, at variance with *temporal* coherence (which is related to the source spectral width), has nothing to do with the physics of the light source, but only with its size and distance from the illuminated surface $S$. Indeed, a source of wavelength $\lambda$ seen from $S$ under a solid angle $\Omega$ generates on $S$ a coherently illuminated region of approximate area $A_c \simeq \lambda^2/\Omega$. The main advantage of using a laser is that it is equivalent to a very brilliant, almost point-like source placed at infinite distance from the illuminated object, thus it is a source with a very large $A_c$. This has also a lot to do with the reason why stars twinkle and planets like Jupiter don't. If you wish to delve more into optical coherence, the reference text is surely J. W. Goodman, *Statistical Optics* (Wiley, 1985).

In both cases, however, we can regard the optical signal in a point $\{x, y\}$ on the observation plane as due superposition and interference of a large number $N$ of individual contributions ("phasors") with different phase and possibly amplitude, which, using the complex notation we can write

$$A_i(x, y, t) = a_i(x, y, t)e^{i\varphi(x,y,t)}.$$

By separating out the real and Imaginary component of the phasors, $A$ can be seen as the sum of $N$ random vectors or, if you prefer, as a random walk in the complex plane (see Fig. 4.4 D). For $N \to \infty$ we know from Sect. 4.4 that the resulting amplitude and phase will have a joint circularly symmetric Gaussian distribution, while the marginal distribution for the amplitude is a Rayleigh distribution,

$$f_A(A) = \frac{A}{\sigma^2} \exp\left(-\frac{A^2}{2\sigma^2}\right) \quad (A \geq 0), \tag{4.81}$$

with $\langle A \rangle = \sqrt{\pi/2}\,\sigma \simeq 1.25\sigma$ and $\sigma_A^2 = (2 - \pi/2)\sigma^2 \simeq 0.43\sigma$.

Let us know evaluate the statistics of the light *intensity*,[22] $I = A^2$. From Eq. (4.5), we get[23]

$$\boxed{f_I(I) = \frac{1}{2\sigma^2} \exp\left(-\frac{1}{2\sigma^2}\right) = \frac{1}{2\langle I \rangle} \exp\left(-\frac{I}{\langle I \rangle}\right) \quad (I \geq 0)} \tag{4.82}$$

which supports the experimental evidence. An exponential distribution of the intensity means that there are more "dark" speckles than "bright" speckles. Indeed, you can easily find that the fraction of speckles with an intensity lower than half the average is almost 40%, while those with an intensity larger than twice the average is less than 14%.

## 4.7  Probability and Information

As we stated in Chap. 2, probability can be interpreted as the degree of certainty we have about the occurrence of an event. Therefore, in this "subjective" approach, there must be a relation between the probability $P(E)$ that we associate with an event $E$, and the amount of *information* we have about $E$. Let us then consider a set of events $\{E_i\}$, representing for instance the possible outcomes of an experiment, which are *mutually exclusive*, namely, such that, for any pair of events $E_1, E_2 \in \{E_i\}$, the

---

[22] Rather than the true radiometric intensity, which is the magnitude of the Poynting vector averaged over many optical cycles, it is customary to call "intensity" what is actually the optical irradiance $I = A^2$.

[23] Note that $f_A(-\sqrt{I}) = 0$.

probability $P(E_1 \cap E_2)$ that *both* of them occur is zero. Suppose also that we can estimate the probabilities $P_i = P(E_i)$ of any of these events, and that $\sum_i P_i = 1$, which, in formal terms, means that this class of events is a complete partition $\mathcal{P}$ of the "event space". Our goal is quantifying the amount of information we *miss* about the problem, if all we know is the set of probabilities $\{P_i\}$.

For example, suppose that, without turning on the light (maybe because my wife is sleeping in the same room) I look for a pair of blue socks, which, however, lie scattered in a drawer together with many other socks of $n$ different colors. In this case, $\mathcal{P}$ is the partition of all the pairs of socks in $n$ groups with a specific color, and the event $E_i$ is the color of the socks I take out from the drawer. Evidently, if I know that all the socks have the same color, I am basically done, whereas selecting a blue pair is surely harder if the socks in the drawer are equally distributed among several colors. This means that the amount of information I need to find the blue socks depends on the specific distribution of the socks among the different colors. Besides, if I am an oddball who likes to change the color of his socks according to the day of the week (blue on Monday, pink on Tuesday, yellow on Wednesday, and so on), I surely need much more information than an ordinary fellow who wears only blue or brown socks. The required information should then depend on the "fineness" of the partition $\mathcal{P}$. Yet, is there a way to gauge the amount of information I need by defining a quantity that depends on the $P_i$'s *alone*?

What we plan to investigate is strictly related to the problem of extracting a meaningful signal from a noisy channel, i.e., a signal mixed with some undesired "random noise", which is a basic problem in communications and, more generally, in information science. Living in the Internet Era, we all know that these disciplines have experienced an incredible boom, thanks to the seminal contributions of great scientists like Harry Nyquist, John von Neumann, and Norman Wiener. However, the true conceptual revolution, leading to the developments in communication systems we are all beneficiaries of, is surely due to the genius of Claude Shannon, and to the concept of *statistical entropy* he developed in Bell Labs in the forties of the past century. Dwelling upon this concept, even at an introductive level, is very useful, both because it provides a criterion to judge the soundness of a given probability distribution, and because of its connection (*nomen omen*) with statistical physics.

### 4.7.1  Statistical Entropy

We first discuss the case of a *discrete* partition of the event space and, by extension, of a random variable that takes on only discrete values. Consider then $n$ mutually exclusive events $\{E_i\}$, whose probabilities $\{P_i\}_{i=1,n}$ satisfy $\sum_i P_i = 1$. We look for a function $S(P_i) = S(P_1, \cdots, P_n)$ that quantifies the "lack of information" originating from the stochastic nature of the problem we are considering, which we shall call *statistical* (or *Shannon*) entropy. If we wish this abstract concept to agree with the common ideas we have about information, $S$ must satisfy some intuitive requirements.

1. Stating that we miss a negative amount of information makes no sense: it simply means that already have more information than we need! Hence, we require $S$ to be a positive definite function, namely $S(P_i) \geq 0$ for all values of $i$. In particular, $S$ should vanish if and only if a specific event $E_j$ happens for sure, i.e., if there is a $j$ such that $P_j = 1$ (and then $P_{i \neq j} = 0$),

$$S(0,\ 0,\ \ldots\ 1,\ \ldots 0) = 0.$$

2. If we slightly change each one of the $P_i$'s, we expect that $S$ does not change too much either. Besides, $S$ should depend only on the whole set of values $\{P_i\}$, but not on the *order* in which they appear in its definition. Therefore, we ask $S$ to be a *continuous* and *symmetric* function of all the variables $P_1, \cdots , P_n$.
3. Looking back at the example of selecting socks, we see that the amount of information we need to find the right pair increases with the number of possible colors, at least when the probability is the same for all colors. Then, if we consider a set of $n$ *equiprobable* events, $S$ must increase with $n$.

We need to add a last, possibly less intuitive requirement, which is however crucial to "characterize" $S$. Coming back to our original example, suppose that, besides the blue socks, I must also find a white shirt, which is in another drawer together with shirts of various colors. From the way I described the scene, you may gather that picking out a pair of socks and choosing a shirt are two independent operations. Then, it is reasonable to assume that the amount of information I need to select a specific combination of shirt and socks is just the *sum* of the information required to perform each one of these two tasks. Again, this also agrees with our intuitive idea of "collecting information". Therefore, we demand the following:

4. Consider two sets of events $\{E_i'\}_{i=1 \cdots n}$ and $\{E_j''\}_{j=1 \cdots m}$, with probabilities $\{P_i'\}$ and $\{P_j''\}$, which are *mutually independent*, and a "double experiment", with the $n \times m$ compound results $\{E_{ij}\}$, which have probabilities $\{P_{ij}\} = \{P_i' \cdot P_j''\}$. Then, we ask $S$ to be *additive*,

$$S(\{P_{ij}\}_{i=1 \cdots n, j=1 \cdots m}) = S(P_1', \ldots, P_n') + S(P_1'', \ldots, P_m'').$$

All the previous requirements, including the last one, sound very general, so we may speculate that they just slightly restrict the class of admissible functions $S$, but the brilliant result obtained by Shannon is truly surprising: these conditions *univocally* define $S$ up to a multiplicative constant $\kappa > 0$. In fact, $S$ is necessarily given by[24]

$$S = -\kappa \sum_{i=1}^{n} P_i \ln P_i. \tag{4.83}$$

---

[24] If some $P_i$'s are zero, we put by convention $P_i \ln P_i = 0$, because $x \ln x \xrightarrow[x \to 0]{} 0$.

Proving that this function is unique is not trivial, but it is easy to show that the statistical entropy defined by (4.83) satisfies all the above requirements.

1. $S$ is clearly a continuous function that does not change by exchanging $P_j \rightleftarrows P_\kappa$, for all possible values of $j$ and $\kappa$.
2. Since $0 \le P_i \le 1$ for all $i$'s, all logarithms are *negative*. Hence, $S \ge 0$.
3. If all the $P_i$'s are the same, so that $P_i = 1/n$, we simply have

$$S = \kappa \ln n, \tag{4.84}$$

   which is a monotonically growing function of $n$.
4. We have

$$S(P_{ij}) = -\kappa \sum_{i=1}^{n} \sum_{j=1}^{m} P_i' P_j'' \ln(P_i' P_j'') = -\kappa \sum_{i=1}^{n} \sum_{j=1}^{m} P_i' P_j'' (\ln P_i' + \ln P_j'') =$$

$$= -\kappa \sum_{j=1}^{m} P_j'' \sum_{i=1}^{n} P_i' \ln P_i' - \kappa \sum_{i=1}^{n} P_i' \sum_{j=1}^{m} P_j'' \ln P_j''.$$

Hence, since both the $P_i'$'s and the $P_j''$'s are normalized,

$$S(P_{ij}) = S(P_i') + S(P_j'').$$

For *generic* (not necessarily independent) events, one can also show that

$$S(P_{ij}) \le S(P_i') + S(P_j'').$$

We can also see that the expression (4.84) for equiprobable events is actually the *maximum* value of $S$.[25] To evaluate this maximum, however, we have to take into account that the $P_i$'s are *constrained* by the normalization condition $\sum_n P_n = 1$, which requires to apply the method of Lagrange multipliers[26], which means minimizing

$$\tilde{S} = S(P_1, \cdots, P_n) - \lambda \left( \sum_{i=1}^{n} P_i - 1 \right). \tag{4.85}$$

---

[25] Since $S \ge 0$, the minimum is clearly $S = 0$, which is attained if and only if there is a certain event $E_j$ with $P_j = 1$.

[26] I guess you are already familiar with Lagrange multipliers. If not, let me just recall that, to find the extrema of the function $f(x_1, x_2, \ldots, x_n)$ subjected to the constraint $g(x_1, x_2, \ldots, x_n) = c$, where $c$ is a constant, we just have to evaluate the *unconstrained* extremal of

$$\tilde{f}(x_1, x_2, \ldots, x_n) = f(x_1, x_2, \ldots, x_n) - \lambda[g(x_1, x_2, \ldots, x_n) - c].$$

The extrema are obtained as a function of the unknown "multiplier" $\lambda$, which is then obtained by applying the constraint condition.

The extrema of $\widetilde{S}$ are found by imposing that, for all values of $j$,

$$\frac{\partial \widetilde{S}}{\partial P_j} = -\kappa(\ln P_j + 1 + \lambda/\kappa) = 0 \implies \ln P_j = -(1 + \lambda/\kappa).$$

Note that, since $P_j$ does *not* depend on $j$, all the $P_j$'s must necessarily be equal to $1/n$. If you are not convinced, just apply the constraint condition,

$$\sum_{j=1}^{n} P_j = 1 \Rightarrow \lambda = \kappa(\ln n - 1) \Rightarrow P_j = 1/n \ \text{ for all } j.$$

When the events $\{E_i\}$ can be associated with the values $\kappa_i$ attained, with probabilities $P_i = P(\kappa_i)$, by a discrete random variable $\kappa$, we say that Eq. (4.83) gives the statistical entropy of the distribution $P(\kappa)$. In this case, we have

$$\boxed{S = -\kappa \, \langle \ln P(\kappa) \rangle} \tag{4.86}$$

**An "objective" view of S.**
We have defined the statistical entropy using a subjective approach, which is fully legitimate in the context of the analysis of random signal developed by Shannon. Yet, we may wonder if $S$ has any intrinsic meaning, namely, whether we can state that a probability distribution can be "objectively" associated with a quantity that gauges its "information content". To see that this is feasible, let us follow a different route, based on an operational approach.

Suppose we try to "construct" a probability distribution $P(k)$ as follows. We divide the total probability in $n$ little packages (which we may call "probability quanta"), each one of value $1/n$, and we ask an "indefatigable monkey" to toss them at random in $r$ boxes, each one labeled with one of the $r$ values that the random variable $k$ can take. Once the monkey is done, we call $k_i$ the number of probability quanta found in box $i$, with $\sum_{i=1}^{r} k_i = n$. When $n \to \infty$, so that the quanta become smaller and smaller, we expect the distribution of the relative frequencies $n_i/n$ to approach a specific probability distribution for $k$. Now we compare this "experimental" distribution with the distribution $P(k)$ we were looking for. If they coincide, we are done, otherwise we ask the (remember, indefatigable!) monkey to repeat the whole procedure, until we get the desired result.

The question is, how long will the monkey have to work? Clearly not a lot, if there are many ways to obtain $P(k)$; much more if $P(k)$ is very hard to obtain with random tosses, exactly as in the problem of finding the right socks. But now we have a way to quantify the statistical entropy of a distribution: we can take it as proportional to the number of rearrangements of probability quanta that yield $P(k)$, in the limit in which these quanta become very small. Note that this is an *intrinsic* property of the distribution, which does not rely at all on the efforts of an indefatigable monkey or of a socks–seeking guy.

From what we have seen in Sect. 2.4, the number of ways to obtain the distribution $\{n_i\}$ is given by the multinomial coefficient

$$M_n^r = \frac{n!}{k_1! k_2! \ldots k_r!}.$$

Then, let us look for the maximum of $M_n^r$, and therefore of the probability of obtaining a given distribution. Of course, we expect this number to grow *very* rapidly with $n$, so we consider the "milder" function $\ln M_n^r / n$ (since this is a monotonically increasing function of $M_n^r$, it must have the *same* maximum). For large $n$, $k_i \to n P_i$, hence

$$\lim_{n \to \infty} \frac{1}{n} \ln M_n^r = \lim_{n \to \infty} \frac{1}{n} \left[ \ln n! - \sum_{i=i}^{r} \ln(n P_i!) \right].$$

Using the Stirling's approximation, it is not difficult to show that

$$\lim_{n \to \infty} \frac{1}{n} \ln M_n^r = \lim_{n \to \infty} \frac{1}{n} \left[ n \ln n - \sum_{i=i}^{r} n P_i \ln(n P_i) \right] =$$

$$= \lim_{n \to \infty} \left[ \ln n - \ln n \sum_{i=i}^{r} P_i - \sum_{i=1}^{r} P_i \ln P_i \right],$$

namely, recalling that the $P_i$'s are normalized,

$$\lim_{n \to \infty} \frac{1}{n} \ln M_n^r = - \sum_{i=1}^{r} P_i \ln P_i,$$

which coincides with (4.83) for $k = 1$. In this view, therefore, $P(k)$ has a large entropy if it is a rather "common" distribution that can be obtained in many ways. Conversely, a distribution with low entropy rarely occurs. Note that, if we consider two distributions $P$ and $P'$ with entropy $S$ and $S'$, and call $M$ and $M'$ the number of ways in which we can respectively obtain $P$ and $P'$, we have $M/M' \sim \exp[N(S - S')]$ which, for $N$ very large, is enormous or negligible depending on whether $S > S'$ or vice versa. In other words, the *vast* majority of distributions generated by the tireless monkey will have a value of statistical entropy close to the maximum. This result, which actually provide the grounds for the method discussed in the following section, has strong similarities with the justification for the existence of the "thermodynamic limit" in statistical mechanics.

**Example 4.11** (*Entropy and data storage*) In Eq. (4.83), we are free to choose any values of $\kappa$, provided it is a positive number, hence in what follows we simply choose $\kappa = 1$. In communication and computer engineering, however, it is usual to choose $\kappa = 1/\ln 2$, so that, using logarithms in base 2, one has

$$S = -\sum_i P_i \log_2 P_i.$$

If we just have two equiprobable events, therefore, $S = 1$: with this choice, the statistical entropy is measured in binary units, better known as *bits*. For example, the entropy associated with the a lottery consisting in the extraction of a number from 1 to 90 (as in the Italian "Lotto") is $S = \log_2 90 \simeq 6.5$ bits.

Consider for instance the "Library of Babel" described in a famous novel by Jorge Luis Borges. In this imaginary library, each book has 410 pages, with 40 lines per page and 40 letters per line, and is written using 25 different characters. If, as Borges writes, the library contains all possible books of this kind (most of them totally meaningless, of course), the information we need to select a specific book is

$$S = \log_2 (25^{40 \times 40 \times 410}) = 6.56 \times 10^5 \log_2 25 \simeq 3\,\text{Mb}.$$

What is the relation between this number and the amount of memory taken by a book of the same kind on the hard disk of a computer? On the one hand, we must observe that, in order to include all the 128 symbols of the ASCII code, each character is encoded with 8 bit $= 1$ Byte (7 bits to choose a character, plus one "parity" bit). On the other, however, we must say that Borges' books are slightly anomalous, since they contain all possible combinations of characters, regardless of any grammar or semantic rules, like they have been written by the previously mentioned indefatigable monkey.[27]

A meaningful book with the same number of characters may require much less memory if it is *compressed*. Compression algorithms can be vary complicated, but they all exploit the fact that, in a real text, letters actually form words with a meaning, and that the number of meaningful words is limited even in the richest tongues. Hence, instead of memorizing all characters, we can, for example, register only the page number and the position in a page where each word is found, with a consistent reduction of the required memory.

### 4.7.2  *The Principle of Maximum Entropy*

In Chap. 8 we will extensively deal with what the so-called *inverse problems*, that is, how from a set of experimental data one can judge the "goodness" of a probability distribution $P(k)$, chosen to describe a quantity with discrete values. However, when only very limited information is available on the distribution, such as the value of $\langle k \rangle$, the concept of statistical entropy can already be useful for deducing some general characteristics of $P(k)$, using a principle based on *Bayesian inference* (an

---

[27] Note that many of these books will fully include Dante's Divine Comedy, which consists of about $4 \times 10^5$ characters.

approach further discussed in Chap. 6), stating that, among the probability distributions compatible with a given amount of information we have obtained, the best possible assumption corresponds to the one with the highest entropy. So, if we know absolutely nothing about $P(k)$, we will assume a uniform distribution as a test distribution. But what happens if we know, for example, that the distribution has a well-defined expectation? In this case, we must maximize $S$ in the presence of *two* constraints, $\sum_i P_i = 1$ and $\sum k_i P_i = \langle k \rangle$. The problem is solved by introducing a *second* Lagrange multiplier $\beta$ in Eq. (4.85), so to maximize

$$\widetilde{S} = -\sum P_i \ln P_i - \lambda \left( \sum P_i - 1 \right) - \beta \left( \sum k_i P_i - \langle k \rangle \right).$$

Therefore, we must have for all $j$:

$$\frac{\partial \widetilde{S}}{\partial P_j} = -(\ln P_j + 1 + \lambda + \beta k_j) = 0 \Longrightarrow P_j = \frac{1}{Z} e^{-\beta k_j},$$

where we have set $Z = [\exp(-1 - \lambda)]^{-1}$. Note that the distribution for $k$ is no longer uniform, but *exponential*. To evaluate $Z$ we could substitute the expression for $P_j$ in the constraint equations and solve for $\lambda$ and $\beta$, but it is more convenient to observe that, for the probabilities $P_j$ to be normalized, we must have

$$Z = \sum_{i=1}^{n} \exp(-\beta k_i). \tag{4.87}$$

Therefore $Z$ is actually a function $Z(\beta)$ of the parameter $\beta$. But what is the meaning of $\beta$? Applying the second constraint equation, we have

$$\langle k \rangle = \frac{1}{Z} \sum_{i=1}^{n} k_i \exp(-\beta k_i) = -\frac{1}{Z} \frac{\partial}{\partial \beta} \left( \sum_{i=1}^{n} \exp(-\beta k_i) \right),$$

wherefrom

$$\langle k \rangle = -\frac{1}{Z} \frac{\partial Z}{\partial \beta} = -\frac{\partial \ln Z}{\partial \beta}, \tag{4.88}$$

an implicit and generally non-invertible analytical relationship, which nevertheless shows that the expectation of $k$ is actually set by the partition function and by the value of the parameter $\beta$.

These results can easily be extended to a situation where it is not $\langle k \rangle$ but rather the expectation of a *function* $\phi(k)$ of the variable to be specified. In this case, setting as constraint $\sum \phi(k_i) P_i = \langle \phi(k) \rangle$, we similarly obtain

$$\begin{cases} P_j = Z^{-1} \exp[-\beta \phi(k_j)] \\ Z = \sum_{i=1}^{n} \exp[-\beta \phi(k_i)] \end{cases} \tag{4.89}$$

so that

$$\langle \phi(k) \rangle = -\frac{\partial \ln Z}{\partial \beta}. \tag{4.90}$$

Therefore, the principle of maximum entropy enables us to significantly reduce the range of possible probability distributions for a random variable. However, the principle only identifies the *broadest* class of distributions compatible with some information, and does not guarantees that $P(k)$ is not provided with a *finer* structure. The following example clarifies this statement.

**Example 4.12** (*A loaded die*) Suppose we know that a die is definitely loaded, because in 60% of the tosses the faces marked with even numbers appear. The single normalization condition is then replaced by the two separate conditions for the probabilities of the "even" and "odd" faces,

$$\begin{cases} P_2 + P_4 + P_6 = \sum_{i=1}^{3} P_{2i} = 0.6 \\[2mm] P_1 + P_3 + P_5 = \sum_{i=1}^{3} P_{2i-1} = 0.4 \,. \end{cases}$$

The maximum entropy principle states that we have to maximize

$$\widetilde{S} = -\sum_{i=1}^{6} P_i \ln P_i - \lambda_f \left( \sum_{i=1}^{3} P_{2i} - 0.6 \right) - \lambda_d \left( \sum_{i=1}^{3} P_{2i-1} - 0.4 \right).$$

By imposing for each variable $\partial S/\partial P_j = 0$, we easily obtain

$$\begin{cases} P_2 = P_4 = P_6 = 1/\exp(1 + \lambda_f) = 0.2 \\ P_1 = P_3 = P_5 = 1/\exp(1 + \lambda_d) = 2/15, \end{cases}$$

where the last equality follows from the constraint conditions, taking into account that the probabilities for the even and the odd faces are equal to each other. However, it would be really strange if the die had actually been loaded in this way, don't you think so? Wouldn't it be simpler to think that the die was unbalanced in such a way that, for example, $P_6 = 4/15$ and $P(1) = 1/15$, while the other probabilities remain equal to 1/6? In Chap. 6 we will see that the possible absence of an important piece of information is a major risk for any inference based on Bayes' theorem.

### 4.7.3  Relative Entropy and Mutual Information

Relative entropy gives a kind of estimate of how "distant", in terms of information content, are two probability distributions $P(k)$ and $Q(k)$ where $k$ takes the same set of values $\{k_i\}$. It is defined as

$$D(P \parallel Q) = -\sum_i P_i \ln \frac{P_i}{Q_i} = \left\langle \ln \frac{P(k)}{Q(k)} \right\rangle_P, \tag{4.91}$$

where $P_i = P(k_i)$, $Q_i = Q(k_i)$, and $\langle \cdots \rangle_P$ is an expectation over the distribution $P(k)$. In (4.91) we put $P_i \ln \frac{P_i}{Q_i} = 0$ if $P_i = 0$, whereas $D(P \parallel Q) = \infty$ if there are one or more values of $i$ such that $Q_i = 0$ and $P_i \neq 0$.

The relative entropy is often called the *Kullback-Leibler distance* (or also *divergence*), although it has not the property of a true distance because it is not symmetric, $D(Q \parallel P) \neq D(P \parallel Q)$, and does not necessarily satisfy the triangle inequality, $D(P + P' \parallel Q) \leq D(P \parallel Q) + D(P' \parallel Q)$. However, using the Jensen inequality it is easy to show that it is a nonnegative quantity and that $D(P \parallel Q) = 0$ only if $P(k) = Q(k)$. Indeed, because of (3.39),

$$-D(P \parallel Q) = \sum_i P_i \ln \frac{Q_i}{P_i} \leq \ln \left( \sum_i P_i \frac{Q_i}{P_i} \right) = \ln \sum_i Q_i = \ln 1 = 0.$$

Let us now introduce *mutual information* as a quantity that gauges the amount of information that two random variables contains about each other, formally defined as the relative entropy between the joint probability distribution $P(k', k'')$ of two variables $k'$ and $k''$ and the product $P(k')P(k'')$ of their marginal distributions,

$$I(k', k'') = \sum_{i,j} P_{ij} \ln \frac{P_{ij}}{P_i' P_j''}, \tag{4.92}$$

where $P_i' = P(k_i')$, $P_j'' = P(k_j'')$, and $P_{ij} = P(k_i', k_j'')$. Mutual information coincides with the reduction of uncertainty of one variable due to the knowledge of the other. Indeed, we can write

$$I(k', k'') = \sum_{i,j} P_{ij} \ln \frac{P(k_i'|k_j'')}{P_i'} = -\sum_i P_i \ln P_i + \sum_{i,j} P_{ij} \ln P(k_i'|k_j'').$$

Hence

$$I(k', k'') = S(k') - S(k'|k''), \tag{4.93}$$

where

$$S(k'|k'') = -\sum_{i,j} P_{ij} \ln P(k_i'|k_j'') \tag{4.94}$$

can be interpreted as the *conditional entropy of $k'$ given $k''$*. Thus the entropy of $P(k')$ is reduced by the amount of information due to the knowledge of $P(k')$. Of, course, this result is perfectly symmetric, and we also have

$$I(k', k'') = S(k'') - S(k''|k').$$

### *4.7.4  The Problem of Continuous Variables*

So far, we have only been dealing with discrete variables. Defining a statistical entropy for a random variable $X$ that takes on continuous values in the interval $[a, b]$ is a much thornier problem. We may try to attack the problem by dividing $[a, b]$ in $n$ subintervals of width $\delta x = (b - a)/n$. The probability that $x$ lies in the $n$-th subinterval can then be written as $P_i \simeq f(x_i)(b - a)/n$, where $f(x_i)$ is the probability density for $x$, evaluated at a generic point $x_i$ inside the subinterval. Hence, we have

$$S(\{P_i\}) = -\sum_{i=1}^{n} P_i \ln(P_i) = -\left[ \sum_{i=1}^{n} \frac{b - a}{n} f(x_i) \ln f(x_i) + \ln \frac{b - a}{n} \right],$$

where we have used $\sum_i P_i = 1$. Now we should take the limit $n \to \infty$. Yet, whereas the first term in brackets tends indeed to $\int_a^b f(x) \ln f(x) \, \mathrm{d}x$, the second one diverges! This has an intuitive explanation: to localize a single point on an interval, we need an *infinite* precision, hence an infinite amount of information.[28]

Can we find a way out? Observing that the diverging term does not depend at all on the specific probability distribution $f(x)$ we are considering, we could simply neglect it, and *define* the entropy of a probability density as

$$S[f(x)] = -\int_a^b f(x) \ln f(x) \mathrm{d}x. \tag{4.95}$$

There are, however, two crucial issues. First, if we consider a variable that is strongly localized around a single value $x_0$, and we take the limit $\epsilon \to 0$ of $f(x) = 1/2\epsilon$, with $|x - x_0| \le \epsilon$, we obtain

$$S[f(x)] = -\frac{1}{2\epsilon} \ln \frac{1}{2\epsilon} \int_{x_0 - \epsilon}^{x_0 + \epsilon} \mathrm{d}x = \ln(2\epsilon) \xrightarrow[\epsilon \to 0]{} -\infty,$$

namely, $S[f(x)]$ *is not positive definite*. Besides this, what may ever be the physical meaning of the logarithm of a quantity such as $f(x)$ that, as we know, is *not* dimensionless? The simplest way to solve both these problems is by introducing a "minimal localization" $\delta x$ for $x$, putting

$$\boxed{S = -\int_a^b f(x) \ln[f(x)\delta x]\mathrm{d}x = - \langle \ln f(x)\delta x \rangle} \tag{4.96}$$

---

[28] Once again, it is useful to recall the warning given in Chap. 1 about the use of continuous quantities in physics.

which corresponds to use a "coarse–grained" probability distribution $f(x)\delta x$. Note that this minimal resolution $\delta x$ does not influence the *difference* between the statistical entropies of two distributions.

The second issue is that we must take great care when we change variables. In fact, if we evaluate $S$ for a random variable[29] $Y = f(X)$, we find, from the "golden rule" (4.1),

$$\int_{f(a)}^{f(b)} p_y(y)\ln[p_y(y)\delta y]\mathrm{d}y = \int_a^b p_x(x)\ln\left[p_x(x)\left|\frac{\mathrm{d}x}{\mathrm{d}y}\right|\delta y\right]\mathrm{d}x.$$

If we wish this result to coincide with Eq. (4.95), we must take $\delta y = |\mathrm{d}y/\mathrm{d}x|\delta x.$, so $\delta y$ and $\delta x$ are not independent. Fixing $\delta x$ fixes $\delta y$ and vice versa: it is up to us deciding *which* is the "reference" variable, and this may not be trivial. For a variable uniformly distributed in $[0, a]$ (with $a \geq \delta x$), Eq. (4.96) gives

$$S = -\frac{1}{a}\int_0^a \ln\frac{\delta x}{a}\,\mathrm{d}x = \ln\frac{a}{\delta x},$$

which *vanishes* for a variable localized with the maximal accuracy $\delta x$.

Nevertheless, while the statistical entropy of a continuous variable cannot be properly defined unless a minimal uncertainty is fixed, the relative entropy between two continuous variables $X$ and $Y$ remains well-defined and nonnegative,

$$D(p \parallel q) = \int_{-\infty}^{+\infty} p(x)\ln\frac{p(x)}{q(x)}\,\mathrm{d}x. \tag{4.97}$$

Furthermore, it is easy to show (try!) that it is invariant upon a variable transformation $x \to y(x)$.

**Example 4.13** (*Entropy of the Gaussian distribution*) For a Gaussian probability density, $g(x) = g(x; \mu, \sigma)$, noticing that

$$\ln[g(x)\delta x] = \ln\left[\frac{\delta_x}{\sigma\sqrt{2\pi}}\right] - \frac{(x-\mu)^2}{2\sigma^2},$$

we have

$$S[g(x)] = \ln\frac{\sigma\sqrt{2\pi}}{\delta_x} + \frac{1}{2\sigma^2}\langle(x-\mu)^2\rangle = \ln\frac{\sigma\sqrt{2\pi}}{\delta_x} + \frac{1}{2},$$

namely,

$$\boxed{S[g(x)] = \ln(\sigma'\sqrt{2\pi\mathrm{e}})} \tag{4.98}$$

where $\sigma' = \sigma/\delta x$ is the standard deviation in units of $\delta x$.

---

[29] For the sake of simplicity, we assume that $f$ is a monotonic function, but the general case is not much more complicated.

We can actually show that the Gaussian distribution has the largest entropy among all the distributions $f(x)$ defined in $x \in (-\infty, +\infty)$ having the same variance. Indeed, since its entropy does not depend on $\mu$, we can choose for $\mu$ the expectation $\langle x \rangle_f$ of $f(x)$, obtaining

$$S[f(x)] = -\int_{-\infty}^{\infty} f(x)\ln[f(x)\delta x]\,\mathrm{d}x = \int_{-\infty}^{\infty} f(x)\ln\frac{g(x)}{f(x)}\,\mathrm{d}x - \int_{-\infty}^{\infty} f(x)\ln[g(x)\delta x]\,\mathrm{d}x.$$

Observing that the logarithm is a *concave* function of its argument, and applying to the first integral the Jensen inequality (3.39) with an inverted sign, we see that this integral is always negative,

$$\int_{-\infty}^{\infty} f(x)\ln\frac{g(x)}{f(x)}\mathrm{d}x = \left\langle \ln\frac{g(x)}{f(x)} \right\rangle_f \leq \ln\left\langle \frac{g(x)}{f(x)} \right\rangle_f = \ln\int_{-\infty}^{\infty} g(x)\mathrm{d}x = 0.$$

Repeating the calculation made to obtain (4.98) with $\mu = \langle x \rangle_f$, the second integral is found to be

$$\int_{-\infty}^{\infty} f(x)\ln[g(x)\delta x]\,\mathrm{d}x = -\ln(\sigma'\sqrt{2\pi\mathrm{e}}),$$

so that, for all densities $f(x)$, $S[f(x)] \leq \ln(\sigma'\sqrt{2\pi\mathrm{e}}) = S[g(x)]$.

When it comes to the principle of maximum entropy, Eq. (4.89) can simply be extended to the case of a continuous variable by writing

$$\begin{cases} f(x) = Z^{-1}\exp[-\beta\phi(x)] \\ Z(\beta) = \int \exp[-\beta\phi(x)]\mathrm{d}x \end{cases} \tag{4.99}$$

where, in analogy with (4.90):

$$\langle \phi(x) \rangle = -\frac{\partial \ln Z(\beta)}{\partial \beta}. \tag{4.100}$$

**Example 4.14** (*Entropy of a random distribution of point-like events*) Let us consider a sequence of point-like events that occur in time according to a law that we do not know a priori, and call $f(t)\mathrm{d}t$ the probability that, if we observe an event at $t = 0$, the next event occurs between $t$ and $t + \mathrm{d}t$. Assuming we *only* know that the average waiting time between two successive events is $\tau$, what is the probability distribution corresponding to the maximum entropy? From (4.99), with $\phi(t) = t$ and $\langle t \rangle = \tau$, we have

$$Z(\beta) = \int_0^{\infty} \exp[-\beta t]\mathrm{d}t = \frac{1}{\beta}.$$

According to (4.100), therefore,

$$\tau = -\frac{\partial \ln(\beta^{-1})}{\partial \beta} = \frac{\partial \ln \beta}{\partial \beta} \implies \beta = \frac{1}{\tau}$$

so that

$$f(t) = \tau^{-1} \exp[-t/\tau],$$

which, with $\tau = 1/\alpha$, is once again a distribution of events in time that follow a Poisson statistics.

### 4.7.5 Statistical Versus Thermodynamic Entropy

Some of you have surely noticed the close relationship between Shannon's statistical entropy and the statistical interpretation of the thermodynamic entropy introduced by Gibbs. The Gibbs entropy of a system at equilibrium can indeed be seen as the statistical entropy associated with the probability distribution of the energy microstates. For an isolated system, this probability distribution is uniform, which immediately leads to the Boltzmann celebrated expression. For a system in equilibrium with a reservoir, one can notice that the reservoir temperature $T$ univocally fixes the average energy of the system as

$$\langle E \rangle = \frac{\sum_{\{i\}} E_i e^{-E_i/k_B T}}{\sum_{\{i\}} e^{-E_i/k_B T}} \tag{4.101}$$

where the summation is over all possible microstates of the system. While in an isolated system the total energy is a fixed parameter, in a system in thermal equilibrium with a reservoir the energy is a fluctuating variable, but with a fixed *average* value.

One can then turn around the question and find the probability distribution for $E$ by maximizing the statistical entropy with the constraint that $\langle E \rangle$ is fixed. As in Sect. 4.7.2, this approach leads to the canonical Maxwell-Boltzmann distribution at a temperature $T$ implicitly *defined* by (4.101). However, this is meaningful only if the value of $T$ obtained by inverting the former relation is *positive*. The maximum entropy approach does not therefore provide an important piece of physical information, the positivity of absolute temperature, which instead comes out directly from a standard approach that balances energy exchanges with a reservoir once one notices that entropy must be a nondecreasing function of energy.

It is worth noting that in our previous discussion, we assumed that the microstates of any finite–size thermodynamic system are *discrete*, which has actually been a stunning result of the novel quantum view of the physical world. In classical mechanics, where the Boltzmann entropy is commensurate to the size of the region of motion in the phase space, all the difficulties in defining the statistical entropy of a continuous variable come back to haunt us. For sure, they haunted Boltzmann all his life long.[30]

---

[30] The Gibbs approach is even more problematic. For instance, defining the entropy as $S = \int_{\mathcal{V}} \rho \ln \rho \, d\mathcal{V}$, where $\rho$ is the density of points in the phase space $\mathcal{V}$, can be rather embarrassing. Indeed, because of the Liouville theorem, $\rho$ is a time–invariant quantity, which means that the entropy would be a constant of motion. Attempts to define a "coarse-grained" density that escapes Liouville's theorem are not usually free from ambiguity. The fact is, while the Boltzmann approach

The Boltzmann view of entropy as a measure of the freedom of motion, or more generally of the richness of different configurations of a macroscopic system is fully in line with the "objective" approach to statistical entropy we presented in Sect. 4.7.1, which can be simply rephrased by saying that a distribution has a large entropy when we are free to build it in many ways.

However, the same caution about the potential lack of information that we mentioned in the Example 4.12 regarding the Shannon entropy applies to thermodynamic entropy. Consider for example the Gibbs paradox about the entropy of mixing $\Delta S$, which is usually solved by showing that, if the atoms of the two species are identical and entropy is defined by counting only their distinguishable rearrangements, $\Delta S = 0$, as it obviously should.

Obviously? Then, if we open up a valve connecting two reservoirs containing $CO_2$ at the same temperature and pressure, we should conclude that $\Delta S = 0$ because, in fact, "nothing happened". Which sounds true, but only because the mixing process did not take thousands of years: otherwise, we would have discovered that, while the $CO_2$ in one reservoir contained the common stable carbon isotope $^{12}C$, the carbon dioxide in the other reservoir contained carbon $^{14}C$, which decays ($\beta^-$) in about 5700 years (so the entropy *did* increase). So, should we speak of entropy or of *entropies*?

---

refers to a *single* system, for which a region of motion is always defined (at least formally), it is not clear what is meant, out of equilibrium, by a distribution of "equally—prepared" systems.

# Chapter 5
# Elements of Random Processes

*It is possible that the movements to be discussed here are*
*identical with the so-called 'Brownian motion'; however, the*
*information available to me*
*regarding the latter is so lacking in precision*
*that I can form no judgement on the matter*

*A. Einstein*

Einstein's 1905 seminal paper,[1] wherefrom the rather skeptical quotation opening this chapter is taken from, is the gateway to the physics of Brownian motion, the archetype of the subject of this chapter: *random* (often called *stochastic*) *processes*, a designation that indicates processes where randomness plays a leading role. By "process" we usually mean a series of events taking place in an *ordered sequence*, so the most common case (but not necessarily the only one) is that of processes happening in time. The crux of the matter is that we can distinguish what *precedes* (in time or space) an event from what *follows* it. This already suggests that random time processes are intrinsically linked with the arrow of time and, consequently, with irreversibility. Indeed, we will see that the link between Brownian motion and diffusion discovered by Einstein necessarily leads to relate random fluctuations to dissipation effects.

Random processes are pervasive in statistical physics, but they are also of paramount importance in other fields ranging from condensed matter to astrophysics. The aim of this chapter is not offering you a complete understanding of random phenomena. I just hope to provide you with an adequate introduction to the methods of stochastic analysis by means of prominent examples in physics. Actually, we have

---

[1] A. Einstein, Ann. Phys. **322**, 549 (1905). Einstein's theory of Brownian motion was explicitly mentioned by Svante Arrhenius, chairman of the Nobel Committee, as a motivation for the 1921 prize for physics (actually awarded to Einstein only one year later).

     233
R. Piazza, *An Invitation to Probability and Data Analysis for Physicists*, UNITEXT for Physics, https://doi.org/10.1007/978-3-031-83856-9_5

already met two important kinds of random processes. The first one is of course random walk in one or more dimension that, as we mentioned, is the simplest abstract model of Brownian motion. An extension to what we will call "continuous-time" random walk will lead us to introduce the concept of *Wiener process* and several very important *stochastic equations*. The other one is related to the waiting time statistics of events that occur according to a Poisson distribution that we studied in Examples 3.18 and 3.20, and consists in the number $N(t)$ of Poisson-distributed events taking place *up to time $t$*. These Poisson processes are an important examples of the more general class of *point processes*.

A random process, or simply RP, can be rigorously defined as a family $X(t) \doteq \{X_t\}_{t \in T}$ of random variables. The index $t$ can either belong to a discrete set (ex., $T = \mathbb{N}$) or to a continuous set (ex., $T = \mathbb{R}$) of values, which allows us to distinguish between discrete and continuous random processes. A further distinction is that the variables $\{X_t\}$ can themselves take discrete or continuous values. The instantaneous value of a RP is called its *amplitude*, because this is usually its interpretation in most physical applications. We can therefore distinguish[2]:

(a)  discrete-index RPs with a discrete amplitude;
(b)  discrete-index RPs with a continuous amplitude;
(c)  continuous-index RPs with a discrete amplitude;
(d)  continuous-index RPs with a continuous amplitude.

A continuous RP can also be regarded as a *random function $X(t)$*. Remember, however, that a RV is *already* a function of the events $\omega$ in the sample space $\Omega$, or more precisely in the $\sigma$-algebra $\Sigma$, so a continuous RP is actually a function $X(\omega, t) : \Sigma \times \mathbb{R} \longrightarrow \mathbb{R}$. Therefore, a RP is actually an ensemble of individual *realizations* obtained by prescribing, for each $t$, the value $x(t)$ taken by $X(\omega, t)$.

In most cases, we will identify the variable $t$ with time, which is the most natural choice, taking into account the intrinsic directionality of a process. But of course one can imagine a more general class of random functions, for example a family of RVs defined on every point of a surface or, in general, of a $d$-dimensional space, $X(\omega, \mathbf{r}) : \Sigma \times \mathbb{R}^d \longrightarrow \mathbb{R}$, where $\mathbf{r} = \{r_1, r_2, \ldots r_d\}$, but, rather than a random "process", this is more properly called a random *field*. Several of the concepts we will develop for RPs can be easily extended to random fields.

**Example 5.1** (*Simple random processes*) Before we introduce those mathematical tools that are required to investigate RPs, let us first discuss some additional simple processes we will often refer to in the following.

WHITE NOISE:    One of the key tasks for an experimentalist in all fields of physics is distinguishing signals from noise. Noise, however, is a rather generic word that refers to several different kinds of random perturbation effects. Different types of noise are usually classified according to their "color", a property that is related, as we will see, to their spectral properties. At the beginning of the 20th century, Lord

---

[2] In the following, discrete-index and continuous-index RPs will be simply called *discrete* and *continuous* RPs.

Rayleigh introduced the concept of *white noise* in acoustic as a disturbance that is "similar to white light, which contains all colors in equal intensity". Although this comparison is of course incorrect (white light spectrum is *not* flat), this concept is central in Shannon's mathematical theory of communications. White noise is actually the simplest example of a discrete-time, continuous amplitude RP, and consists of a sequence of $n$ independent, identically distributed (i.i.d.) random variables $\epsilon(t_1), \epsilon(t_2), \ldots, \epsilon(t_n)$, where the time increment $\Delta t = |t_j - t_i|$ (with $i \neq j$) is assumed to be constant, all of which have a zero average, $\langle \epsilon(t_i) \rangle = 0$. In the limit $\Delta t \to 0$ one obtains a continuous RP that is usually called "pure" white noise. As we will see, although presenting several nonphysical properties like a spectrum infinite bandwidth, this limiting process is extremely useful to develop a consistent physical model of Brownian motion.

RANDOM TELEGRAPH:    A special kind of "digital" noise, originally discovered in reverse–biased semiconductor p-n junctions, is *burst noise*, which consists of sudden transitions between two or more voltage or current levels at random times.[3] While its physical origins remain uncertain, burst noise lends itself to a simple modeling as *random telegraph signal*. A continuous-time RP $\{X(t), t \geq 0\}$ is called a random telegraph signal if (i) its amplitude can take only the values $\pm 1$, and (ii) the times at which amplitude switching occurs are Poisson–distributed. Then, if $N(t)$ is the number of switchings occurring up to time $t$, we have $X(t) = (-1)^{N(t)} X(0)$, where the initial amplitude $X(0)$ is $\pm 1$ with the same probability.[4] Besides noise in MOSFETs and operational amplifiers, a random telegraph signal has recently been used to model dephasing in spin-qubit systems and fluorescence intermittency in quantum dots.

RANDOM OSCILLATIONS:    Oscillating quantities where either the amplitude or the phase display a (partially or fully) random behavior are RPs that occur in fields ranging from acoustic to statistical optics. Several noise sources can for instance introduce a random modulation of the amplitude of a sinusoidal signal. A random–amplitude signal can then be written as

$$X(t) = a \cos(\omega t + \phi) = a \operatorname{Re}\left[e^{i\omega t + \phi}\right] \tag{5.1}$$

and its statistical properties will then depends on those of the random variable $a$. Random amplitude oscillations play an important role in random signal analysis based on Fourier decomposition. In other cases, noise affects the signal generated by an oscillator, introducing a random phase modulation $\varphi(t)$ that, as we will see, has an important effect on the spectrum of the signal. A random–frequency oscillation can then be represented as

---

[3] When the noise source is connected to an audio system it produces a popping sound, which is why burst noise is also known as "popcorn noise".

[4] Note that every random process $R_{xx}(t)$ whose amplitude switches randomly (i.e., according to a Poisson distribution) between two values $R_1$ and $R_2$, can be expressed using a random telegraph process $X(t)$ as $R_{xx}(t) = \overline{R} + \Delta_R X(t)$, where $\overline{R} = (R_1 + R_2)/2$ and $\Delta_r = |R_2 - R_1|/2$.

$$X(t) = a_0 \cos[\omega t + \phi] = a_0 \operatorname{Re}[e^{i(\omega t + \phi)}] \qquad (5.2)$$

where the either the frequency $\omega$, or the initial phase $\phi$, or both are random variables.[5]

## 5.1  Mathematical Tools for Random Processes

A RP is a collection of random variables that, in general, are neither independent nor identically distributed. Therefore, a full description of a RP would in principle require to know the joint probability distribution of all the involved variables. For a discrete RP, this amounts to state the $n$-point joint probability distribution

$$P(X_1 \le x_1 \cap X_2 \le x_2 \cap \ldots \cap X_n \le x_n)$$

for all values of $x_1, x_2, \ldots x_n \in \mathbb{R}$. For a continuous RP, one should rather specify the $n$-point joint probability density

$$f_n(x_1, t_1; x_2, t_2; \ldots x_n, t_n), \qquad (5.3)$$

which gives the probability density that the RV has an amplitude $x_i$ at time $t_i$ for $i = 1, 2, \ldots n$ (or the corresponding joint probability densities $f_n(x_1, t_1; x_2, t_2; \ldots x_n, t_n)$ for a RP with continuous amplitude), hoping that, for a rather small value of $n$, they already yield a reasonable representation of the process. Except in very simple cases, however, this is a prohibitive task. Luckily, many features of RPs can be captured by means of a simpler approach, grounded in the concepts of mean and correlation functions, which require only the lowest order distributions $P(x, t)$ (or $f(x, t)$) and $P_2(x_1, t_1; x_2, t_2)$ (or $f_2(x_1, t_1; x_2, t_2)$).

### 5.1.1  The Mean Function

The mean function $\mu_x(t)$ of a random process is simply the expectation of $X$ as a function of $t$, provided of course that the latter exists. Thus, for a continuous RP, it is given by

$$\boxed{\mu_x(t) = \langle x(t) \rangle = \int_{-\infty}^{+\infty} x f(x, t) \mathrm{d}x} \qquad (5.4)$$

---

[5] According to our previous convection, we should actually call $\Omega$ and $\Phi$ the frequency and initial phase *variables*, reserving $\omega$ and $\phi$ for their *values*, but this would make the notation a bit clumsy.

while for a discrete RP we have, for example, $\mu_x(n) = \langle x_n \rangle$ for all $n \in \mathbb{N}$. The mean function provides us with a picture how the RP evolves on average in time.

**Example 5.2** (*Mean function of simple random processes*) Let us figure out the mean functions of the RPs that we have introduced in Example 5.1:

WHITE NOISE:     Since the white noise process consists of identically distributed random variables with zero expectation, $\langle \epsilon(t_i) \rangle = 0$, the mean function identically vanishes for all values of $t$.

RANDOM OSCILLATIONS:     In a signal of the form (5.1), the random amplitude $a$ is decoupled from the (deterministic) time dependence $\cos(\omega t + \phi)$. Therefore, the mean function oscillates in time between $\pm\langle a \rangle$,

$$\mu_x(t) = \langle a \rangle \cos(\omega t + \phi).$$

For a random phase–modulation signal $X(t) = a_0 \cos[\omega_0 t + \phi]$, where $\phi$ is uniformly distributed in $[0, 2\pi]$, while the amplitude $a_0$ and the frequency $\omega_0$ are fixed, we have

$$\mu_x(t) = a_0 \int_0^{2\pi} \cos[\omega_0 t + \phi]\, d\phi = 0,$$

which shows that, as expected, the mean function of a superposition of oscillation with a uniformly distributed initial phases vanishes for any $t$. Let us instead consider a random-frequency signal $X(t) = a_0 \cos[\omega t]$, where $\omega$ is uniformly distributed over an interval $[0, \Omega]$. We have

$$\mu_x(t) = \frac{a_0}{\Omega} \int_0^{\Omega} \cos(\omega t)\, d\omega = a_0 \frac{\sin(\Omega t)}{\Omega t} = a_0 \operatorname{sinc}(\Omega t),$$

where $\operatorname{sinc}(x) = \sin(x)/x$. Figure 5.1A shows that the mean function is characterized by a peak of width $\omega t \simeq 2\pi$ centered on the time origin (we are taking $-\infty < t < +\infty$) with oscillations on both sides progressively decreasing in amplitude.

RANDOM WALK:     The position $X(t)$ in a 1-dimensional random walk of unitary length is $X(t) = 2k(t) - N(t)$, where $k(t)$ is the number of steps in the positive $x$-direction at time $t$ over a total of $N(t)$. If the probability of taking a step along $+x$ is $p$, then $\langle k(t) \rangle = pN(t)$, so that $\mu_x(t) = (2p - 1)N(t)$. Hence, if $N(t) = t/\tau$, where $\tau$ is the time for a single step, $\mu_x(t)$ grows linearly in time (unless $p = 1/2$, when it vanishes for any $t$).

POISSON PROCESS:     The average number of events occurring up to time $t$ is $\mu_x(t) = \langle N(t) \rangle = \alpha t$, where $\alpha$ is the event frequency. So, the mean values grows linearly with time in this case too. Note, however, that if the Poisson process consists in considering the number of event $\Delta N_{ij}(t)$ taking place *between $t_j$ and $t_i$*, where $\delta_{ij}(t) = t_j - t_i$ is kept constant, then $\mu_x(t) = \langle \Delta N_{ij}(t) \rangle$ is constant too.

RANDOM TELEGRAPH:    Take $X(0) = 0$. At time $t$ the value of $X(t)$ will be 0 or depending on whether the number $n$ of switches in $[0, t]$ is even or odd, respectively, i.e., $X(t) = n \mod 2$. The probability that exactly $n$ switches occur in $[0, t]$ is $(\alpha t)^n \exp(-\alpha t)/n!$, where $\alpha$ is the frequency of the switches. Since

$$\mu_x(t) = 1 \times P[X(t) = 1] + 0 \times P[X(t) = 0] = P[X(t) = 1],$$

the mean function is given by

$$\mu_x(t) = P[X(t) = 1] = e^{-\alpha t} \sum_{n \text{ odd}} \frac{(\alpha t)^n}{n!}.$$

Recalling that

$$\sum_{n \text{ odd}} \frac{(x)^n}{n!} = \sinh(x) = \frac{1}{2}\left(e^x - e^{-x}\right),$$

we finally have

$$\mu_x(t) = \frac{1}{2}\left(1 - e^{-2\alpha t}\right),$$

which show that the mean function approaches exponentially the constant value $\mu_x(\infty) = 1/2$. You can easily show that, if we choose $X(0) = 1$, the mean function turns out to be

$$\mu_x(t) = \frac{1}{2}\left(1 + e^{-2\alpha t}\right)$$

which also approaches exponentially $\mu_x(t) = 1/2$. Therefore the random telegraph looses any memory of the initial state on a time scale of the order of $\tau = 1/\alpha$. Both cases are shown in Fig. 5.1B.

For the white noise, but also for the RW with $p = 1/2$ and for the "incremental" Poisson process we mentioned, the mean function does not depend on time, and said to be *stationary*. However, stating that the mean function is stationary does not mean that the *random process* is stationary, which is a much stronger condition. A RP is said to be *stationary in the strict sense* if and only if *all* joint probability distributions are invariant upon an arbitrary time shift $\Delta t$, that is, for every $n$,

$$f_n(x_1, t_1; x_2, t_2; \ldots x_n, t_n) = f_n(x_1, t_1 + \Delta t; x_2, t_2 + \Delta t; \ldots x_n, t_n + \Delta t). \quad (5.5)$$

However, many RPs can be considered stationary in a less restrictive sense, which we will later specify.

The definition of mean function rises another important issue. In real terms, evaluating $\mu_x(t)$ requires to perform an average over all possible realizations of the RP, weighted according to the probability distribution of (in general time-dependent) $X(t)$. Experimentally, however, it is often too demanding to collect a large number of realizations of a RP, so to obtain a good approximation of an ensemble average.

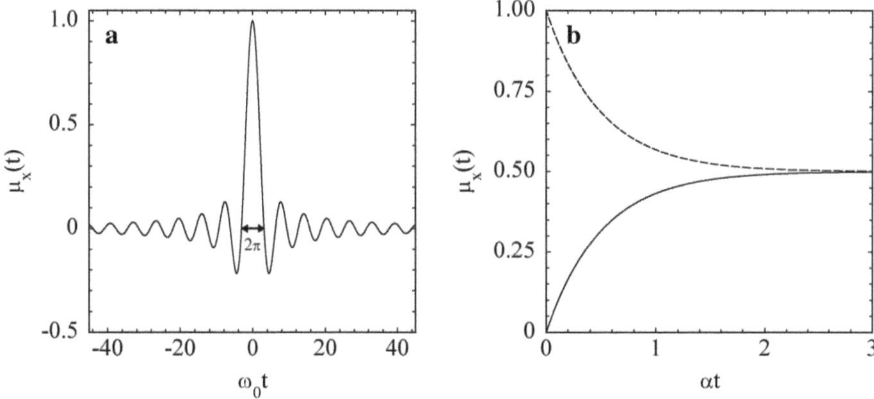

**Fig. 5.1**   Mean function $\mu(t)$ for a random-frequency signal with $\omega$ uniformly distributed in $[0, \omega_0]$ (A) and for the random telegraph signal (B)

It is much easier to measure the *time* average of a single realization

$$\bar{x}_T = \frac{1}{T} \int_0^T x(t)\mathrm{d}t. \tag{5.6}$$

For example, to obtain the Brownian diffusion coefficient $D$ one can profit from optical techniques that, by measuring the correlation in time of the light scattered by a colloid, intrinsically provide an ensemble average over the motion of many particles. However, when the particle displacement is observed under a microscope, acquiring the trajectories of a large number of particles is quite demanding, and one rather settles with measuring the displacement $X(t)$ of a single particle for a sufficiently long time $T$, which yields $D$ as the slope of $X^2(t)$ versus $t$ (which, as we will see, should be a straight line).

But the question is, provided that we choose $T$ long enough, is $\bar{x}_T$ close to $\mu_x(t) = \langle x(t) \rangle$? A first condition is that, obviously, $\mu_x(t)$ should not appreciably vary on a time $T$, i.e., that on this time scale the mean function is quasi-stationary. But this is generally not enough. Some of you may have already spotted the similarity between this question and the *ergodic problem* in statistical mechanics.[6] However, the analogy

---

[6] In statistical mechanics, the issue of ergodicity concerns the relation between the time average of the trajectory of the representative point of a system in the phase space $\Gamma$ and the ensemble average made over the generalized coordinates and momenta of the allowed region of motion on $\Gamma$. Ergodicity is a key issue in dynamical systems theory, but for what concerns statistical mechanics is has probably been a rather misleading problem. Indeed, even if a system were ergodic (which is generally not the case), how long it would take for trajectory of the representative point to cover sufficiently well the whole region of motion? The equivalence between time and ensemble averages is rather suggested by the typicality of the trajectory, which almost always passes through regions of the phase space where the values of the physical quantities approximately coincide with

is only partial and, as we will see, the conditions for the time and ensemble average of a RP to be (almost always) equal are much less stringent than for a complex dynamical system.

### 5.1.2   Correlation Functions

In the first introductory chapter we have stressed the paramount importance of the concept of correlation in physics. In Chap. 4 we have then shown that correlations between two random variables $X, Y$ are essentially embodied in their covariance $\sigma_{xy}$, which does not vanish if and only if $\langle xy \rangle \neq \langle x \rangle \langle y \rangle$. Extending these notions to RPs, which are sequences of random variables, is almost immediate and allows us to introduce the fundamental concept of *n-point correlation function* of a (discrete or continuous) random process. This is defined as the average of the product of $X(t)$ evaluated at $n$ distinct times $t_1, t_2, \ldots, t_n$,

$$R_{xx}(t_1, t_2, \ldots, t_n) = \langle x(t_1)x(t_2) \ldots x(t_n) \rangle. \tag{5.7}$$

We will mainly focus on 2-point (or second order) correlation functions, usually called "correlation function" with no further specification, which can be expressed in terms of the mutual (2-point) distribution function as

$$\boxed{R_{xx}(t_1, t_2) = \langle x(t_1)x(t_2) \rangle = \int_{-\infty}^{+\infty} \int_{-\infty}^{+\infty} x_1 x_2 f(x_1, t_1; x_2, t_2) \, \mathrm{d}x_1 \mathrm{d}x_2} \tag{5.8}$$

In particular, for a continuous RP, putting $t_1 = t$ and $t_2 = t + \tau$, we can write

$$\boxed{R_{xx}(t, \tau) = \langle x(t)x(t + \tau) \rangle} \tag{5.9}$$

In general, $R_{xx}(t, \tau)$ is a function of both $t$ and $\tau$. Quite often, however, the correlation function is only dependent on the time difference $\tau$, not on $t$. If this is the case, we say that the correlation function is stationary and, by shifting the time axis by $-t$, we can write $R_{xx}(\tau) = \langle x(0)x(\tau) \rangle$. Note that a stationary correlation function is also symmetric, $R_{xx}(-\tau) = R_{xx}(\tau)$ (just shift $R_{xx}(\tau)$ by $-\tau$). Stationarity of the correlation function, however, does not imply that the mean function $\mu_x(t)$ is stationary too.

We will only consider RPs where correlations progressively "die out", that is, RPs with a finite temporal correlation range. This means that $x(t_1)$ and $x(t_2)$ become independent after a sufficiently long time, so that

---

their ensemble averages (see, for instance, J. Bricmont, *Science of chaos or chaos in science?*, arXiv:chao-dyn/9603009).

$$\lim_{|t_2 - t_1| \to \infty} \langle x(t_1) x(t_2) \rangle = \langle x(t_1) \rangle \langle x(t_2) \rangle,$$

or, if you prefer, that the *covariance* $\sigma_{12}^2$ of the two variables vanishes. This is the reason why, in many fields (in particular in statistical optics), physicists often prefer to define a "net" correlation function

$$\boxed{C_{xx}(t_1, t_2) = \sigma_{12}^2 = \langle x(t_1) x(t_2) \rangle - \langle x(t_1) \rangle \langle x(t_2) \rangle} \qquad (5.10)$$

which is zero in the limit $|t_2 - t_1| \to \infty$. One can finally make $R_{xx}(t_1, t_2)$ and $C_{xx}(t_1, t_2)$ *dimensionless* by dividing them by the product $\sigma_1 \sigma_2$ of the standard deviations at $t = t_1$ and $t = t_2$,

$$r_{xx}(t_1, t_2) = \frac{R_{xx}(t_1, t_2)}{\sigma_1 \sigma_2} \; , \quad c_{xx}(t_1, t_2) = \frac{C_{xx}(t_1, t_2)}{\sigma_1 \sigma_2}. \qquad (5.11)$$

Observe that $c_{xx}(t_1, t_2)$ is the equivalent for a RP of the correlation coefficient of two random variables.

**Example 5.3**  (*Simple correlation functions*)

WHITE NOISE:  Since white noise is a sequence of *independent* and identically distributed random variables, we have

$$R_{xx}(t_i, t_j) = \langle \epsilon^2 \rangle \delta_{ij},$$

where $\delta_{ij}$ is the Kroneker delta. Thus $R_{xx}(t_i, t_j)$ vanishes for $t_j \neq t_i$ and for $t_i = t_i$ coincides with the average squared amplitude of the signal. Since $\langle \epsilon \rangle = 0$, $\langle \epsilon^2 \rangle$ is also equal to the variance identically distributed variables, so we can write

$$R_{xx}(t_i, t_j) = C_{xx}(t_i, t_j) = \sigma_\epsilon^2 \delta_{ij}.$$

In the continuum limit $\Delta t \to 0$, the correlation function becomes

$$R_{xx}(t, t + \tau) = \langle \epsilon^2 \rangle \delta(\tau),$$

which only depends on the time difference $\tau$.

RANDOM OSCILLATION:  For a random–amplitude oscillation $X(t) = a \cos(\omega_0 t)$ we have

$$R_{xx}(t, \tau) = \langle a^2 \rangle \cos(\omega_0 t) \cos[\omega(t + \tau)] = \frac{\langle a^2 \rangle}{2} [\cos(\omega_0 \tau) + \cos(2\omega_0 t + \tau)],$$

which for $\tau \ll t$ is a frequency-doubled signal with a small phase and amplitude shift. For a random–phase signal $X(t) = a_0 \cos(\omega_0 t + \phi)$, with $\phi$ uniformly distributed in $[0, 2\pi]$ and $a_0, \omega_0$ fixed, we have to evaluate

$$R_{xx}(t, \tau) = a_0^2 \langle \cos(\omega_0 t + \phi) \cos[\omega_0(t + \tau) + \phi] \rangle$$

which, using again the Werner's product-to-sum identity, becomes

$$R_{xx}(t, \tau) = \frac{a_0^2}{2} [\cos(\omega_0 \tau) + \langle \cos(2\omega_0 t + \omega_0 \tau + 2\phi) \rangle].$$

When averaged over the random phase $\phi$, however, the second term vanishes, so we simply have

$$R_{xx}(t, \tau) = \frac{a_0^2}{2} \cos(\omega_0 \tau),$$

which, notably, does not depend on time $t$.

RANDOM TELEGRAPH:    The average $\langle x(t)x(t + \tau) \rangle$ does not vanish if and only if $x(t)$ and $x(t + \tau)$ are both equal to one, thus it is given by

$$P[(x(t + \tau) = 1) \cap (x(t) = 1)].$$

That is, we must have an odd number of switches in $[0, t]$ followed by an even number of switches in $[t, t + \tau]$. Reasoning as we have done to find the mean function, we easily find

$$R_{xx}(\tau) = \frac{1}{4} \left(1 - e^{-2\alpha t}\right) \left(1 + e^{-2\alpha \tau}\right).$$

which does depend on $t$, so the "standard" random telegraph signal is not stationary in the correlation function either.

Let us however consider a modified version of the RP, which starts at $x(0) = 1$ and switches back and forth, according to a Poisson distribution of switching times, between $+1$ and $-1$. Then $x(t)x(t + \tau)$ is $+1$ or $-1$ depending on whether the number of switches in $[t, t + \tau]$ is, respectively, even or odd. So, assuming that $\tau$ can be positive or negative, we have

$$R_{xx}(t, t + \tau) = e^{-\alpha|\tau|} \left[ \sum_{n \text{ odd}} \frac{(\alpha|\tau|)^n}{n!} - \sum_{n \text{ even}} \frac{(\alpha|\tau|)^n}{n!} \right] =$$

$$= e^{-\alpha|\tau|} \sum_{n=0}^{\infty} \frac{(-\alpha|\tau|)^n}{n!} = e^{-2\alpha|\tau|},$$

which does not depend on $t$, but only on the time difference $\tau$.

RANDOM WALK:    We want to find the correlation between the displacement after $n$ and $m > n$ steps of unitary length. Writing $x_k$ for the displacement at step $k$,

$$R_{xx}(m, n) = \langle x_n x_m \rangle = \left\langle \sum_{k=1}^{n} x_k \sum_{k'=1}^{m} x_{k'} \right\rangle = \sum_{k=1}^{n} \sum_{k'=1}^{m} \langle x_k x_{k'} \rangle.$$

But the single steps are independent and, since $\langle x_i^2 \rangle = 1$, $\langle x_k x_{k'} \rangle = \delta_{k,k'}$. Therefore, only the $n$ terms with $k' = k$ are left, and we have

$$R_{xx}(m, n) = \sum_{k=1}^{n} \langle x_k^2 \rangle = n,$$

which, rather surprisingly, does not depend on the number of steps $m - n$ made from the initial position $x_m$. Since the variance of a single unit-length step is $\sigma_i^2 = 4p(1 - p)$, where $p$ is the probability of making a step in the positive $x$-direction, a similar calculation gives

$$C_{xx}(m, n) = n\sigma_i^2 = 4np(1 - p).$$

Then you can easily find that the normalized net correlation function is simply

$$c_{xx}(m, n) = \frac{n}{m},$$

which approaches one if $m \simeq n$. In particular, if we write $k = m - n$ and assume $k \ll n$ we have $c_{xx}(k, n) \simeq 1 - k/n$.

POISSON PROCESS:    Adding and subtracting $N^2(t_1)$, we have

$$R_{xx}(t_1, t_2) = \left\langle N(t_1)N(t_2) + N^2(t_1) - N^2(t_1) \right\rangle = \langle N(t_1)[N(t_2) - N(t_1)] \rangle + \left\langle N^2(t_1) \right\rangle,$$

which, observing that $N(t_2) - N(t_1) = N(t_2 - t_1)$, becomes

$$R_{xx}(t_1, t_2) = \langle N(t_1)N(t_2 - t_1) \rangle + \left\langle N^2(t_1) \right\rangle = \langle N(t_1) \rangle \langle N(t_2 - t_1) \rangle + \left\langle N^2(t_1) \right\rangle,$$

because $N(t_1)$ and $N(t_2 - t_1)$ are independent. For a Poisson process, we have $\langle N(t) \rangle x = \alpha t$, and

$$\left\langle N^2(t) \right\rangle = \sigma_N^2 + \langle N(t) \rangle^2 = \alpha t + \alpha^2 t^2 = \alpha t(1 + \alpha t).$$

Hence we obtain

$$R_{xx}(t_1, t_2) = \alpha^2 t_1(t_2 - t_1) + \alpha t_1(1 + \alpha t_1) = \alpha t_1(1 + \alpha t_2),$$

while the net correlation function is

$$C_{xx}(t_1, t_2) = R_{xx}(t_1, t_2) - \langle N(t_1) \rangle \langle N(t_2) \rangle = \alpha t_1,$$

which does not depend on the time difference $t_2 - t_1$.

We have introduced $R_{xx}(t_1, t_2)$ and $C_{xx}(t_1, t_2)$ to quantify the degree of correlation of $X(t)$ at different values of $t$. This idea can be extended to two distinct RPs $X(t)$ and $X$ by introducing the *mutual* correlation functions

$$\begin{cases} R_{xy}(t_1, t_2) = \langle x(t_1)y(t_2) \rangle \\ C_{xy}(t_1, t_2) = \langle x(t_1)y(t_2) \rangle - \langle x(t_1) \rangle \langle y(t_2) \rangle \end{cases} \tag{5.12}$$

More generally, for a set of $n$ RPs, one can introduce the *correlation matrix*

$$\mathbf{R} = [R_{ij}(t_1, t_2)] = \left[ \langle x_i(t_1)x_j(t_2) \rangle \right], \tag{5.13}$$

with of course its net counterpart $\mathbf{C} = [C_{ij}]$, where the diagonal elements $C_{ii}$ are the "auto"-correlation functions of a process $X_i(t)$ with itself.


### 5.1.3   *Wide-Sense Stationarity and Ergodicity*

As we mentioned, stationarity of a RP strictly requires that all joint probability distributions are invariant upon an arbitrary time shift, which is a very stringent condition. However, many RPs satisfy two weaker requirements, that is, (i) their mean function is stationary, that is $\mu_x(t)$ is a time-independent constant, and (ii) $R_{xx}(t_1, t_2)$ only depends on the difference $|t_2 - t_1|$, which for a continuous process means that the correlation function is a function of $\tau$ alone. If this is the case, the process is said to be *stationary in the wide sense*, or WSS. An example of a WSS process is a random phase–modulation signal with $\phi$ is uniformly distributed in $[0, 2\pi]$.

If a RP is stationary in wide sense, the question of ergodicity that we introduced in Sect. 5.1.1 is easier to address. We will formally state that a WSS process is *ergodic in the mean* if

$$\overline{x(t)} = \lim_{T \to \infty} \bar{x}_T = \lim_{T \to \infty} \frac{1}{T} \int_0^T x(t)\, dt \tag{5.14}$$

is equal to the ensemble average $\langle x(t) \rangle$. But what about the correlation function? Under what conditions the time–average of $R_{xx}(t, \tau)$,

$$\overline{x(t)x(t + \tau)} = \lim_{T \to \infty} \frac{1}{2T} \int_{-T}^T x(t)x(t + \tau)\, dt, \tag{5.15}$$

coincides with $\langle x(t)x(t + \tau) \rangle$? If this is the case, we say the process is ergodic *in the correlation function*. Ergodicity in the mean does not imply ergodicity in correlation, and vice versa.

Actually a RP is said to be ergodic in the strict sense if a single realization contains *all* the statistical information about the process, which is a very stringent condition. As for stationarity, however, one can say that a RP is ergodic in the wide sense if it is ergodic in the mean and in the correlation function. For example, the random–phase oscillation we have previously mentioned is an example of a RP that is ergodic in the wide sense. Indeed, $\overline{x(t)} = a_0\overline{\cos(\omega_0 t + \phi)} = 0$, like $\langle x(t)\rangle$), because the time-average of a simple oscillation vanishes. Moreover, if we write

$$
\overline{x(t)x(t+\tau)} = a_0^2 \overline{\cos(\omega_0 t + \phi)\cos\left[\omega_0(t+\tau)+\phi\right]}
$$

$$
= \frac{a_0^2}{2}\left[\cos(\omega_0\tau) + \overline{\cos(2\omega_0 t + \omega_0\tau + 2\phi)}\right],
$$

the last term in brackets vanishes for the same reason, so the time-averaged correlation function coincides with the ensemble average we previously found.

For WSS processes, one can find a simple condition on $R_{xx}(\tau)$ ensuring that the mean of a RP is ergodic. Notice first that the finite-time average $\bar{x}_T$ defined in (5.6 is a random variable with an expectation that is equal to the mean function,

$$
\langle \bar{x}_T \rangle = \left\langle \frac{1}{T}\int_0^T x(t)\,\mathrm{d}t \right\rangle = \frac{1}{T}\int_0^T \langle x(t)\rangle\,\mathrm{d}t = \langle x(t)\rangle = \mu_x,
$$

where we have exchanged the order of expectation and integration because they are linear operators. However, the question is: will actually $\bar{x}_T$ approach $\mu_x$ for $T \to \infty$? Requiring that this happens for *all* possible realizations of $X(t)$ is a bit too much, as we will better see in the next chapter. Let us settle with asking that the variance $\sigma_T^2$ of $\bar{x}_T$ vanishes in the limit of large T,

$$
\lim_{T\to\infty}\sigma_T^2 = \lim_{T\to\infty}\left\langle (\bar{x}_T - \mu_x)^2 \right\rangle = 0.
$$

We have

$$
\sigma_T^2 = \left\langle (\bar{x}_T - \mu_x)^2 \right\rangle = \left\langle \left[\frac{1}{2T}\int_{-T}^T (x(t) - \mu_x)\mathrm{d}t\right]^2 \right\rangle =
$$

$$
= \frac{1}{4T^2}\int_{-T}^T\int_{-T}^T \langle (x(t') - \mu_x)(x(t'') - \mu_x)\rangle\mathrm{d}t'\mathrm{d}t'' = \frac{1}{4T^2}\int_{-T}^T\int_{-T}^T C_{xx}(t',t'')\,\mathrm{d}t'\mathrm{d}t''.
$$

Since for a WSS process $C_{xx}(t',t'') = C_{xx}(|t''-t'|)$, changing the variables from $(t',t'') \to (t = t', \tau = t'' - t')$, we obtain[7]

$$
\sigma_T^2 = \frac{1}{2T}\int_{-T}^T \left(1 - \frac{|\tau|}{2T}\right) C_{xx}(\tau)\,\mathrm{d}\tau = \frac{1}{T}\int_0^{2T}\left(1 - \frac{\tau}{2T}\right) C_{xx}(\tau)\,\mathrm{d}\tau
$$

---

[7] Note that, changing the variables, the integration region changes from a square in the $t', t''$) plane to a tetragon in the $(t, \tau)$ with two sides parallel to the $\tau$-axis and the other two to the $\tau = -t$ line, so that the elementary area is $(1 - |\tau|/2T)\mathrm{d}\tau$.

A continuous WSS process will then be ergodic if

$$\lim_{T \to \infty} \frac{1}{T} \int_0^{2T} \left(1 - \frac{\tau}{2T}\right) C_{xx}(\tau)\, d\tau = 0.$$

Since $|1 - \tau/2T| < 1$, this clearly implies that

$$\int_0^{\infty} |C_{xx}(\tau)|d\tau < \infty, \tag{5.16}$$

which in turn means that *the net correlation function must vanish for $\tau \to \infty$*.

A final property of the correlation function of a WSS process, is that it has its maximum at the origin, that is, for all $\tau$

$$|R_{xx}(\tau)| \le R_{xx}(0).$$

To show this, remember that, according to Eq. (4.15), for any pair $(X, Y)$ of random variables, $\langle xy \rangle^2 \le \langle x^2 \rangle \langle y^2 \rangle$. Therefore, for *ergodic* RPs, where $C_{xx}(\tau)$ vanishes for $\tau \to \infty$, $R_{xx}(\tau)$ decays[8] from $\langle x^2 \rangle$, for $\tau = 0$, to $\langle x \rangle^2$ in the limit $\tau \to \infty$. One can also show that condition (5.16) entails that, by increasing $T$, there is less and less correlation between the RP mean up to time $T$, $x_T$, and the last observation $x(T)$.[9]

Finding a necessary and sufficient condition for the ergodicity of the correlation function is harder, and surely beyond the scope of this book. In several cases, however, it is possible to *check* directly whether a WSS process is also ergodic in the wide sense by comparing the time and ensemble averages of the correlation function.

### 5.1.4  *Spectral Density and the Wiener-Khinchine Theorem*

Harmonic analysis in the frequency domain is a powerful tool to provide valuable information about a time-process. Regrettably, for a *random* process this is not at all a straightforward task. To understand why, let us first recall how one proceeds for a *deterministic* process, that is, a known (real) function of time $u(t)$.

We can distinguish two classes of deterministic time processes. If

$$\int_{-\infty}^{+\infty} u^2(t)dt < \infty \tag{5.17}$$

---

[8] Not necessarily monotonically: sometimes $R_{xx}(\tau)$ decays with damped oscillations.
[9] See E. Parzen, *Stochastic processes*, Dover, New York (2015).

then $u(t)$ admits a Fourier transform,

$$u(\omega) = \mathfrak{F}[u(t)] = \int_{-\infty}^{\infty} u(t)\, e^{-i\omega t}\, dt$$

which provides a direct route to spectral analysis. Since the energy of a signal is usually proportional to the square of its amplitude (think for instance of $V(t)$ or $I(t)$ in an electric circuit), the condition in (5.17) is equivalent to state that the total energy content of the signal is finite. Using Parseval's theorem, which loosely states that the integral of the square of a real function is equal to the integral of the magnitude squared of its (usually complex) transform, we have

$$\int_{-\infty}^{+\infty} u^2(t)dt = \int_{-\infty}^{+\infty} |u(\omega)|^2 d\omega,$$

which tells us that $|u(\omega)|^2$ is an energy density per unit frequency, or an *energy spectral density*.

In most cases, however, the signals of interest do not have a finite energy content, maybe simply because they don't vanish for $t \to \infty$, a necessary (but not sufficient) for condition (5.17) to hold. Nevertheless, a *truncated* signal,

$$u_{\Delta t}(t) = \begin{cases} u(t) & |t| \le \Delta t/2 \\ 0 & \text{otherwise} \end{cases} \tag{5.18}$$

*does* admit a Fourier transform $u_{\Delta t}(\omega)$. Thus, its normalized energy spectrum

$$S_{\Delta t}(\omega) = \frac{|u_{\Delta t}\omega)|^2}{\Delta t},$$

has the dimension of power per unit frequency, which suggest to define the *power spectral density* (PSD, or simply *power spectrum*) of $u(t)$ as[10]

$$S_u(\omega) \triangleq \lim_{\Delta t \to \infty} \frac{|u_{\Delta t}(\omega)|^2}{\Delta t}. \tag{5.19}$$

For a random process, however, definition (5.19) makes little sense. Indeed, even if we can define a truncated version $X_{\Delta t}(t)$ of the process, $S_T(\omega)$ would not be a unique function of $\omega$, but a random process *itself*, whose value usually fluctuates

---

[10] Note that $S_u(\omega)$ can be a function in the generalized sense. For example the constant signal $u(t) \equiv 1$ has a PSD $S_u(\omega) = \delta(\omega)$.

erratically as $T$ grows without limits.[11] A reasonable modification of (5.19), consists in defining the PSD as an *average* over all the realizations of $X(t)$,

$$S_x(\omega) \triangleq \lim_{\Delta t \to \infty} \frac{\langle |X_{\Delta t}(\omega)|^2 \rangle}{\Delta t}, \tag{5.21}$$

provided of course that this limit exists (which is luckily true in most cases of practical interest). Definition (5.21) clearly entails that $S_x(\omega)$ is real and positive. Besides, since $X_{\delta T}(t)$ is real, its Fourier transform is hermitian, $X_{\delta T}(-\omega) = X_{\delta T}^*(-\omega)$. Consequently, $S_x(\omega)$ is an even function of $\omega$, $S_x(-\omega) = S_x(\omega)$. Finally, if we consider the integral of $S_x(\omega)$ over all frequencies, we obtain

$$\int_{-\infty}^{+\infty} S_x(\omega)\,\mathrm{d}\omega = \int_{-\infty}^{+\infty} \lim_{\Delta t \to \infty} \frac{\langle |X_{\Delta t}(\omega)|^2 \rangle}{\Delta t}\,\mathrm{d}\omega = \lim_{\Delta t \to \infty} \frac{1}{\Delta t} \left\langle \int_{-\infty}^{+\infty} |X_{\Delta t}(\omega)|^2\,\mathrm{d}\omega \right\rangle,$$

so that, using Parseval's theorem,

$$\int_{-\infty}^{+\infty} S_x(\omega)\,\mathrm{d}\omega = \lim_{\Delta t \to \infty} \frac{1}{\Delta t} \left\langle \int_{-\infty}^{+\infty} X_{\Delta t}^2(t)\,\mathrm{d}t \right\rangle = \lim_{\Delta t \to \infty} \frac{1}{\Delta t} \int_{-\Delta t/2}^{+\Delta t/2} \langle X_{\Delta t}^2(t) \rangle\,\mathrm{d}t.$$

If $X(t)$ is a WSS process, however, $\langle X_{\Delta t}^2(t) \rangle$ does *not* depend on time, and we simply have

$$\int_{-\infty}^{+\infty} S_x(\omega)\,\mathrm{d}\omega = \langle X^2 \rangle, \tag{5.22}$$

which shows that $S_x(\omega)$ is indeed the average signal power per unit frequency.

The power spectrum is a very useful tool for operating on a RP, for example with the aim of extracting a signal from a noisy background.[12] Obtaining $S(\omega)$ is usually not trivial, but is made considerably easier by an important result, due to Norbert Wiener and Aleksandr Khinchine, which basically states that the power spectrum and the autocorrelation function of a WSS process are a Fourier transform pair,

---

[11] This is equivalent to say that a Fourier transform of the RP defined as

$$\widetilde{X}(\omega) = \mathfrak{F}[X(t)] = \int_{-\infty}^{\infty} X(t)\,\mathrm{e}^{-\mathrm{i}\omega t}\,\mathrm{d}t \tag{5.20}$$

cannot exists for *all* possible realizations of $X(t)$.

[12] This is usually done by *linear filtering*. A time-invariant (TI) filter is conceptually a "black box" that, in response to an input function $x(t)$, produces an output $y(t) = F[x(t)]$ such that $F[x(t + \tau)] = y(t + \tau)$. The filter is linear if $F[ax_1(t) + bx_2(t)] = ay_1(t) + by_2(t)$, where $y_1(t) = F[x_1(t)]$ and $y_2(t) = F[x_2(t)]$. Example of linear filters are for instance $y(t) = \mathrm{d}x(t)/\mathrm{d}t$ or a convolution $y(t) = \int_{-\infty}^{t} w(t - t')x(t')\mathrm{d}t'$. It is easy to show that the response of a time-invariant linear filter to a complex harmonic of unit amplitude, $x(t) = \exp(\mathrm{i}\omega t)$, is a complex harmonic with the same frequency, $y(t)A(\omega)\exp(\mathrm{i}\omega t)$, where $A(\omega)$ is called the *frequency response* of the filter. If $X(t)$ is a WSS random process, one can show that the spectral density of the output $Y(t)$ of a TI linear filter is related to $S_x(\omega)$ by $S_y(\omega) = |A(\omega)|^2 S_x(\omega)$..

$$\boxed{S_x(\omega) = \mathfrak{F}[R_{xx}(\tau)] = \int_{-\infty}^{+\infty} R_{xx}(\tau)e^{i\omega\tau}} \qquad (5.23)$$

The demonstration of this theorem is rather technical and entails delicate passages, so is not presented here.[13] Nevertheless, the next examples show how the WK theorem easily allows the PSD of some basic RPs to be obtained.[14]

**Example 5.4** (*Power spectra of selected processes*)

WHITE NOISE (PURE):  Since in the continuum limit the correlation function of white noise with amplitude $\epsilon$ is $R_{xx}(\tau) = \langle\epsilon^2\rangle\delta(\tau)$, the PSD is *constant* over the whole frequency domain,

$$S_x(\omega) = \langle\epsilon^2\rangle \int_{-\infty}^{\infty} \delta(\tau)e^{i\omega\tau} = \langle\epsilon^2\rangle, \qquad (5.24)$$

and therefore the total power associated with pure white noise is *infinite*. The spectrum of a discrete-time white noise with constant time increment $\Delta t$ has instead a bandwidth of the order of $(2\Delta t)^{-1}$ with a spectrum that is approximately flat for $\omega \in [-B, +B]$.[15]

RANDOM OSCILLATOR:  Consider again a process $X(t) = a\cos(\omega_0 t + \phi)$, where both $a$ and $\phi$ are random uncorrelated variables. If the phase is uniformly distributed in $[0, 2\pi]$, the correlation function is

$$R_{xx}(t, t+\tau) = \langle a^2 \cos(\omega_0 t + \phi)\cos[\omega_0(t+\tau)+\phi]\rangle =$$
$$= \left\langle \frac{a^2}{2}[cos(\omega_0 t) + \cos(2\omega_0 t + \omega_0\tau + 2\phi)]\right\rangle = \frac{\langle a^2\rangle}{2}\cos(\omega_0\tau),$$

which does not depend on $t$ (note that the last passage *requires* $a$ and $\phi$ to be uncorrelated). Therefore, since the Fourier transform of $\cos(\omega_0\tau)$ is half of the sum of two Dirac deltas at $\pm\omega_0$, we simply have

$$S_x(\omega) = \int_{-\infty}^{+\infty} R_{xx}(\tau)e^{i\omega\tau}\,d\tau = \frac{\langle a^2\rangle}{4}[\delta(\omega - \omega_0) + \delta(\omega + \omega_0)] \qquad (5.25)$$

MODIFIED RANDOM TELEGRAPH:  We have seen that the correlation function of a "modified" random telegraph that switches between $\pm 1$ is $R_{xx}(\tau) = \exp(-2\alpha|\tau|)$. The Fourier transform of this "time-symmetric" exponential decay has a *Lorentzian* shape

---

[13] See, for instance, the book by BALAKRISHAN in the Suggested Readings.

[14] In problems concerning electric networks, it is much more common using the ordinary frequency $\nu$, considered as a *positive* real number, then the angular frequency $\omega = 2\pi\nu$. The "one-sided" power spectral density, defined for $f \geq 0$, is simply $S_x(f) = 2S_x(\omega)$.

[15] This estimate is based on the Nyquist-Shannon sampling theorem.

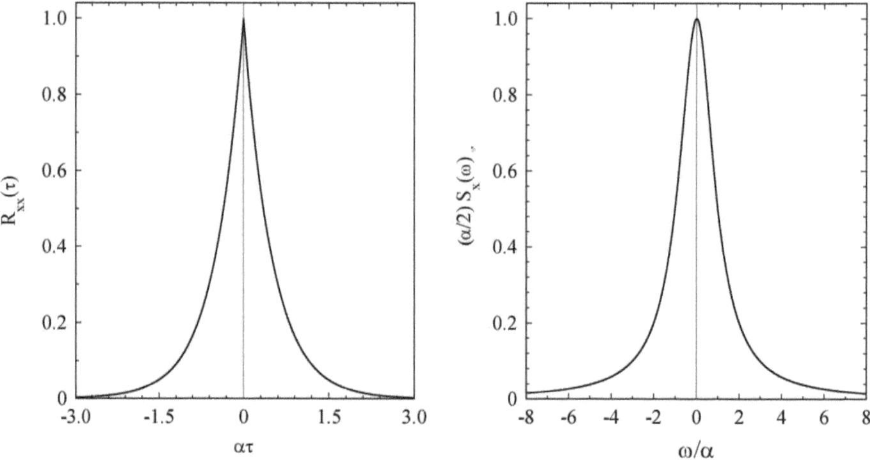

**Fig. 5.2** Correlation function (left) and power spectrum (right) for the modified random telegraph

$$S_x(\omega) = \frac{2\alpha}{\alpha^2 + \omega^2} \tag{5.26}$$

The correlation function and the power spectrum of the modified random telegraph signal are shown in Fig. 5.2.

## 5.1.5   Random Pulse Sequences

The statistical properties of a process that consists of a sequence of random pulses are particularly interesting because of their relation with several types of noise that we will discuss in the next section. Suppose then that a "noise source" emits a stationary train of independent sharp pulses at random instants of time $t_i$ that are distributed according to a Poisson with a mean frequency $\nu$. All pulses have the same shape, but can have a different amplitude. Figure 5.3 shows for instance the signal

$$\xi_T(t) = \sum_{i=1}^{N} a_i \exp\left(-\frac{t - t_i}{\tau}\right) \theta(t - t_i) \tag{5.27}$$

obtained by superposing $N$ exponentially decaying pulses that start at a random time $t_i$ within a time interval $T$ and an amplitude $a_i$ uniformly distributed in $[0, 1]$. The figure shows that increasing by increasing the frequency $\nu$ or, equivalently, the record length $T$, the single pulses progressively merge, generating a very irregular signal

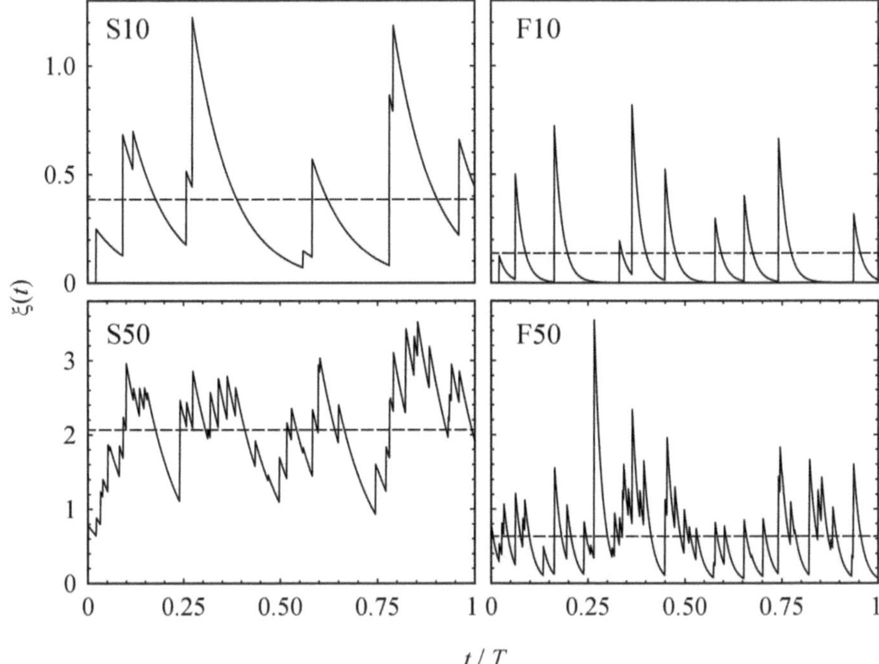


**Fig. 5.3**  Sum of a finite number of exponentially decaying pulses for two values of the decay time $\tau$ and of the frequency $\nu$ (or, equivalently, of the record length $T$)

$\xi_T(t)$. It is then convenient to consider the limit $\lim_{T\to\infty} \xi_T(t)$, that is, the statistically stationary process[16]

$$\xi(t) = \sum_{-\infty}^{+\infty} a_i\, f(t - t_i), \tag{5.28}$$

where all the amplitudes $a_i$ have the *same* probability distribution, with expectation $\langle a \rangle$ and variance $\sigma_a^2$, and the probability that $k$ pulses are emitted in a time interval $t$ is $(\nu t)^k \exp(-\nu t)/k!$.

A rather surprising result about these unceasing pulse sequences is that *all the statistical properties of the stationary process $\xi(t)$ are fixed by the pulse shape and by the statistics of the random amplitude*. This is the content of two very general theorems, respectively due to Norman Campbell[17] and John Carson,[18] which we simply state without demonstration.

---


[16] This reference model for random pulse processes ha been introduced by Stephen Rice, see WAX, Suggested Readings, pp. 133–294.

[17] N. Campbell, Proc. Camb. Phil. Soc. **15**, 117 (1909).

[18] J. R. Carson, Bell Syst. Techn. J. **10** 374 (1931).

CAMPBELL'S THEOREM:    The cumulants of $\xi(t)$ are related to the pulse shape $f(t)$ and to the moments $\langle a^n \rangle$ of the amplitude distribution by

$$\kappa_n = v\langle a^n \rangle \int_{-\infty}^{+\infty} f^n(t)\mathrm{d}t. \tag{5.29}$$

Thus, the mean function and the variance of $\xi(t)$ are

$$\langle \xi(t) \rangle = v\langle a \rangle \int_{-\infty}^{+\infty} f(t)\mathrm{d}t \; ; \;\; \sigma_\xi^2 = v\langle a^2 \rangle \int_{-\infty}^{+\infty} f^2(t)\mathrm{d}t. \tag{5.30}$$

CARSON'S THEOREM:    The power spectrum of $\xi(t)$ is

$$\boxed{S_\xi(\omega) = 2v\langle a^2 \rangle |\tilde{f}(\omega)|^2 + 4\pi \langle \xi \rangle^2 \delta(\omega)} \tag{5.31}$$

where $\tilde{f}(\omega)$ is the Fourier transform of the shape function $f(t)$ and $\langle \xi(t) \rangle$ is given by the Campbell theorem. We see that the DC (zero-frequency) term $4\pi \langle \xi \rangle^2 \delta(\omega)$ disappears only if probability distribution of the amplitude pulse amplitude is symmetric about zero.

These results, however, hold only if the pulses start at times $t_i$ that are Poisson-distributed, and should be corrected in the presence of correlations between the pulses.

**Example 5.5** (*Simple pulse shapes*) As a first example, consider a sequence of rectangular pulses of unit area, fixed duration $\tau$, and mean frequency $v$

$$f(t) = \begin{cases} 1/\tau & 0 \le t < \tau \\ 0 & \text{otherwise} \end{cases}$$

Campbell's theorem gives $\langle \xi(t) \rangle = v$ and $\sigma_\xi^2 = v/\tau$. Since the Fourier transform of the shape function is

$$\int_{-\infty}^{+\infty} f(t)\mathrm{e}^{-i\omega t}\,\mathrm{d}t = \frac{1}{\tau} \int_0^\tau \mathrm{e}^{-i\omega t}\,\mathrm{d}t = \frac{1 - \mathrm{e}^{-i\omega\tau}}{i\omega\tau} = \frac{\sin(\omega\tau/2)}{\omega\tau/2}\mathrm{e}^{-i\omega\tau/2},$$

we obtain

$$S_\xi(\omega) = 2v\,\mathrm{sinc}^2\left(\frac{\omega\tau}{2}\right) + 4\pi v^2 \delta(\omega). \tag{5.32}$$

Consider instead a sequence of exponentially decaying pulses with

$$f(t) = \mathrm{e}^{-t/\tau}\theta(t).$$

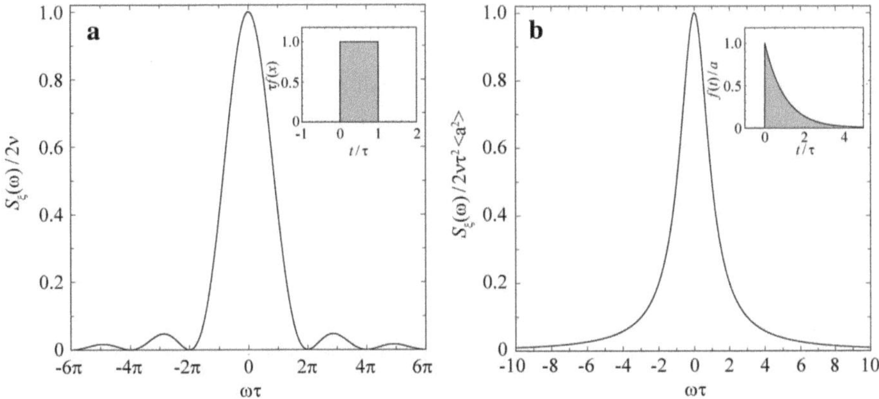

**Fig. 5.4** Power spectra of a rectangular (A) and exponential (B) pulse sequence

and a random, time independent amplitude $a$. You can immediately check that $\xi(t) = \nu\tau\langle a\rangle$ and $\sigma_\xi^2 = (\nu\tau/2)\langle a\rangle^2$. Since

$$\tilde{f}(\omega) = \int_{-\infty}^{+\infty} e^{-(1/\tau+i\omega)t}\theta(t)\mathrm{d}t = \int_0^{+\infty} e^{-(1/\tau+i\omega)t}\mathrm{d}t = \frac{\tau}{1+i\omega\tau},$$

we obtain

$$S_\xi(\omega) = 2\nu\langle a^2\rangle \frac{\tau^2}{1+\omega^2\tau^2} + 4\pi\langle a\rangle^2\delta(\omega), \tag{5.33}$$

which is the sum of a Lorentzian plus a DC term. The power spectra (5.32) and (5.33) are shown (without the DC terms) in Fig. 5.4.

## 5.2 Noise

Noise is any kind of signal that, for some reason, we find "disturbing". But is noise only nuisance? Should not we be annoyed, if not nauseated,[19] by the pervasiveness of noise? No doubt, this omnipresent buzz of Nature fixes the ultimate limit of the accuracy of our measurements. Moreover, noise is a hindrance to signal detection, so that one of the main tasks of scientists and engineers is separating the wheat (the signal) from the chaff (the noise). Nevertheless, a great physicist like Rolf Landauer (one of my heroes) dared to claim that noise *is* the signal,[20] by which he meant that scrutinizing the statistical properties of noise we may learn a lot of new physics.

---

[19] After all, the English word "noise" comes from the Latin *nausea*....

[20] R. Landauer, Nature **392**, 658 (1998).

We have already discussed the properties of continuous (pure) white noise, which can be summarized as follows.

1. The mean function identically vanishes for all $t$, $\mu_\epsilon(t) \equiv 0$;
2. The statistical fluctuations of the process are constant, $\sigma_\epsilon^2(t) = \langle \epsilon^2 \rangle$;
3. The process is $\delta$-correlated, $R_{xx}(t) = \langle \epsilon^2 \rangle \delta(t)$.

Actually, properties 1.-3. fully characterize a pure white noise process, in the sense that nothing more can be said about it: *every* process that satisfy them is a white noise RP. As we said, however, the continuous white noise is an idealized process, for no real signal can have a spectrum that is flat over all possible frequencies (hence with a total infinite power). However, there are two prominent *physical* processes that resembles very much pure white noise up to very high frequencies. Their importance stems from the two main messages they convey, namely, a) the reality of thermal agitation, and b) that matter is not continuous, but made of distinct particles. The first observation provide the links between macroscopic thermodynamics and statistical mechanics, the second is, according to Feynman, the only sentence to be passed down to future generations if a cataclysm destroys all scientific knowledge (I fully agree). These two statements are also the cornerstones of Einstein's theory of Brownian motion, which also means the pillars on which the whole theory of random processes has been built.

Noise became a leading word in physics with the advent of radio communications. Therefore, it is not surprising that the cradle of noise investigation has been electronics. Indeed, it was Walter Schottky, one of the founders of semiconductor physics, who, in an attempt to identify the performance limits of vacuum tube amplifiers, accounted for *both* these kinds of noise in a *single* landmark paper.[21] Let us then examine these two "whitish" noises, starting with thermal noise.

### 5.2.1   *Thermal Noise*

Back in 1906, in his second paper on the Brownian motion, Einstein suggested that the thermal motion of electrons should generate a fluctuating voltage at the ends of a resistor, concluding however that

> Since I could not find any additional experimentally verifiable consequences, any treatment of further special cases seems useless to me.[22]

Indeed, it took two decades before the instrument sensitivity increases enough to allow John Bertrand Johnson, at Bell Labs, to discover the occurrence of a fluctuating voltage generated at the terminals of the input resistance of a vacuum-tube amplifier.[23]

---

[21] An English translation of the original paper, written in German, can be found in J. Micro/Nanolith. MEMS MOEMS **17**, 041001 (2018).

[22] See *The collected papers of Albert Einstein*, Vol. 2, doc. 32, J. Stachel ed., Princeton Univ. Press (1989).

[23] J. B. Johnson, Phys. Rev. **32**, 97 (1928).

Johnson correctly interpreted this noise voltage as the result of thermal agitation, and indeed showed that it grows linearly with $T$, but the full explanation of the effect is due to Harry Nyquist (also at Bell labs) who, in a very elegant paper published the same volume of Physical Review, used the 2nd Law of thermodynamics and the equipartition theorem to show that the electromotive force due to thermal agitation in conductors must be a universal function of (ordinary) frequency $\nu$, resistance $R$, and temperature $T$ alone, and that average squared the noise voltage $\langle V_n^2 \rangle$ in a frequency interval $\Delta \nu$ is

$$\langle V_n^2 \rangle = 4 k_B T R \Delta \nu,$$

which is equivalent to state that the noise spectral density is frequency independent.[24] The magnitude of the Johnson-Nyquist noise (as it is currently known) is usually small. A convenient rule of thumb is that at room temperature the noise voltage from a $1\,\text{k}\Omega$ resistor, integrated over a $1\,\text{MHz}$ bandwidth, is about $4\,\mu\text{V}$, but at much higher load or bandwidth it becomes relevant. In the same paper, Nyquist also generalized noise this result to networks containing capacitors or inductors, thus described by a complex electrical impedance $Z(f)$, for which the PSD becomes

$$S_{v_n} = 4 k_B T \text{Re}[Z(f)] \tag{5.34}$$

However, we know that if this result holds up to any value of the circuit bandwidth $\Delta \nu$, the power associated with the Johnson-Nyquist noise would diverge. Nyquist was fully aware of this, and pointed out that, if quantum effects are taken into account, the result should be modified as[25]

$$S_{v_n} = 4 h \nu \, Re[Z(f)] \, \frac{1}{e^{h\nu/k_B T} - 1}. \tag{5.35}$$

Do you see something familiar in this formula? I am sure that many of you have spotted the same correction factor to the classical Rayleigh-Jeans formula that allowed Planck to avoid the "ultraviolet catastrophe" in the blackbody radiation. In fact, it is *exactly* the same problem. In a close circuit, the energy associated with the noise voltage is distributed to a set of standing e.m. waves like those existing in a cavity in thermal equilibrium with the walls. It must be noticed, however, that the quantum correction factor $(h\nu/k_B T)[\exp(h\nu/k_B T) - 1]^{-1}$ is important only when $\nu/T$ is of the order of $k_B/h \simeq 0.02\,\text{THz/K}$, that is, for very low temperatures or very high frequency. At room temperature, for example, $k_B T/h$ is in the far infrared range.

The PSD of Johnson-Nyquist noise voltage only depends on the resistance (or, more generally, on the real part of the impedance) of the circuit and on the absolute

---

[24] H. Nyquist, Phys. Rev. **32**, 110 (1928). Nyquist's result is actually a direct consequence of the fluctuation-dissipation theorem that we will later discuss.

[25] When Nyquist wrote his paper, the concept of zero-point energy had just been introduced by Pascual Jordan in 1926, who was actually skeptical about its reality. Taking it into account would modify Eq. (5.34), but whether it really contributes to the PSD of thermal noise is still strongly debated.

temperature. Thus, it can be exploited to make probes that do not need any external temperature calibration source. Besides, since changes of the circuit impedance can be detected and compensated non-invasively with very high precision, these sensors are immune from aging and environmental effects that affect most temperature probes. Today, *thermal noise thermometry* is applied in environments where standard temperature probes would easily be damaged and could not easily be recalibrated, but also for remote measurements of distant sources like the cosmic microwave background.[26]

### 5.2.2  Shot Noise

Shot noise is solely due to the discrete nature of matter, and more precisely to the transport of particles. Indeed, while thermal noise is an equilibrium effect that does not entail any net particle transport, electronic shot noise is *always* associated with a direct current flow. In fact, we will se that it is basically due to the *fluctuations of the number* of transported particles. Schottky key intuition was indeed that what macroscopically looks as a DC current is actually a train of uncorrelated pulses, each one containing a single elementary charge that, if the pulses are uncorrelated, will show typical Poisson number fluctuations $\Delta N = \sqrt{\langle N \rangle}$.

For usual values of the current, the relative fluctuation $\Delta N / N$ is very small. For instance, with a current of 1 nA, current fluctuations due to shot noise will be of the order of 10 parts per million, far smaller than the effects of thermal noise, unless $T$ is very close to $0\,\mathrm{K}$. But there are situations, such as those induced by "tunnel junctions" (thin insulating layers separating two electrically conducting materials), where the electron transport is strongly hindered, allowing shot noise to be detected.

The spectral properties of shot noise in an electric circuit can be easily calculated using the results we obtained for a random pulse sequence. Since the current intensity is the charge that flows per unit time, the mean rate of arrival of the charge carriers (electrons, but also holes in semiconductors) is $v = I/e$, where $e$ is the elementary charge. Assume that each charge carries corresponds to a current pulse of infinitesimal duration, $e\delta(t - t_i)$, where $t_i$ is the carrier arrival time. Since the Fourier transform of the pulse is simply $e$, we have

$$S_\xi(\omega) = 2eI + 4\pi I^2 \delta(\omega) \tag{5.36}$$

which, apart from the DC term, is pure white noise. However, this result is entirely dependent on the assumption of current pulses that are infinitely narrow. If we drop this nonphysical assumption and consider pulses with a finite width or decay time, such as those discussed in the Example 5.5, the shot noise bandwidth is limited to frequencies of the order of $\tau^{-1}$, where $\tau$ is the finite width or decay time of the pulse.

---

[26] For a review, see D. R. White *et al.*, Metrologia **33**, 325 (1996).

Equation (5.36) is only valid for Poisson-distributed pulses. Electrons, however are fermions, and therefore show correlations, basically due to the Pauli principle, that appear as an effective repulsive interaction. As we may expect, shot-noise fluctuations are then partly quenched compared to uncorrelated pulses. This reduction can be quantitatively estimated[27] using a simple model of the conductivity of solid due to Landauer (once again my hero!) that yields the so-called *Fano factor $F$*, which is the ratio between the variance and the expectation of the distribution of the arrival times (hence for a Poisson process $F = 1$).[28]

### 5.2.3  Colored Noise

Noise is not only white, but can be attributed a "color" based on the frequency dependence of its power spectrum. Let us then examine some hues of this color palette (which is very popular among acoustic engineers).

#### 5.2.3.1  Red Noise (Wiener Noise)

This kind of noise is also called "brown", not with a chromatic meaning, but rather as a tribute to Robert Brown. Indeed, it is deeply related to Brownian motion, which we will extensively discuss in Sect. 5.3. So far, we just have to remember that the total displacement in a random walk with $p = 1/2$ is the result of the sum of uncorrelated random steps with zero expectation, i.e., of a white noise process. Thus, the Wiener process $X(t)$ describing continuum random walk in one dimension is, in some sense,[29] the time-integral of the white noise, which can then be written as $dX(t)/dt$. If we remember that taking the derivative of a function corresponds, in the Fourier space, to multiply its Fourier transform by $i\omega$,

$$\mathfrak{F}\left[\frac{dX(t)}{dt}\right] = i\omega\mathfrak{F}[X(t)],$$

then, from Eq. (5.24),

$$S_x(\omega) = |\mathfrak{F}[X(t)]|^2 = \frac{\langle \epsilon^2 \rangle}{\omega^2}, \tag{5.37}$$

---

[27] See C. Beenakker and C. Schönenberger, Physics Today **56**, 37 (2003).

[28] Shot noise effects are not limited to electron transport, but play an important role in the optical signal detection too, and in particular in photon counting experiments. See, for instance, L. Mandel and E. Wolf, *Optical Coherence and Quantum Optics*, Cambridge Univ. Press (1995).

[29] Actually, the definition of the integral (or of the derivative) of a RP is a rather delicate topic that unfortunately goes beyond the scope of this introductory chapter. For a good introduction, see the book by JACOBS in the Suggested Readings.

where $\langle \epsilon^2 \rangle$ is the average squared step length (or, more generally, the constant PSD of the associated white noise). The power spectrum of the Wiener noise decreases then as $f^{-2}$, and is therefore dominant at low frequencies. That's why in acoustic it is also known as "red noise" (a kind of sound that many people consider as relaxing).

### 5.2.3.2  Pink Noise (Flicker, or $1/f$, Noise)

One of the most mysterious kind of noise is "pink" noise, better known as flicker noise in electronics or, in a more general context, as $1/f$ noise. The last expression means that, usually, its power spectrum is inversely proportional to the frequency, although the definition is often extended to cover all kinds of noise with a PSD proportional to $f^{-\alpha}$, with $0.5 \lesssim \alpha \lesssim 1.5$. Noise with these spectral properties, intermediate between those of white and red noise (thus duly called "pink"), has been detected for an impressive gamut of random phenomena in science, technology, even music and human cognition. In physics, $1/f$ spectra are widely found in condensed matter, astronomy, and geophysics. However, no general and exhaustive explanation of flicker noise has been proposed, thus its pervasiveness is still an open question.

The origin of flicker noise is actually a very old puzzle. Indeed, it was first observed by Johnson in 1925[30] (so, *before* he discovered thermal noise) while trying to detect shot noise. Johnson found that the power spectrum of the process was indeed flat at sufficiently high frequency. However, at low frequency he also find an unexpected additional contribution that approximately scaled as the inverse of the frequency. A first tentative explanation was given by Schottky[31] by assuming that this contribution to the vacuum tube current electrons comes from electrons released from the cathode surface according to a simple exponential relaxation law,

$$n(t; t_i) = \begin{cases} n_0\, e^{-(t-t_i)/\tau} & \text{for } t \geq t_i \\ 0 & \text{for } t < t_i \end{cases}$$

In Example 5.5, we found that this kind of current pulses yield a Lorentzian power spectrum (5.33) that, if $\omega \gg \tau^{-1}$ decrease as the inverse *square* of the frequency. So, the agreement of Schottky's model with Johnson's data is only qualitative. However, by assuming a power-law *distribution* of relaxation times, one can always find a frequency region where the power spectrum approximately decreases as $\omega^{-(1+\beta)}$, with $\beta \geq 1$.[32] Later, more and more sophisticated attempts were made to explain resistivity fluctuations with a $1/f$ spectrum that occur in most conducting materials. In particular, a primary role seems to be played by the random motion of impurities.[33] Since these approaches are based on specific solid state models, however, they cannot account for the ubiquity of pink noise in totally different contexts.

---

[30] J. B. Johnson, Phys. Rev. **26**, 71 (1925).

[31] W. Schottky Phys. Rev. **28**, 74 (1926).

[32] See E. Minotti, arXiv:physics/0204033.

[33] For a review, see M. B. Weissman, Rev. Mod. Phys. **60**, 537 (1988).

A distinctive feature of $1/f$ noise, however, is that, as a power law, it is scale invariant. This suggests that it may be a general manifestation of scale-free natural phenomena. Along these lines is the self-organized criticality "paradigm" introduced by Bak, Tang and Wiesenfeld[34]. Self-criticality is a general property of dynamical systems that may explain many features of complex systems such as the ubiquity of fractal geometry, power laws and, above all, $1/f$ noise, which has been applied to a disparate topics, generating thousands of publications. Yet, this ambitious approach has its shortcoming and several of its applications have been harshly debated.[35] For sure, like many other theoretical approaches to complexity, self-organized criticality can provide at best a qualitative explanation of scale-free noise (and no *prediction*).

As far as we are concerned, however, a very alarming feature of $1/f$ noise is that it diverges at *low* frequency. For an experimentalist, this means that taking long time-averages of a signal affected by this kind of "infrared divergence" does not necessarily mean improving the accuracy of the results. All the way around. Commenting the apparent low-frequency divergence in the fluctuation spectrum of the Bermuda sea level, William Press[36] remarked that

> If you postpone your Bermuda vacation for too long,
> the island may be underwater!

### 5.2.3.3 Blue Noise (Cherenkov Noise)

I guess you have hardly had the chance to look inside a nuclear reactor. But in the Internet era, of course, you can. Just take a look at this page of the International Atomic Energy Agency:

https://www.iaea.org/newscenter/news/what-is-cherenkov-radiation.

What you will see is that the water surrounding the fuel, which acts as a neutron moderator, beautifully shines in blue. This is due to Cherenkov radiation. The first reported observation comes from the notebook of when Marie Skłodowska Curie, who in 1910 noticed that a concentrated radium solution glowed with a strange blue light[37] However, the first quantitative study was made only in 1934 by Pavel Cherenkov, whose PhD thesis concerned the irradiation of uranium salt solutions with $\gamma$ rays. With meticulous experiments, which were almost incredible for the times and earned him the 1958 Nobel Prize in Physics, Cherenkov was able to show that the effect is not due to fluorescence, and that the radiation is emitted with a very asymmetric pattern.

---

[34] P. Bak, C. Tang and K. Wiesenfeld, Phys. Rev. Lett. **59** (1987) 381.

[35] See N. W. Watkins *et al.*, arXiv:1504.04991.

[36] W. H. Press, Comments Astrophys. **7**, 103 (1978).

[37] About the discovery of the Cherenkov radiation, see A.A: Watson, https://arxiv.org/abs/1101.4535..

Cherenkov radiation is generated by charged particles[38] moving in a medium with a speed larger than the speed of light *in the medium*, $c/n(\omega)$, where $n(\omega)$ is the medium refractive index (which depends on frequency). The effect has *some* analogy with the shock waves produced by an object moving through the air faster than the speed of sound, which are responsible for the crack of a supersonic bullet or the sonic boom of a fighter plane, but is *much* harder to be accounted for in classical electrodynamics. This task was first accomplished in 1937 by Ilya Frank and Igor Tamm, who shared with Cherenkov the Nobel Prize. Their main results can be summarized as follows:

1. The radiation is emitted by the particle in a forward cone[39] with an apex angle $\vartheta$ such that

$$\cos(\vartheta) = \frac{c}{n(\omega)v},$$

   where $v$ is the particle velocity. Note that, in the ultrarelativistic limit $v \simeq c$, $\cos(\vartheta) \to n(\omega)^{-1}$, which for water is an angle of about $40°$ in the visible.
2. The radiated spectrum is continuous, with a PSD proportional to

$$S_x(\omega) \propto \frac{v}{c^2} \left( 1 - \frac{c^2}{v^2 n^2(\omega)} \right) \omega \tag{5.38}$$

The power spectrum is then linearly proportional to $\omega$, thus *diverges* at short wavelengths, which is a kind of ultraviolet *super* catastrophe! If this were really true, taking a picture of a glowing nuclear tank would be rather unsafe, to use a strong understatement.... What saves the poor photographer is the dispersion, that is, the frequency dependence of the refractive index $n(\omega)$ which for all materials approaches one for $\omega \to \infty$ and can even becomes negative close to high frequency absorption bands.

Cherenkov radiation detectors are used in several fields, ranging from high energy and nuclear physics to cosmic ray and neutrino astronomy. In particular, Cherenkov telescopes that detect the radiation emitted by electrons and positrons produced by energetic $\gamma$-rays penetrating the atmosphere, are fostering important advances in high-energy astrophysics. Yes, but what has Cherenkov radiation to do with *noise*? In fact Cherenkov radiation is the main disturbance in fiber-optic radiation sensor for radiotherapy dosimetry, a field where calling it "noise" is perfectly appropriate. Nevertheless even in this case it seems to be possible to turn chaff into wheat, exploiting Cherenkov radiation in novel radiation sensor for proton therapy.[40]

---

[38] Which means that in nuclear plants the culprits are *not* the neutrons, but rather the electrons generated by the ionization of the water molecules.

[39] Of course, the particle sees it as a *backward* cone.

[40] K. Jang *et al.* Opt. Express **20**, 13907 (2012).

### 5.2.3.4  Crackling Noise ((Barkhausen Noise)

The noise color palette used by acoustic engineers is actually richer, but the other tints have little to do with physics. There is however a kind of noise that is not given a color, but has a well defined *sound*: "crackling" noise, whose relevance to physics and science in general is steadily growing. It is then worth giving a look at (or better, listen to) this colorless but noisy noise.

Crackling noise was first (literally) heard in a landmark experiment performed by Heinrich Barkhausen in 1919, which provided the first experimental evidence of the existence of ferromagnetic domains. Barkhausen showed that the magnetization of a ferromagnetic sample subjected to an increasing external field changes discontinuously, by tiny but sizeable jumps. The experimental trick used by Barkhausen to discover this effect, which still bears his name, was brilliant: each of these tiny jumps generated a feeble current in a coil surrounding the test material that, suitably amplified, produced, precisely, a crackling noise in an earphone. Since then, the Barkhausen effect has been investigated in a large number of magnetic systems. Notably, the PDS of the Barkhausen noise often has a Lorentzian shape, which is consistent with a simple model that is very like the one used by Schottky for flicker noise.[41]

As we anticipated, however, crackling noise is a much more general effect, and typically arises when a system responds to an external force with a cascade of restructuring events that appear very similar over different magnitude scales. A pivotal case are earthquakes, whose magnitude, as we have seen in 1.15, is distributed as a power law according to the Gutenberg-Richter law. By mapping the magnitude of the earthquakes in figure 1.18 to a sound amplitude and creating a soundtrack, the resulting effect would indeed be a kind of crackling noise[42] But crackling noise is observed in many other physical phenomena, ranging from friction to fracture in disorder materials, structural rearranging in foams, the dynamics of superconductors and superfluids, the occurrence of solar flares, fluctuations in the stock market.[43]

Scale invariance naturally suggests that crackling noise have some relation with critical phenomena, a connection originally put forward by John Rundle to justify the Gutenberg–Richter law.[44] Most of the later work in this direction took inspiration from the concept of self-organized criticality we mentioned while discussing $1/f$ noise. The basic argument goes as follows. There are systems that respond to external forces with a random sequence of small event with a similar magnitude, like popcorn popping when heated, and others that give way in one single event, like a snapping piece of chalk. Crackling is in between snapping and popping, and occurs when the connections between parts of the system are stronger than in systems that pop but

---

[41] For a review, see B. Alessandro *et al.*, J. Appl. Phys. **64**, 5355 (1988).

[42] See, for instance, the sound produced by the US earthquake records, https://en.wikipedia.org/wiki/Crackling_noise..

[43] For a review, see J. P. Sethna, K. A. Dahmen, and C. R. Myers, Nature **410**, 242 (2001).

[44] J. B. Rundle, J. Geophys. Res., **94**(B9), 12,337 (1989).

weaker than in systems that snap. A crucial condition for crackling, however, is the presence of disorder, that is, lack of homogeneity between the individual parts the system is made of.

Consider for instance an ensemble of ferromagnetic domains. One can account for disorder by introducing a "random field" that acts differently on each domain, in addition to an interaction term between neighbour domains. Let us now apply a slowly increasing external magnetic field to the system. If the interaction between the domains is negligible with respect to the random field, most domains flip independently, and we get popping noise. Conversely, if the disorder is small compared to the mutual interaction between the domains, most of the domains will reorient as a group, so the system snaps. Cracking occurs when the disorder approaches a well-defined critical magnitude. Models developed along this line have been able to explain many features of crackling noise. Once again, however, to explain does not mean to be able to predict. Unfortunately, predicting large earthquakes using a scaling approach is still a long way off. Assuming that it will ever be possible...

## 5.3   From Random Walk to Brownian Motion

As we stresses, Brownian motion is the archetype of random processes. All stochastic methods and most important results in nonequilibrium statistical mechanics, derive from Einstein's explanation of the roots of Brown's pollen dance. In the previous chapters, we already have introduced random walk as a basic model for Brownian motion. Now we will generalize our investigation and frame it in the general context of random processes. This can of course be done at different levels of sophistication. I have chosen to adopt a simple "classical" approach, related to the historical development of the subject, trying to emphasize the most important physical issues. We will see that even this basic plain of attach allows some advanced topics to be introduced.

### 5.3.1   The Diffusion Limit

In this chapter we have seen that the displacement $X(t)$ in a random walk is a discrete RP obtained as the sum of independent increments. On the other hand, Example 3.22 shows that, when the number of steps becomes very large, the probability distribution for $X(t)$ becomes a Gaussian with a variance $\sqrt{L^2 t/\tau}$, where $\tau$ is the time it takes to make a step of length $L$. We also said that a displacement growing as $\langle x(t) \rangle \propto \sqrt{t}$ is the typical signature of a *diffusion* process, with a diffusion coefficient $D = L^2/2\tau$.

One may then wonder if, in the limit $\tau \to 0$, the probability distribution satisfies an equation similar to the macroscopic diffusion equation obtained from the phenomenological Fick's law stating that mass flux is directly proportional to concentration gradients. For a random walk, however, passing to the continuum is a major challenge. Since the directions of two consecutive steps are totally uncorrelated, when $L \to 0$ a single path becomes a line that, albeit continuous, is *nowhere differentiable*: a true mathematical freak! But a freak with a name, because this is exactly that Wiener process we introduced in Sect. 5.2 as a kind of integral of white noise, where the words "kind of" actually conceals the whole intricate subject of stochastic calculus.

The feature of a Wiener process that most clashes with physics is that the derivative of $X(t)$ is discontinuous for every $t$, which makes Newton roll in his grave. Einstein was fully aware of this crucial problem when, in his first landmark paper on Brownian motion, he wrote[45]

> We introduce a time interval $\tau$ into consideration, which is very small compared to the observable time intervals, but nevertheless so large that in two successive time intervals $\tau$, the motions executed by the particle can be thought of as events which are independent of each other

Let us then consider a particle undergoing a one-dimensional random walk and follow Einstein suggestion by assuming that $\tau$ is very short compared to the time scale $T$ over which we want to describe the particle motion, but not *truly* infinitesimal. For generality, we will also assume a generic single-step probability $0 < p \leq 1$. To find the probability density $f(x, t + \tau)$ that the particle is in $x$ at $t + \tau$, we write:

$$f(x, t + \tau) = f(x - L, t)p + f(x + L, t)(1 - p),$$

that is, either the particle at $t$ was a step back, and then makes a step forward with probability $p$, or it was a step forward, and makes a step back with probability $1 - p$. Since $\tau \ll T$, we approximate $f(x, t + \tau)$ to first order as

$$f(x, t + \tau) \simeq f(x, t) + \frac{\partial f}{\partial t}\tau.$$

We can expand also the right side, but here, for reasons that will soon be clear, we better expand $f(x \pm L, t)$ to second order,

$$f(x \pm L, t) \simeq f(x, t) \pm \frac{\partial f}{\partial x}L + \frac{1}{2}\frac{\partial^2 f}{\partial x^2}L^2.$$

Substituting in the original equation we have:

$$\frac{\partial f}{\partial t} = (1 - 2p)\frac{L}{\tau}\frac{\partial f}{\partial x} + \frac{L^2}{2\tau}\frac{\partial^2 f}{\partial x^2}.$$

---

[45] This statement may also be the first explicit introduction of a *coarse-graining approximation* in physics.

Therefore

$$\boxed{\frac{\partial f}{\partial t} = (1 - 2p)\frac{L}{\tau}\frac{\partial f}{\partial x} + D\frac{\partial^2 f}{\partial x^2}}.$$

(5.39)

This is an example of what we will later call a *Fokker–Planck equation*. If then we consider a large number $N$ of particles, the fraction of them that is found between $x$ ed $x + \mathrm{d}x$ at time $t$ is:

$$\rho(x, t) = Nf P(x, t),$$

and therefore obeys *the generalized diffusion equation*

$$\boxed{\frac{\partial \rho(x, t)}{\partial t} = (1 - 2p)\frac{L}{\tau}\frac{\partial \rho(x, t)}{\partial x} + D\frac{\partial^2 \rho(x, t)}{\partial x^2}}$$

(5.40)

When $p = 1/2$, the first term at the right side vanishes (this is why we needed a second order expansion) and we obtain the standard macroscopic diffusion, whose solution for an initial condition $f(x, 0) = \delta(x)$, corresponding to a particle distribution concentrated at the origin, is a Gaussian with $\sigma_x^2 = \langle x^2 \rangle = 2Dt$. Physically, this could describe the spreading in time of an ink drop in *still* water. However, the quantity $\rho(x)$ in (5.40) is not necessarily a mass or particle number density: other physical quantities, such as heat or momentum in a moving fluid, also spread out diffusively.

But what is the meaning of the first term at the r.h.s. of (5.40)? If $p \neq 0.5$, we may expect each particle to drift in time along the positive (if $p > 0.5$) or negative (when $p < 0.5$) $x$-direction. Dimensionally, the quantity $(1 - 2p)L/\tau$ must then correspond to the stationary drift velocity $v_d = F/f$, where $f$ is a friction coefficient, that a particle takes on in the presence of a constant external force $F$ like gravity.[46]

Einstein exploited the previous observation to derive a fundamental relation between the diffusion and friction coefficients. Consider for instance a colloidal suspension contained in a bottle and subjected to gravity, which makes particle settle under their own weight with a stationary speed $v_d = mg/f$. The particles will progressively accumulate at the bottom of the container, so that a downward–directed concentration gradient $\nabla n$ along the vertical forms. This will in turn generate a pressure gradient that, if the suspension is sufficiently diluted, can be written as[47] $\nabla P = k_B T \nabla n$. Equilibrium is reached when

---

[46] This can also be seen by noticing that, if the whole distribution moves rigidly with a constant speed $v_d$, $n(x, t)$ cannot be an arbitrary function of position and time, but only of the combined variable $z = x + v_d t$. Indeed, if we neglect the Brownian diffusive term, Eq. (5.40) is satisfied by any arbitrary function $n(z)$.

[47] Actually, this is the *excess* hydrostatic pressure (the "osmotic" pressure) due to the particles, which collectively behave as an ideal gas.

1. The total force per unit volume due to gravity and to the pressure gradient vanishes, $mgn = k_B T \nabla n$;
2. The downward particle flux due to sedimentation balances the upward diffusive flux due to the concentration gradient, $n v_d = D \nabla n$.

These two equations together with $v_d = mg/f$ yield

$$\boxed{D = \frac{k_B T}{f}} \tag{5.41}$$

Remarkably, the Einstein relation (5.41) relates a quantity like $D$, which accounts for particle fluctuations in Brownian motion, to $f$, which quantifies energy dissipation due to friction. In fact, it is a token of the *fluctuation-dissipation theorem*, a general result in nonequilibrium statistical mechanics.


### 5.3.1.1   Diffusion in 3$d$

The close relation between random walk and diffusion can also be assessed using the characteristic function $\tilde{f}_x(k)$ that we introduced in Chap. 4. Let us consider, more generally, a random walk in three dimensions, and write the displacement vector after $n$ independent equal steps $\mathbf{s}_i$ of length $L$ as $\mathbf{r} = \sum_{i=1}^{n} \mathbf{s}_i$. Notice first that the average displacement squared is

$$\langle r^2 \rangle = \sum_{i,i=1}^{n} \mathbf{s}_i \cdot \mathbf{s}_j = L^2 \sum_{i,i=1}^{n} \langle \cos \theta_{ij} \rangle,$$

where $\theta_{ij}$ is the angle between the directions of step $i$ and $j$. But since the steps are uncorrelated, we have $\langle \cos \theta_{ij} \rangle = \delta_{ij}$. Therefore we still have $\langle r^2 \rangle = nL^2$, like in the one-dimensional case. Since $\mathbf{r}$ is the sum of independent random variables, Eq. (4.45) gives $f_{\mathbf{r}}(\mathbf{k}) = [f_{\mathbf{s}}(\mathbf{k})]^n$, where $f_{\mathbf{s}}(\mathbf{k})$ is the characteristic function of a single step. We assume that the probability density (in 3$d$) for a step $\mathbf{s}$ does not depend on the step orientation,[48] with a fixed magnitude $|\mathbf{s}| = L$. To be correctly normalized, it must then be

$$f_{\mathbf{s}}(r) = \frac{1}{4\pi L^2} \delta(r - L).$$

Thus, $f_{\mathbf{s}}(\mathbf{k})$ is given by (inverse) Fourier transform (in *three* dimensions)

$$f_{\mathbf{s}}(\mathbf{k}) = \int e^{i\mathbf{k}\cdot\mathbf{r}} f_{\mathbf{s}}(r)\, d^3 r,$$

---

[48] Thus excluding any drift along a specific direction.

which does not seems to be immediate. However, for symmetry reason, the characteristic function will also be circularly symmetric, i.e., will depend only of $k = |\mathbf{k}|$. We can then perform the integration in spherical coordinates, taking the direction of $\mathbf{k}$ as the polar axis. We have

$$f_{\mathbf{s}}(k) = \frac{1}{2L^2} \int_0^\pi d\theta \, \sin\theta \int_0^\infty dr \, r^2 e^{ikr\cos\theta} \delta(r - L).$$

The radial integral is simply given by

$$\int_0^\infty r^2 e^{ikr\cos\theta} \delta(r - L) \, dr = \frac{1}{2} \int_{-\infty}^{+\infty} r^2 e^{ikr\cos\theta} \delta(r - L) \, dr = \frac{L^2}{2} e^{ikL\cos\theta},$$

so that, putting $\cos\theta = z$,

$$\tilde{f}_{\mathbf{s}}(k) = \frac{1}{2} \int_0^\pi e^{ikL\cos\theta} \sin\theta \, d\theta = \frac{1}{2} \int_{-1}^1 e^{ikLz} dz = \frac{\sin kL}{kL}, \tag{5.42}$$

which gives[49] $\tilde{f}_{\mathbf{r}}(k) = \left(\frac{\sin kL}{kL}\right)^n$. The displacement p.d.f. after a time $n\tau$ is then[50]

$$f_{\mathbf{r}}(n\tau) = \frac{1}{(2\pi)^3} \int d^3k \, e^{-i\mathbf{k}\cdot\mathbf{r}} \left(\frac{\sin kL}{kL}\right)^n.$$

We are interested in the continuum limit $n \to \infty$, $\tau \to 0$, keeping $n\tau = t$ constant. However, $\sin kL / kl < 1$ for all $kL \neq 0$, which means that, for $n \to \infty$, $f_{\mathbf{r}}(n\tau)$ vanishes *unless* $L^2 \propto n^{-1} \propto \tau$. In this case, expanding at second order $\sin(x) \simeq x - x^3/6$, the quantity

$$\left(\frac{\sin kL}{kL}\right)^n = \left(\frac{\sin kL}{kL}\right)^{t/\tau} \simeq \left(1 - \frac{k^2 L^2}{6}\right)^{t/\tau}$$

remains indeed finite in the limit $\tau \to 0$. Therefore, if we define the diffusion coefficient in 3d as

$$D = \lim_{\tau \to 0} \frac{L^2}{6\tau}, \tag{5.43}$$

---

[49] Using the same method, you can easily show that the Fourier transform of any spherically symmetric function $f(r)$ is given by

$$\tilde{f}(k) = \frac{4\pi}{k} \int_0^\infty r \sin(kr) f(r) \, dr.$$

[50] Note that the normalization factor in $3D$ is $(2\pi)^{-3}$.

we have

$$\left(1 - \frac{k^2 L^2}{6}\right)^{t/\tau} = (1 - Dk^\tau)^{t/\tau} \xrightarrow[\tau \to 0]{} e^{-Dk^2 t}$$

Therefore, the probability density for the displacement is

$$f_{\mathbf{r}}(n\tau) = \frac{1}{(2\pi)^3} \int d^3 k \, e^{-i\mathbf{k}\cdot\mathbf{r}} \left(\frac{\sin kL}{kL}\right)^n.$$

Writing $\mathbf{k}\cdot\mathbf{r} = k_x x + k_y y + k_z z$ and evaluating the three Gaussian integrals separately, one finally obtains

$$f_{\mathbf{r}}(t) = \frac{1}{(4\pi Dt)^{3/2}} e^{-r^2/4Dt} = \frac{e^{-x^2/4Dt}}{2\pi\sqrt{Dt}} \times \frac{e^{-y^2/4Dt}}{2\pi\sqrt{Dt}} \times \frac{e^{-z^2/4Dt}}{2\pi\sqrt{Dt}} \qquad (5.44)$$

which is then the product of the p.d.f. of three independent 1d diffusive displacements along $x, y, z$. The mean square displacement is then given by

$$\langle r^2 \rangle = \langle x^2 \rangle + \langle y^2 \rangle + \langle z^2 \rangle = 6Dt \qquad (5.45)$$

One can finally show that the particle density $\rho(r, t) = N f_{\mathbf{r}}(t)$, with $f_{\mathbf{r}}(t)$ given by (5.44), is the solution of the $3d$ diffusion equation

$$\frac{\partial \rho(\mathbf{r}, t)}{\partial t} = D\nabla^2 \rho(\mathbf{r}, t) \qquad (5.46)$$

with the initial condition $\rho(\mathbf{r}, 0) = \delta(x)\delta(y)\delta(z)$.


### 5.3.2   *The Langevin Equation*


By "smoothing" the random walk over a timescale $\tau$, Einstein managed to unravel the fundamental link between Brownian motion and diffusion and the close connection between fluctuations and dissipation effects. However, he did not provide any clue about the meaning and magnitude of $\tau$, which is crucial to avoid a crucial limit of the diffusion model. Indeed, although Einstein "coarse-graining" approach avoid the thorny mathematical problems of the Wiener process, a mean displacement proportional to $\sqrt{t}$, if extrapolated to $t = 0$, would entail a particle speed diverging as $t^{-1/2}$. Below a minimal timescale, arguably of the order of $\tau$, we should reasonably expect the r.m.s. displacement to be linear in time, so that at very short time the motion is rather "ballistic" (that is, with a finite constant speed) than diffusive.

Giving a physical meaning to $\tau$, however, necessarily requires to formulate a *dynamical equation* that suitably takes into account the effect of the random collisions with the solvent molecules on the particle motion. It was Paul Langevin who, in 1908[51] paved the way for a truly dynamic theory of Brownian motion by deriving an equation that was bound to become a paradigm in nonequilibrium physics.

It is very instructive to examine how Langevin himself derived and discussed the equation that bears his name. Langevin's stroke of genius was separating the total force due to the solvent into a fluctuating plus a mean contribution. The only assumptions made by Langevin made about the fluctuating part $\eta(t)$ is that it had to be "indifferently positive and negative" (that is, $\langle \eta(t) \rangle = 0$) and totally uncorrelated with the particle position. If this were the only force, however, there would be a steady transfer of kinetic energy to the particle, and its mean square velocity $\langle v^2(t) \rangle$ might increase without limits, while, as Langevin clearly states, it has to reach an equilibrium value given by Boltzmann's equipartition theorem, $\langle v^2 \rangle = \langle v_x^2 + v_y^2 + v_z^2 \rangle = 3k_B T/m$, where $m$ is the particle mass. Therefore, there must be a mechanism for transferring back energy to the fluid that is more efficient, the larger the particle speed. According to Langevin, this is nothing but the Stokes friction force $F_v = -f\mathbf{v}$, where the friction coefficient $f$ is proportional to the particle size and to the viscosity coefficient $\eta$ of the suspending fluid.[52]

Thus, Newton's equation for the particle displacement in time along a given direction $x$ is

$$m\frac{\mathrm{d}^2 x(t)}{\mathrm{d}t^2} = -fv(t) + \eta(t), \tag{5.47}$$

where $v(t) = \mathrm{d}x(t)/\mathrm{d}t$. Multiplying both sides by $x(t)$, Eq. (5.47) can be written as

$$m\frac{\mathrm{d}^2}{\mathrm{d}t^2}[x^2(t)] - 2mv^2(t) = f\frac{\mathrm{d}}{\mathrm{d}t}[x^2(t)] + 2\eta(t)x(t),$$

Since $\langle \eta(t)x(t) \rangle = 0$, if we take the average of both sides the random force cancels out. Then, setting $\frac{\mathrm{d}}{\mathrm{d}t}\langle x^2(t) \rangle = z$, and noting that $m\langle v^2 \rangle = k_B T$, we obtain

$$m\frac{\mathrm{d}z}{\mathrm{d}t} + fz = 2k_B T,$$

which has the general solution

$$z = 2\frac{k_B T}{f} + Ce^{-\gamma t} \tag{5.48}$$

where $C$ is a constant and $\gamma = f/m$ is called the *dissipation coefficient*. For $t \gg \gamma^{-1}$, therefore, $z$ becomes does not depend on time anymore, so that

---

[51] P. Langevin, C. R. Acad. Sci. (Paris) **146**, 530, 1908 (translated in English by A. Gythiel, Am. J. Phys. **65**, 1079, 1997).

[52] Besides these internal forces of the particle/solvent system, there might also be some external forces $F_{ext}$ like gravity or an electric/magnetic field.

$$\langle x^2(t) \rangle = 2\frac{k_B T}{f} t = 2Dt,$$

Therefore, the *Brownian relaxation time* $\tau_B = \gamma^{-1}$ plays exactly the role of the "coarse-graining time" $\tau$ in the Einstein model, and for $t \gg \tau_B$ the mean square displacement grows *diffusively*, with a diffusion coefficient $D$ given by the Einstein relation (5.41).

Let us carry on Langevin's task using the tools we have introduced in this chapter. For generality, we will discuss Brownian motion in the $3d$-space. For symmetry reasons, different components of the particle displacement and velocity will be uncorrelated, so we can focus on a specific axis, say $x$. the Brownian particle is subjected to a net dissipative viscous force $F_x^v = -m\gamma v_x$, where $v_x = \dot{x}$ is the component of the particle velocity along $x$, plus a randomly fluctuating force $\eta(t)$, with $\langle \eta(t) \rangle = 0$ and $\langle \eta^2 \rangle$ constant in time. Equation (5.47) can then be written

$$\boxed{\frac{dv_x}{dt} = -\gamma v_x + \frac{1}{m}\eta(t)} \tag{5.49}$$

Taking the Fourier transform of both members, we have:

$$v_x(\omega) = -\frac{1}{\gamma + i\omega}\frac{\eta(\omega)}{m}.$$

Therefore, the power spectrum of the particle velocity is related to the power spectrum of the random force by

$$S_{v_x}(\omega) = \frac{1}{\gamma^2 + \omega^2}\frac{S_r(\omega)}{m^2}.$$

Following Langevin, we now assume that the random force is not correlated with the particle position. This is surely guaranteed if $\eta(t)$ is $\delta$-correlated, i.e., if $\langle \eta(t)\eta(t') \rangle = \langle \eta^2 \rangle \delta(t - t')$. In other words, we assume that the random force has all the properties of white noise. The power spectrum of $v_x$ is then Lorentzian,

$$S_{v_x}(\omega) = \frac{\langle \eta^2 \rangle}{m^2}\frac{1}{\gamma^2 + \omega^2}, \tag{5.50}$$

and the autocorrelation function of a velocity component is

$$R_{v_x v_x}(t_1, t_2) = \frac{\langle \eta^2 \rangle}{2\pi m^2}\int_{-\infty}^{\infty} d\omega \frac{e^{i\omega t}}{\gamma^2 + \omega^2} = \frac{\langle \eta^2 \rangle}{2m^2\gamma}e^{-\gamma\tau}, \tag{5.51}$$

where $\tau = |t_2 - t_1|$. Setting $\tau = 0$, we see that

$$\langle v_x^2 \rangle = \frac{\langle \eta^2 \rangle}{2m^2\gamma}.$$

On the other hand, after a sufficiently long average time the particle velocity gets fully "thermalized" with the surrounding solvent, hence $m\langle v_x^2\rangle = k_B T$ and

$$\boxed{\langle \eta^2\rangle = 2\,m\gamma k_B T} \tag{5.52}$$

Therefore *the viscous dissipation coefficient is directly related to the intensity of the fluctuating random force*, which is an alternative form of the Einstein relation more closely related to the general fluctuation-dissipation theorem.

Let us now find the particle displacement $x(t)$, which given by

$$x(t) = \int_0^t v_x(t')\mathrm{d}t'.$$

Since $R_{v_x v_x}(t_1, t_2)$ only depends on $\tau = |t_2 - t_1|$, we can write:

$$\langle x^2(t)\rangle = \int_0^t \mathrm{d}t_2 \int_0^t \mathrm{d}t_1 \langle v_x(t_1)v_x(t_2)\rangle = 2 \int_0^t \mathrm{d}t_2 \int_0^{t_2} \mathrm{d}\tau\, \langle v_x(0)v_x(\tau)\rangle.$$

Integrating by parts:

$$\langle x^2(t)\rangle = \left[2t_2 \int_0^{t_2} \mathrm{d}\tau\, \langle v_x(0)v_x(\tau)\rangle\right]_0^t - 2 \int_0^t \mathrm{d}t_2\, t_2\langle v_x(0)v_x(\tau)\rangle,$$

which, substituting $t_2 \to \tau$ in the second term at the r.h.s., yields:

$$\langle x^2(t)\rangle = 2 \int_0^t \mathrm{d}\tau\,(t - \tau)\langle v_x(0)v_x(\tau)\rangle. \tag{5.53}$$

Using now Eq. (5.51) in the form $\langle v_x(0)v_x(\tau)\rangle = (k_B T/m)\exp(-\gamma\tau)$, and introducing the diffusion coefficient defined again as $D = k_B T/m\gamma$ we finally have

$$\langle x^2(t)\rangle = 2Dt - \frac{2D}{\gamma}\left[1 - \mathrm{e}^{-\gamma\cdot t}\right]. \tag{5.54}$$

which coincides with the integral of (5.48) if we take $C = 2D$.

Because different components of the displacement are independent, extension to the $3d$ case is immediate,

$$\boxed{\langle \mathbf{r}^2(t)\rangle = 6Dt - \frac{6D}{\gamma}\left[1 - \mathrm{e}^{-\gamma t}\right]} \tag{5.55}$$

with the limiting behavior

$$\langle \mathbf{r}^2(t)\rangle \xrightarrow[t\ll\tau_B]{} \langle \mathbf{v}^2\rangle t$$

$$\langle \mathbf{r}^2(t)\rangle \xrightarrow[t\gg\tau_B]{} 6D(t - \tau_B),$$

where we have again introduced the Brownian relaxation time $\tau_B = 1/\gamma = m/f$ that, as we see, *coincides with the relaxation time of the velocity correlation function*. For $t \ll \tau_B$, the particle motion is then *ballistic*, and the particle moves with a constant r.m.s. velocity

$$\langle v^2\rangle^{1/2} = \langle v_x^2 + v_y^2 + v_z^2\rangle^{1/2} = \sqrt{3k_B T/m}.$$

The mean square displacement given by Eq. (5.55)is shown in Fig. 5.5.

But what is the typical order of magnitude of $\tau_B$? For a spherical particle of radius $a$, the Stokes friction coefficient is $f = 6\pi\eta a$. Hence

$$\tau_B = \frac{m}{f} = \frac{m}{k_B T}D = \frac{2}{9}\frac{da^2}{\eta}, \tag{5.56}$$

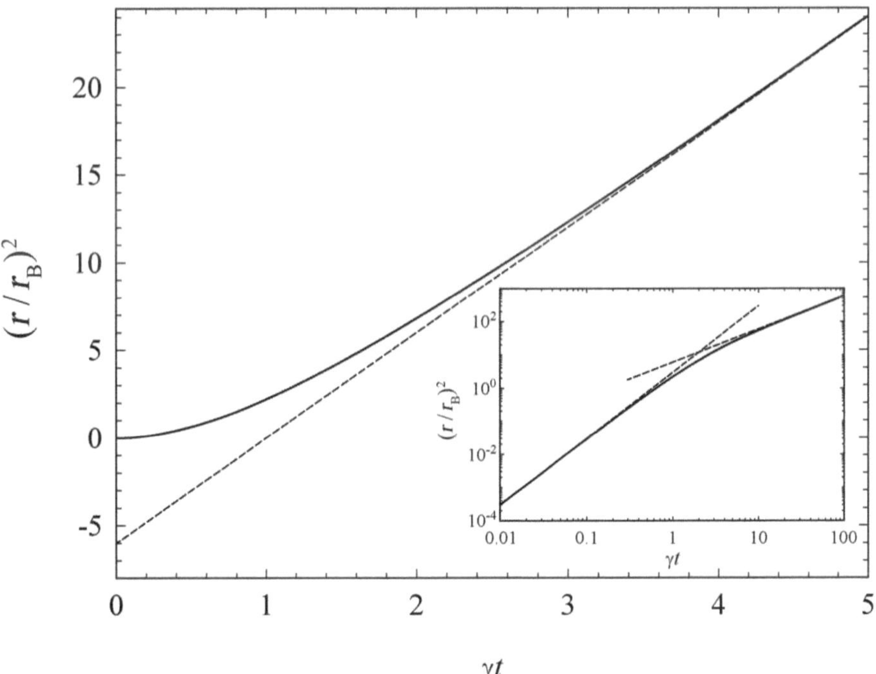

**Fig. 5.5** Langevin's result for the mean square displacement $(r/r_B)^2$, where $r_B = D/\gamma$, versus the scaled time $\gamma t$. The inset shows that the r.m.s. square displacement changes from ballistic ($\propto t^2$) at short times to Brownian ($\propto t$) in the long time limit

where $d$ is the particle material density. For a particle of radius $a = 1\,\mu$m, taking $d \simeq 1000\,$kg/m$^3$ and $\eta \simeq 0.85 \times 10^{-3}\,$Pa $\cdot$ s (the value for water at $T = 300\,$K), one finds $\tau_B \simeq 0.2\,\mu$s. How much does the particle move? From (5.55), the particle mean square displacement in $\tau_B$ is $\langle \mathbf{r}^2(\tau_B)\rangle = 6D\tau_B/$e. Since $D \simeq 2.6\,\mu$m$^2$s$^{-1}$, we have

$$\frac{\langle \mathbf{r}^2(\tau_B)\rangle^{1/2}}{a} \simeq 10^{-3},$$

which is really a negligible fraction of the particle size!

From (5.51) and (5.52), one finally obtains

$$D = \int_0^\infty \langle v_i(0)v_i(t)\rangle \, \mathrm{d}t,$$

where $v_i$ is any component of $\mathbf{v}$. Therefore, since $\langle \mathbf{v}(0)\mathbf{v}(t)\rangle = 3\langle v_i(0)v_i(t)\rangle$ because of independence, we have

$$\boxed{D = \frac{1}{3}\int_0^\infty \langle \mathbf{v}(0)\mathbf{v}(t)\rangle \, \mathrm{d}t} \tag{5.57}$$

that gives the diffusion coefficient in terms of the time-integral of the velocity autocorrelation function.

Unfortunately, at this point I have to give you unpleasant news: all we have done is highly questionable because our starting point, Eq. (5.49) is basically meaningless. Indeed, the occurrence of the white-noise term at the r.h.s. implies that the particle acceleration is discontinuous *for all values of* $t$. That is, $v(t)$ does not, after all, really exist! Nevertheless, let me reassure you, that Eq. (5.55) is *almost* correct.[53] The way the Langevin equation should be properly interpreted and dealt with will hopefully become clearer in the next section. Let me however point out that, in his original treatment, Langevin dodged this deadly bullet tracing out the noise term by *averaging*, by which most of the delicate issues related to stochastic differential equations can be sidestepped.

---

[53] Actually, it turns out that the particle displacement is not a fully "memoryless" process (a feature that we will better specify in the next section). Because of this, before entering the diffusive regime the mean square displacement exhibits a "long time tail" proportional to $t^{-3/2}$. See, for instance, G. L. Paul and P. N. Pusey, J. Phys. A **14**, 3301 (1981).

## 5.4 Markov Processes

A distinctive feature of random walks is that the position after the $n$-th step solely depends on the starting point, i.e., the position attained after step $n - 1$.

In other words, random walkers have no memory of the path they have followed. Most of the RPs of interest in physics share this feature, which is the key defining property of a *Markov process*.[54]

In Sect. 5.1 we stated that a RP can be fully characterized by means of the single and multiple-time probability distributions

$$P_1((x_1, t_1), P_2(x_1, t_1; x_2, t_2), P_3(x_1, t_1; x_2, t_2; x_3, t_3) \ldots$$

However, $P_n(x_1, t_1; \ldots x_n, t_n)$ can be obtained from the joint probability distribution of order $n - 1$ using the "chain rule"

$$P_n(x_1, t_1; \ldots x_n, t_n) = P_n(x_n, t_n | x_{n-1} t_{n-1}, \ldots, x_1, t_1) P_{n-1}(x_1, t_1; \ldots x_{n-1}, t_{n-1}),$$

where the $n$-times *conditional* probability $P_n(x_n, t_n | x_{n-1} t_{n-1}, \ldots, x_1, t_1)$ is the "repository" of the process memory of earlier times. Iterating this scheme, one eventually obtains $P_2(x_1, t_1; x_2, t_2) = P_2(x_2, t_2 | x_1, t_1) P(x_1, t_1)$. Therefore, we can also specify a stochastic process by giving the single-time probability $P_1(x_1.t_1)$ and a hierarchy of multiple-time conditional probabilities.

The "memory" of a Markov process is limited, at any instant of time, to the immediately preceding value of the time argument, that is, for all $n \geq 2$,

$$P_n(x_n, t_n | x_{n-1} t_{n-1}, \ldots, x_1, t_1) = P_2(x_n, t_n | x_{n-1}, t_{n-1}), \tag{5.58}$$

Like for a random walk, the future state of a Markov process only depends on its present state, and not on the history of *how* the present state has been reached. All the joint probability distributions can therefore be obtained as products of the single-time probability $P_1(x, t)$ with two-times conditional probabilities $P_2(x', t' | x, t)$, with $t' > t$. If the process is stationary in the wide sense, we can also write $P_1(x, t) = P_1(x)$ and $P_2(x', t' | x, t) = P(x', t' - t | x, 0)$.

In principle, the stationary probability distribution $P(x)$ and the conditional probability $P(x', t' - t | x, 0)$ are independent quantities. However, in real physical processes the memory of the initial state $x$ is often lost as $\tau = t' - t$ increases, so the conditional probability approaches the stationary value,

$$\lim_{\tau \to \infty} P(x', \tau | x, 0) = P(x'). \tag{5.59}$$

---

[54] This brief outline is limited to *continuous* Markov processes. There is also a vast literature on discrete-time Markov processes, usually called *Markov chains*, but, in the physical processes we will deal with, time is always a continuous variable.

As a consequence, the single conditional probability distribution (which we will simply write $P(x', \tau \mid x)$ leaving out the indication of the time origin) completely determines all the statistical properties of a WSS Markov process.

### 5.4.1 The Master Equation

The amplitude of Markov processes can be either discrete or continuous. Let us begin with the former case, indicating for short the amplitude values $x_j$ (also called *states*) as $j$. The evolution of a Markov process can often be described using a so-called "master" equation.

We start observing that, because of the "memoryless" nature of Markov processes, the probability of "propagating" in a time $t$ from an initial state $i$ to a final state $j$ can be written as the product of the probabilities of making first a transition from $i$ to an intermediate state $k$ in a time $0 < t' < t$ followed by a transition from $k$ to $j$, summed over all possible intermediate states,

$$P(j, t \mid i) = \sum_k P(j, t - t' \mid k) P(k, t' \mid i) \tag{5.60}$$

which is called the *Chapman-Kolmogorov equation*, an identity that is very general but with the strongly annoying property of being *quadratic* in $P$, that is, nonlinear. In many cases the Chapman-Kolmogorov equation can be reduced to a linear equation by introducing a *transition rate* $w(k' \mid k)$ from a state states $k$ to a state $k' \neq k$, which can be properly defined if and only if, in the limit of a very short time interval $\delta t$, the transition probability becomes proportional to $\delta t$,

$$P(k', \delta t \mid k) = w(k' \mid k)\delta t. \tag{5.61}$$

*If* this holds true (which is not always the case), the Chapman Kolmogorov equation leads to the following differential equation for $P(j, t \mid i)$

$$\frac{\mathrm{d}}{\mathrm{d}t}[P(j, t \mid i)] = \sum_{k \neq j} [w(j \mid k) P(k, t \mid i) - w(k \mid j) P(j, t \mid i)] \tag{5.62}$$

which is called a *master equation* and is satisfied for any initial state $i$ with the initial condition $P(j, 0 \mid i) = \delta_{ij}$. A physical interpretation of the master equation can easily be spotted by noticing that each term at the r.h.s. is the difference between a "gain" term, expressing the instantaneous probability of a transition $k \to j$, and a "loss" term, giving the same for a "backward" transition $j \to k$.

Note that, as a direct consequence of the Markov nature of the RP, (5.62) is a *first-order* differential equation, which only requires the specification of the initial distribution $P(j, 0 \mid i)$ to be solved, like the Hamilton equations of motions in classical mechanics or the Schrodinger equation in quantum mechanics. In some sense, the master equation is then the "stochastic analog" of the fundamental equations of deterministic dynamics.[55]

A very important issue is whether or not the master equation has a stationary solution, that is, if $dP(j, t \mid i)/dt \to 0$ for $t \to \infty$. From (5.62) and the definition (5.59) for a stationary distribution, we see that this happens if

$$\sum_{k \neq j} [w(j \mid k)P(k) - w(k \mid j)P(j)] = 0, \tag{5.63}$$

which, if there are $n$ accessible final states $j$, is a set of $n$ linear equations, which can be extremely hard to solve if $n$ is large or even infinite. The problem is however much simpler if the transition rates satisfy detailed balance,

$$w(j \mid k)P(k) = w(k \mid j)P(j) \tag{5.64}$$

for every pair of distinct states $k, j$. The principle of detailed balance, which originates from the time-reversibility of the microscopic equation of motion, states that in a system *at thermodynamic equilibrium* every process and its own reverse process runs at the same average rate. If (5.64) holds, the Markov process is then reversible, and the stationary solution $P$ is actually the *equilibrium*[56] solution $P^{eq}(j)$. Since $\sum_k P(k) = 1$, the equilibrium solution (if it exists!) is simply found to be

$$P^{eq}(j) = \left[ 1 + \sum_{k \neq j} \frac{w(k \mid j)}{w(j \mid k)} \right]^{-1} \tag{5.65}$$

**Example 5.6** (*A two-state Markov process*) Let us consider a stationary Markov process where, like for the random telegraph, the system randomly and suddenly switches between two states 1 and 2 with constant transition rates $\alpha_1 = w(2, t \mid 1)$ and $\alpha_2 = w(1, t \mid 2)$. It is easy to show that the Markov equations for the time derivatives of $P_1(t) = P(1, t \mid 2)$ and $P_2(t) = P(2, t \mid 1)$ can then be written in a compact form as

---

[55] You may counter that the simple Newton's equation $m\ddot{x} = F(x)$ is *second* order. However, it can be written as a *pair* of first-order equations, $m\dot{x} = p$ and $\dot{p} = F(x)$. Similarly, several RPs that seems to have a "longer memory" than a Markov process can often be reduced to a set of coupled Markov processes.

[56] One has always to be very careful in distinguishing between equilibrium states and *nonequilibrium stationary* states (NESS), where energy or momentum fluxes are present. For example, a system in a time-independent thermal gradient can reach a stationary condition, but this is a NESS, because a heat flux must be maintained. In a NESS of this kind, detailed balance cannot hold, otherwise the 2nd Law of Thermodynamics would be violated!.

$$\frac{d\mathbf{P}(t)}{dt} = \mathbb{W}\mathbf{P}(t), \tag{5.66}$$

where $\mathbf{P}(t)$ and the *transition matrix* $\mathbb{W}$ are given by

$$\mathbf{P}(t) = \begin{bmatrix} P_1(t) \\ P_2(t) \end{bmatrix} \;;\quad \mathbb{W} = \begin{bmatrix} -\alpha_1 & \alpha_2 \\ \alpha_1 & -\alpha_2 \end{bmatrix} \tag{5.67}$$

This system of first-order differential equations can be formally solved as

$$\mathbf{P}(t) = e^{\mathbb{W}}\mathbf{P}(0),$$

where, by definition

$$e^{\mathbb{W}t} = \sum_{n=0}^{\infty} \frac{(\mathbb{W}t)^n}{n!}$$

Finding the exponential of a matrix is not at all an easy task, but in the specific case we are discussing it is not incredibly hard, as we will see.

Before we do it, however, it is useful to find a solution using an approach solely based on the physics of the problem. The probability that the system makes a transition from state 1 to state 2 in $\delta t$ is $\alpha_1 \delta t$. Since $\delta t$ is infinitesimal, the probability that it does *not* make a transition is $1 - \alpha_1 \delta t$. So, the probability that the system *remains* in state 1 for a time interval $t$ is

$$(1 - \alpha_1 \delta t)^{t/\delta t} \underset{\delta t \to 0}{\Longrightarrow} \exp(-\alpha_1 t),$$

which is what we expect if the switching times are distributed according to a Poisson. Applying the same argument to 2, the probability of remaining in the other state for a time interval $t$ is of course $\exp(-\alpha_2 t)$. The average "residence time" in states $1, 2$ is therefore $\tau_{1,2} = (\alpha_{1,2})^{-1}$. Then, if we consider a long interval of time, the steady-state probabilities $P(1)$, $P(2)$ will be given by the fraction of the total time that the system spends in ech of the two states,[57]

$$P(1) = \frac{\tau_1}{\tau_1 + \tau_2} = \frac{\alpha_2}{\alpha_1 + \alpha_2} \;;\quad P(2) = \frac{\tau_2}{\tau_1 + \tau_2} = \frac{\alpha_1}{\alpha_1 + \alpha_2},$$

To find the entire time dependence of $P_1(t)$ and $P_2(t)$, we have to solve the system of equations (5.66). The exponential of the transition matrix can be calculated by first observing that $\mathbb{W}^2 = -2\alpha\mathbb{W}$, where $\alpha = \frac{1}{2}(\alpha_1 + \alpha_2)$ is the mean transition rate, so that $\mathbb{W}^n = (-2\lambda)^{n-1}\mathbb{W}$ for all $n \geq 1$. Then, you should not find too difficult to show that

---

[57] This can also be directly obtained from Eq. (5.63) using $P(1) + P(2) = 1$.

$$\mathrm{e}^{Wt} = 1 - \frac{1}{2\alpha}\left(1 - \mathrm{e}^{-2\alpha t}\right)\mathbb{W} = \frac{1}{2\alpha}\begin{bmatrix} \alpha_2 + \alpha_1 \mathrm{e}^{-2\alpha t} & \alpha_2(1 - \mathrm{e}^{-2\alpha t}) \\ \alpha_1(1 - \mathrm{e}^{-2\alpha t}) & \alpha_1 + \alpha_2 \mathrm{e}^{-2\alpha t} \end{bmatrix}.$$

Using as initial condition either $\mathbf{P}(0) = \begin{bmatrix} 1 \\ 0 \end{bmatrix}$ or $\mathbf{P}(0) = \begin{bmatrix} 0 \\ 1 \end{bmatrix}$ we obtain

$$P(1, t\,|\,1) = \frac{\alpha_2 + \alpha_1 \mathrm{e}^{-2\alpha t}}{\alpha_1 + \alpha_2} \;\; ; \;\; P(2, t\,|\,1) = \frac{\alpha_1(1 - \mathrm{e}^{-2\alpha t})}{\alpha_1 + \alpha_2}$$

$$\tag{5.68}$$

$$P(1, t\,|\,2) = \frac{\alpha_2(1 - \mathrm{e}^{-2\alpha t})}{\alpha_1 + \alpha_2} \;\; ; \;\; P(2, t\,|\,2) = \frac{\alpha_1 + \alpha_2 \mathrm{e}^{-2\alpha t}}{\alpha_1 + \alpha_2}$$

In the limit $t \to \infty$ we then have the stationary probabilities

$$P(1, t\,|\,1) = P(1, t\,|\,2) = \frac{\alpha_2}{\alpha_1 + \alpha_2}$$
$$P(2, t\,|\,1) = P(2, t\,|\,2) = \frac{\alpha_1}{\alpha_1 + \alpha_2},$$

which do not depend on the initial state and coincide with the probabilities $P(1)$ and $P(2)$ guessed from the fractions of the residence time.

Suppose now that the amplitudes of the process in states 1 and 2 are $x_1$ and $x_2$. Then, at steady-state, the expectation and variance of the RP are

$$\langle x \rangle = x_1 P(1) + x_2 P(2) = \frac{x_1 \alpha_2 + x_2 \alpha_1}{\alpha_1 + \alpha_2} \tag{5.69}$$

$$\sigma^2(x) = \langle x^2 \rangle - \langle x \rangle^2 = \frac{\alpha_1 \alpha_2 (x_1 - x_2)^2}{(\alpha_1 + \alpha_2)^2} \tag{5.70}$$

If we now consider the specific case of the "symmetrized" random telegraph, with $x_1 = 1$, $x_2 = -1$ and $\alpha_1 = \alpha_2$, we have $\langle x \rangle = 0$, $\sigma_x^2 = 1$, as already found.

### 5.4.2  The Fokker-Planck Equation

What we have said so far can be easily extended to Markov processes with a continuous set of states, that is, with a continuous amplitude. A WSS Markov process is therefore fully characterizes by the conditional probability density $f(x, t\,|\,x_0)$, which becomes in the $t \to \infty$ limit the steady-state p.d.f. $f(x)$. Correspondingly, the Chapman-Kolmogorov equation becomes

$$f(x, t\,|\,x_0) = \int f(x, t - t'\,|\,x') f(x', t'\,|\,x_0)\,\mathrm{d}x', \tag{5.71}$$

where the integration is over the range of values of $x$. By introducing the *transition probability density* $w(x \mid x')$, so that $w(x \mid x')dx$ gives the probability per unit time of a transition from $x_0$ to any state in the range $(x, x + dx)$, the master equation (5.62) becomes

$$\frac{\partial}{\partial t} f(x, t \mid x_0) = \int \left[ f(x', t \mid x_0) w(x \mid x') - f(x, t \mid x_0) w(x' \mid x) \right] dx' \qquad (5.72)$$

with the initial condition $f(x, 0 \mid x_0) = \delta(x - x_0)$.

However, solving the integro-differential equation (5.72)is generally a prohibitive task. The key strategy to make it manageable is assuming that the transition densities are slowly varying function of $x$. Expanding then $w(x \mid x') = w(x' + \delta x, \mid x)$ at *second* order in $\delta x$ one obtains, with some manipulation,[58] the *Fokker–Planck equation*

$$\boxed{\;\frac{\partial}{\partial t} f(x, t \mid x_0) = -\frac{\partial}{\partial x}[A_1(x) f(x, t \mid x_0)] + \frac{1}{2}\frac{\partial^2}{\partial x^2}[A_2(x) f(x, t \mid x_0)]\;} \qquad (5.73)$$

where the quantities

$$A_1(x) = \int x' w(x + x' \mid x)dx', \quad A_2(x) = \int (x')^2 w(x + x' \mid x)dx',$$

which basically are the first two moments of the transition rate, are respectively called the *drift* and *diffusion* coefficients because of the close relationship between Eq. (5.73) and Eq. (5.39). Indeed, in the simplest case where both coefficients do not depend on $x$, the Fokker-Planck equation reduces to the generalized diffusion equation Eq. (5.39) for the probability density once we identify $A_1(x) \to v_d$ and $A_2(x) = D$.

If we define a probability current density $j(x, t \mid x_0)$ as

$$j(x, t \mid x_0) = A_1(x) f(x, t \mid x_0) - \frac{\partial}{\partial x}\left[\frac{1}{2} A_2(x) f(x, t \mid x_0)\right],$$

the Fokker-Planck equation can be written in the form of a *continuity equation*

$$\frac{\partial}{\partial t} f(x, t \mid x_0) + \frac{\partial}{\partial x} j(x, t \mid x_0) = 0. \qquad (5.74)$$

When a stationary density $f(x)$ exists, $j(x, t \mid x_0)$ tends for $t \to \infty$ to a stationary current $j^{st}$ that, since $f(x)$ does not depend on $t$, must also satisfy, according to 5.74,

---

[58] For a complete derivation, see for instance the book by BALAKRISHNAN in the Suggested Readings.

$$\frac{\partial j^{st}}{\partial x} = 0,$$

that is, $j^{st}$ does not depend on $x$ either. The stationary density $f(x)$ can then be found by solving the *first*-order, *ordinary* differential equation

$$\boxed{\frac{1}{2}\frac{\mathrm{d}}{\mathrm{d}x}[A_2(x)f(x)] - A_1(x)f(x) = c} \tag{5.75}$$

where the constant $c$ is fixed by the boundary conditions.

### 5.4.3  The Ornstein-Uhlenbeck Process

Consider now a slightly more complicated form of the Fokker-Planck equation, where we assume that the diffusion coefficient is still constant, but the drift coefficient depends linearly on $x$, so that

$$\frac{\partial}{\partial t}f(x,t\,|\,x_0) = -c_1\frac{\partial}{\partial x}[xf(x,t\,|\,x_0)] + \frac{c_2}{2}\frac{\partial^2}{\partial x^2}[xf(x,t\,|\,x_0)] \tag{5.76}$$

where $c_1$ and $c_2$ are positive constants (so that the drift coefficient is in the *opposite* direction of $x$). If we choose as natural boundary conditions $f(x,t\,|\,x_0) \xrightarrow[x\to\pm\infty]{} 0$, with the initial condition $f(x,0\,|\,x_0) = x_0$, the solution of (5.76) can be shown to be a *Gaussian* with expectation and variance[59]

$$\mu_x(t) = \int_{-\infty}^{+\infty} xf(x,t\,|\,x_0)\mathrm{d}x = x_0 e^{-c_1 t}$$

$$\sigma_x^2 = \int_{-\infty}^{+\infty} x^2 f(x,t\,|\,x_0)\mathrm{d}x - \mu_x^2(t) = \frac{c_2}{2c_1}(1 - e^{-2c_1 t}),$$

that is,

$$f(x,t\,|\,x_0) = \frac{1}{\sigma_x\sqrt{2\pi}}\exp\left[-\frac{(x - \mu_x(t))^2}{2\sigma_x^2}\right] \tag{5.77}$$

At $t = 0$ the conditional density $f(x,t\,|\,x_0)$ is therefore a $\delta$-function in $x_0$, while for all $t > 0$ is a Gaussian that widens in time with a peak value that drift monotonically to zero and a variance that grows attaining a constant value for $t \to \infty$. One can finally show that the process autocorrelation decays exponentially to zero,

$$R_{xx}(\tau) = \langle x(0)x(\tau)\rangle = \frac{c_2}{2c_1}e^{-c_1\tau}.$$

---

[59] Note that $\mu_x(t)$ and $\sigma_x^2$ are *conditional* on the given initial condition $x(0) = x_0$.

Notably, if we put $c_1 = \gamma$ and $c_2 = \langle \eta^2 \rangle / m^2$, *this coincides with the velocity auto-correlation function*, Eq. (5.51), *obtained from the Langevin equation*. Therefore, the Ornstein-Uhlembeck process gives the time-dependence of the probability distribution of the velocity of a Brownian particle, as described by the Langevin stochastic equation. This is just a special case of a more general result, according to which the probability density of a random process $\xi(t)$ that satisfies the stochastic differential equation

$$\frac{d\xi}{dt} = g(\xi, t) + h(\xi, t)\eta(t)$$

where $g$ and $h$ are regular functions of their arguments, and $\eta(t)$ is a Gaussian white noise, obeys the partial differential equation[60]

$$\frac{\partial}{\partial t} f(\xi, t | x_0, t_0) = -\frac{\partial}{\partial \xi} \left[ g(\xi, t) f(\xi, t | x_0, t_0) \right] + \frac{1}{2} \frac{\partial^2}{\partial \xi^2} \left[ h^2(\xi, t) f(\xi, t | x_0, t_0) \right].$$

There would be much more to say about random processes. As I anticipated, however, this chapter cannot pretend to be more than a simple introduction. Many additional applications to physics can be found in the books mentioned in the Suggested Readings.

---

[60] See the excellent book by BALAKRISNAN in the Suggested Readings.

# Chapter 6
# What Is Probability?

> *Probability is the most important concept*
> *in modern science, especially as nobody*
> *has the slightest notion of what it means*
>
> B. Russell (quoted by E. T. Bell)

Let me say this straight. If your sole goal is to learn about probability applications you won't need this chapter at all, so you are permitted to skip the coming lengthy digression. Before jumping to this conclusion, however, consider the following. Like myself, you have probably spent some time as a budding physicist speculating about the dramatic revolution in our perception of reality that the quantum revolution has brought about: we have to accept that God *does* play dice. Do you think we can really hope to figure out the rules of this game of (weird, and perhaps loaded) dice without trying to understand what we mean by "probability"?

IF you do, THEN move on to the next chapters, which will provide you with the basic tools for the (esoteric) art of data fitting, although several of the notions that follow provide a good primer in statistics. ELSE...stay with me.

Before we start, however, let me stress that the aim of this chapter is neither providing you with a review of the different interpretations of probability (a job for philosophers), nor to discuss the foundations of hypothesis testing (a task for professional statisticians). In the local dialect of my native town of Milan, we say

Offelee, fa el tò mestee

which means something like "Confectioner, stick to your own job!". And I am neither a philosopher, nor a professional statisticians. My goal is just to pique your curiosity by presenting questions about probability that have no straightforward answers, or maybe have no final answers at all.

© The Author(s), under exclusive license to Springer Nature Switzerland AG 2025     281
R. Piazza, *An Invitation to Probability and Data Analysis for Physicists*, UNITEXT for Physics, https://doi.org/10.1007/978-3-031-83856-9_6

To begin, it is useful to summarize and further categorize the three main interpretations of the concept of probability presented in the examples that open Chap. 2:

OBJECTIVE (EXPERIMENTAL) PROBABILITY:   Probability is seen as some kind of unspecified *element of reality*, whose value can be figured out by determining a "limiting value" for the relative frequency with which a certain event occurs. It is then a purely *empirical* approach to probability.

THEORETIC (SPECULATIVE) PROBABILITY:   Probability can be *deduced* from abstract reasoning alone, for example (but not necessarily) because a class of events can be regarded as equiprobable. (Mis)quoting Beethoven, probability has a given value because *so muss es sein*. Here we are then speaking of a purely *speculative* approach to probability, rather popular among theoretical physicists.

SUBJECTIVE (LOGIC) PROBABILITY:   Probability gauges our predictive power. It tells us the degree of certainty we can have about the outcome of an event based on the information we already have. Objective reality does not know anything about chance: it is all about the way we *learn* about things (philosophers would call this an *epistemic* approach). In some sense, it is a subjective approach, but *not* personal caprice: it is an extension of classical logic that provides the foundation for rational belief.

It is worth noting that previous categorization is not standard at all. Most contemporary books on statistics, particularly those focused on economists and social scientists, tend to distinguish just between the first and third approach to probability, commonly known as "frequentist" and "Bayesian" (with a strong preference for the last one). However, forgetting about a priori probability, or considering it just as a starting point for the Bayesian analysis is not only historically inaccurate, but also disrespectful of what theoretical physicists actually do.

As a matter of fact, till the thirties of the last century, physicists tended to ignore the Bayesian interpretation, with their main focus being on comparing the frequentist versus the "Laplace" (deductive) interpretation of probability.[1] Since the latter actually provided the first consistent method of assessing probabilities, we better start with the deductive approach.

## 6.1   Probability for the Theorist: Equiprobability and Beyond

Laplace is usually credited with proposing the definition of probability as the proportion of favorable to total cases, "so long as all these cases are equally probable", but the idea of equiprobability was actually already familiar to Leibnitz, Jacob Bernoulli, and other scientists almost one century earlier.[2]

---

[1] See, for instance, R. B. Lindsay and H. Margenau, *Foundations of Physics*, Dover Publ. (1956).

[2] Even Galileo, when discussing the problem of dice (see Example 2.21), spoke of events happening "with the same ease".

At that time, however, whether the equiprobability of a set of events should be assigned on objective (experimental) or epistemic bases was rather vague. Leibnitz stated that equiprobability should be based on the principle of "insufficient reason" (or of "indifference"), which in a nutshell means that two events have to be consider as equiprobable if we see no reason why one of them should happen more than the other, a clearly epistemic view that Laplace later adopted too. But Bernoulli already questioned this approach by stating that, when we say that it is equally likely that a child should be born a boy or a girl only because we know *from experience* that the number of boys and girls is nearly equal all over the world.

Perhaps, Bernoulli had already perceived that equiprobability is mostly, if not only, useful when dealing with games of chance. For life insurers or physicians it is surely of no use, and our discussion about classical vs. quantum dice suggests that sorting out equiprobable events may not be trivial even for games of chance. However, it is when the number of possible events is continuous that assigning equal probabilities becomes really ambiguous, as revealed by the next example due to Richard[3] von Mises, a scientist we will cover a lot in the following section.

**Example 6.1** (*von Mises' cheating barman*) Imagine you are having a happy time with a friend in a cozy Italian tavern, and that you ask the barman, who is not fully committed to honesty, to bring you some good white wine, for example a glass of Soave. Your friend, who has already been in that tavern, warns you that the glass you will be given will surely contain at least as much water as wine, but not more than twice as much water as wine (after all, the barman is not *so* bold). Assuming that any ratio of water ($W$) to Soave ($S$) between these two extremes is equally likely, you would arguably conclude that there is a 50% chance that the watered down wine glass you will be served has a ratio $W/S \geq 3/2$. But the problem can also be viewed "from the wine perspective", in which of course we have $1/2 \leq S/W \leq 1$. If we consider equiprobable all *these* ratios, we might expect that, with a 50% chance, the content of the glass will have a ratio $S/W \leq 3/4$, i.e., a ratio of water to wine $W/S \geq 4/3$, which is a *different* answer![4]

Using the results we obtained in Chap. 4 for the probability density of a function $\phi(x)$ of a random variable, you should already have figured out the origin of this paradox. Indeed, if $X = W/S$ is uniformly distributed, $S/W = 1/X$ is *exponentially* distributed (and vice versa). Actually, the inconsistency arises solely because we are treating $W/S$ as a *continuous* variable. Suppose indeed that the dishonest bartender is actually a deceitful vending machine, which mixes water and Soave only in (equally likely) proportions of $W/S = (5 + n)/5$, with $n = 0$ to 5. Then, the reciprocals of these numbers are *not* evenly spaced in the interval [0.5, 1] (see the top panel in Fig. 6.1) and therefore can hardly be equiprobable.

Another classic problem that warns us of the possible ambiguities that can arise from taking a continuous RV variable as uniform is the famous paradox that Joseph

---

[3] Not to be confused with his elder brother Ludwig, one of the fathers of classical liberalism.

[4] Check that if you calculate the ratio of water to the total liquid content, $W/(W + S)$, you still get another answer, different from the others.

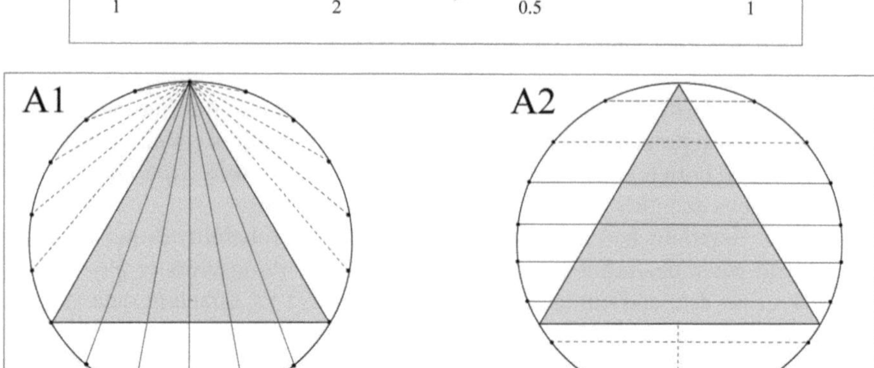


**Fig. 6.1** Top panel: Discrete version of the von Mises problem showing that equally spaced $W/S$ ratios do not correspond to equally spaced $S/W$ ratios. Bottom panel: Discrete version of the Bertrand's paradox, with the constructions corresponding to answers A1 and A2. Chords that are shorter than the triangle side are shown with broken lines

Bertrand presented in 1889 in his book *Calcul des probabilités*. The problem posed by Bertrand goes as follows:

> Q: In a circle, what is the probability that a cord[5] selected at random is longer than the side $\ell$ of an inscribed equilateral triangle?

Bertrand showed (and you can easily check) that one obtains at least three different answers according to the procedure used to select the chord:

A1:  Select a point $P$ at random on the circumference as a vertex of the triangle and draw the chord from $P$. The chord will be longer than $\ell$ if it cuts across the triangle, which happens with probability $p = 1/3$.

A2:  A chord that is not a diameter is completely defined by its midpoint $C$ and will be longer than $\ell$ if $C$ lies closer to the center than half the circle's radius, which happens with $p = 1/2$.

A3:  All the midpoints of chords longer than $\ell$ fall within a concentric circle with a diameter equal to the radius of the original circle. The ratio of the areas of the two circles yields $p = 1/4$.

There have been several attempts to show that these three conflicting solutions cannot be put on equal footing, and that only one of them is actually *the* correct one. For example, according to Edwin Jaynes, another scientist we will shortly meet, only

---

[5] More precisely, one should say *a chord that is not a diameter*. This is because any point in the circle is the midpoint of a single chord, except for the center, which is the midpoint of infinite diameters. Nevertheless, the diameters are a subset of measure zero of the set of all possible chords.

solution A2 can be accepted if we require the solution to be invariant by translations, rotations, and scale change.[6]

However, this results still depends on what we mean by "random", and specifically on the experimental protocol we devise to generate a "random" distribution of chords. Indeed, solution A2 is obtained if chords are selected by throwing long straws at a circle drawn on the floor (as Jaynes actually successfully tested). However, if for instance one selects the midpoints of the chords by throwing darts at the circle, then the only solution that satisfies symmetry invariance is A3.[7] Note once again that the paradox only arises if the chord endpoints are located anywhere on the circumference. As shown in the bottom panel of Fig. 6.1, if we restrict ourselves to a finite number of endpoints, the distribution will be different depending on whether we choose strategy A1 or A2. In the latter case they are the intersections with the circumference of equispaced *parallels*, in the former they are on the vertexes of a regular polygon (so they are *angularly* equispaced).

Stating equiprobability for continuous variables can therefore be rather ambiguous, unless there is a clear prescription to select *which* events should be regarded as equiprobable.[8] But the major problem is that, as we said, the principle of indifference has in practice very limited applications. Does this preclude any chance of assessing a priori probabilities, or at least some general features that a probability distribution must satisfy? Surely not! Remember for instance the Example 2.4. There we suggested, on the footsteps of Galileo, that there are several quantities which are not equiprobable on a linear, but rather on a logarithmic scale, namely, different *orders of magnitude* are equiprobable. Similarly if some events are reasonably expected to follow a geometric distribution $P(k; p)$, their probabilities should satisfy $P(k'; p)/P(k''; p) = (1 - p)^{k'-k''}$.

In fact, it is exactly when there are good reasons to *reject* the principle of indifference that powerful conjectures about a probability distribution can be made. One of the leitmotifs of this book has been, since the first chapter, the occurrence of frequency and probability distributions that are scale-free. Some of the most important advancements in statistical physics of the past decades, for instance the modern theory of critical phenomena, are actually based on the key observation that some fluctuating quantities possess scale invariance, so their probability distribution *has* to be, at least in a given range, a power law. Using a well-grounded model to claim that the fluctuations of a random variable must satisfy some general requirements is part of the job of good theorists: do we really want to take it away from them?

---

[6] E. T. Jaynes, *Found. Phys.* **3**, 477 (1973).

[7] A. Drory, *Found. Phys.* **45**, 439 (2015).

[8] This is the case of classical statistical mechanics, where the set of available microstates for an isolated system coincides with the region (the "hypersurface") $\Sigma(E)$ of the phase space where the energy has a given value $E$. The reason why one *must* use the phase space, where the coordinates are the positions and momenta of the particles, and not a "$\mu$-space" where the coordinates are the particle positions and *velocities* is that in the latter the Liouville theorem is *not* valid, and therefore volume (and then probability) is not conserved in the time evolution. Anyway, in quantum mechanics, where the microstates of energy of a finite-size system are discrete, an unambiguous definition of equiprobability is much easier.

## 6.2   Probability for the Experimentalist: the Tormented Story of Frequentism

Assessing probabilities as the values that relative frequencies approximate in a long series of observations is likely as old as, if not older than, the a priori approach. The primacy of empirical estimates over (poor) a priori conclusions is already endorsed by Galileo, when he states (see Example 2.21) that "it is nevertheless known that *long observation* has made players consider 10 and 11 to be more advantageous than 9 and 12". Validating frequentism was arguably in the back of the mind of Bernoulli, when he derived the "law of large numbers" that we will shortly discuss, and even Laplace admitted that, as more and more experimental data build up, "their *true possibility* is known better and better".[9]

However, the first earnest defense of frequentism is due to John Venn, who conceived and popularized the eponymous diagrams, so useful to teach elementary set theory. Venn's essay,[10] is a severe (although rather verbose) rebuttal of any "objective" probability that can be theoretically predicted a priori. Assessing probability values is only possible because of the empirical evidence that on the long run the experimental frequencies settle on rather constant values (although he admits that the expression "on the long run" is highly questionable for those we now call non-stationary processes).[11] Yet, when trying to give a more precise definition of what he means by "the proportions in the long run" he just states that

> The proportion, in fact, will gradually approach towards some fixed numerical value, what mathematicians term its *limit*.

And that's the point! For what we also have nonchalantly called "the limit of relative frequencies" has *nothing* to do with a mathematical limit. Consider, for example, the relative frequencies $f_k$ of the digits in $\pi$ we discussed in Chap. 1. To speak of a limit in a mathematical sense, we should be able to say that the difference between $f_k$ and 0.1 becomes smaller than any chosen value $\varepsilon$ any time we consider a number of decimals $N$ greater than a certain value $N_0(\varepsilon)$. Yet, if $\pi$ is a normal number, this cannot be strictly true for *all* values of $N > N_0$, since in the sequence of decimals of a normal number there can always be an arbitrarily long sequence that does not contain the digit $k$ and that ruins our party.

But there is a second, much more important difference, because the "frequency endpoint" we talk about is conceptually very different from the ordinary limit of a sequence, which obviously depends on the *order* of the terms. This is not what we think of when we consider the trend of relative frequencies: "shuffling" the decimals of $\pi$, or randomly choosing a decimal "every so often" and considering only the subsequence of the extracted decimals, we expect that almost always the relative

---

[9] See I. Hacking, *The emergence of probability*, Cambridge Univ. Press (2006).

[10] J. Venn, *The Logic of Chance*, MacMillan and Co., Cambridge 1866 (freely available on Google Books).

[11] Venn also makes a clear distinction between what we now call experimental mean and expectation (which he respectively calls "average" and "mean") using as an example the St. Petersburg paradox.

frequencies still all converge to 0.1 in a fairly continuous way. In some way, then, the set of decimals of $\pi$ (or any set of statistical data) is a mathematical entity much "richer" than a sequence.

It was von Mises, the real father of modern frequentism, who took the herculean effort to define and characterize these "aggregates" of a large number of observations that he called "collectives" (*Kollectiven*). Actually, the frequency interpretation was mostly a bonus of his profound inquiry, whose main goal was to comprehend what is really meant by a "random" sequence. We already know from Chap. 1 that randomness is a subtle concept, extremely hard (or even impossible) to delineate properly. Von Mises' effort, however, has a more restricted scope, namely, finding the requisites a sequence of observations satisfy to *univocally* define probabilities "on the long run". Clearly, just giving the correct frequencies is not enough. For example, a sequence of coin flips that yields a pattern like HTHTHTHTHTHTHTHT...*seems* to indicate that the coin is fair, but you'd surely strongly suspect that the game is rigged.

To examine von Mises' theory[12] it is crucial to remember that he has not only been an applied mathematician but also (and chiefly) a physicist and an engineer, who gave important contributions to aerodynamics and solid mechanics. In fact, he firmly believed that probability theory does not belong to the realm of pure mathematics, but is rather a *physical* theory like hydrodynamics. Therefore he starts from considering a physical random variable that can assume a set of possible values $R = \{r_1, \ldots r_m\}$. In a real experiment consisting in $N$ measurements we then obtain a finite sample $\{x_1, \ldots, x_j, \ldots, x_N\}$, with $x_j \in R$. A collective is an ideal infinite extension of this finite sample with two properties:

P1 (STATISTICAL STABILIZATION):    The relative frequency of any possible result $r_j$ approaches a limit as $N \to \infty$, which is called the probability $p_j$ of $r_j$,

$$\lim_{N \to \infty} f_j = \frac{n_j(N)}{N} = p_j.$$

You might say, "Here we go again, that's where the shoe pinches!", since we already say that there cannot be a true mathematical limit of the frequencies. But you (like most of the critics of von Mises' theory) would be mistaken, because von Mises', as we will see, is not a mathematical but an *empirical* concept of limit.

P2 (RANDOMNESS):    The limits of relative frequencies must not change if the original sequence is substituted by a subsequence obtained by any *preassigned* place selection, where by "preassigned" we mean that the selection rule *before* the value of the element to be selected or rejected is known. Possible selection rule could then be choose those $x_j$ for which $j$ is prime, exclude every other $x_j$, flip a coin and choose $x_j$ if it's heads, but surely not a rule like "select $x_j$ if $x_j = r_k$", which would give $p_k = 1$! More generally, one can accept that the decision to retain or

---

[12] R. von Mises, *Probability, Statistics and Truth*, Dover Publ. (1981).

reject an element $x_j$ of the original sequence depends on the preceding elements $x_1, \ldots, x_{j-1}$ but not on $x_j$ or on the following ones.[13]

By mans of collectives and of their properties, axioms K1-K3 of Kolmogorov's theory and all their consequences can be fully justified. It is also interesting that conditional probability, which must be *defined* in the measure–theoretical approach, is a *derived* concept in von Mises' theory, once one defines an operation on collectives called "partition" that is very similar to the idea of conditional probability as a projection we introduced in Chap. 2.

Unfortunately, property P2, which can be considered as the first quantitative definition of randomness, cannot hold for *all* possible preassigned selection. Extensive mathematical studies have actually shown that the more selection rules are allowed, the smaller is the class of possible collectives. Nevertheless, this is not true if we restrict the class of place selections, which von Mises suggested to select on the bases of the physical problem in question.

Coming back to von Mises' concept of "limit", he clearly links it to an ultimate experimental accuracy, for *in physics* there is no room for infinite precision (as we learned discussing statistical entropy, infinite precision requires infinite information). In other words, if we introduce an "elementary uncertainty" $\varepsilon_0$, we can actually *stop* the experiment after a *finite* number of measurements $N(\varepsilon_0)$, because a difference $f_j - p_j < \varepsilon_0$ has no experimental meaning (which is what scientist actually do). For a physicist beware of the continuum illusion, this factual idea of limit can be perfectly adequate, but I understand that this is the main reason why most mathematicians are skeptical about frequentism. Still, it is nevertheless hard to deny that, in our starting example on the decimals of $\pi$, all relative frequencies converge in some sense to $1/6$. We would almost say that the probability that the opposite happens is arbitrarily small…if we were not using frequencies *a la* von Mises to *define* probability!

But if we *do* have an independent way of defining probability (maybe simply a priori, or better with an axiomatic approach) then an important theorem, the "law of large numbers" (LLN), first derived in a weaker form by Jakob Bernoulli in 1713, and then in a stronger and progressively more general form by several "top guns" of the 20th century mathematics like Émile Borel, Aleksandr Khinchin, and Kolmogorov. Discussing in detail the LLN is beyond our scope, but let us enjoy at least the taste of this remarkable result.

### 6.2.1  The Law of Large Numbers

The simplest approach to the LLN justifies the progressive approach of the experimental frequencies to the theoretical probability values from the convergence of the binomial distribution to a normal distribution, as the number $n$ of trials goes to infinity. Thus, we have to analyze the behavior of unlimited Bernoulli sequences, which

---

[13] This is equivalent to stating that, in a game of chance, there is there is no winning system unless we know about the future. See FELLER vol.1, pp. 198–200.

we will generally indicate with $110010111001\ldots$, where 1 indicates a "success" and 0 a "failure". Venturing into a space where elementary events are infinite sequences can, however, be treacherous, as we already noticed discussing the St. Petersburg paradox. Nevertheless, if we regard each unlimited sequence as the binary representation of the decimal expansion of a real number[14] between 0 and 1, the sample space can be taken as the interval [0, 1] of the real axis, which however means the events of are not countable. This requires some care. Luckily, as with many drugs, the law of large numbers can be given in small amounts, or in stronger and more effective doses. Like for medicines, the strong formulation of the law of large numbers is a bit harder to digest. So, let us begin with a small dose.

The "weak" statement

We first show that, by increasing the number of trials $n$, the relative frequency of successes $k/n$ approaches $p$ *in probability*, by which we mean that

$$\forall \epsilon > 0 : \quad P\left(\left|\frac{k}{n} - p\right| \le \epsilon\right) \xrightarrow[n \to \infty]{} 1. \tag{6.1}$$

Therefore, putting $k_m = n(p - \epsilon)$ and $k_M = n(p + \epsilon)$, we should evaluate

$$P\left(\left|\frac{k}{n} - p\right| \le \epsilon\right) = P\left(k_m \le k \le k_M\right) = \sum_{k=k_m}^{k=k_M} \binom{n}{k} p^k (1 - p)^{n-k}.$$

As $n$ increases, the binomial is better and better approximated by a Gaussian. Therefore, using (3.65) and recalling the definition of $z$, we can write

$$P\left(\left|\frac{k}{n} - p\right| \le \epsilon\right) \simeq G\left(\frac{k_M - np}{\sqrt{np(1 - p)}}\right) - G\left(\frac{k_n - np}{\sqrt{np(1 - p)}}\right) =$$
$$= 2G\left(\frac{n\epsilon}{\sqrt{np(1 - p)}}\right) - 1.$$

But for every $\epsilon$, the argument of $G$ at the r.h.s. diverges,

$$z = \frac{n\epsilon}{\sqrt{np(1 - p)}} = \epsilon\left[\frac{n}{p(1 - p)}\right]^{1/2} \xrightarrow[n \to \infty]{} \infty,$$

---

[14] Actually, more than a single sequence can sometimes represent the same real number. For example, "1/2" can be written indifferently as $0.1000\ldots$ or $0.01111\ldots$ But this will not affect our conclusions.

so we can use Eq. (3.66) and write

$$P\left(\left|\frac{k}{n} - p\right| \le \epsilon\right) \simeq 1 - 2\frac{\exp(-z^2/2)}{z\sqrt{2\pi}} \xrightarrow[n\to\infty]{} 1.$$

Operationally, this means that if, for example, we flip a fair coin $n = 100$ times, the probability of obtaining a number of heads between 40 and 50 (i.e. $\epsilon = 0.1$, and therefore $z = 2$) is approximately

$$P\left(\left|\frac{k}{100} - 0.5\right| \le 0.1\right) \simeq 1 - \frac{\exp(-2)}{\sqrt{2\pi}} \simeq 0.95,$$

that is, if we repeat the "experiment" many times, in 95% of cases the relative frequency will not differ by more than 20% from the theoretical probability.

### 6.2.1.1   The "strong" Statement

As reassuring as the weak formulation is, it does not fully correspond to what we hoped to discover. Suppose, for example, that in the previous example you flip the coin another 1000 times. Even if after the first 100 tosses the relative frequency is between 0.4 and 0.6 (which is very likely), we do not know if this will *continue*. True, the probability that this will *not* happen in any of the following flips is very small, but the probability that this may happen sooner or later is obtained by summing *a lot* of small probabilities! In other words, Eq. (6.1) tells us that for a fixed number of attempts $n$ the relative frequency of successes is almost always equal to $p$, but it does not tell us that it *remains* so, i.e., it does not assure us that it will be valid for all $k > n$. If we remember our discussion of the decimals of $\pi$, this is actually what we are really interested in.

A much more stringent condition is guaranteed by the strong form of the law of large numbers. Indeed, calling $f_n = k/n$, it can be shown that for every $\epsilon > 0$ the values of $n$ such that $|f_n - p| > \epsilon$ are, with probability $P = 1$, a *finite* number. Thus, since the number of these values is finite, for every $\epsilon$ and $\delta$ we can choose a value $n_0$ such that $P\left(|f_n - p| < \epsilon\right) > 1 - \delta$ *for every* $n > n_0$, i.e., the difference between $f_n$ e $p$ becomes small and *remains* so.

To obtain the strong LLN, we need a result by Émile Borel and Francesco Paolo Cantelli of paramount importance in measure theory. In simple words the *Borel-Cantelli* (BC) *lemma* states that if the sum of the probabilities $P(E_n)$ of an infinite sequence of events $E_n$ is finite, then the probability that infinitely many of these events occur is zero. More formally, let us introduce the event

$$A = \limsup_{n\to\infty} A_n \doteq \bigcap_{n=1}^{\infty}\left(\bigcup_{k=n}^{\infty} A_k\right),$$

i.e., we first look for the smallest event $B_n = \bigcup_{k=n}^{\infty} A_k$ that contains all $A_k$ with $k > n$, and then find the interception of all the events $B_n$. Event $A$ is then the equivalent for a sequence of sets of the *limit inferior* of a numerical sequence. Because of this definition, an elementary event $x$ belongs to $A$ *if and only if* $x \in A_n$ for infinite values of $n$. The BC lemma then states that $A$ is "almost always" empty,

$$\sum_{n=1}^{\infty} P(A_n) < \infty \Rightarrow P(A) = 0. \tag{6.2}$$

Indeed, since $A \in B_n$ for all values of $n$, we have from (2.4)

$$P(A) \leq P(B_n) = P\left(\bigcup_{k=n}^{\infty} A_k\right) \leq \sum_{k=n}^{\infty} P(A_k).$$

Because $\sum_{n=1}^{\infty} P(A_k)$ is finite, however, the last term on the terms vanishes for $n \to \infty$, therefore $P(A) = 0.$[15]

The BC lemma very easily yields the strong LLN (and much more). Indeed, for any $a > 0$ consider the event

$$A_n : \left\{ \left| \frac{k_n - np}{\sqrt{np(1-p)}} \right| \geq \sqrt{2a \ln n} \right\},$$

which amounts to say that the standard variable $z$ defined in Chap. 3 takes a value larger than $z_0 = \sqrt{2a \ln n}$ of probability $1 - G(z_0)$. For large $n$ we have, using Eq. (3.63),

$$P(A_n) \simeq \frac{\exp(-z_0^2/2)}{z_0\sqrt{2\pi}} = \frac{1}{4\pi a \ln n} \frac{1}{n^a} < \frac{1}{n^a}, \tag{6.3}$$

---

[15] One can show (see FELLER, vol. 1) that for *independent* events $A_n$ the BC lemma becomes

$$\sum_{n=1}^{\infty} P(A_n) = \infty \Rightarrow P(A) = 1,$$

i.e., if the sum of the probabilities diverges, then *infinite* events $A_k$ will surely (with probability one) take place, no matter how small are the $P(A_k)$.

When considering infinite Bernoulli sequences, it is easy to construct a string of independent events. Just divide the sequences into "blocks" of attempts and consider those events $A_k$ that are only related to the attempts contained in block $k$: in this way, it is immediate to show that any finite sequence of successes and failures will occur infinitely many times. For example, consider the string "101" (success-failure-success), and choose as $A_k$ the events "string 101 will happen at attempts $3k, 3k + 1, 3k + 2$". These events refer to disjoint blocks of attempts and are therefore independent. Besides, each of them has probability $p^2(1 - p)$, so $\sum_{k=1}^{\infty} P(A_k)$ diverges. Or consider a book of any length translated into Morse code, which is a finite Bernoulli sequence. Then, the BC lemma becomes the "theorem of the tireless monkey", according to which one of these relatives of ours, placed in front of a keyboard for a sufficient time, will sooner or later write (infinite times) the all the books in Borges' library of Babel.

where the last inequality derives from $4\pi a \ln n < 1$ for $n > 1$. Since $a > 1$, la series $\sum_n P(A_n) = \sum_n n^{-a}$ converges and therefore, because of the BC lemma, only a *finite* number of events $A_n$ can take place. On the other hand, $|k_n/n - p| > \epsilon$ implies

$$\left| \frac{k_n - np}{\sqrt{np(1-p)}} \right| > \frac{\epsilon}{p(1-p)} \sqrt{n}.$$

But, by increasing $n$, the r.h.s. becomes *larger* than $\sqrt{2a \ln n}$. Thus, Eq. (6.3) also implies that the event $|k_n/n - p| > \epsilon$ takes place for a finite number of values $n$ too, which is the strong LLN.

A particularly interesting corollary of this result, not hard to prove rigorously, is that any real number, which can be thought of as the set of all infinite Bernoulli sequences in which the possible outcomes are the values of the individual digits, is normal with probability one. Indeed, the frequency with which a string of digits like "7523" appears in the distribution of almost all real numbers approaches (and remains) close to its theoretical probability p $= 10^{-4}$. In fact, since the result we obtained does not depend on the specific base in which we represent a number, almost all real numbers are normal in any base $b > 1$, that is, they are *absolutely normal*.

Thus, using the measure–theoretical definition of probability, the idea that "on the long run" the relative frequencies approach constant values, which coincides with the probabilities, can be given an explicit (probabilistic) meaning. As a matter of fact, Kolmogorov developed his theory precisely to provide sound foundations for frequentism. Discussing the relation to experimental data of his axiomatic approach, he indeed writes[16]:

> Under certain conditions, which we shall not discuss here, we may assume that to an event $A$ which may or may not occur under conditions $\mathfrak{S}$, is assigned a real number $P(A)$ which has the following characteristics:
>
> (a) One can be practically certain that if the complex of conditions $\mathfrak{S}$ is repeated a large number of times, $n$, then if $m$ be the number of occurrences of event A, the ratio $m/n$ will differ very slightly from $P(A)$.
>
> (b) If $P(A)$ is very small, one can be practically certain that when conditions $\mathfrak{S}$ are realized only once, the event $A$ would not occur at all.

Statement (a) is nothing but the von Mises frequency interpretation of probability, with the explicit requirement of fixed experimental conditions. However, this also means that "single shot" events cannot be attributed a meaningful probability value. This is an insurmountable limit for frequentism, but is also, as we will shortly see, a substantial problem for Kolmogorov theory. Before that, let's go a little deeper into the concept of randomness.

---

[16] KOLMOGOROV, pp. 3–4.

### 6.2.2 *Randomness and Algorithmic Complexity*

We have seen that in order to properly define probability, von Mises considered it crucial to provide first an answer to the question "what is a random sequence?". The invariance by place selection he conceived was meant to identify "erratic" sequences that no gambler could predict, regardless of their betting strategy. As we mentioned, however, it is rather difficult to limit the class of place selections to those that preserve the limit of relative frequencies.

A different and more effective approach is scrutinizing not so much the unpredictability, but rather the "typicality"of a sequence. It was Laplace that pointed out that among all very long sequences only a few have a "rule that is easy to grasp", which he attributed to the fact that these sequences have "a regular cause". The other sequences, "incomparably more numerous", are irregular: these typical sequences are the ones that must consider random. Laplace's suggestion was formalized for infinite sequences by Per Martin-Löf. Focusing on infinite binary sequences, whose class is usually indicated as $[0, 1]^*$, Martin-Löf called "special" a property of a sequence $s \in [0, 1]^*$ if the probability that the property holds is zero, while a property is called "typical" if the probability that it holds is one (namely, if it holds for almost all $s \in [0, 1]^*$). Sequences that have a special property cannot be random, so an infinite binary sequence is random if, roughly speaking, there is no effective way to specify a special property that the sequence possesses. A special property could for example be that, within an infinite sequence, there is a finite number of blocks of seven consecutive 0's: this won't happen for a truly random sequence. By extension, it is not difficult to see that every random sequence must be the binary expansion of a *normal* number, but the opposite is not true. This one, for example,

$$C_2 = 0.110111001011101111000100110101011110011011110111100001000\ldots$$

ia a normal number in base 2, but as a sequence it is definitely *not* random: it is nothing but the Champernowne constant we met in Chap. 1, obtained in this case by concatenating the natural numbers in base 2 one after the other in order.[17] Martin-Löf was able to provide a rigorous definition, based on measure theory, of a typical sequence. As Laplace suspected, almost all sequences in $[0, 1]^*$ are typical, but establishing whether a *particular* sequence is typical, i.e., if it passes *all* tests of "immunity" any special property that can be conceived, is another story.

If deciding whether an infinite sequence is random is arduous, asking the same question for a *finite* sequence (a string) is almost meaningless. Consider for instance the following string:

$$001001000011111101101010100010001000010110100001100$$

---

[17] Note that, as a normal number, the Champernowne constant contains all possible binary strings of any length, including those corresponding to everything that has or will ever be written: Borges would have loved this Universal Library of Babel (actually, I have to confess that I found the book you are reading by patiently looking inside the Champernowne constant).

It could definitely be a sequence of results obtained by flipping a fair coin: the fraction of 1's is 21/50, which is slightly lower than 0.5, but this is perfectly admissible. However, if prefixed by a dot, these are just the first 50 digits in the binary expansion of the fractional part[18] of $\pi$, which is rather suspicious. Nevertheless, whether random or not, this string can surely be *described* with a compact sentence, for instance by asking a computer to write the first 50 digits of $\pi$. Similarly, these two strings

$$A = 00110100010111001011010101010101000110001011010101100$$

$$B = 0101010101010101010101010101010101010101010101010101$$

have the same probability $P = 2^{-50}$ of being obtained in a series of Bernoulli trials with $p = 1/2$, but B can surely be described in a much more compact form. This suggest that string that looks more "irregular" require a longer description.

However, what do we exactly mean by "description"? To avoid confusion and paradoxes, an unequivocal procedure is mandatory,[19] and this is far from being easy. The solution to the problem was suggested by Ray Solomonoff and later independently found by Kolmogorov and Gregory Chaitin, who introduced the concept of *algorithmic complexity*, which can be defined as follows. Let us first distinguish between a computing algorithm, which is a general step-by-step procedure to solve a problem, and a program, which is a specific implementation of an algorithm on a computer architecture with some programming language. Then, for a given algorithm $A$, the algorithmic complexity $K_A(s)$ of a binary string $s$ is the length of the shortest program, based on $A$ and itself coded as a binary string $s^*$, that reproduces $s$. Of course $K_A(s)$ depends on the algorithm $A$, but Solomonoff and Kolmogorov managed to show that there is an optimal algorithm $A_o$ such that, for any algorithm $A$ and all strings $s$,

$$K_{A_o}(s) \leq K_A(s) + c(A), \tag{6.4}$$

where $c(A) \geq 0$ is a constant that depends in general on $A$.[20]

A more precise definition of algorithmic complexity (also known as Kolmogorov's complexity) would require rather advanced concepts of the theory of computation, and in particular Alan Turing's iconic idea of a "Universal Computing Machine", that cannot unfortunately fit in this book.[21] Nevertheless, the little we have said already allows us to roughly define a random string as a string with an algorithmic complexity

---

[18] The fractional part of a real number $x$, i.e., the excess beyond its integer part, is usually written as $\{x\}$.

[19] How ambiguous is the verb "define" can be appreciated by considering the question posed by G. G. Berry, a junior librarian in Oxford, who asked Bertrand Russell "what is the smallest positive integer that cannot be defined by an English sentence with fewer than 1000 characters?". You can see yourself the paradoxical nature of this question.

[20] Such an optimal algorithm $A_o$ is not necessarily unique.

[21] A simple but very precise introduction, which includes many useful references, can be found in M. Davis, *What is computation?*, in *Mathematics Today Twelve Informal Essays*, L. A. Steen eds., Springer, New York (1978).

comparable to its length.[22] More precisely, we will say that a string of length $n$ is $c$-random if, for some small integer $c$,

$$K_{A_o}(s) \geq n - c. \tag{6.5}$$

But how many strings are random? The number of distinct programs whose binary code has a length $r$ is clearly $2^r$. If we sum all programs with a length less than $n - c$ we find, according to (6.5), the total number of strings that are *not* $c$-random. Using the expression for a geometric sum

$$\sum_{r=1}^{n-c-1} 2^r = \sum_{r=0}^{n-c-1} 2^r - 1 = 2^{n-c} - 2 < 2^{n-c},$$

thus, the fraction $f$ of the $2^n$ strings of length $n$ that are not $c$-random is less than

$$f = \frac{2^{n-c}}{2^c} = 2^{-c}.$$

Even for $c = 10$, $f$ is less than $10^{-3}$: hence almost all strings with $n \gg c$ are random. Therefore, random strings are *typical* in the sense of Martin-Löf.

But can we say if a *specific* string is random? The answer, if the string is long, is definitely no. More precisely, it is not possible to decide whether there exist a binary string whose algorithmic complexity is larger than an integer $m$, if $m$ is large enough. A rough demonstration of this important theorem, due to Chaitin, goes as follows. Suppose that we can make a program that, for every $m$, finds out a string whose complexity is larger than $m$. What is the length $\ell(s^*)$ of the binary code of that program? We need a fixed number of bits $\ell_0$, which does *not* depend on $m$, to provide the instruction for computing the complexity of all the strings of a given length, plus approximately $\log_2 m$ bits[23] to input the value $m$ we wish to check. So we have, approximately, $\ell(s^*) \simeq \ell_0 + \log_2 m$. But then $\ell(s^*)$ grows only as the *logarithm* of $m$, so it will eventually be smaller than $m$. We reach then the contradicting conclusion that a string whose complexity is greater than $m$ is calculated by a program whose length is *less* than $m$. This implies that such a program cannot exist (more precisely, that there is no program that executes this task on a universal Turing machine and eventually halts having completed its job).

Thus, of a long string we can only state that it is very likely random, but we can never be sure: it could contain a pattern that we are not able to pick out and that, if grasped, may allow the string to be reproduced by a short program. It is much easier

---

[22] Note that, with this definition, the string $s^*$ corresponding to the shortest program that codes for $s$ *must* be random. Otherwise, one could write a program $s^{**}$ that just says "generate $s^*$ and use it to produce $s$" that would be a *shorter* description of $s$.

[23] You can easily check that a natural number $2^k \leq m \leq 2^{k+1}$ requires a number of bits $k \simeq \log_2 m$ to be specified. More precisely, $k = \lfloor \log_2 m \rfloor + 1$, where $\lfloor x \rfloor$ is the largest integer $n$ such that $x \geq n$ (the "floor" of $x$).

to write a program that generates the first 1000 digits of $\pi$ than a program that *detects* them within a string. For example, this string

$$1001000100110011101100101010000011010000110001110 0 \ldots$$

may have been obtained by flipping a fair coin, but nothing precludes that it is part of a quantum gravity fundamental constant discovered by the Klingons. Hard to fly to Kronos and check. Similarly, while it might not be impossible to spot the first (digitized) bars of a simple Baroque counterpoint, I bet it won't be so easy to figure out the beginning of a Beethoven sonata.

There are similarities but also important differences between the algorithmic complexity and the Shannon entropy. Both are, in some sense, a measure of the information content. But while $S$ gauges the degree of complexity of a whole probability distribution, algorithmic complexity focuses on a single sequence of events (for example, a Bernoulli string), measuring the effort required to describe it with *no reference* to probability concepts. As I tried to highlight, both of them have little to say about *meaning*. The string coding for the Beethoven sonata we mentioned is neither too simple, like a trivial pattern such as 01001000100001000001 . . ., not too complex like a random sequence. Meaningful pattern are in between and, as far as I know, we have not find a general way to describe them.

## 6.3  Subjective *ma Non Troppo*

Consider these statements:

(a) "Heavy rain is expected today";
(b) "The roof of my house is going to leak";
(c) "I'll have to collect buckets of water from the floor".

It comes naturally to try to evaluate the probability of (c) based on those of (a) and (b), which are obviously not independent, since it could be the rain itself that damages the roof. But in what kind of sample space can these three events be framed? Which are the "elementary events"? It does not seem trivial to get away with Kolmogorov's theory, which indeed struggles quite a bit to accommodate problems where the degree of probability of a logical proposition must be assessed.

As a practical matter, if the question that are difficult to deal with Kolmogorov's theory were few or trivial, we could simply ignore them, stating, as many supporters of frequentism do, that they are outside the scope of a probability theory.[24] But actually, in everyday life the cases where a well-defined ensemble clearly exist that are rather exceptional. Therefore, it is certainly important to inquire about the possibility of

---

[24] Von Mises for instance maintained that questions such as 'Is there a probability of Germany being at some time in the future involved in a war with Liberia?' has nothing to do with his probability theory. Nevertheless, I'm not sure if I'd be content if there was absolutely no way to estimate the chance of a (last) World War....

inferring the probability of an event from previously obtained information. This is the task faced by the so-called subjective approach to probability, which however has nothing to do with private opinions about chance, but rather aims to interpret probabilistic reasoning as an extension of classical logic. So, it is indeed subjective, because it is based on "reasonable expectation", *ma non troppo*, not too much so.

As we will see, the key to introduce reasonable expectation as a primary concept in probability is the Bayes theorem on conditional probabilities. Before expatiating on the subjective and extended logic approaches to probability, it is then worthwhile discussing the power of the Bayes theorem (which, remember, holds *whatever* interpretation we give to probability) but also the dangers of using it carelessly.

### 6.3.1  *The Bayes Theorem, Cross and Delight of the Predictor*

It was (once again) Laplace who first realized that conditional probability, and specifically the Bayes theorem, can be used to improve the estimate of the likelihood of an event. In a landmark paper of 1774, he indeed wrote[25]:

> The uncertainty of human knowledge is concerned with events or with causes of events. If one is assured, for example, that an urn only contains white and black tickets in a given ratio, and one asks the probability that a ticket drawn by chance will be white, then *the event is uncertain but the cause upon which the probability of its occurrence depends, the ratio of white to black tickets, is known.* In the following Problem, an urn is supposed to contain a given number of white and black tickets in an unknown ratio; if one draws a ticket and finds it white, determine the probability that the ratio of white to black tickets is that of $p$ to $q$. *The event is known and the cause is unknown.*

To solve the problem, he introduces a principle:

> If an event can be produced by a number n of different causes, the probabilities of these causes given the event are to each other as the probabilities of the event given the causes, and *the probability of the existence of each of these is equal to the probability of the event given that cause, divided by the sum of all the probabilities of the event given each of these causes.*

If we call $C_i$ the causes and $E$ the event, the sentence in Italic reads

$$P(C_j|E) = \frac{P(E|C_j)P(C_j)}{\sum_i P(E|C_i)P(C_i)},$$

which is exactly Bayes' theorem (2.12). Even the simple problem posed by Laplace, however, shows how delicate can be the application of Bayes' theorem to what we will later call "inverse" probability. Suppose indeed that the urn contains only two tickets, which can either be white ($q = 2$), black ($q = 0$), or of different colors

---

[25] P.-S. Laplace, *Memoir on the Probability of the Causes of Events*, reprinted in Laplace's Oeuvres completes **8** 27–65 (translation by S. M. Stigler).

($q = 1$). Calling $W$ the event of drawing a white ticket, the probability that both tickets are white is then given by

$$P(2\,W) = \frac{P(W|2)P(2)}{P(W|0)P(0) + P(W|1)P(1) + P(W|2)P(2)}.$$

Evidently, $P(W|2) = 1$, $P(W|1) = 1/2$, and $P(W|0) = 0$. But what about the *unconditional* probabilities $P(2)$, $P(1)$, and $P(0)$? Laplace, using the principle of indifference, would have set $P(2) = P(1) = P(0) = 1/3$, obtaining $P(2|W) = 2/3$, but we have to stress that this choice of equiprobability a priori is totally arbitrary. No one knows the protocol with which the urn was prepared: maybe the staff who set it up did not care a bit about impartiality!

You may say, but that was only a first guess that we can surely improve if we draw *another* ticket, and our estimate will arguably become more reliable the more tickets we draw. Right, but this only works if the urn contains *more* than two tickets. Otherwise, it is a take it or leave it game (exactly as predicting if it will rain tomorrow).

This simple example already highlights two aspects of the inductive approach that we will delve into: the need for an initial assumption (a "prior") and the problem of quantifying the amount of information we have to gather before an estimate becomes truly reliable. But the most serious problems with inductive guessing arise when we *miss* crucial information, as the following sad story highlights.

**Example 6.2** (*A hapless bright turkey*) Once upon a time, a clever Kentucky turkey was pondering her life expectancy, asking herself: 'Tomorrow, I will have been living for 13 weeks, but will I still be alive?'. She had actually spent a lot of time trying to learn probability by reading Laplace's *Oeuvres complètes*,[26] eventually mastering the subject better than most of the writer's students, so argues the following

Let's call $x$ the probability that I stay alive in a certain day. Of course, if $x$ takes a *specific* value $p$, the probability $P(n)$ that I survive for $n$ days in a row is (assuming that these are independent events)
$$P(n|x = p) = p^n.$$

Then, if I knew that $x$ can take only a few discrete values $p_i$ with probabilities $P(x = p_i)$, I might estimate
$$P(n) = \sum_i P(n|x = p_i)P(x = p_i).$$

However, since I do not know anything about $x$, the principle of indifference forces me to assume, as a first guess, that $x$ is a continuous RV uniformly distributed in $[0, 1]$. Thus I'd better write
$$P(n) = \int_0^1 P(n|x = p)\mathrm{d}p = \int_0^1 p^n \mathrm{d}p = \frac{1}{n+1}$$

But then, the probability $P(n + 1|n)$ of remaining alive for another day if I have already lived for $n$ days is
$$P(n + 1|n) = \frac{P[(n + 1) \cap n]}{P(n)} = \frac{P(n + 1)}{P(n)} = \frac{n+1}{n+2},$$

---

[26] In French, because she was multilingual too, despite being American.

since of course $P[(n+1) \cap n] = P(n+1)$. So, for $n = 90$ (thirteen weeks minus one day), $P(n+1|n) = 91/92 \simeq 0.99$, and I can sleep soundly!

The reasoning of our bright turkey is flawless. The only problem is that she makes this argument on the fourth week of November…

This little joke, where the turkey's estimate is actually based on Laplace's "rule of succession",[27] already tells us that missing a relevant piece of information (the fact that Thanksgiving is coming) may lead to disastrous conclusions. But the real puzzle is that the impeccable turkey solution is valid *for all n* $\geq 1$[28]. Let us indeed use the same logic to predict the next outcome of the toss of a coin, which, for all we know, could be heavily rigged, so that we can only say that the probability $x$ of getting heads is a number between 0 and 1. Suppose we flip the coin and observe 48 consecutive heads. Then, we should conclude that the probability that the next throw will still result in heads is 49/50, or about 98%. I believe that the vast majority of you would find this result entirely reasonable, as in this case it is hard to imagine a dramatically disruptive effect like Thanksgiving.

Good. Now apply the same reasoning to another coin that, flipped only *once*, shows heads. Do you think it is a fair statement to say that the probability of getting heads on the next throw is 2/3? Something tells us that (as long as there are no hidden pieces of information) Bayesian inference can work so much better (i.e., depend less on initial assumptions) the more it is supported by observations.[29] But at what point can we really feel 'safe'?

---

[27] P.-S. Laplace, *A philosophical essay on probabilities* (translated by F. W. Truscott and F. L. Emory, Frederick Lincoln. Chapman & Hall, 1902). In this landmark essay, Laplace used the rule of succession to calculate the probability that the Sun will rise tomorrow, given that it has risen every day for the past 5000 years. Of course, he would have reached the same conclusion even if the following day the Sun and the entire solar system had been sucked into a black hole….

[28] One can find it a bit disturbing that the turkey took for granted that she could, in principle, live forever, and may wonder what would change had she realized that $n$ must be upper–limited. How is the rule of succession modified if the number of attempts is finite? Suppose for instance that we have already drawn $n$ white balls in a row from an urn that contains an *assigned* number $N$ of white and black balls in an unknown proportion. What is the chance the next draw will also be a white ball? Assuming that all black/white proportions are equiprobable, one surprisingly finds again $p = (n+1)/(n+2)$ *independently* of N (see, for instance, S. L. Zabell, *Erkenntnis* **31**, 1989). You can easily find however, that the probability that all the remaining balls are white *does* depend on $N$, and is given by $(n+1)/(N+1)$. Note that when $n \ll N$ this is still a small number. For example, in Switzerland there are almost five million sheep. Many of then are of the Blacknose breed. Suppose that, wandering through the mountain pastures of Valais you have already seen 98 sheep, all with a black nose. Then you should conclude that there is a 99% chance that the next sheep you see will still be Blacknose, but also that the chance that *all* sheep in Switzerland have a black nose is about 20 in a million.

[29] Besides, if we call $H_1$ ($H_2$) the event heads at the first (second) toss, using the estimate $P(H_2|H_1) = 2/3$, we get $P(H_2 \cap H_1) = P(H_2|H_1)P(H_1) = 1/3$, while if we consider the two events independent, we obtain $P(H_2 \cap H_1) = 1/4$. This means that, if we accept using the succession rule, $H_2$ *cannot* be considered independent of $H_1$ (which of course, is a natural consequence of the fact that $H_1$ changes our estimate of $H_2$!).

Yet, the problem of missing crucial data can be even more severe: as shown in the next example, we may realize that we do miss something, but we are unable to pinpoint what we really missed.

**Example 6.3** (*A baffling genie*) You are sipping on a beer with a friend while discussing ancient stories about genies and agreeing that there must be some truth to them. Suddenly, a genuine genie materializes and warmly thanks you for your trust in his existence. He actually wants to reward both of you, but since he is an avant-garde pragmatic spirit, instead of granting you the usual three wishes, he decides to give you two checks.

He shuffles the checks, puts them in sealed envelopes, and gives them to you and your friend in random order. However, just before disappearing in a puff of smoke, he whispers: "Sorry, I forgot to tell you that the two checks are not for the same amount of money: one is twice as large as the other. But, if you wish, you can swap them".

You open the envelope you picked, check the check, and wonder whether you should propose to your friend to make the trade. Since you have read this book up to this point, you surely know how to proceed. Call $x$ the sum you found on your check. The sum on your friend's check is either $x/2$ or $2x$, both with probability 1/2, because the checks were distributed at random. Then if you both accept to trade the checks, you expect to end up with an amount of money

$$x' = \frac{1}{2}\left(\frac{x}{2} + 2x\right) = \frac{5}{4}x,$$

so you conclude that you *should* propose a swap. The problem is, your friend will follow exactly the same logic and necessarily come to the *same* conclusion!

So, should you make the swap proposal or not? The fact is, you have no way to tell because you can't make a guess about the *total* amount of money that the genie donated. But let me modify the story as follows. Before the genie vanishes, he declares that one of the two checks has a *thousand times* more money than the other. What would you do if you open your envelop and find a ten dollar check? And if you instead find a check for ten *thousand* dollars, would you still like to swap? Note that there is no reason to make different choices. It only depends on whether you picture the genie as a billionaire, generously giving away millions of dollars, or as a cheapskate who is not ashamed of writing a check for one cent. Where is the catch? Actually, I don't know.

Coming back to our hapless turkey, to find the probability that she lives tomorrow she actually used only a limited version of the Laplace succession rule, which is much more powerful. Let us then discuss it in general by considering a sequence of Bernoulli trials where, however, the probability of success in a single trial is not a fixed value, but is rather distributed according to a (non necessarily uniform) probability density $f(p)$ with support in [0,1].

Suppose that in the first $N$ trials we had $n$ successes. What is the probability $P(n+1|n)$ that the $N+1$ trial is a success too? For a *specific* value of $p$ we have

$$P(n+1\mid n, p) = \frac{p^{n+1}(1-p)^{N-n}}{p^{n}(1-p)^{N-n}},$$

so that, averaging over $f(p)$,

$$P(n+1\mid n) = \frac{\int_0^1 p^{n+1}(1-p)^{N-n} f(p)\mathrm{d}p}{\int_0^1 p^{n}(1-p)^{N-n} f(p)\mathrm{d}p} \tag{6.6}$$

If $p$ is uniform, $f(p)$ simplifies and, using the definition of the beta function $B(\alpha, \beta)$ and its relation with the gamma function in the appendix, you can easily find that

$$P(n+1\mid n) = \frac{n+1}{N+2}. \tag{6.7}$$

Notably, $P(n+1\mid n)$ converges for large $N$ and $n$ to the relative frequency $\nu = n/N$ of successes. So, we see that in this limit the Bayesian and the frequentist approach go hand-in-hand (and we can finally "feel safe").

However, you have the right to be skeptical about using an uniform distribution for $p$, in particular because we know that, when $n = N$, the conclusions drawn from the succession rule must be taken with care (and the same of course is true for $n = 0$, for your failures are your friend's successes). If these cases are excluded, the Laplace result (6.7) can be derived with much weaker assumptions than equiprobability. Indeed, substituting $n = \nu N$ in Eq. (6.6),

$$P(n+1\mid n) = \frac{\int_0^1 p\left[p^{\nu}(1-p)^{1-\nu}\right]^N f(p)\mathrm{d}p}{\int_0^1 \left[p^{\nu}(1-p)^{1-\nu}\right]^N f(p)\mathrm{d}p}. \tag{6.8}$$

If $0 < p < 1$, then $p^{\nu}(1-p)^{1-\nu}$ has a maximum for $p = \nu$. Hence, for sufficiently large $N$, $[p^{f}(1-p)^{1-f}]^N$ has a strong and very narrow peak around the same value of $p$. Thus, unless $f(p)$ is extremely small when $p = f$, its only important values in the integral are those for $p \simeq \nu$, and we can approximate $f(p) \simeq f(\nu)$, obtaining again (6.7). Therefore, except in very special cases, *the equiprobability assumption is not really needed if $N \gg 1$.*

### 6.3.2  *From Bayes' Theorem to Subjective and Logical Probability*

So far, we have discussed the pros and cons of Laplace's application of Bayes' theorem, with no need of assigning any specific meaning to probability. As we anticipated, however, Bayesian methods are at the core of the view of probability as inference, whose development has been an alternation of approaches that stressed its subjective aspects with others that appealed more to universal, impersonal reasons. This ambivalence is clearly evident when considering some distinguished scholars who have followed this approach.


### 6.3.2.1    Keynes: More Practicality, Less Math!

You probably know John Maynard Keynes as one of the intellectual giants of the twentieth century. His ideas, developed within the intellectual ferment of the Bloomsbury group, radically changed the economic policies of governments and the contemporary views of the human society. His struggle for world peace, although less successful, was exemplary too. Regrettably, his legacy in probability ranks significantly lower. So, he would not have found place in this short summary if not for the fact that he was arguably the first to explicitly claim that probability is an extension of logic that goes beyond the reach of mathematics.

The first page of his treatise[30] leaves no doubt about his central idea: we need probability because there are issues, not only in philosophy, science, and ethics, but also in practical life that cannot be tackled with a yes-or-no logic. Doubting conclusive claims is a cornerstone of pragmatism, a lifelong guiding light of Keynes' deeds.

This does not means that probability reasoning is subjective. A proposition is not probable because we think it so. What is probable or improbable (within certain prescribed circumstances) is independent of our opinion. Indeed, Keynes views probability as an extension of logic and the foundation of rational belief, which he hopes to elaborate on the footsteps of the masterpieces in formal logic by Frege and Russell. As such, probability is a measure of the likelihood of *propositions*, not of the occurrence of "events", a term that he considers vague and ambiguous.[31]

According to Keynes, however, there are cases when a quantitative assessment of probability is hardly possible for several reasons:

> Either [because] in some cases there is no probability at all; or probabilities do not all belong to a single set of magnitudes measurable in terms of a common unit; or these measures always exist, but in many cases are, and *must remain*, unknown; or probabilities do belong to such a set and their measures are *capable* of being determined by us, although we are not always able so to determine them in practice.

In other words, sometimes we cannot attribute a numerical "truth-value" to a proposition: in many practically important situations we can only *compare* two propositions, just stating which one is more probable than the other, but not how much. This reflection drives Keynes to present a harsh criticism of the principle of indifference and to dismantle Laplace's rule of succession. He is a bit kinder with frequentism, believing however that it is almost always useless.

---

[30] J. M. Keynes, *A Treatise on Probability*, Macmillan and Co., London, 1921 (freely available on the Project Gutenberg). The treatise is actually a development and a refinement of a work that went on for many years, starting from 1904 as an undergraduate paper for the 'Cambridge Apostles'.

[31] He actually declares that "With the term 'event', which has taken hitherto so important a place in the phraseology of the subject, I shall dispense altogether".

Keynes' appeal for a comparative view of probability is certainly rooted in practical life, but unfortunately his treatise does not provide us with rigorous rules for it. The logical axioms he introduces sound somewhat arbitrary and too sophisticated, and could hardly be conceived without having in the back of his mind those devices associated with the concept of sample space such as boxes, dice, coins. It was still a long way to a consistent epistemic probability. A leap forward was definitely made by the next scholar we consider.

### 6.3.2.2 De Finetti: "Probability Does Not Exist"

Bruno de Finetti is regarded by most of the statisticians working in economics and social sciences as the founding father of Bayesian inference.[32] Yet, the provocative statement that opens this section reveals him as a lifelong wrangler too, engaged in an endless crusade against any objective conception of probability. Indeed, de Finetti joins Poincaré in his battle against positivism, fully agreeing with him that:

(a) A reality completely independent of the mind which conceives it is an impossibility, and an external world, even if it exists, would for us be forever inaccessible[33];
(b) The laws of science do not belong to this inscrutable world. They are merely human mental constructions. Scientific hypotheses must however confront with the evidence that comes from this mysterious external entity, and this requires probability calculus. In fact, all the sciences are only "unconscious applications" of the calculus of probabilities.

However, de Finetti comments (about Poincaré):

> But then we must go beyond: why does he stop? Because his point of view, like perhaps any point of view alive, intelligent, subtle, leads - to think it all the way - to relativism and absolute subjectivism. There are many who have horror of such a conclusion and stop halfway. So the Poincaré. [Instead] if we do not fear this conclusion, if on the contrary it is precisely on it that our conception is based, we can develop unscrupulously and up to the extreme consequences Probabilism.[34]

For de Finetti, therefore, only *subjective* probabilities exist, that is, the *degree of belief* in the occurrence of an event attributed by a given person at a given instant and with a given set of information. However, if we wish that these subjective assessments satisfy the basic rules of probability calculus, they must be *consistent*.[35]

To clarify what de Finetti means by consistency we better use, as he did, a parallel with wagers. Consider again Laplace's two-tickets urn, and suppose that I propose

---

[32] Contrary to what the English version of Wikipedia states, De Finetti's *alma mater* was not Politecnico di Milano: indeed, he began studying as an engineer, but soon decided to switch to mathematics at the University of Milan, encouraged by the great mathematician Tullio Levi-Civita.

[33] Those of view fond of philosophy will recognize in this the Kantian *noumenon* (although for Kant the noumenon must *necessarily* exist).

[34] B. de Finetti, Probabilismo, F. Perrella Publ., Napoli, 1931 (my translation).

[35] Or 'coherent', as de Finetti's original term 'coerente' is usually, and rather inaccurately, translated.

you the following deal: I make available to you a sum of money $S$ which I undertake to deliver to you provided that the urn contains two white tickets (hypothesis $ww$). I ask you now what is the maximum price $P_{ww}$ you are willing to offer me in exchange for the contract I have proposed. Bear in mind, however, that whatever price $P_{ww}$ you set, you must be willing on my request to pay *me* the sum $S$, against my payment to you of the price $P_{ww}$, if the hypothesis in question is true. Then I propose similar contracts for hypothesis $bb$ (two black tickets) and $bw$ (one black, one white), asking for your corresponding offers $P_2$ and $P_3$. What can be shown is that, unless you choose $P_{ww}/S + P_{bb}/S + P_{bw}/S = 1$, so that the ratios $P_i/S$ sum as probabilities, you will *surely* loose. More generally, de Finetti proved that the subject who expresses a set of values quantifying his degrees of belief is immune from any combinations of bets that, whatever event occurs, he is certain to lose[36] *if and only if these values combine according to the rules of probability theory.*

Using the "Dutch book theorem", which can actually be reformulated with no reference to betting, Kolmogorov's axiom are recovered,[37] and subjective probability is not anymore "individual caprice". Having managed to quantify subjective probabilities, de Finetti points then out two key differences between the traditional approach and his one:

- While in the classical approach an hypothesis is either true of false, here the aim is to evaluate its *likelihood*[38] using the available information.
- Two propositions or hypotheses cannot be judged as independent or not in the absolute sense, but only subject to a prescribed condition. For instance, in the example that opens Sect. 6.3, whether propositions (a) and (b) are independent *does* depend on whether proposition (a) is verified, i.e., on whether it will rain or not! Similarly, two successive extractions from an urn that contains $n$ white and $m$ black balls are independent. However, if we just know that the urn contains $n + m$ balls, *but we don't know which are white and which black*, knowing which color is drawn makes more likely that the urn contains more balls of that color. In other words, probabilities should always be regarded as *conditional*, or subordinate to a set of premises, as de Finetti writes.

The last observation raises however an alarming problem. The usual way all experimental scientists make estimates is by taking many measurements in (reasonably) identical conditions and consider the results as a set of values of independent and identically distributed (i.i.d.) variables. But Bayesian inference relies precisely on using each result to update the estimate of the following ones, and if the measurements are assumed to be *independent*, this is clearly "mission impossible" (remember what

---

[36] This is called a "Dutch book". Why Dutch betting books should be so unsafe is not clear. Anyway, I found this expression not too kind to the Dutch.

[37] Although what is meant by a "complete set" of possible hypotheses, which would be the equivalent of the sample space, is not clear to me. Evidently, our turkey omitted at least one of them.

[38] Here "likelihood" is used with the current technical meaning in statistics: one evaluates the probability of an event, but the *likelihood* of a proposition or an hypothesis. For the latter, in his Italian writings, de Finetti often uses the word "plausibilità". Maybe, even in English, 'plausible' could have been a better choice than 'likely', which is substantially a synonym for probable.

we have said about the effects of the succession rule in footnote 29)! Independence is really the Beelzebub of Bayesians: they will *never* accept independence a priori (lest they lose their job). De Finetti is conscious of the crucial importance of multiple observations made in identical conditions to make estimates, but also realizes that, to this aim, assuming independence is not crucial: we can settle for a weaker condition, namely, that the *order* of the results plays no role. This observation leads de Finetti to further develop the concept of *exchangeable variables*, which we already discussed in passing while introducing the hypergeometric distribution, and turn it into a viable alternative to independence. It is worth taking a brief detour to clarify this concept, one of the few groundbreaking notions in probability after Laplace.

Before that, let me mention that, regrettably, it took a long time before de Finetti's radically new approach spread within the international community. This is of course partly due to the fact that he mostly wrote in Italian, but a reason of no less importance is that his writing style cannot be considered linear, to use an understatement, and is often laden with ideological considerations that are rather out of place. In the English-speaking world, he became first known because of the work of his friend Leonard Savage, who explicitly proposed to establish statistics on de Finetti's subjective approach. The work of Savage,[39] an excellent mathematician who during WW2 had worked as an assistant to John von Neumann, put de Finetti's ideas on firmer bases and introduced several important concepts in decision theory. But his ideas were still regarded as rather controversial, so it took several more decades before the Bayesian method pioneered by de Finetti prevailed among economists, psychologists, and social scientists. However, not among physicists: we will later inquire why. For now, let's return to the most lasting contribution by de Finetti.

### 6.3.2.3  Exchangeability and the Representation Theorem

As a basic notion, exchangeability is arguably easier to grasp than independence. Two events $E_1$, $E_2$ are "exchangeable" if their subscripts them can be swapped without affecting the results. For example, two flips of a coin are exchangeable because the order in which they are done is irrelevant for the probabilities of possible outcomes[40] More formally, this means that the sequence of events $(A, B)$ and $(B, A)$ have the same probability, because both sequences describe the occurrence of exactly one of the two events, either at the 1st or at the 2nd trial. This is easily extended to any finite sequence of events $(E_1, \ldots E_n)$, which are said to be exchangeable when the probability of occurrence of exactly $k$ of them in any order is the same for all $k \leq n$, and to an infinite sequence if $(E_1, E_2, \ldots, E_n)$ is exchangeable for every $n \geq 1$.

---

[39] L. Savage, *Foundations of Statistics*, John Wiley & Sons, New York (1954).

[40] For a subjectivist, this means that observers realized that they have no way to distinguish between them. As physicists used to deal with indistinguishable objects, however, we may agree no "observer" is really needed.

Independent events are obviously exchangeable,

$$P(E_1, E_2) = P(E_2|E_1)P(E_1) = P(E_2)P(E_1) = P(E_1|E_2)P(E_2) = P(E_2, E_1).$$

However, exchangeable events are not necessarily independent. Consider for example an extraction *without replacement* from an urn that contains 5 blue balls and twice as much red balls. The results of two subsequent extractions are clearly interchangeable,

$$P(R_1, B_2) = \frac{2}{3}\frac{5}{14} = \frac{1}{3}\frac{10}{14} = P(B_1, R_2)$$

On the other hand, the probability $P(R_2)$ of drawing a red ball at the second extraction, whatever the result of the first, is

$$P(R_2) = P(R_1, R_2) + P(B_1, R_2) = \frac{2}{3}\frac{9}{14} + \frac{1}{3}\frac{10}{14} = \frac{2}{3}$$

while the probabilities of drawing a red ball at the second extraction, conditional to the result of the first, are
$$\begin{cases} P(R_2|R_1) = \dfrac{9}{14} \\ P(R_2|B_1) = \dfrac{10}{14} \end{cases}$$

which are both *different* from $P(R_2)$. Therefore, the results of the second draw is *not* independent from the first, which would obviously be true for two extractions *with* replacement. Note however that, even in the latter case, if the urn content is *unknown*, a Bayesian would object to consider the draws as independent a priori, and would rather use the information obtained from the first extraction to improve the estimate of the ratio red/blue balls with respect to an initial (arbitrary) assumption.

The concept of exchangeability is readily extended to random variables. A collection of $n$ random variables $\{X_1, X_2, \ldots, X_n\}$ is exchangeable if the label we attach to them is not informative. That is, their joint probability does not change upon any permutations $\{i\} \to \pi(\{i\})$ of their labels,

$$P(X_1 = x_1, \ldots, X_n = x_n) = P(X_{\pi(1)} = x_1, \ldots, X_{\pi(n)} = x_n)$$

which, for continuous variables can be written using their joint densities as

$$f_{X_1, X_2, \ldots X_n}(x_1, x_2, \ldots x_n) = f_{X_{\pi(1)}, X_{\pi(2)}, \ldots X_{\pi(n)}}(x_1, x_2, \ldots x_n) \tag{6.9}$$

for all permutations $\pi$ of the subscripts $\{1, 2, \ldots, n\}$. Exchangeability therefore entails a complete symmetry between the variables. If the variables are independent they are clearly exchangeable, because the joint density factorized into the marginal densities,

$$f_{X_1,X_2,\ldots X_n}(x_1, x_2, \ldots x_n) = f_{X_1}(x_1) f_{X_2}(x_2) \ldots f_{X_n}(x_n),$$

hence independence implies exchangeability, but not vice versa.

The most important result about exchangeable variables, which is actually a milestone of inferential statistics, is that their joint probability can be expressed in a few simple and very general forms. This result, which is the content of the so-called representation theorem obtained in its simplest form by de Finetti in 1930, is more easily obtained by extending the previous definition to include *infinitely* exchangeable variables. We than say that

*An infinite collection of random variables* $\mathfrak{C} = \{X_i\}_{i=1}^{\infty}$ *is said to be infinitely exchangeable if every finite subsequence* $\{X_1, \ldots X_n\} \in \mathfrak{C}$ *is exchangeable.*[41]

The representation theorem, derived by de Finetti for a sequence of Bernoulli variables, i.e., variables that can attain only the values 0 ("failure") and 1 ("success"), can be stated as follows[42]

**Theorem 6.4** *Let* $\{X_i\}_{i=1}^{\infty}$ *be an infinitely exchangeable sequence of Bernoulli variables, then for every subsequence* $\{X_i\}_{i=1}^{n}$ *with joint probability distribution* $P(x_1, \ldots, x_n)$ *and density* $p(x_1, \ldots, x_n)$ *there exists a probability distribution* $F(\theta)$ *such that*

$$p(x_1, x_2, \ldots, x_n) = \int_0^1 \prod_{i=1}^{n} \theta^{x_i}(1-\theta)^{1-x_i} dF(\theta), \tag{6.10}$$

*where*

$$\theta = \lim_{n \to \infty} \frac{\sum_{i=1}^{n} x_i}{n} \quad \text{and} \quad F(\theta) = \lim_{n \to \infty} P\left(\frac{\sum_{i=1}^{n} x_i}{n} \leq \theta\right).$$

The main message of the representation theorem is that, for exchangeable sequences, the relative frequency of successes, $\sum_i x_i/n$, defines in the limit $n \to \infty$ a random variable $\Theta$ with a cumulative probability distribution $F(\theta)$, itself the limiting value for the cumulative distribution of the frequencies. More important, the theorem says that *the joint probability distribution of exchangeable variables can be seen as a superposition (a "mixture") of identical distributed independent variables, "weighted" according to* $F(\theta)$. Indeed, the joint probability density *conditional* to a given value of $\Theta = \theta$) is

$$p(x_1, x_2, \ldots, x_n \mid \theta) = \prod_{i=1}^{n} \theta^{x_i}(1-\theta)^{1-x_i} = \prod_{i=1}^{n} p(x_i \mid \theta).$$

---

[41] One may think that *every* finite sequence of exchangeable variables can be embedded in an infinitely exchangeable collection of variables of the same kind, but the answer is definitely no. See BERNARDO- SMITH, Sect. 4.3.

[42] For a detailed proof, see BERNARDO AND SMITH, Sect. 4.3.

In fact, the representation theorem is a generalization of the Law of Large Numbers, which we proved in Sect. 6.2.1 for an unlimited sequence of i.i.d variables, to exchangeable and not necessary independent sequences. When a probability density $f(\theta) = \mathrm{d}F(\theta)/\mathrm{d}\theta$ can be defined, we can also write

$$p(x_1, x_2, \ldots, x_n) = \int_0^1 \prod_{i=1}^n \theta^{x_i}(1-\theta)^{1-x_i} f(\theta)\mathrm{d}\theta. \tag{6.11}$$

For a Bernoulli sequence, we can write

$$p\left(\sum_{i=1}^n x_i = k\right) = \binom{n}{p} p(x_1, x_2, \ldots, x_n),$$

because $k$ coincides with the number of "successes", which can be chosen in $\binom{n}{k}$ ways from $n$ trials. Therefore, from (6.10),

$$p\left(\sum_{i=1}^n x_i = k\right) = \binom{n}{k} \int_0^1 \prod_{i=1}^n \theta^k(1-\theta)^{1-k}\mathrm{d}F(\theta) \tag{6.12}$$

The representation theorem, which can actually be extended to a much larger class of random variables, provides the basic justification for the Bayesian analysis. For example, suppose that we wish to estimate a parameter $\theta$ in a statistical model. While parameters are considered by a frequentist unknown but *fixed* quantities, in the Bayesian approach they can take different values depending on the degree of belief of the appraiser, so they actually are random variables with a probability distribution. So, one makes an initial hypothesis $f(\theta)$ about the distribution, a *prior* in the current jargon,[43] and then used the collected information, i.e., the probability of getting a set of data for a given value $\theta$ (the *likelihood function*) to "update" the estimate of $f(\theta)$ using the Bayes theorem, written as

$$f(\theta|x_1, x_2, \ldots x_n) = \frac{p(x_1, x_2, \ldots, x_n|\theta)f(\theta)}{\int_0^1 p(x_1, x_2, \ldots x_n|\theta)f(\theta)\mathrm{d}\theta}, \tag{6.13}$$

where $p(x_1, x_2, \ldots x_n) = \int_0^1 p(x_1, x_2, \ldots x_n|\theta)f(\theta)\mathrm{d}\theta$ is the total probability of obtaining the data set $\{x_i\}$, called the *marginal distribution* because $\theta$ has been "marginalized out" by integration. For instance, using (6.10) and (6.13), it is not difficult to show that

---

[43] Expression that de Finetti, however, never liked. In his words, "So far as the terms 'prior' and 'posterior' are concerned, they simply signify "before" and "after" the acquisition of the information A. One should avoid giving too much weight to this, lest the impression is given that "prior" refers to some mysterious circumstance of being prior to any experience, or to a state of absolute ignorance, or total indifference [...] as was the case with the old terminology relating to a priori and a posteriori probabilities".

$$p(x_{n+1} = 1|x_1, x_2, \ldots, x_n) = \int_0^1 \theta f(\theta|x_1, x_2, \ldots, x_n)\mathrm{d}\theta = \langle f(\theta|x_1, x_2, \ldots, x_n)\rangle \, .$$

which is a generalization of Laplace's succession rule.

The representation theorem ensures that a consistent "prior" estimate for $f(\theta)$ can be made, but does not provide a practical method to choose it. When no information is available, one better adopts an "uninformative" prior. The less compromising choice is a uniform distribution, justified by the indifference principle, but other options, based for instance on the information content (see the next section) are possible. The story is different if we have definite information about a variable, either because previous data exist or because there are theoretical reason to select a class of prior distributions. In this case, it is worth considering *conjugate priors*, that is, distribution pairs such that, for a given likelihood function $p(\mathbf{x}; \theta)$, the prior and posterior are distribution of the same kind, but with different parameters.[44] For example, if the likelihood function is a binomial or a geometric distribution, the family of beta distributions provide a class of conjugate priors, while the same role is played by the class of gamma distributions for a Poisson likelihood. When the likelihood function is a normal distribution, then the prior/posterior pair depends on the kind of parameter $\theta$ one wishes to estimate. Notably, if $p(\mathbf{x}; \theta)$ is a Gaussian with known variance, and we wish to estimate its expectation, then the conjugates are normal too. Even though statistical inference is not our focus in this chapter, the previous brief introduction should be sufficient for you to start exploring the tortuous paths of Bayesian jargon.

### 6.3.2.4 Physicists Enter the Scene: From Jeffreys to Jaynes

Starting with the work of Maxwell and Boltzmann, and even more so with the advent of quantum mechanics, probability has become a key tool in physics since the last decades of the 19th century. Yet, until much later, physicists have never been particularly eager to examine its "meaning" because its basic rules were more than sufficient for their purposes. Most of them were de facto frequentists, not only when they analyzed experimental data, but also when they use experimental frequencies to develop models: think for instance of the "rediscovery" of the Poisson distribution by Rutherford and coworkers.

A notable exception has been Harold Jeffreys, a prominent geophysicist[45] but also a skilled mathematician.[46] His thoughts on probability originate from the work of

---

[44] A "prior pair" because a posterior of an estimate is the prior of the next one.

[45] Among his main results, he was the first to show that the core of the Earth is liquid, he performed an analysis model for the travel times of seismic waves that is still a reference standards, he gave seminal insight concerning wind circulation. Perhaps, his only flaw was staunchly opposing till his death the idea of continental drift, which he refused (curiously for him, a priori) because no force even remotely strong enough to move continental plaques was conceivable.

[46] In 1924 he found an approximate solution of differential equations of the form $y'' = k\varphi(x)y$ for large $k$ that essentially coincide with the Wentzel–Kramers–Brillouin (WKB) solution of the Schrodinger equation in the semiclassical limit.

Keynes, which however he tries to put on firmer logical bases, aiming to disclaim Keynes' belief that, in many cases, probabilities can only be compared, and not put on an absolute scale. His idea of probability is based on the concept of *rational* degree of belief, where the qualifier "rational" marks his distance from the subjectivists. Jeffreys was totally unaware of the work of de Finetti, but he was strongly critical of the subjectivist approach of Frank Ramsey, who anticipated many of the ideas of the Italian.[47] Jeffreys indeed asserts that, given a set of data,

> a proposition $q$ has in relation to these data one and one only probability. If any person assigns a different probability, he is simply wrong.[48]

This means that there is a *unique* reasonable degree of belief that any satisfactory theory must account for. Nothing can be farther from de Finetti, who on the contrary stated that probability does not correspond to a self-proclaimed rational belief, but to the effective personal belief of anyone. Jeffreys admits that fixing a probability scale according to the subjective expectation of benefit is something that "a business man often has to make, whether he wants or not, or whether it is legitimate or not", but this concession suggests that, for him, subjective probability is just stuff for businessmen.

A point of connection between Jeffreys and subjectivists is that the degree of personal belief is not aimed at unraveling some unknown, hidden probability, as Keynes had speculated: objective probability does not exist. Except, maybe for physics…Jeffreys had indeed to admit that

> On the other hand, some scientific laws may contain an element of probability that is intrinsic to the system and has nothing to do with our knowledge of it. Carnap in particular has maintained that there are two kinds of probability. I am inclined to think that there may be such a thing as intrinsic probability, without accepting Carnap's optimistic attempt at a frequency determination of it.[…] *If we accept the notion of intrinsic probability, however, we have a pigeon-hole where we can put any differences that there may be between classical determinism and quantum indeterminacy.*[49]

Here the ghost that haunts any physicists that sticks to a subjective or purely epistemic idea of chance comes out: quantum probability is baffling, and seems to defy any commonly accepted logic. We'll come back to this crucial issue.

In his main work *Theory of Probability* Jeffreys dedicated particular attention to the choice of priors, and in particular to a problem that we have already anticipated when we discussed the von Mises cheating barman in Example 6.1 and the Bertrand paradox. Within the Bayesian approach, this means that if we assume that a uniform prior for a parameter $\theta$, we cannot do the same for a parameter $\varphi$ that is a nonlinear function of $\theta$. Jeffreys showed that one can nevertheless define uninformative priors that are *invariant* upon a change of variable. Regrettably, these are frequently

---

[47] Ramsey's work "Truth and probability", written in 1926 but publishes posthumous in 1931, present a very interesting discussion of the difference between probability in physics and in logic. Besides, Ramsey has given important contributions in formal logic and in economics, concerning taxation and saving. Unfortunately, this brilliant polymath died in 1930, when he was only 26.

[48] H. Jeffreys, Statistical Inference, Cambridge Univ. Press (1931).

[49] H. Jeffreys, Brit. J. Phil. Sci., **5**, 275 (1955).

"improper" priors, which means distributions $f(\theta)$ that cannot be normalized or lack finite moments. Jeffreys' approach was controversial and rather disregarded at the time, but surely put the bases for the development of a quantitative Bayesian analysis.

**Example 6.5** (*Jeffreys' priors and the Poisson distribution*)
Suppose that, in a unit time, we have detected $k$ decays from a radioactive sample, and that we want to use this information to estimate the decay rate $\alpha$. We know that radioactive decays are distributed according to a Poisson, so the probability of observing $k$ decays in a unit time *if* the decay rate equals $\alpha$ is

$$P(k|\alpha) = \frac{\alpha^k e^{-\alpha}}{k!},$$

which plays the role of a likelihood function. Then, the Bayes theorem tells us that

$$P(\alpha|k) = \frac{\alpha^k e^{-\alpha} P(\alpha)}{\int_0^\infty \alpha^k e^{-\alpha} P(\alpha)\, d\alpha},$$

where $P(\alpha)$ is the initial probability distribution we assume for $\alpha$ (the prior). Since $\alpha$ can take all values between $0$ and $+\infty$, $P(\alpha)$ cannot be strictly uniform. However, if $P(\alpha)$ does not vary too rapidly in the region where $P(\alpha|k)$ is appreciable, we know from Eq. (6.8) that we can still assume $P(k) \simeq$ const., which then cancels out. Therefore

$$P(\alpha|k) = \frac{\alpha^k e^{-\alpha}}{\int_0^\infty \alpha^k e^{-\alpha}\, d\alpha} = \frac{\alpha^k e^{-\alpha}}{k!},$$

At a first glance, this result may seem reasonable, because $P(\alpha|k)$ (where, remember, the random variable is now $\alpha$ and $k$ a parameter of the distribution!) is concentrated, at least for large $k$ around $\alpha = k$. Yet, the expectation of $\alpha$ is

$$\langle \alpha \rangle = \int_0^\infty \alpha P(\alpha|k) d\alpha = \frac{1}{k!} \int_0^\infty \alpha^{k+1} e^{-\alpha}\, d\alpha = \frac{(k+1)!}{k!} = k+1,$$

which is a bit weird, since we would rather expect $\langle \alpha \rangle = k$. Note for instance that if we have observed *no* decays, this result would suggest that we have an average frequency of *one* decay per unit time! You may say, aha!, but is not the frequency, but the average waiting time $\tau = 1/\alpha$ that should be taken as uniform! If $P(\tau)$ is uniform, $P(\alpha) = 1/\alpha^2$ (again the cheating barman!), and if you try to repeat the previous calculation using $P(\alpha) = 1/\alpha^2$ you easily get

$$P(\alpha|k) = \frac{\alpha^{k-2} e^{-\alpha}}{(k-1)!} \implies \langle \alpha \rangle = k-1,$$

which is incorrect in the opposite way!

However, remember our examination of the slide rule in the Example 2.4 and the following discussion of Galileo's comparison between two estimates of the price of a horse: there are quantities that can be expected to be uniform on a *logarithmic* scale, which means assuming, for example, that the probability of observing a number of decays between 0.1 and 1, or between 1 and 10, is equal.[50] Thus, we may guess $P(\ln\alpha) = C$, which implies $P(\alpha) = C/\alpha$. Unfortunately, $P(\alpha)$ is not a "good" probability distribution, because it cannot properly be normalized, that is, it is an improper prior. However, as in the case of the uniform distribution, it can still be used if it does not vary too rapidly where $P(\alpha|k)$ is appreciable. With this choice, we easily obtain

$$P(\alpha|k) = \frac{\alpha^{k-1}e^{-\alpha}}{(k-1)!}$$

and therefore

$$\langle\alpha\rangle = \frac{1}{(k-1)!}\int_0^\infty \alpha^k e^{-\alpha} = k.$$

Notably, if $P(\alpha) = C/\alpha$, $P(\tau) = C/\alpha$ too: $P(\alpha)$ is therefore a Jeffreys prior.

*"Dedicated to the memory of Sir Harold Jeffreys, who saw the truth and preserved it."* These words opens *Probability Theory*, the book that summarizes the struggle of Edwin Thompson Jaynes to endorse Jeffreys' approach to probability and to apply it to statistical mechanics and physics in general.[51] Jaynes most acknowledged contribution to physics is the model he developed with Fred Cummings of a two-level atom interacting with a quantized electromagnetic field in a cavity. However, his primary interest was in statistical mechanics, which is the subject of most of his early papers. His goal is fully evident in the abstract of his first general paper on the subject, *Information Theory and Statistical Mechanics*,[52] where he writes

> If one considers statistical mechanics as a form of statistical inference rather than as a physical theory, it is found that the usual computational rules, starting with the determination of the partition function, are an immediate consequence of the maximum-entropy principle. In the resulting 'subjective statistical mechanics', the usual rules are thus justified independently of any physical argument, and in particular independently of experimental verification; whether or not the results agree with experiment, they still represent the best estimates that could have been made on the basis of the information available.

From the outset, Jaynes regards the use of probability in statistical mechanics just as expression of human ignorance. This is particularly evident in his position about entropy that he does not regard as a property of a physical system, but rather of the experiment performed on it. According to him, entropy is just "an anthropomorphic

---

[50] In general this is a reasonable choice for any random variable with support $(0, +\infty)$.

[51] E. T. Jaynes, *Probability theory, the Logic of Science*, Cambridge Univ. Press (2003). The book, left unfinished, was edited after Jaynes demise by G. Harry Bretthorst.

[52] E. T. Jaynes, Phys. Rev. **106**, 620 (1957), Curiously, this paper was issued despite the (rather generic) objections to publications raised by the referee (see https://bayes.wustl.edu/etj/report.html), something that would never happen today. Unless you find an attentive editor.

concept".[53] The most important contribution of the paper is surely the introduction of the maximum-entropy principle, which we extensively discussed in 4.7.2 and that, as we have seen, allows all basic probability distributions of the microstate energy to be derived.

Jaynes became interested in probability theory when he realized that Jeffrey's viewpoint made all the problems of theoretical physics appear in a different light. But it was the work of Richard Cox, who formally proved that the algebra of symbolic logic allows probability rules to be derived from a few primitive notions that directly appeal to common sense,[54] that Jaynes set out on a 40-year quest to create a fully coherent and comprehensive framework for logic probability.

One of the issues that he considered more attentively was that of consistently defining priors, which he identifies as those that maximize the Shannon entropy, subjected to any constraints provided by the prior information. As we have shown in 4.7.2, this is indeed a powerful method to estimate the parameters of a probability distribution for a discrete variable but, as Jaynes clearly realized, defining a consistent statistical entropy for a probability density $f(x)$ is a much harder task. However, discretizing $S$ by introducing an accuracy limit, however, is not within the comfort zone of a Bayesian, thus Jaynes chooses to introduce an arbitrary function $m(x)$ and consider the relative (Kullbach–Leibler) entropy

$$S_{rel} = D(f\|m) = -\int f(x) \ln\left(\frac{f(x)}{m(x)}\right) dx.$$

However, maximizing $S_{rel}$ subject only to the normalization condition yields

$$f(x) = \frac{m(x)}{\int m(x)dx},$$

so that, aside from a normalization factor, $m(x)$ is the prior distribution describing "complete ignorance". So, choosing $m(x)$ requires to state what we mean by complete ignorance…and we are back to square. Jaynes is aware of the lack of invariance under a change of parameters of a prior simply based on the indifference principle, but does not seek a general (but improper) solution as Jeffreys had done. Rather he requires only on invariance under those changes that merely convert the original problem into an equivalent one. In other words, Jaynes defines the "ignorance prior" as a distribution that is invariant upon a suitable group of transformations. As we discussed (and criticized) in Sect. 6.1, this is the way he dealt with Bertrand's paradox, but it is hard to see how this approach can be applied in general.

---

[53] There is surely some truth in stating that entropy, considered as information, depends of what you are able to detect - remember the puzzle about mixing of $CO_2$ with different isotope composition closing Chap. 4? - but stating that it is only an anthropomorphic notion is a preposterous offence to thermodynamics!.

[54] R. T. Cox, Am. J. Phys. **14**, 1 (1946). Cox was also involved in a measurement of the polarization of scattered electrons, which actually showed the parity violation in $\beta$-decay later predicted by Lee and Yang.

While the target of Jeffreys criticism was frequentism,[55] Jaynes had to confront with the solid framework of measure-theoretical probability. One of his main goals was indeed showing that Kolmogorov's K1-K3 axioms can be derived from a purely logical approach. Which indeed he manages to do, but leaving some ambiguity about the concept of "hypothesis space" which should substitute for sample space in hypothesis testing. Once more, how can we be confident that we have taken into account all possible hypotheses?

Partly because of his fully unconventional approach, and partly because of his gigantic and rather annoying immodesty[56] Jaynes ideas were practically ignored for decades. Only recently they have been reappraised by philosophers and by a rather restricted group of physicists. Regardless of the criticisms of his ideas about physics,[57] I truly believe that *Probability Theory* is a rich book that deserves to be read and pondered.

### 6.3.2.5   Why Isn't Every Physicist a Bayesian?

According to Dennis Lindley, who wrote the foreword for Theory of Probability, De Finetti used to say that "we shall all be Bayesian by 2020". While I am writing this book, four years after that unlucky year, the situation is rather different. Currently, Bayesian methods are widely used in social and economic sciences, and are often regarded as a crucial tool in decision theory. Furthermore, their significant contribution to the advancement of image recognition, machine learning, and AI should not be overlooked. But applications in physics are mostly limited to astrophysics, cosmology and, to a lesser extent, experimental particle physics, while the Bayesian approach is almost ignored in fields like statistical and quantum mechanics where probability concepts are pivotal.

What prevents most physicists from being inspired by Bayes' disciples? Back in 1995, the point of view of a high-energy experimentalist, Robert Cousins, has been expressed in a nice paper wherefrom I borrowed the title of this section.[58] Cousins mostly focuses on data analysis and concludes that Bayesian methods have so far had a limited diffusion mainly because, as students, they were usually taught traditional concepts like confidence intervals that are easier to grasp than priors and likelihood

---

[55] In the 1930s, Kolmogorov's axioms were not widely known among physicist, most of which ignored by the way measure theory.

[56] If you consider my judgement too harsh, just take a look at Jaynes' autobiographical paper *A Backward Look to the Future*, in Physics and Probability, edited by W. T. Grandy and P. W. Milonni, Cambridge Univ. Press (1993).

[57] In particular, he believed that the use of probability in quantum mechanics was a clear indication that the present theory is "incomplete". Besides, he was convinced that the concept of photon is useless, and that all problems in quantum optics should and can be solved by treating the electromagnetic field as a purely classical quantity.

[58] R. D. Cousins, Am. J. Phys. **63**, 398 (1995).

functions. Besides, frequentists[59] refrain to renounce to the idea that estimation consists in finding the probability that the "true" value of a quantity (which *exists*) lies within a confidence region.[60] This is probably true, but I think there is more to it, mostly related to the inseparable link between the Bayesian approach and the epistemic (either subjective or logical) conception of probability. Let me give you a list of possible reasons, which I humbly ask you to view as my own personal thoughts.

A LACK OF COMPELLING MOTIVATION.    Most of the results in physics can be obtained by repeating an experiment with a long sequence of trials that are reasonably assumed to be identical and independent. When a prior based on experimental frequencies can be used, the results of classical and Bayesian analysis mainly differ in the interpretation. One-off situations are mostly limited to cosmology. For example, statements like "at present, dark matter constitutes 26.7% of the mass-energy content of the Universe" cannot be tested in repeated experiments. Another case in astrophysics or high-energy physics is that of very rare events. To evaluate the probability that a supernova explodes in our Galaxy, we cannot (and luckily should not) wait for many events of this kind. Similarly, the search for the Higgs boson required finding a needle event in a haystack of fake evidence. But to solve these questions require starting from a solid theory, not from a generic, uninformative prior. And this brings us to the next point.

THEORY HAS THE LEAD.    For a good physicist, even if a poor experimentalist, theory comes first. Bayesian inference assumes that an initial hypothesis, even if elementary and unspecific, can be progressively improved by repeated use of the likelihood function. I doubt. Relativity would have never come out from improvements of Newtonian physics: following Thomas Kuhn, this required a "change of paradigm". To me, this also means that AI machines, which work on a Bayesian scheme, will never replace theoretical physicists.[61] Besides (following now Karl Popper), we are not chiefly seeking methods to improve weak theories, but rather to *disprove* bad ones. Primacy of theory also means primacy of probability over statistics. Consider for example critical phenomena. Theory predicts that density fluctuations in a fluid close to the critical point diverge according to a power law. Driven by the likelihood of a set of data, a Bayesian experimentalist may try to "improve" this conjecture by introducing, for example, logarithmic corrections. But no serious reviewer will ever trust these adjustments: corrections to scaling laws exist, but their form is rigorously prescribed by theory.

A SIMPLIFIED LANDSCAPE?    Nevertheless, there are fields where rigorous theories are lacking, and experimental data can be fitted only using complicated models that depend on a large number of parameters. Think for instance of models of global

---

[59] Bayes' supporters usually refer to their opponents as "frequentists". But today, of course, there are no more (or very few) frequentists *a la* von Mises. Modern "frequentists" are scientists that, following Kolmogorov, regard probability as a Lebesgue measure with the distinctive feature of independency, while the frequency limit in the long run is ensured by the strong LLN (or, for a finite collection of experimental data, by introducing a resolution limit).

[60] In the frequentist jargon, this is called *coverage*.

[61] The case may be different for psychotherapists.

carbon emissions, oceanic circulation, or any other processes in geophysics. In this case, one may expect Bayesian "updating" to be effective. There is however a big open question: would the set of values of the model parameters that maximize the likelihood of the observation be *unique*? May the scenery be instead similar to the free energy "landscape" of a glassy material, peppered with a myriad of almost equivalent minima? Considering the nonlinearity of the Bayes updating, this won't be a surprise.

IGNORANCE BEGETS ONLY IGNORANCE.  Starting from Laplace's indifference principle, we have extensively discussed the thorny question of "uninformative priors". The choice of a prior for the frequency of events following a Poisson distribution has taught us that choosing a correct prior is anything but easy, despite the efforts of Jeffreys and Jaynes. But apart from technical reasons, I think that what disturbs many physicists (including me) is the idea that starting from a state of total ignorance can help find a learned solution. Knowledge never stems from ignorance.

BLACK SWANS SHOULD NOT BE SNUBBED.  A weak point in the logical approach is the definition of a "hypothesis space", which is much vaguer that the measure-theoretical sample space. As the turkey learned the hard way, taking into account all possible hypotheses is extremely hard. These are what Nassim Talem calls "black swans". But more than overlooking black swans, the danger is to *snub* them.

This is illustrated by a clever strip that you can see on the webcomic XKCD created by Randall Munroe.[62] A frequentist and a Bayesian dispute about the chance that the Sun has just exploded. Since it is night, they cannot decide whether this has happened or not, so they place their trust in a neutrino detector, which however is addict to gambling. Thus, before giving a response, the detector uses to throw two dice, and to tell the truth if and only if they do not both come six. When asked if it has detected a large neutrino flux, the detector says "YES". The frequentist immediately concludes that the chance that the detector has told a lie is only 1/36, less than 3%, so the Sun has very probably become a nova. But the Bayesian replies: "Bet you $ 50 it hasn't".

What reasoning provided the Bayesian with this confidence? He surely assumed that, according to the theory, the (unconstrained) probability $p$ that the Sun will becomes a nova before a few billion years is very low, for sure much lower than 1/36. So, according to the Bayes theorem, the probability $P(E|Y)$ that the Sun has exploded if the detector has given a positive answer is

$$P(E|Y) = \frac{35/36\, p}{35/36\, p + 1/36(1-p)} \simeq 35\, p,$$

which is still a very low figure. The Bayesian reasoning is flawless…pity that just that day Darth Vader was passing by, eager to try an enhanced version of the Death Star capable of obliterating not only planets, but stars too (so that losing

---

[62] https://xkcd.com/1132.

$50 is anyway the lesser evil). This may sound like a joke, but the same argument - "This event is so improbable that no additional evidence will make it possible" - has been used several times by economists to disregard warnings of a financial collapse (usually stating that "this time is different").

Actually, we made a similar unjustified assumption of negligible probability in Example 2.10, but with the only aim of showing that *even if* the chances that life develops on a planet are extremely low, it is almost certain that at least one planet in our Galaxy hosts life. On the contrary, in Example 2.15 the probability of a rare disease was not regarded as an assumption, but rather as an experimental evidence. Even in this case, however, one has to be careful, because no reference is made to the way clinical tests were conducted. While tests on a generic population give a very low probability for a disease, testing a sample that is more similar to the patient who underwent the check-up, such as male patients over 60 years old, may yield quite different results.

NOISE IS NOT NUISANCE    All we have said so far contributes to the uneasiness of many physicists towards subjective or logical probability. But the real problem is that most physicists believe in their heart that it cannot be all in their mind, that there is an element of reality in randomness. This is not to imply that they (I'd rather say, we) are old-fashioned positivists. Not at all. I am simply saying that we hold the belief that our will cannot fully control, let alone eliminate, chance. This basically means that there is some objectiveness in chance, of which noise, whether due to thermal fluctuations, to the granularity of matter, or to quantum effects, is an undeniable evidence.

In fact, a classic argument of Bayesians against objective probability is that there is nothing like a "random flip" of a coin. Were the initial conditions and the tossing protocol carefully controlled, they say, one could force a "fair" coin to show only heads or only tails: randomness is just a nuisance that we can get rid of. For a 1 € piece, or even for the smallest coin ever minted, this may work. But it won't be effective if the coin is the size of a cell, because of thermal noise. Of course, a 1 € coin has a thermal energy too, but it is a ludicrous fraction of the gravitational energy that drives its fall. Stating it differently, you can convince a coin to show only heads, but not a Brownian particle to always go straight! In the previous chapter we learned that thermal fluctuations are intrinsically linked to dissipation, that is, to irreversibility. Then, if subjectivists claim that noise is not an element of reality, they must also conclude that the arrow of time is just a human mental construction.

Yet, the main challenge for a purely epistemic probability is the enigmatic world of quantum mechanics with its baffling paradoxes. Which is the subject of the last section of this chapter.

## 6.4   Probability in Quantum Mechanics

Quantum mechanics, arguably the most successful theory about physical reality, cannot do without probability. All attempts to circumvent its constitutional stochastic nature, for instance by conjecturing "hidden variables" that would make it an incomplete theory, either failed or led to labyrinthine alternatives. Yet, the way probability is incorporated into quantum mechanics is quite unique. If physical properties are given probabilities instead of definite values, it may be expected their time evolution to be a random process, similar to those we discussed in Chap. 5. Not at all. Theory does not provide any stochastic equation for the evolution of quantum probabilities. Rather, the physical state of a system is described by a probability *amplitude* (the wave function or, more generally, the state vector), which evolves in time according to the *deterministic* Schrödinger equation.

However, the wave function is generally a complex quantity both in the coordinate and in the momentum representation, hence its only physical meaning is given by the Born rule that relates probability densities to its modulus squared. The "shadowy" nature of the wave function is at the roots of most of the puzzling aspects of quantum mechanics, such as interference effects, non-separability, non-locality. Besides, it opens up the main question in quantum mechanics: what does measuring something means? Today, many physicists consider the answer given by the standard ("Copenhagen") interpretation, which by the way introduces time-asymmetry and irreversibility in the temporal evolution, as very unsatisfactory.[63]

Yet, there might be deeper reasons for the disturbing features of the quantum world. Conceptually, the state vector is an element of a Hilbert space $\mathcal{H}$, a function space equipped with a distance induced by an inner product. The problem is, compared to ordinary Euclidean vector spaces, Hilbert spaces are *huge*! This simple but crucial difference has paramount effects on the basic assumptions of probability and on classical logic itself, definitely ruling out the possibility of describing the quantum world using concepts borrowed from the macroscopic world.

Let us see why.[64] Classically, we describe a physical system with $f$ degrees of freedom by means of the phase space $\Sigma$ with $2f$ dimensions (all the generalized coordinates and momenta). A property of the system is in general a subset $\mathcal{P} \subset \Sigma$, which can be located by introducing an "indicator" $\mathcal{P}_\mathbf{x}$, where $\mathbf{x}$ is a point in the phase space, defined as

$$\mathcal{P}_\mathbf{x} = \begin{cases} 1 & \text{if } \mathbf{x} \in \mathcal{P} \\ 0 & \text{otherwise} \end{cases}$$

Consider for example an harmonic oscillator of frequency $\omega$. The property

$$\mathcal{P} = \text{"the total energy } E \text{ of the oscillator is less than } E_0\text{"}$$

---

[63] Note that an "instantaneous" collapse of the wave function may also conflict with relativity, if $\psi$ extends over a large spatial region.

[64] This section relies heavily on the book by GRIFFITHS in the Suggested Readings, which I consider to be the best pedagogical account of the foundations of quantum mechanics.

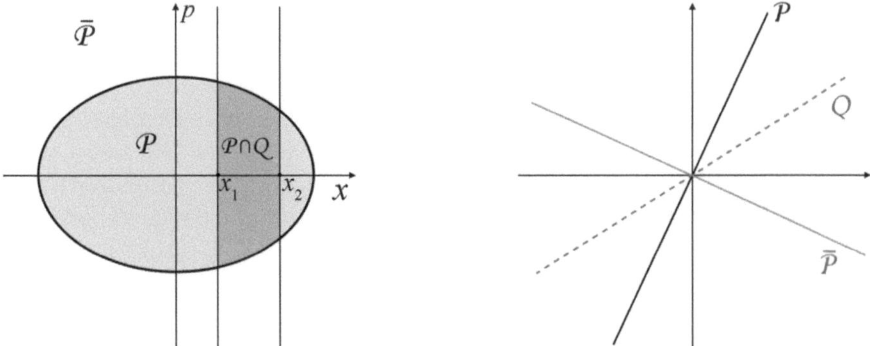

**Fig. 6.2**  Left: Phase space for a one-dimensional harmonic oscillator. The ellipse corresponds to a total energy of the oscillator $E = E_0$. The property $P : E < E_0$ is the region shaded in light grey, while $P \cap Q$ is the region where the oscillator lies between $x_1$ and $x_2$ and its energy is less than $E_0$. Right: A ray $\mathcal{P}$ and its orthogonal complement $\bar{\mathcal{P}}$ in a two dimensional Hilbert space $\mathcal{H}$, which can for instance represent the states of a spin-half particles as the two possible values of the spin component along a selected direction. Evidently, $\mathcal{P} \cup \bar{\mathcal{P}}$ does not contain any vector like $Q$ that is not orthogonal to $P$.

corresponds to the inside of the ellipse of equation $\frac{m\omega^2}{2} x^2 + \frac{1}{2m} p^2 = E$ in the $(x, p)$ space (see Fig. 6.2A). It is a natural choice to identify the opposite property, $\bar{\mathcal{P}} = E \geq E_0$, with the indicator $\bar{P}_{\mathbf{x}} = 1 - P_{\mathbf{x}}$. Note that $\mathcal{P} \cap \bar{\mathcal{P}} = \emptyset$ and, more importantly, $\mathcal{P} \cup \bar{\mathcal{P}} = \Sigma$, which means that *every point of the phase space (every physical state of the system) is either in $\mathcal{P}$ or in $\bar{\mathcal{P}}$*.

Classically, defining the probability $P(a)$ that some physical quantity $A(\mathbf{x})$ of the system has a given value $a$ means taking $\Sigma$ as the sample space and setting $P(a)$ as the Borel measure of the subset $\mathcal{P} : \{\mathbf{x} \in \Sigma : A(\mathbf{x}) = a\}$. One can then define the probability that a system has both properties $\mathcal{P}$ and $\mathcal{Q}$ as the measure of $\mathcal{P} \cap \mathcal{Q}$, and the probability that the system has either the property $\mathcal{P}$ or the property $\mathcal{Q}$ as $\mathcal{P} \cup \mathcal{Q}$. For example, the joint property

$\mathcal{P} \cup \mathcal{Q} =$ "the energy of the oscillator is $E < E_0$ *and* it is found within $[x_1, x_2]$"

corresponds to the area in Fig. 6.2A. shaded in dark grey.

Generating a probability measure in the phase space that satisfies Kolmogorov's axioms is therefore rather straightforward. However, this is much harder to do in quantum mechanics. Let us briefly recall that in quantum mechanics the allowed values of a physical quantity (an *observable*) $A$ are the eigenvalues $a_i$ of a self-adjoint operator $\hat{A}$, which allow to decompose $\hat{A}$ as

$$A = \sum_i a_i \hat{P}_i,$$

where $\hat{P}_i$ is a projector on the subspace $\mathcal{P}_i$ where $A = a_i$. Projector referring to distinct eigenvalues $a_j, a_k$ are orthogonal, that is, $\hat{P}_j \hat{P}_k = \hat{P}_k \hat{P}_j = \hat{P}_j \delta_{jk}$ or, equivalently, $\mathcal{P}_j \cap \mathcal{P}_k = \emptyset$ if $j \neq k$. Given a complete set of eigenvalues $\{a_i\}$ (a basis), the relative projectors form a decomposition of the identity $\hat{I}$, $\sum_i \hat{P}_i = \hat{I}$. Equivalently, one has $\bigcup \mathcal{P}_i = \Sigma$.

It is then natural to assume that the property $A = a_i$ corresponds to the subspace $\mathcal{P}_i$ onto which $\hat{P}_i$ projects, so that $\hat{P}_i$ plays the same role as the indicator function in the phase space. The negation $\bar{\mathcal{P}}$ of a quantum property $\mathcal{P}$ can be consistently defined as the *orthogonal complement* of $\mathcal{P}$, the collection of the state vectors which are orthogonal to all those in $\mathcal{P}$, which is the projector $1 - \hat{P}$. Consider for example a simple quantum oscillator with energy eigenvalues $\epsilon_n = (n + 1/2)\hbar\omega$, and call $\hat{P}_0$ and $\hat{P}_1$ the projectors on the ground and first excited state, respectively. Then the property

$$\mathcal{P} = \text{``The energy of the oscillator is more than } 2\hbar\omega\text{''}$$

is associated with the projector $\bar{\hat{P}} = \hat{I} - \hat{P}$, where $\hat{P} = \hat{P}_0 + \hat{P}_1$.

However, with this definition, a wide gap opens between classical phase and quantum Hilbert space, for it is *not* true that $\bar{\mathcal{P}} \cup \mathcal{P} = \mathcal{H}$. For instance suppose that if $\mathcal{P}$ is made of a *single* state. As shown in Fig. 6.2B, in the Hilbert space, this corresponds to a *ray*, that is, to all the state vectors $\alpha\psi$, where $\psi$ is a normalized state and $\alpha$ a complex number. Then, $\bar{\mathcal{P}}$ is the perpendicular ray passing through the origin, but, evidently, $\mathcal{P} \cup \bar{\mathcal{P}}$ is a very small subset of $\mathcal{H}$ (actually, a subset of measure zero!), which does not include the ray corresponding to any state $\mathcal{Q}$ that is not orthogonal to $\mathcal{P}$. More generally, in a Hilbert space there is a vast number of states that neither lie in the subspace associated with a property $\mathcal{P}$, not in its orthogonal complement $\bar{\mathcal{P}}$. For these states, $P$ is neither true nor false: it is simply *undefined*.

This observation consistently makes far more challenging to regard a Hilbert space as a single probability space, since $\mathcal{H}$ can actually host an *infinite* number of mutually "disconnected" sample spaces. In fact, two properties $\mathcal{P}$ and $\mathcal{Q}$ are compatible, that is, can be given a common set of probability values, if and only if their associated projectors commute, $\hat{P}\hat{Q} = \hat{Q}\hat{P}$. Besides mutually exclusive events, we have events that are *incompatible*, because the properties they refer to do not possess a common sample space.

A further difference with the classical case emerges if we consider the conjunction $\mathcal{P}$ AND $\mathcal{Q}$ of two properties. The indicator for a classical conjunction is the product $PQ$ of the indicators, but in the quantum case the product $\hat{P}\hat{Q}$ of two projectors is a projector if and only if they commute. Of course, a similar problem exists for the disjunction $\mathcal{P}$ OR $\mathcal{Q}$, whose indicator is $\hat{P} + \hat{Q} - \hat{P}\hat{Q}$. Instead of the product of two non-commuting projectors, one may think of simply using the intersection $\mathcal{P} \cap \mathcal{Q}$ of the corresponding subspaces for and, correspondingly, their union $\mathcal{P} \cup \mathcal{Q}$ for $\mathcal{P}$ OR $\mathcal{Q}$, but this leads to rather embarrassing physical conclusions. Consider for instance an electron. The statement

$$S = (s_z = +\hbar/2) \text{ AND } (s_x = +\hbar/2),$$

where $s_z, s_x$ are the spin components along $x$ and $z$, is always *false*, which means that $\mathcal{P}_{++} = (\mathcal{P}_{s_z} = +\hbar/2) \cap (\mathcal{P}_{s_x} = +\hbar/2) = \emptyset$. But then we have

$$\overline{\mathcal{P}}_{++} = (\mathcal{P}_{s_z} = -\hbar/2) \cup (\mathcal{P}_{s_x} = -\hbar/2) = \mathcal{H},$$

so that $\bar{S} = (s_z = -\hbar/2)$ OR $(s_x = -\hbar/2)$ is always *true* (which is wrong!).

There are basically two ways to escape these inconsistencies. The first is rebutting one or more principles of classical logic, by claiming that quantum mechanics does not comply with them. This is the route originally followed by Garrett Birkhoff and John von Neumann,[65] who suggested that quantum oddities can be accounted for by assuming that the distributive laws,

$$A \cup (B \cap C) = (A \cup B) \cap (A \cup C) \text{ and } A \cap (B \cup C) = (A \cap B) \cup (A \cap C),$$

where $A$, $B$, $C$ are quantum properties, hold only if these properties are compatible. This approach has led to a proliferation of alternative quantum logics that would require a full book to be discussed.

Although not necessitating new logic principles, the second route, pioneered by Robert Griffiths and later developed by Murray Gell-Mann, Pierre Hohenberg, and Roland Omnès, requires a radical change in our picture of reality. Indeed, the central idea of consistent quantum theory, as this approach is called, is that a unique exhaustive description of a physical system does not exist. Instead reality is such that it can be described in diverse incompatible ways using descriptions, called "frameworks", which cannot be combined.[66]

To put it simply, this means that we are always bound to view reality from a limited perspective. For example, if we decide to focus on a property that we call "the $z$-component of a particle spin', $S_z$ is all we can know about something that *behaves* as a classical internal angular momentum besides its magnitude $S^2$. And that is not because "we cannot measure" $S_x$ and $S_y$, rather because we have no right to assert that an element of reality called "spin" can be pinned on the particle. This of course happens in the classical word too. The Matterhorn, as seen from Zermatt, looks very different from Mount Cervino viewed from Breuil-Cervinia, but we know that these are just different viewpoints of the same mountain, as much as in tomography we know that each single image is a different projection of the same organ. In the quantum world, such a "tomographic reconstruction" is probably an illusion.

Some of you may have visited the splendid Ryōan-ji temple in Kyoto, known for its famous *karesansui* (dry landscape) rock garden. The garden contains fifteen stones of varying sizes arranged in such a way that from any point you look, sitting on the veranda of the abbot's residence, you cannot see all of them at once. This is believed to be possible only for the few ones who reach Enlightenment. Those

---

[65] G. Birkhoff and J. von Neumann, Ann. Math. **37**, 823 (1936).

[66] By framework one means a projective decomposition of the identity generated by the projectors of a physical quantity. Two frameworks $\{\hat{P}_j\}$ and $\{\hat{Q}_k\}$, generated by the physical properties $\mathcal{P}$ and $\mathcal{Q}$, are compatible if and only if every $\hat{P}_j$ commutes with every $\hat{Q}_j$, i.e., $\hat{P}_j\hat{Q}_k = \hat{Q}_k\hat{P}_j$ for all $j, k$.

who have not achieved perfection, like most of us, must be content with their limited perspective. This is possibly one of the most profound lessons that physicists should learn from Zen Buddhism.

We can either accept this new vision of reality or see the quantum world as the realm of a different logic. In both cases, however, it is evident that the effort by Jeffreys, Cox, and Jaynes to reduce probability to logic, claiming that this reduction is *unique*, faces an insurmountable obstacle.

**Example 6.6** (*The Bell inequality*) One of the most astonishing feature of quantum mechanics is the prediction of long-range correlations that give the theory a nonlocal flavour. Consider for example a singlet state of two electrons, 1 and 2, with zero total spin, and suppose we measure the spin component along $z$ of electron 1. There will be 50/50 chance of finding $S_z^{(1)} = \pm\hbar/2$, but once we know that $S_z^{(1)} = +\hbar/2$ we know *for sure* that $S_z^{(2)} = -\hbar/2$, and vice versa.[67] Remarkably, this kind of correlation persists even if the two electrons fly apart and cease to interact, provided that in their spin states do not change. So, if Alice measures $S_z^{(1)}$, she can predict the outcome of the measurement of $S_z^{(2)}$ by Bob. Worse yet, it is as if Alice's measurement instantly *determines* what Bob will find, regardless of his distance.

You may think that this kind of correlations is not, after all, peculiar to quantum mechanics. As John Bell pointed out, if I know that a colleague of mine always wears socks of two different colors, say red or green, and I see that one sock is red, then I can be certain that the other is green. Alas, quantum reality is much more intricate! Because Alice can choose to measure $S_x^{(1)}$ instead of $S_z^{(1)}$, and this drastically changes the set of Bob's possible results. This is like stating that I also know that one of my colleague socks, but not the other one, always has a hole in it, but inspecting his socks magically makes me color blind.

Many physicists have felt uncomfortable with the explanation of these correlations given by the standard interpretation of quantum mechanics, which leads to paradoxical conclusions pointed out by an influential paper by Einstein Podolsky and Rosen.[68] This led Einstein to declare resolutely that quantum mechanics is an incomplete theory, because what happens in a system $S_1$ *must* be independent of what happens in another system $S_2$ that is spatially separated from $S_1$.[69] But does this Einstein's "locality principle" make predictions different from those of quantum mechanics? An affirmative answer to this key question was given for the first time by John Bell,[70] who concocted an experimentally testable inequality relation among the observables of spin-correlation experiments that is an inescapable consequence of the locality principle, but disagrees with the predictions of quantum mechanics.

---

[67] This is not easily realized with electrons or other material particles, but is not uncommon in quantum optics to find processes where a source simultaneously emits two photons with orthogonal polarization states.

[68] The relation between spin correlations and the EPR paradox (Phys. Rev. **47**, 777, 1935) is clearly discussed in D. Bohm and Y. Aharonov, Phys. Rev. **108**, 1070 (1957).

[69] Einstein meant "spatially separated" in the relativistic sense that no information can be transmitted from $S_1$ to $S_2$ within the time lapse between two compared events in $S_1$ and $S_2$.

[70] J. S. Bell, Physics **1**, 195 (1964).

We will discuss Bell's inequality in the context of a simple model by Eugene Wigner. Consider again a series of measurements by Alice and Bob of two particles with zero total spin that have been spatially separated. This time, however, Alice and Bob agree to take measurements along the directions set by *three* unit vectors $\hat{\mathbf{a}}$, $\hat{\mathbf{b}}$, and $\hat{\mathbf{c}}$ that are, in general, not mutually orthogonal.

According to Einstein's locality principle, Alice's results should not depend on what Bob chooses to measure, so let us focus on her. In Example 2.3, we found that *any* three events $E_1$, $E_2$, and $E_3$ must satisfy.

$$P(E_1 \cap E_2) \leq P(E_1 \cap E_3) + P(\overline{E}_3 \cap E_2),$$

Let us put

$E_1 = \quad$ {Taking a measurement of the spin projection along direction $\hat{\mathbf{a}}$, Alice finds a *positive* result};

$E_2 = \quad$ {Taking a measurement of the spin projection along direction $\hat{\mathbf{b}}$, Alice finds a *negative* result};

$E_3 = \quad$ {Taking a measurement of the spin projection along direction $\hat{\mathbf{c}}$, Alice finds a *negative* result}.

Then, indicating with $\hat{\mathbf{v}}+$ ($\hat{\mathbf{v}}-$) a positive (negative) spin component along $\hat{\mathbf{v}}$, the probability $P_A[(\hat{\mathbf{a}}+) \wedge (\hat{\mathbf{b}}-)]$ that Alice finds *both* $\hat{\mathbf{a}}+$ and $\hat{\mathbf{b}}-$ satisfies

$$P_A[(\hat{\mathbf{a}}+) \wedge (\hat{\mathbf{b}}-)] \leq P_A[(\hat{\mathbf{a}}+) \wedge (\hat{\mathbf{c}}-)] + P_A[(\hat{\mathbf{c}}+) \wedge \hat{\mathbf{b}}-)].$$

for any choices of the three axes $\hat{\mathbf{a}}$, $\hat{\mathbf{b}}$, $\hat{\mathbf{c}}$. Since Bob's results are always the opposite of those obtained by Alice, the joint probability $P(A = \hat{\mathbf{a}}+; B = \hat{\mathbf{b}}+)$ that Alice finds $\hat{\mathbf{a}}+$ and Bob $\hat{\mathbf{b}}+$ satisfies therefore the *Bell inequality*

$$P(A = \hat{\mathbf{a}}+; B = \hat{\mathbf{b}}+) \leq P(A = \hat{\mathbf{a}}+; B = \hat{\mathbf{c}}+) + P(A = \hat{\mathbf{c}}+; B = \hat{\mathbf{b}}+), \quad (6.14)$$

which is a direct consequence of Einstein's locality principle and classic logic.

Let us however check whether in quantum mechanics (6.14) is always satisfied. Since Alice has observed that the spin of one of the two particles is directed along $+\hat{\mathbf{a}}$, the spin of the other particle, measured by Bob, *must* be directed along $-\hat{\mathbf{a}}$. Then, by decomposing the spin in components parallel and perpendicular to a direction $\hat{\mathbf{b}}$ that forms an angle $\theta_{ab}$ with the *positive* $\hat{\mathbf{a}}$ direction, it can be shown[71] that

$$P(A = \hat{\mathbf{a}}+; B = \hat{\mathbf{b}}+) = \frac{1}{2} \sin^2 \left( \frac{\theta_{ab}}{2} \right),$$

---

[71] For details, see J. J. Sakurai and J. Napolitano, *Modern quantum mechanics* (3nd ed.) Cambridge University Press, Cambridge, 2020.

and similar relations for the other two terms in (6.14). Therefore, the Bell's inequality is equivalent to

$$\sin^2\left(\frac{\theta_{ab}}{2}\right) \leq \sin^2\left(\frac{\theta_{ac}}{2}\right) + \sin^2\left(\frac{\theta_{bc}}{2}\right). \tag{6.15}$$

Choosing for simplicity $\theta_{ac} = \theta_{bc} = \theta$ and $\theta_{ab} = \theta_{ac} + \theta_{bc} = 2\theta$ (so that $\hat{\mathbf{c}}$ bisects the directions defined by $\hat{\mathbf{a}}$ and $\hat{\mathbf{b}}$, this would imply that

$$\sin^2\theta \leq 2\sin^2\left(\frac{\theta}{2}\right) \implies \cos\left(\frac{\theta}{2}\right) \leq \frac{1}{\sqrt{2}},$$

which is violated for *all* values of $\theta$ in $(0, \pi/2)$, as currently confirmed by conclusive experimental evidence.[72]

Therefore, Einstein's locality principle does not hold. Does this imply that quantum mechanics is a nonlocal theory? Not necessarily. Indeed, as clearly stated in the EPR paper,

> One would not arrive at our conclusion if one insisted that two or more physical quantities can be regarded as simultaneous elements of reality *only then they can be simultaneously measured or predicted*…

…which is exactly what the consistent QM approach asserts. Besides, using logical $\wedge$ connectors as we did between properties belonging to incompatible frameworks, as we did, is not allowed.

Before you proceed reading this book, let me make a final distinction. Stating that, for the reasons discussed above, a subjective approach to probability can hardly meet the demands of a physicist does not mean that all the refined statistical methods devised by Bayes' supporters should be thrown away! Therefore, using a purely utilitarian approach, in the final chapters I will take the liberty to borrow any concept from Bayesian statistics that I consider beneficial for our purposes.

---

[72] For a review, see A. Aspect, Nature **398**, 189 ((1999).

# Chapter 7
# Noise and Bias in Measurements

*La science, mon garçon, est faite d'erreurs,*
*mais d'erreurs qu'il est bon de commettre,*
*car elles mènent peu à peu à la vérité*

*J. Verne*

To the delight of the my budding experimentalist readers, in this chapter we come back to the opening theme of this book, the subtle art of digging into experimental data, hoping that the heavy theoretical burden we took upon our shoulders will help up to assess quantitatively the accuracy and precision of a measurement. We are indeed going to introduce what is usually called "the theory of errors", although in what follows you will hardly find the word "error", which seems to suggest that fluctuations in the values of a physical quantity are solely caused by a medley of the setup drawbacks with the ineptitude of the experimenter. This is unjust to experimentalists and definitely wrong. As we stressed in Chap. 5, noise is not just nuisance, and fluctuations are a constitutional element of reality. The word "error" will mainly be limited to what are usually called *systematic* errors, that is, deviations of the single measurements from the correct value that (see Sect. 7.1.1) are consistently of the same sign. These are still "noise contributions", but with the specific feature of having a non-zero average, which means that the results are *biased*. Based on their origin, we can distinguish three 'flavours' of noise:

INTRINSIC NOISE:    As extensively discussed in Chap. 5, thermal agitation, matter granularity, and quantum effects can generate intrinsic fluctuations in a signal, but additional noise can also be introduced during the signal propagation from the source to the detection unit. For example, atmospheric scattering or turbulence introduce fluctuations in a light or radio signal.

© The Author(s), under exclusive license to Springer Nature Switzerland AG 2025
R. Piazza, *An Invitation to Probability and Data Analysis for Physicists*, UNITEXT for Physics, https://doi.org/10.1007/978-3-031-83856-9_7

INSTRUMENTAL NOISE:    The various stages of signal detection and elaboration, which we present in the next section, add further noise, not only of the types we already discussed (thermal, flicker, shot), but also due, for example, to signal transduction and digitalization. Besides, as we will see, the measurement process is often a source of errors, with the meaning given above.

HUMAN NOISE:    This should better be called human *error*, since apart from some trivial cases like mistakes in sample preparation, manual measurements (which are becoming rarer and rarer), or miscalculations, it usually generate systematics deviations in the results. These errors, which can be insidious and hard to discover, range from the incorrect unit conversion of a master student to the flawed setup design by the group leader.

This chapter focuses on the most common strategies to cope with noise and errors in order to obtain a reliable assessment of the value of a physical variable, and of the precision of this estimate. Then, we will discuss how noise transforms in indirect measurements (what is usually, but not here, called "error propagation"). As mentioned in the introduction, the level of presentation will be really introductory, since there are already excellent textbooks on these subjects that I won't be able to compete with. Thus, more than on the specific techniques, I will mainly dwell upon the crucial assumptions underlying them. Besides, I firmly believe that an adequate introduction to data analysis can only be made by maintaining a constant link with concrete experimental practice. Thus, contradicting the incipit of this chapter, the first section of this chapter is mainly dedicated to my fellow theoretical colleagues, who often have a rather naive idea of the general structure and functions of an experimental setup. So let us start by figuring our what "taking a measurement" concretely means.

## 7.1   A Theorist's Little Guide to the Lab

### 7.1.1   Measurement Setup

The instrumentation used for physical measurements has an extremely variable degree of complexity, which ranges from simple scales to particle accelerators. However, every apparatus must ultimately provide data. We can therefore try to outline at least the essential features of a data acquisition process through the layout in Fig. 7.1. Any experimental apparatus contains at least one sensor-transducer (which we will also simply call *detector*) and a reading system, while the signal adaptation system (signal conditioning) may or may not be present and the processor may consist of only…a student provided with pen and paper. Let us analyze the individual blocks.

DETECTION: The purpose of the sensor is to detect the quantity to be measured, which constitutes the input of the measurement system, and to provide a response signal proportional to the value of the measured quantity. Furthermore, the sensor should ideally be insensitive to external stimuli of a different nature than those of interest, i.e., it must be *selective*. The signal provided by the sensor is generally of

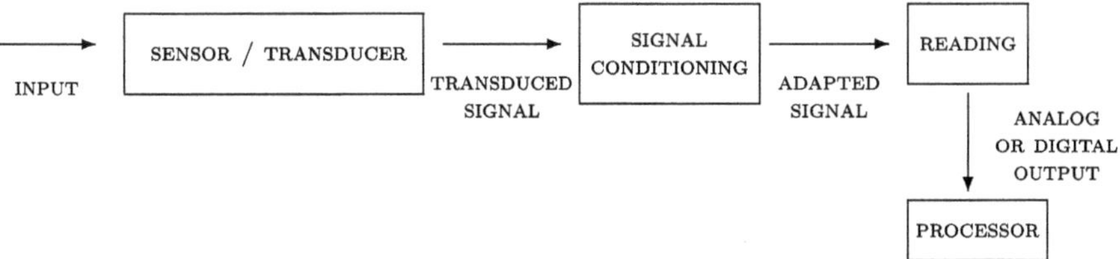


**Fig. 7.1** Outline of a measurement apparatus

a different nature from the input signal, and therefore we say that the signal related to the original quantity is TRANSDUCED into a signal of another (usually electrical) type, more easily adjustable and controllable than the original one.

Consider for example a simple pressure gauge for measuring tire pressure on cars. The nozzle applied to the valve is connected to a cylinder containing a sealed piston. The piston is pushed upwards by the tire pressure until the elastic force of a return spring balances the product of pressure force on the piston. Finally, the spring compression is gauged by the displacement of a movable rod that rotates an index on a graduated scale. Here, the sensor is the spring, and the pressure signal is transduced into a linear displacement of the rod. As an example of the "selective" purpose of a sensor, consider a photocell of an alarm system that must detect the presence of infrared light. If we do not want the ambient light to affect the detection, we can place an optical filter in front of the photocell that excludes the other chromatic components. The nature of the sensitive element of a detector obviously depends on the physical quantity we wish to measure. Thus, for example:

- springs, torsion pendulums, or materials that provide an electrical signal when compressed (piezoceramics) are force sensors;
- floats, liquid columns, elastic membranes are hydraulic sensors;
- the mercury column of a thermometer, or components with temperature-sensitive electric components such as thermocouples and thermistors are thermal sensors;
- films, photocells, photomultipliers, avalanche photodiodes, CCD and CMOS cameras, are optical sensors.

And this is just a small part of the list we could compile, since the specific nature of the signal to be detected often requires to design "dedicated" detectors like scintillators and bubble chambers in particle physics.

CONDITIONING: This block has the function of transforming the signal into a form suitable for the final stage of reading and processing. Signal adjustments can be of several kinds, but two processes are particularly important:

AMPLIFICATION: The signal is often too weak to be recorded by the reading system and therefore requires amplification. An amplifier must be able to provide a "faithful copy" of the input signal with a larger amplitude. Normally, amplification requires providing power to the system, but we can also "amplify" in the broad sense a mechanical or light signal using, for example, a hydraulic or an optical lever. Particularly interesting is when the amplification already occurs within the detector, like in a photomultiplier, where the electrons emitted by a photosensitive material, the photocathode, are multiplied by a "cascade" emission from internal photosensitive surfaces the dynodes.

FILTERING: We have already seen that filtering of the signal can sometimes be done upstream of the detection, as in the case of an alarm photocell, or of a spectrophotometer, where the wavelength of the light incident on a photodetector is first selected through a prism or a grating. Often, however, especially when the output signal from the detector is electrical, filtering is performed *after* detection. Filtering is generally used to reduce noise by limiting the information content of the input signal, preserving only what is of interest.

READING AND PROCESSING: The signal, possibly adapted, can be read directly, for example by means of a gauge that moves on a graduated scale or a photographic film, or through a reading instrument such as an oscilloscope, a pulse counter, a multichannel analyzer, a digital acquisition module. A general characteristic of a reading block is that its sensitivity is usually related to the maximum value of the reading scale ("full scale"). Almost always, a preliminary processing stage is entrusted to a unit consisting of a computer with peripherals, which add further noise due to digital processing. Of course, the final stage of an acquisition process is left to the experimenters themselves, who collect and analyze the data, a delicate stage that should never be underestimated.[1]

### 7.1.2  Specs and Features of an Experimental Setup

We can try to identify the origins of the deviations of a measured value from the "true" value of a physical quantity,[2] bearing however in mind a basic empirical principle: no list of possible causes of uncertainty is ever complete, and neglected items are almost never negligible. To examine instrumental noise, some general specifications of the experimental setup should be taken into account.

SENSITIVITY: Any detector has a minimum *sensitivity threshold* $s_0$ that, in addition to making it unfeasible to measure signals below $s_0$, it also establishes an upper limit to the accuracy with which we can estimate the value of the measured quantity. The minimum sensitivity of a device is often related to the amplitude of the noise generated internally by the system. Many detectors show for example a *background noise*, so that the sensor provides a non-zero output even in the absence of an input signal. This is the case of the eye, where the spontaneous generation of a small amount of

---

[1] I always tell new students who join my group that, despite all the instrumentation they see around the lab, the most important instrument is always within their heads.

[2] Pointless to repeat that I do not belong to the Bayesian party, thus I firmly believe that true values do exist.

photoreception nerve impulses sets the sensitivity limit discussed in Example 3.14, or of photomultipliers, where background noise is due to spontaneous electron emission from the dynodes due to several causes, including cosmic rays.

In general, signal amplification does not improve sensitivity, since the threshold $s_0$ is amplified too. Moreover, the amplification process generally introduces additional electronic noise. The most favorable situation is that of a system with "internal" amplification, such as a photomultiplier, where the amplification noise is generally reduced to a minimum theoretical value. Signal conditioning procedures can enhance sensitivity, particularly when they are equivalent to averaging the signal over a specific set of values, as in the case of an integrating system or, in the case of a periodic signal, when *synchronous detection* is performed. The goal is to keep only those signal components that have the frequency of the signal, like a lock-in amplifier does, but at the cost of slowing down the detection process. As we already mentioned, the minimum value that a reading system can detect is often related to its measurement range, i.e., to the full scale of the instrument, with a larger full scale corresponding to a lower sensitivity. As a trivial example, while to measure the diameter of a small hole we can use a caliper, which allows a resolution of $10^{-1} - 10^{-2}$ mm, the same sensitivity is hard to achieve when measuring the length of a room, even with a laser meter.

SIGNAL- TO- NOISE RATIO: The most important parameter determining the resolution of a device is not so much the sensitivity itself, but rather the signal-to-noise ratio (SNR). For example, we all know that turning up the volume of the signal received from a very noisy radio station can only cause our eardrums to shatter, since we also amplify the noise, keeping the SNR constant. A separate discussion can be made on the effects of digitizing a signal on the SNR. In the recent past, analog stereo systems have been replaced first by CDs and then by MP3s.[3] The main reason for the success of these sound reproduction systems is the transition to a digital writing and reading system, which essentially transcribes a complex sound into a binary code. This is a bad blow for noise, since in a binary system made up of ones and zeros there is no room for things like "one plus some noise". Most data transmission and processing systems have faced the same fate, mostly because at the end of the chain there is usually a computer that "thinks digitally". But there is a trade-off, of course. Digital processing requires splitting a continuous range in subintervals of minimum size, the "bits" and this limits the resolution to the minimum value of one bit. So, for example, a 16-bit acquisition card that measures electrical signals up to an amplitude of 10 V dividing the measurement range into $2^{16}$ parts has a minimum resolution of about 0.15 mV.

REPRODUCIBILITY: No measurement setup provides *exactly* the same value in two measurements made under identical conditions of the same quantity. This is something different from sensitivity, and differences due to lack of reproducibility can often be larger than the minimum resolution. For example, if we are performing a position measurement with a micrometric-drive translator driven by an electric

---

[3] Although we are all witnessing a revival of the old vinyl records: never say never!.

motor, position resolution, obtained for instance with an electro-optical system called encoder, can be much more precise than position reproducibility, which is limited by the play of the micrometer screw that controls positioning.

CALIBRATION: Calibrating an experimental apparatus involves taking measurements in conditions where the value of the investigated quantity value is already known. As simple examples, the spring of a scale needs to be adjusted with a set of standard reference weights, and the grayscale of a camera must be checked. An important aspect of the calibration of each block of the setup (sensor, amplification stage, signal adaptation, reading) is the kind of link relating the output to the input. Sometimes the latter is linear, but it is crucial to establish within what limits this is valid, which permits to set a maximum range of values (referred to as *dynamic range*) within which the signal is not distorted. However, many sensors, for example sensitive temperature probes like the thermistors, have a response that is far from linear, which requires a complete calibration curve obtained by comparing the measured values with a reference. Another issue is the zero-point calibration, since often, particularly in the presence of amplification stages, devices have a non-zero output value even in the absence of a signal, which is called an *offset*.

BANDWIDTH: A concept of the utmost importance when the collected data refer to the same quantity measured at different times or locations in space is *bandwidth*, a property that actually concerns *all* components of an acquisition system, and what is "upstream" of the acquisition process too. The concept of bandwidth is better grasped in the time domain, observing that a detector can faithfully follow a signal up to a maximum frequency. In the detection process, higher frequencies are then cut out and the detected signal is partially distorted because of the elimination of the high frequency components. Similar effects occur in the amplification process. It can indeed be shown that the "bandwidth times gain" product of an amplifier is constant. In other words, the more we amplify a signal, the more the frequency band of the amplified signal is reduced. Any measurement system has then a finite "passband", which obviously also contributes to establishing the linearity of the response.

As different as it may seem, the same problem arises for signals that vary in space, for example in imaging, where the combination of optical components (lenses, mirrors, diaphragms, and so on) sets a maximum spatial resolution with which the image can be captured and reconstructed. Leaving details to specialized and authoritative texts,[4] I just want to point out that the methods used to analyze the "resolving power" of an optical device (for example, the resolution limit of a microscope, or the smallest details that can be distinguished in a photographic image) are still based on Fourier analysis using spatial variables called spatial frequencies. Although the problem is formally more complex, there is still a close parallelism with the frequency analysis of a temporal signal, also for what concerns the effects of the limited passband on the fidelity of the acquired signal.

Everything we have said so far just concerns a simplified setup dedicated to data acquisition. But of course an experimental apparatus generally includes many other

---

[4] See, for instance, J. W. Goodman, *Introduction to Fourier Optics*, McGraw-Hill (Singapore).

components that *precede* signal detection. Each of them may have shortcomings, deriving for instance from mechanical defects, vibrations, temperature fluctuations, which limit the sensitivity and reproducibility of a measurement. Finally, we should never underestimate human mistakes such as, for example, a wrong instrument reading or an inaccurate data transcription. There is no universal remedy for blunders of this kind, which are perhaps rare events but often with devastating consequences, but it is always a good idea to keep a detailed record of all lab operations, better using a classical lab notebook.[5]

### 7.1.3   *Noise Versus Errors*

While instrumental electronic and digital noise average to zero, issues of reproducibility, incorrect calibration, limited bandwidth are the main sources of those "systematic errors" that here we call instrumental errors or simply *bias*, which cause all measurements to deviate in the same direction (deficit or excess) from the expected value. Instrumental bias can be difficult to detect at times. For example, the gain or the dark noise of an accurately calibrated photomultiplier can change in the long run because the photocathode deteriorates, or because gas slowly infiltrates the vacuum tube. In fact, the sensitivity of many instruments *drifts* over time, and these drifts are often the main cause of errors. Detecting *human* errors, caused for instance by a poor experimental plan, can be even more challenging.

In order to limit noise, is may sometimes be beneficial to *coexist* with errors. For example, suppose that we want to measure the intensity of light scattered by a sample at a given angle $\vartheta$ from the incident beam, using a photomultiplier mounted on a rotating arm that, due to the mechanical play of a screw pitch, can be positioned with a precision $\Delta\vartheta$. If we take measurements at the angles $\vartheta_1 < \vartheta_2 < \ldots < \vartheta_N$, starting at $\vartheta_1$ and moving the arm always in the direction of increasing $\vartheta$, all angles will be systematically in excess by $\Delta\vartheta$ of the value we set. Instead, if move the arm from one angle to another at random, trying to balance positive and negative deviations, the systematic error is significantly reduced, but the measured values randomly fluctuate by $\pm\Delta\vartheta/2$ about the set point. If we are more interested in angular *differences* than in the absolute value of $\vartheta$, the first strategy is surely better.

### 7.1.4   *Precision Versus Accuracy*

So far we have used terms like "precision" and "accuracy" somewhat superficially, without defining exactly what we mean by these terms. We can be more accurate (or precise) by identifying a fluctuating physical quantity $X$ with a random variable

---

[5] I know firsthand that relying exclusively on computer logs, as my younger students often do, guarantees that many results will soon be lost, given the fast pace of digital media changes!.

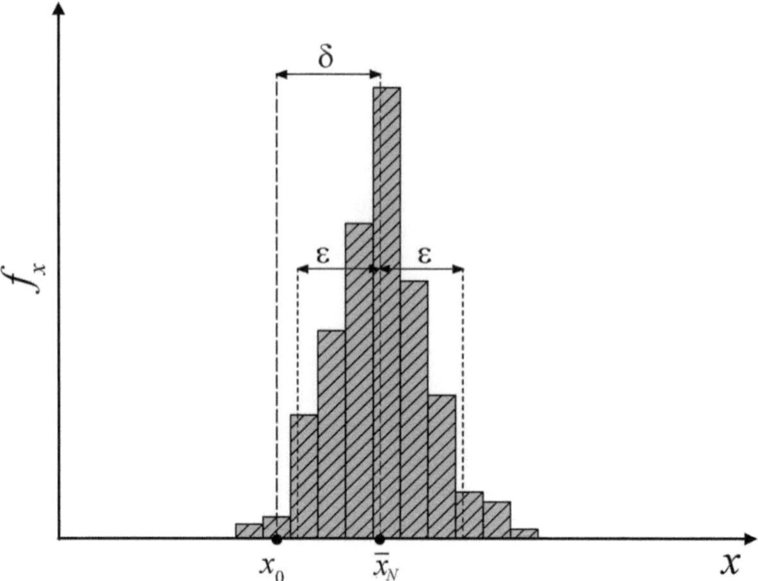

**Fig. 7.2** Schematic frequency distribution of measured values. $\delta$ and $\varepsilon$ are the bias and noise contributions, respectively

whose probability distribution quantifies the amount of noise and errors associated with the measurement process. By making a series of $N$ identical independent measurements of $X$, we can expect the distribution of the measured values to look like the histogram in Fig. 7.2, where $x_0$ is the theoretical value of $X$ and $\bar{x}_N$ is the mean of the results.

As you see, the bias contribution $\delta$ shifts the average with respect to the theoretical value, while the noise contribution $\varepsilon$ widens the distribution around the mean. The *precision* of a measurement is the uncertainty with which the mean is known. However, in the presence of bias, $\bar{x}$ will differ from $x_0$ even for a (hypothetical) infinite sequence of measurements. The *accuracy* of a measurement is then assumed to be the limit difference

$$\lim_{N \to \infty} |\bar{x}_N - x_0|.$$

Precision and accuracy are then two distinct concepts, and there can be cases of precise but not accurate measurements, as well as cases of accurate but not precise measurements.

### *7.1.5  Is Noise Gaussian?*

My answer to this question is simple: it would be nice, and in many cases it is true. Sometimes, however, it is just wishful thinking. The usual explanation for the Gaussian distribution of noise is rooted in the CLT. Suppose indeed that a *large number* of sources contribute to the overall noise level $\epsilon$. Writing $\epsilon = \sum_k \varepsilon_k$, each contribution $\varepsilon_k$ can be considered as a particular value of a random variable that summarizes the effects of the $k$-th source. Then, it is usually claimed that, regardless of the probability distributions of the individual noise sources, when their number is large, $\epsilon$ will be Gaussian. This argument may sound straightforward and faultless, but it actually conceals some crucial assumptions, namely that:

i.    The noise sources are *independent*;
ii.   Each of them has a probability distribution with a *finite variance*;
iii.  There is not a *dominant* noise source.
iv.   Last but not least, that the noise contributions sum up, that is, that noise is *additive*.

The first three assumptions have already been discussed in relation to the CLT. The last, apparently "innocent" one, is not necessarily guaranteed in a signal acquisition process. Therefore, it is worth discussing each of these issues on relation to the diverse origins of noise pointed out in the previous sections. Before proceeding, it is important to acknowledge that this "official" explanation actually reveals the wishful thinking of any experimentalist. Most of the statistical tests we will describe rely on the assumption that errors are Gaussian, which greatly simplifies the development of highly effective analysis criteria. Methods and tests much less sensitive to noise distribution do exist, but require more advanced statistical tools. They are generically known as robust statistics, and I will later provide a brief introduction to them.

Let's start by examining intrinsic noise, i.e., noise inherent to the quantity being measured. In most cases this is thermal noise that luckily, as discussed in Chap. 5, is in some sense the paradigm of Gaussian white noise. However, large non-Gaussian fluctuations often show up out of thermodynamic equilibrium, like in geophysical and climate phenomena, which often display the extreme events statistics typical for instance of crackling noise.

Even if the source intrinsic fluctuations are Gaussian, non-Gaussian noise can add to a signal that, as a radio or optical wave, has to propagate through a medium before being detected. For example, I believe most you are all familiar with the static that can be heard on a weak AM radio signal during a storm, which is caused lightning. Similarly, atmospheric turbulence, the main disturbance in astronomical observations from earth-based telescopes, adds non-Gaussian noise.

Detection can be a further source of non-Gaussian noise, in particular when dealing with optical signal measurements or image capture/reconstruction. Low-intensity signals are subjected to photon-counting noise that, as a type of shot noise, has a *Poisson*, and not a normal, distribution. As a consequence, in a picture taken with a CCD or CMOS sensor, the SNR will be higher in brighter regions than in dark regions, where photon counting noise is more relevant. The resolution of images obtained with partially coherent illumination can also be degraded by speckle noise that, as we know, has an *exponential* intensity distribution.[6]

Electronic noise in signal processing is usually the main source of instrumental disturbance, but here we are often on the safe side. Indeed, even if active devices often generate non-Gaussian flicker noise, the overall noise is an incoherent sum of many independent noise sources, the individual electronic components. Flicker noise, however, is a concern not so much for its distribution, but because of its power spectrum. After all, the dominant contribution of very low frequencies suggests that taking long-term averages may not be as beneficial as one might think (remember Press' comment about postponing vacation in Bermuda?). Digitalization is usually another source of non-Gaussian noise. It should finally be stressed that amplification stages introduce *multiplicative*, not additive noise. This is actually true for speckle noise too.

Summing up, assuming noise Gaussianity, as we will do, will make our life simpler, but we should never forget to tread carefully. Just a final comment about the supporter of Bayesian statistics. For them, the problem is not so compelling, since they believe that we should not strive to find "true" values (which may even not exist) but just to figure out our best subjective guess, given the available information. If we don't have any information about the noise except its mean square value, they will simply pick a Gaussian distribution that, as we now know, maximizes the statistical entropy of a p.d.f. on $\mathbb{R}$ with a pre-assigned variance.

### 7.1.6   Outliers

What is an outlier? Two distinguished scholars as Vic Barnett and Toby Lewis defined it as "an observation (or subset of observations) which appears to be inconsistent with the remainder of that set of data".[7] But what is the appropriate criterion for claiming that a result is inconsistent? Unassumingly, I just prefer say that outliers are "strange data": many scientists would acknowledge that a result is peculiar, but only a few would concur that it is inconsistent. To the best of my knowledge, every "rigorous" criterion proposed so far to identify outliers is based on assuming a particular data distribution, typically Gaussian. However, this is a serious issue: just remember that the width of a distribution is estimated using $s_x$, to which deviations due to "peculiar" data contribute *quadratically*.

Where do outliers originate from? I'd say that there are at least three explanations:

- They are *proper* outliers, due to an atypical distribution of data;
- They are *blunders*;
- They originate from data *contamination*,

---

[6] Let me recall that speckles are due to *spatial* coherence. So, speckles do not require laser light to be seen: images obtained in white light though a narrow aperture (for instance, pictures taken with an objective stopped-down by a diaphragm) *do* show speckles.

[7] V. Barnett and T. Lewis, Outliers in Statistical Data, 2nd ed., John Wiley & Sons, New York, 1984.

We are already familiar with the first explanation: distributions with long tails are intrinsic sources of extreme, totally unexpected data. Black swans, as Taleb would say. As we have seen, large deviations are indeed ubiquitous in fields ranging from climatology to financial market trading. Proper outliers are then *apparent* outliers. If there are good reasons to expect a long-tailed distribution, the issue can be turned around. One should not be worried by the occurrence of (apparent) outliers, but rather by their absence.[8] Unfortunately, this explanation is often neglected in those statistical studies that stick to a Gaussianity assumption (which includes of course Bayesians).

To elucidate the dangers of a "Gaussian bias", consider the following example. Suppose that I tell you that, in a series of measurements of a quantity $X$, I have obtained the following results

$$0.32, 0.17, 0.30, \mathbf{0.55}, 0.09, 0.15, 0.03, 0.28, 0.13, 0.31.$$

I am pretty sure that most of you will find the fourth data point, $x_4$, rather strange. Indeed, the mean and the standard deviation of the other nine data are $\bar{x} = 0.20$ e $s_x = 0.11$, so $x_4$ deviates from the mean by more than three standard deviations. This is surely very unlikely, *provided that* the data have a normal distribution. As a matter of fact, however, I obtained the previous data by randomly sampling a *Cauchy* distribution, which has much longer tails than a normal distribution, so much so that it does not have a finite variance. With this data distribution, apparent outliers like $x_4$ can occur quite often. Besides, if I had submitted this second group of results to you,

$$13.70, 13.95, 13.71, 13.72, 13.68, 13.49, 13.55, 13.43, 13.53, 13.57$$

it is probable that you didn't notice any anomalies. Actually, I just added a constant $x_0 = 13.40$ to the previous data and shuffled them a bit. This teaches us something about our ability to identify outliers at a glance.

About blunders, let me make a simple example. This sequence of readings I personally made of the voltage signal provided by a photomultiplier, read on a multimeter:

$$0.002\,\text{V}, 2.334\,\text{V}, 2.310\,\text{V}, 2.275\,\text{V}, 2.290\,\text{V}, \ldots$$

Obviously, the first result is more than strange. Well, the reason is simply that in the first measurement I had forgotten to power the photodetector, which was clearly a blunder. However do not take blunders lightly: gross mistakes in reading, copying, communicating are more common than you may expect!

By "contamination" (or "infection") I mean that some data related to a different phenomenon with a different statistics "slip" into our measurement. For example, a photomultiplier sometimes generates current pulses of significantly lower amplitude

---

[8] Sometimes, this is a way to suspect that a set of data (maybe obtained by your own students) has been "adjusted".

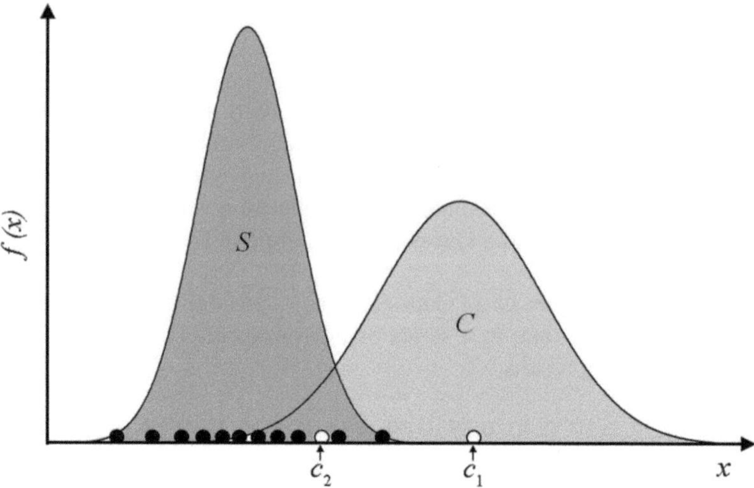

**Fig. 7.3**  Example of data "contamination" (see text)

that the standard ones, which are not due to light absorption by the photocathode, but to electrons generated by thermal emission from the internal dynodes. These pulses have their own distribution that has nothing to do with the phenomenon we are studying. An example of contamination is shown in Fig. 7.3, where the black dots represent genuine data from the sample, with a distribution $S$ of values, while the white dots are contaminants, with a distribution $C$. However, note that while measurement $c_1$ appears to be an outlier, a contaminant like $c_2$ seems to be fully compatible with the distribution of "honest" data. A contaminant does not necessarily have a value external to the typical interval where genuine results fall.

In more complex situations, a piece of data may appear to be an outlier even if it does not have a value that is too high or low. In particular, this may happen while measuring a quantity $Y$ as a function of another variable $X$, that is, a set of data pairs $(x_i, y_i)$. For example, many of you will find "strange" the data point indicated by an arrow in Fig. 7.4, but why? Certainly not because of the value of $Y$, which is well within the normal range of variation. In fact, the value of the fourteenth data point looks much more suspicious, but I do not think it bothers any of you. The fact is that the measurement in question is not consistent with an evident oscillating pattern of the other data. Yet, quantifying this observation is not trivial at all. How would you train a computer (or even a generative AI chatbot) to track down strange data of this kind?

The crucial issue is: should we accept or reject a result that looks like an outlier? Finding a general criterion has long been a kind of "Holy Grail" for experimental scientists. The first and most famous suggestion, made by the American astronomer Willian Chauvenet, can be summed up as follows.

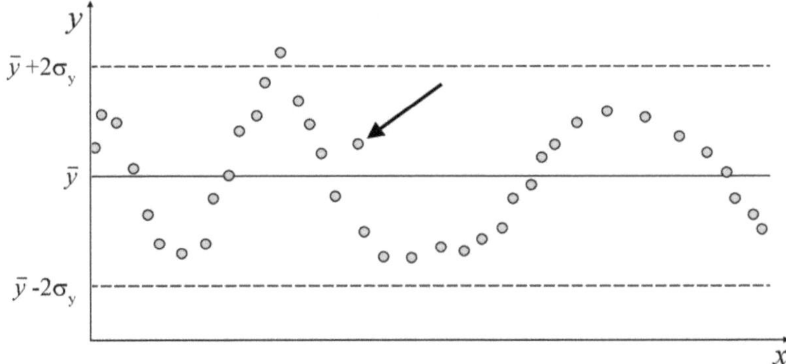

**Fig. 7.4**  Accept or dismiss? This is the problem

Analyze the data statistics and make a guess about their probability distribution. Calling $P(\varepsilon)$ the probability that a deviation from the expected value is greater than $\varepsilon$, the number of data point deviating from the experimental mean by more than $\varepsilon$, out of $N$ measurements, must be of the order of $NP(\varepsilon)$. Hence, if we find a value $\varepsilon_0$ such that $NP(\varepsilon_0) = 1/2$, a result deviating by $\epsilon > \epsilon_0$ is more likely extraneous to the distribution than consistent with it, and can therefore be rejected.

A simple example may clarify this statement. Suppose we have collected $N = 500$ data points of a quantity $X$, and found $\bar{x} = 3, s_x = 0.6$. For a sufficiently large number of measurements, as we will shortly see, the standard deviation can be considered a good estimate of $\sigma_x$. If we assume that noise has a Gaussian distribution (and that there are not errors) we can then look for that value of $z_0$ such that the residual probability $1 - G(z_0) = 1/2N = 10^{-3}$. Using the table in the appendix, we find $z_0 \simeq 3.08$. With a sample of 500 data points, the probability of finding at least one result corresponding to a value $z > z0$ is then less than 50%. Hence, Chauvenet's criterion tells us to reject a data point $x$ if

$$x > \bar{x} + s_x z_0 \simeq 4.85.$$

Chauvenet's criterion is simple, reasonable, and effectively rejects "bad" data. Pity that, as it is easy to prove,[9] the probability of rejecting *good* data is about 40% ! Since the seminal work by Chauvenet, more advanced rejection criteria have been proposed, but, to a greater or lesser extent, they all have potential bugs and should be used with caution.

Vice versa, one of the aims of robust statistics is trying to *coexist* with outliers. The basic idea using parameters like the median that are much less dependent on the

---

[9] The probability of erroneously rejecting a valid data point is $1 - (1 - \frac{1}{2N})^N$, which, for large $N$, is approximately $1 - e^{-1/2} \simeq 0.39$.

shape of a probability distribution than the mean or the standard deviation. Just to give a simple example, consider this set of data

$$2.7, 2.7, 2.8, 2.8, 2.9, 3.0, 3.2, 3.3, 3.6, 2.9, 5.0$$

where the last data point is definitely strange. The mean of all the data is $\bar{x} = 3.2$, while if we exclude the last one we have $\bar{x} = 3.0$. If we include it, the median, which for the first nine measurements is $x_m = 2.9$, stays instead between 2.9 and 3.0. If you remember that the median is also the parameter that minimizes the sum $|s|_x$ of the *absolute values* of the deviations, you can understand why, in robust statistics, $|s|_x$ takes on the role usually played by the standard deviation.

## 7.2  Parameter Assessment

### 7.2.1  Why Is It Worth Taking More Measurements?

Let us get to the core of the measurement problem. Our goal is to reduce the imprecision and inaccuracy in determining the value of one or more physical quantities as much as possible. Accuracy, as we already know, can only be improved by eliminating the causes of errors, but the unavoidable presence of noise requires addressing the issue of measurement precision in depth. By taking a *single* measurement of a quantity, it is rather hard to guess how different the obtained value is from a theoretical prediction. Besides, we have any clue about the width of the data distribution. Intuitively, we expect that better estimates can be obtained by taking several measurements and calculating averages, but we need some evidence to support this guess. So let's ask ourselves:

i.   What is the benefit of taking many measurements and *how much* is it convenient?
ii.  Why does averaging improve precision?
iii. Can we use the standard deviation $s_x$ of the data to estimate the precision of a measured average?

Let us start with some qualitative ideas, for example by comparing the result $x_0$ of a single measurement of a quantity $X$, to which we associate a random variable with expectation $\langle x \rangle$ and variance $\sigma_x^2$, with a second series of measurements in which we have obtained $N$ values $x_1, ..., x_N$. In the first case we may tentatively guess that $\delta_0 = x_0 - \langle x \rangle$ is of the order of $\sigma_x$, either with a positive or a negative sign, and the same guess should be done for *each one* of the results of the second experiment. But what about the average? Writing $x_i = \langle x \rangle + \delta_i$, with each $\delta_i$ of the same order as $\delta_0$, the experimental average is equal to $\bar{x} = \langle x \rangle + \delta$, with $\delta = (\delta_1 + \delta_2 + ... + \delta_N)/N$.

Had most of the $\delta_i$'s the same sign, the deviation would still be of the order of $\delta_0$, but they will likely be evenly distributed between positive and negative, hence $\delta$ will be *significantly smaller* than $\delta_0$. If we assume for simplicity that each $\delta_i$ is exactly equal to $\delta_0$ in magnitude, and completely random in sign, the problem is equivalent to that of a one-dimensional random walk with $N$ steps of length $\delta_0$. We then expect that $\langle \delta \rangle = 0$, that is, that the average does not present *systematic* deviations from the expected value, and that, for large $N$, $\sigma_\delta^2 \sim \sigma_x^2/N$, namely, that the deviation of $\bar{x}$ from $\bar{x}$ is reduced compared to that of a single data by a factor equal of $\sqrt{N}$.

If $N$ is large, using the average as an "estimator" of $\langle x \rangle$ improves the precision in proportion to the square root of the number of collected data. This is the simple (but not trivial) reason why taking more measurements is a winning strategy. The reasoning we have followed is quite approximate, but contains the essence of what we now want to work out in detail.

### 7.2.2  The Mean as an Estimator of the Expectation

When we calculate the mean as $\bar{x} = N^{-1} \sum_{i=1}^{N} x_i$, we are actually summing the values obtained for $N$ random variables $X_i$ that are in fact identical, but that we can conceptually distinguish. The experimental result for the average can therefore be thought of as a single value of the variable $\overline{X} = \sum_i Y_i$ obtained by summing $N$ random variables $Y_i = N^{-1} X_i$, each one with expectation and variance

$$\langle y_i \rangle = \frac{\langle x_i \rangle}{N} = \frac{\langle x \rangle}{N} \ , \quad \sigma^2(y_i) = \frac{1}{N^2} \left\langle (x_i - \langle x \rangle)^2 \right\rangle = \frac{\sigma_x^2}{N^2}.$$

The experimental average is then given a novel and different interpretation. Looking at the mean as a random variable means that, if we repeat the entire set of $N$ measurements several times, we would get a slightly different value for $\bar{x}$, with a distribution of values that reflects the probability distribution for $\overline{X}$. Then, if $N$ is large, the Central Limit Theorem ensures us that this distribution is *Gaussian*, with expectation

$$\boxed{\langle \bar{x} \rangle = \sum_{i=1}^{N} \langle y_i \rangle = \sum_{i=1}^{N} \frac{\langle x \rangle}{N} \implies \langle \bar{x} \rangle = \langle x \rangle} \tag{7.1}$$

and variance

$$\boxed{\sigma^2(\bar{x}) = \sum_{i=1}^{N} \sigma^2(y_i) = \frac{1}{N^2} \sum_{i=1}^{N} \sigma_x^2 \implies \sigma^2(\bar{x}) = \frac{\sigma_x^2}{N}} \tag{7.2}$$

Equation (7.2) means that the distribution of the mean is narrower (by a factor of $\sqrt{N}$) than the distribution of values of individual measurements, that is, *the mean $\overline{X}$ shows much smaller fluctuations than the physical variable $X$*. In the vast majority of cases, the average value we calculate from experimental data will then approximate the "exact" value of $X$ with a precision of a few $\sigma_{\bar{x}} = \sigma_x/\sqrt{N}$. Note that this result holds *whatever the distribution of the variables $X_i$*, provided that they satisfy the requirements of the CLT. Therefore, *the noise does not necessarily have to be Gaussian for the mean to be normally distributed*, the only requirement being that $N$ is large enough.[10]

### 7.2.3  Estimating $\sigma_x$: The "Adjusted" Standard Deviation

You may not have noticed, but the results we have just found are totally *useless* if we don't know the variance of $X$! Obtaining from the experimental data a satisfactory estimate of $\sigma_x$ is then crucial. Since the variance is a sum of the squares of deviations with respect to $\langle x \rangle$, we might think that a good estimate is given by the sum of the squares of deviations from $\bar{x}$, that is, by the square of the standard deviation,

$$s_x^2 = \overline{(x - \bar{x})^2} = \frac{1}{N} \sum_{i=1}^{N} (x_i - \bar{x})^2 = \frac{1}{N} \sum_{i=1}^{N} \left( x_i^2 - \bar{x}^2 \right).$$

As much as the mean, however, $s_x^2$ must be regarded as a random variable, which takes on distinct values in different series of measurements. Thus, our guess is correct if and only if the expectation of $s_x^2$ is equal to $\sigma_x^2$. Let's see.

$$\langle s_x^2 \rangle = \frac{1}{N} \left\langle \sum_{i=1}^{N} \left( x_i^2 - \bar{x}^2 \right) \right\rangle = \frac{1}{N} \sum_{i=1}^{N} \left( \langle x_i^2 \rangle - \langle \bar{x}^2 \rangle \right) = \langle x^2 \rangle - \langle \bar{x}^2 \rangle.$$

The expectation of $s_x^2$ is than the difference between $\langle x^2 \rangle$ and the expectation of the *mean* of $X$, but this is *different* from $\sigma_x^2$. Indeed, adding and subtracting $\langle \bar{x} \rangle^2 = \langle x \rangle^2$, we have

$$\langle s_x^2 \rangle = \left( \langle x^2 \rangle - \langle x \rangle^2 \right) - \left( \langle \bar{x}^2 \rangle - \langle \bar{x} \rangle^2 \right) = \sigma_x^2 - \sigma_{\bar{x}}^2,$$

that is, using (7.2):

$$\langle s_x^2 \rangle = \sigma^2 \left( 1 - \frac{1}{N} \right) = \frac{N-1}{N} \sigma_x^2. \tag{7.3}$$

Hence, the standard deviation as defined in Eq. (**??**) *underestimates* the variance of the limiting distribution by a factor $(N-1)/N$. Where does this rather unexpected result come from? In Chap. **??** we have shown that the root mean square deviation from a generic value $\mu$, that is, the second moment $M_2(\mu)$, is minimal when $\mu = \bar{x}$. But since in general $\langle x \rangle \neq \bar{x}$, we may expect $M_2(\langle x \rangle)$ to be *greater* than $M_2(\bar{x})$. To obtain a correct estimate, we must then *redefine* the standard deviation as

---

[10] How to deal with small samples will be discussed later.

$$S_x = \sqrt{\frac{\sum_{i=1}^{N} (x_i - \bar{x})^2}{N - 1}} \qquad (7.4)$$

Using this new definition, we have $\langle s_x^2 \rangle = \sigma_x^2$, so that the square of the "adjusted" standard deviation yields indeed the best estimate of the variance. Note that in the limiting case when the data set consists of a *single* measurement, equation (7.4) provides an indefinite value for $s_x$ (which is sensible), while for very large data samples definitions (7.4) and (**??**) are almost identical. Using the new definition of $s_x$, we also have

$$s_x^2 = \frac{N}{N - 1} \sum_{j=1}^{r} f_j \left( x_j - \bar{x} \right)^2 = \frac{N}{N - 1} \left( \overline{x^2} - \bar{x}^2 \right).$$

### 7.2.4  *How to Report a Result*

Using (7.4), we can restate in "operational terms Eq. (7.2) by means of the *standard deviation of the mean*,[11]

$$s(\bar{x}) = \frac{s_x}{\sqrt{N}}, \qquad (7.5)$$

which is the best estimate of the deviation of the mean from the expected value that we can get from the experimental data. We also note that as the number $N$ of measurements increases the uncertainty on the mean decreases as $N^{-1/2}$ (while the uncertainty of each individual data point does *not* change).

We now have all the ingredients to state how the final result of a set of experimental measurements of a physical quantity $X$ should be reported. Indeed, since for sufficiently large $N$ the probability distribution of the mean is Gaussian, if we write

$$\boxed{x = \bar{x} \pm s(\bar{x})} \qquad (7.6)$$

we mean that, with a chance of about 68%, the "true" value of $X$ is within an interval of width $\pm s(\bar{x})$ around $\bar{x}$.

Yet, there is something strange in what we have just state. Supporters of the Bayesian approach would indeed raise a strong objection, which I fully agree with, though I mostly consider it as a question of interpretation. The "true" value of $X$ is a *fixed* quantity, not a fluctuating one. What sense does it make writing that the "true" value of $X$ is, with some probability, in a given interval around the mean? The probability that $X$ has a specific value can only be 1 if the value is correct, or 0 otherwise! In fact, we are actually using an "inverse probability" reasoning, yielding

---

[11] This is usually—but not here—called the "standard error".

the *estimate* we can give of the parameters of the distribution of the mean (whose specific experimental value is known). It is a subtle but significant difference. This is the kind of reasoning that is done whenever we try to adapt a theoretical distribution to a set of experimental data.

The standard deviation of the mean can then be taken as the half-width of what is usually called the "error bar" for our best estimate of $X$, given by the experimental mean. This is the simplest way to provide the result of a repeated measurement of a physical quantity. However, noting that the variable

$$z = \frac{x - \bar{x}}{s(\bar{x})} = \sqrt{N}\,\frac{x - \bar{x}}{s_x} \tag{7.7}$$

has, for large $N$, a standard normal distribution, we can then also define more generally a *confidence interval at p %* by evaluating for which value $z_0$ of $z$ at least $p$ % of the area subtended by a unit Gaussian falls within $[-z_0, +z_0]$.

As a simple example, suppose that 100 measurements have given, for the mean and the (adjusted) standard deviation of a quantity $X$,

$$\bar{x} = 3.565; \; s_x = 0.124.$$

so that $s(\bar{x}) = 0.124/\sqrt{100} = 0.0124$. We may report the result as

$$x = 3.565 \pm 0.0124.$$

Alternatively, we may wish to give a result for $X$ with a confidence interval of, say, 95%. This requires to find the value $z$ such that the area subtended by the tails of the Gaussian outside the interval $(-z, +z)$ is less than 5%. From the table in the appendix, we find $z = 1.96$, thus

$$x = 3.565 \pm 1.96\,s(\bar{x}) = 3.565 \pm 0.024$$

with a confidence of 95%. Note however that, so far, we have no idea about the *precision* with which $s(\bar{x})$ is known, and therefore of the number of digits that can be considered significant in this result.

Pray, do not place excessive or improper confidence in confidence intervals! Eq. (7.7) is strictly based on the conditions that a) the number $N$ of measurements is *large enough* to guarantee that the distribution of the mean is Gaussian, and b) $s_x$ is *calculated*, not *assumed*. What happens if condition a) is not met due to the small data sample will be discussed in Sect. 7.6. See below for the more serious consequences of not meeting condition b). But even if both conditions are satisfied, we must remember that, according to the TLC, the convergence to the normal distribution is *not* uniform. Thus, even if $N$ is large enough to fairly grant that $\bar{x}$ is Gaussian in $[\bar{x} - s_x, \bar{x} + s_x]$, this does not mean at all that the probability that $X$ lies within $\bar{x} \pm 3s_x$ is 99.7%.

In many situations (many more than you can guess!), however, one does not take several, but just *one* measurement of a physical quantity, relying for the uncertainty of the results on the pre-calibrated accuracy of the instrumental setup. This is commonly done for example when a specimen is weighted on a balance, using the reading on the scale as the measured value and the accuracy of the balance stated by the producer as it uncertainty. For quantities that can take only positive values, so that their distribution can *never* be exactly Gaussian, this "shorter route" can lead to outrageous claims. For instance, if you add a pinch of salt to a water-containing flask placed on a balance, and the reading of the balance, which has a stated resolution of $\pm 10$ mg, increases by 18 mg, you can fairly state that the true amount of salt you poured is $w = 18 \pm 10$ mg, but what does it mean stating that there is a probability of 2.5% that $w < -2$ mg ($w < \bar{w} - 2\sigma$)? That probability is clearly zero!

Things get worse if we cannot be sure that the instrumental noise is Gaussian, for example because it is shot or amplification noise, or even more when the uncertainties are not due to instrumental, but to *intrinsic* noise. If $X$ has not a normal distribution, not only the probability values associated with a confidence interval are different, but the concept itself of confidence interval must be deeply reconsidered. For instance, if the probability distribution for $X$ is not symmetric around $\langle x \rangle$,

$$P\left[(\langle x \rangle - \sigma) \le X \le \langle x \rangle\right] \ne P\left[\langle x \rangle \le X \le (\langle x \rangle + \sigma)\right].$$

In this case, there can be several different definitions of "confidence interval". And of course, if the distribution of $X$ has not a finite variance, like a Cauchy, the situation is really desperate…Therefore, beware of excessive confidence!

### 7.2.5  *Estimating Correlations*

Since the mean is the best estimate of $\langle x \rangle$, and the standard deviation of $\sigma_x$, we may expect the experimental correlation coefficient,

$$r_{xy} = \frac{s_{xy}}{s_x s_y} = \frac{\overline{xy} - \bar{x}\bar{y}}{s_x s_y},$$

to be a good estimator of the correlation coefficient of the limiting distribution,

$$\rho_{xy} = \frac{\langle xy \rangle - \langle x \rangle \langle y \rangle}{\sigma_x \sigma_y}.$$

This is true, provided that $r_{xy}$ is adjusted, like we did for $s_x$, using a correction factor $N/(N-1)$. However, while the probability distributions for the mean and the standard deviation approximately has a Gaussian shape even for moderate $N$, the probability distribution for $r_{xy}$ approaches a normal distribution very slowly, that is, only for a very large number of measurements. This is why assessing the degree of correlation of two variables using a few experimental data is very dangerous.

## 7.3   Indirect Measurements

In most cases, the physical quantity $Y$ that we want to determine is actually obtained from data on one or more other variables $X_1, X_2, \ldots X_N$ that are actually measured, through a known functional relationship $Y = f(X_1, X_2, \ldots X_N)$. Besides, preparing the experiment may require to set some experimental parameters, a procedure that can introduce errors. For example, we could determine the mass of an unknown particle by measuring its momentum after a collision with a particle of known mass, using the incident momentum of the incident particle as an adjustable experimental parameter.

How can we fix a confidence interval for a quantity $Y$ obtained as a function of a directly measured quantity $X$? Of course, if we could determine the entire probability density $f_x(x)$ for $X$, we could obtain $f_y(y)$ using the "golden rule" obtained in Chap. 4. Usually, however, all we know is an estimate of the expectation and variance of $X$. Nevertheless, we can still give an approximate estimate for the confidence interval for $Y$ when the distribution of measured values for $X$ is narrow. What we plan to do in this section is showing how the confidence interval for $Y = f(X)$ is related to that for $X$. This is what is usually called "propagation of errors", but I definitely prefer the expression *noise rescaling*. The following observations are also particularly useful while designing an experiment to provide an a priori estimate of the precision of an indirect measurement.

### 7.3.1   *Expectation of $Y = f(X)$*

Remember that, in general, $\langle y \rangle$ cannot be obtained be calculating the function $f(x)$ for $X = \langle x \rangle$, i.e., $\langle f(x) \rangle \neq f(\langle x \rangle)$. However, this can be shown to be a good approximation for small errors, that is, up to terms of the order of $(\sigma_x)^2$. Indeed, using a Taylor expansion,

$$f(x) = f(\langle x \rangle) + \left(\frac{df}{dx}\right)_{\langle x \rangle} (x - \langle x \rangle) + \frac{1}{2}\left(\frac{d^2 f}{dx^2}\right)_{\langle x \rangle} (x - \langle x \rangle)^2 + \ldots$$

and taking the expectation of both sides, we have

$$\langle f(x) \rangle = f(\langle x \rangle) + \left(\frac{df}{dx}\right)_{\langle x \rangle} \langle (x - \langle x \rangle) \rangle + \frac{1}{2}\left(\frac{d^2 f}{dx^2}\right)_{\langle x \rangle} \langle (x - \langle x \rangle)^2 \rangle + \ldots$$

However, $\langle (x - \langle x \rangle) \rangle = 0$. Hence noting that the last term on the r.h.s. is nothing but the variance of $X$, we obtain:

$$\langle f(x) \rangle = f(\langle x \rangle) + \frac{1}{2} \left( \frac{\mathrm{d}^2 f}{\mathrm{d}x^2} \right)_{\langle x \rangle} \langle f(x) \rangle = f(\langle x \rangle) + \left( \frac{\mathrm{d}f}{\mathrm{d}x} \right)_{\langle x \rangle} \sigma_x^2 + \dots$$

Therefore, at *1st order in* $\sigma_x$,

$$\langle f(x) \rangle \simeq f(\langle x \rangle).  \tag{7.8}$$

Note that the small term we are neglecting actually introduces a slight systematic deviation between $\bar{y}$ and $\langle y \rangle$.

### 7.3.2   *Functions of a Single Variable*

Consider first a simple linear relation between $Y$ and $X$, $Y = aX + b$. In this case, the standard deviation of $Y$ can be exactly related to $\sigma_x$, since

$$\sigma_y^2 = \left\langle (ax + b)^2 \right\rangle - \langle (ax + b) \rangle^2 = a^2 \left\langle x^2 \right\rangle + 2ab \langle x \rangle + b^2 - a^2 \langle x \rangle^2 - 2ab \langle x \rangle - b^2,$$

wherefrom

$$\sigma_y^2 = a^2 \left\langle x^2 \right\rangle - a^2 \langle x \rangle^2 = a^2 \sigma_x^2,$$

that is

$$\sigma_y = |a| \sigma_x.  \tag{7.9}$$

note that the absolute value of $a$ acts as indeed as a "rescaling factor" between $\sigma_x$ and $\sigma_x$, while the constant term $b$ does not contribute to $\sigma_y$.

Let us now deal with a generic relationship $Y = f(X)$, assuming that most measured values $x_i$ of $X$ are expected to fall within a narrow interval around $\langle x \rangle$. Namely, we assume that the *relative* deviations are small, $|x_i - \langle x \rangle| / \langle x \rangle \ll 1$. Then, if we expand at *first* order $f(x)$,

$$f(x) \simeq f(\langle x \rangle) + \left( \frac{\mathrm{d}f}{\mathrm{d}x} \right)_{\langle x \rangle} (x - \langle x \rangle),$$

we are actually back to the linear case, with

$$\begin{cases} a = \left( \dfrac{\mathrm{d}f}{\mathrm{d}x} \right)_{\langle x \rangle} \\[2ex] b = f(\langle x \rangle) + \left( \dfrac{\mathrm{d}f}{\mathrm{d}x} \right)_{\langle x \rangle} \langle x \rangle. \end{cases}$$

We have therefore the general formula

$$\sigma_y \simeq \left| \frac{df}{dx} \right|_{\langle x \rangle} \sigma_x. \tag{7.10}$$

Let's now apply Eq. (7.10) to two particularly significant cases.

POWER LAW:   When $Y = AX^{\alpha}$, we have $f'(x) = \alpha A x^{\alpha - 1}$, hence

$$\sigma_y = |\alpha A \langle x \rangle^{\alpha - 1}| \sigma_x.$$

Dividing by $\langle y \rangle$, we obtain

$$\boxed{\frac{\sigma_y}{|\langle y \rangle|} = |\alpha| \frac{\sigma_x}{|\langle x \rangle|}} \tag{7.11}$$

When $\alpha = -1$, $Y$ is inversely proportional to $X$, and we have

$$\boxed{\frac{\sigma_y}{|\langle y \rangle|} = \frac{\sigma_x}{|\langle x \rangle|}} \tag{7.12}$$

namely, the two variables have the same *relative* standard deviation.

LOGARITHMIC LAW:   If $Y = \ln(X)$, $f'(x) = -1/x$, hence

$$\boxed{\sigma_y = \frac{\sigma_x}{\langle x \rangle}} \tag{7.13}$$

This expression is particularly useful when the amplitude of the fluctuations of $X$ is approximately proportional to the value of $X$ itself (for example, if the measurement is made with an instrument whose precision is proportional to the full scale). In this case, the error on the *logarithm* of $X$ turns out to be a constant.

**Remark 7.1**  Bear in mind that Eq. (7.10) is only valid for *small* deviations. In some cases, one needs to go beyond this first-order approximation. For example, if $Y = X^2$, Eq. (7.10) yields $\sigma_y = 2|\langle x \rangle| \sigma_x$, which *vanishes* when $\langle x \rangle = 0$. This does not mean that $Y$ is determined exactly, but only that deviations are of order $\sigma_x^2$ or more. If $X$ has a Gaussian distribution of variance $\sigma^2$ centered at the origin, $\sigma_y$ is exactly given by

$$\sigma_y^2 = \langle y^2 \rangle - \langle y \rangle^2 = \langle x^4 \rangle - \langle x^2 \rangle^2 .$$

Since $\langle x \rangle = 0 = 0$, $\langle x^2 \rangle^2$ coincides with $(\sigma^2)^2 = \sigma^4$, whereas (see the appendix) $\langle x^4 \rangle = 3\sigma^4$. This yields $\sigma_y = \sqrt{2}\,\sigma_x^2$, hence $\sigma_y$ is proportional to $\sigma_x$ *squared*.

**Example 7.2**  (*Friction coefficient*) We want to find the viscous friction coefficient of a liquid of density $\rho_f$ by measuring the final stationary speed of a mass of density $\rho$

and volume $V$ falling through the fluid under the effect of its own weight. Assuming a downward directed $z$ axis, and taking into account the buoyancy force $-\rho_f V g$, the equation of motion is

$$V\rho\ddot{z}(t) = (\rho - \rho_f)Vg - k\dot{z}(t),$$

thus the terminal velocity $v_\infty$, obtained by setting to zero the acceleration $\ddot{z}$, is

$$v_\infty = \frac{(\rho - \rho_f)Vg}{k}.$$

Then, if $v_\infty$ is known with an accuracy $\sigma(v_\infty)$, we have

$$\frac{\sigma(k)}{k} = \frac{\sigma(v_\infty)}{v_\infty},$$

that is, the estimate of $k$ has the same relative precision as that of $v_\infty$.

**Example 7.3** (*Elastic scattering*) A particle of mass $m$ is scattered elastically by a second particle of mass $M \gg m$. The change in of $m$ can be obtained by measuring the angle that the direction of motion of the particle after the collision makes with the incident direction by observing that the kinetic energy of $m$ does not appreciably change in the collision because of the small mass ratio $m/M$. Therefore for the momentum of $m$ does not change in the collision, $|\mathbf{p}_f| = |\mathbf{p}_i| = p$, so that (see Fig. 7.5)

$$\frac{\mathrm{d}}{\mathrm{d}\vartheta}\Delta p = p\cos\left(\frac{\vartheta}{2}\right).$$

From (7.10) we obtain

$$\frac{\sigma(\Delta p)}{\Delta p} = \frac{1}{2}\cot\left(\frac{\vartheta}{2}\right)\sigma(\vartheta),$$

where $\sigma(\vartheta)$ is the precision with which we measure the scattering angle. Since $\cot x$ diverges for $x \to 0$, this shows that the relative precision in the estimate of the momentum change becomes arbitrarily small for $\vartheta \to 0$.

### 7.3.3  Functions of Several Variables

More generally, the physical quantity we are interested in can depend on several directly measured variables. Consider first a simple bilinear law, $Z = aX + bY$. Expanding $\sigma_z^2 = \langle(ax + by)^2\rangle - \langle(ax + by)\rangle^2$ we obtain

$$\sigma_z^2 = a^2\left(\langle x^2\rangle - \langle x\rangle^2\right) + b^2\left(\langle y^2\rangle - \langle y\rangle^2\right) + 2ab\left(\langle xy\rangle - \langle x\rangle\langle y\rangle\right),$$

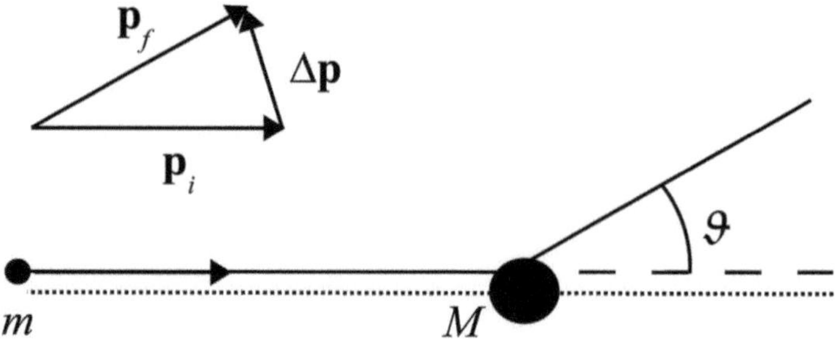

**Fig. 7.5**   Elastic particle scattering

that is,

$$\sigma_z^2 = a^2\sigma_x^2 + b^2\sigma_y^2 + 2ab\sigma_x\sigma_y\rho_{xy}$$

(7.14)

What is the effect of the presence of the correlation coefficient $\rho_{xy}$? Note that, taking $a = b = 1$, if $X$ and $Y$ are completely

(a) CORRELATED ($\rho_{xy} = 1$):   $\sigma_z^2 = (\sigma_x + \sigma_y)^2$;
(b) ANTI- CORRELATED ($\rho_{xy} = -1$):)   $\sigma_z^2 = (\sigma_x - \sigma_y)^2$;
(c) UNCORRELATED ($\rho_{xy} = 0$):)   $\sigma_z^2 = \sigma_x^2 + \sigma_y^2$.

with $\sigma_z$(case a) $\geq \sigma_z$(case c) $\geq \sigma_z$(case b). When $x$ and $y$ are fully correlated fluctuations in $Z$ are maximal, and coincide with the sum of those of $X$ and $Y$. Vice versa, if the two variables are *anti*correlated, fluctuations in $X$ and $Y$ tend to compensate each other, and $\sigma_z$ is minimal. In the intermediate case of two *uncorrelated* variables, which we will mostly deal with, the fluctuations add up in quadrature.

Since the best estimate of $\rho_{xy}$ is the experimental correlation coefficient $r_{xy}$, $\sigma_z$ is correctly estimated by

$$s_z = \left(a^2 s_x^2 + b^2 s_y^2 + 2ab s_x s_y r_{xy}\right)^{1/2}$$

(7.15)

Hence, for fully uncorrelated variables (but *only* in this case!),

$$s_z = \sqrt{a^2 s_x^2 + b^2 s_y^2},$$

(7.16)

Consider now a generic function of two variables, $Z = g(X, Y)$. In analogy with what we have done for a function of a single variable, for small deviation we can expand $g(x, x)$ at first order in $x - \langle x \rangle$ and $y - \langle y \rangle$,

$$g(x, y) \simeq g(\langle x \rangle, \langle y \rangle) + \left(\frac{\partial g(x, y)}{\partial x}\right)(x - \langle x \rangle) + + \left(\frac{\partial g(x, y)}{\partial y}\right)(y - \langle y \rangle)$$

where the partial derivatives are evaluated in $x = \langle x \rangle$, $y = \langle y \rangle$. Once again, at first order $g(x, y)$ is linear in the small deviations from $\langle x \rangle$ ed $\langle y \rangle$, hence

$$\sigma_z^2 = \left( \frac{\partial z}{\partial x} \right)^2 \sigma_x^2 + \left( \frac{\partial z}{\partial y} \right)^2 \sigma_y^2 + 2 \left( \frac{\partial z}{\partial x} \right) \left( \frac{\partial z}{\partial y} \right) \sigma_x \sigma_y \rho_{xy}. \tag{7.17}$$

For uncorrelated variables, the variance of $Z$ is simply

$$\sigma_z^2 = \left( \frac{\partial z}{\partial x} \right)^2 \sigma_x^2 + \left( \frac{\partial z}{\partial y} \right)^2 \sigma_y^2. \tag{7.18}$$

The previous results are easily extended to functions of more than two variables. For example, if $Z = X_1 + X_2 + \ldots + X_N$, where all the variables $X_i$ are uncorrelated, we simply have

$$\sigma_z^2 = \sigma_{x_1}^2 + \sigma_{x_1}^2 + \ldots \sigma_{x_N}^2 \tag{7.19}$$

a result that we have already obtained in a more general form in Chap. 4. Let us consider two additional useful examples.

PRODUCT:    If $Z = XY$, where $X$ and $Y$ are uncorrelated, we have $\dfrac{\partial z}{\partial x} = y, \dfrac{\partial z}{\partial y} = x$,

so that

$$\sigma_z^2 = \langle x \rangle^2 \sigma_x^2 + \langle y \rangle^2 \sigma_y^2. \tag{7.20}$$

Dividing both members by $\langle z \rangle^2 = \langle xy \rangle^2 = \langle x \rangle^2 \langle y \rangle^2$ we obtain

$$\frac{\sigma_z^2}{\langle z \rangle^2} = \frac{\sigma_x^2}{\langle x \rangle^2} + \frac{\sigma_y^2}{\langle y \rangle^2}. \tag{7.21}$$

To appreciate the effect of correlations, just try and apply Eq. (7.20) to the case $Y = X$, so that $Z = X^2$, which gives

$$\sigma_z^2 = 2 \langle x \rangle^2 \sigma_x^2,$$

while we know that, for $Z = X^2$,

$$\sigma_z^2 = 4 \langle x \rangle^2 \sigma_x^2,$$

a discrepancy that, since $\rho_{xx} = 1$, is removed by introducing the correlation contribution $2 \langle x \rangle^2 \sigma_x^2$.

For $N$ uncorrelated variables, expression (7.21) simply generalizes to

$$\frac{\sigma_z^2}{\langle z \rangle^2} = \frac{\sigma_{x_1}^2}{\langle x_1 \rangle^2} + \frac{\sigma_{x_2}^2}{\langle x_2 \rangle^2} + \ldots + \frac{\sigma_{x_N}^2}{\langle x_N \rangle^2}. \tag{7.22}$$

This "composition law", which is particularly useful when dealing with multiplicative noise, has the same form as the one we derived for the sum, provided that the $\sigma_i$'s are replaced with the relative variances $\sigma_i / \langle x_i \rangle$.

RATIO:   If $Z = X/Y$, $\dfrac{\partial z}{\partial x} = \dfrac{1}{y}$ and $\dfrac{\partial z}{\partial y} = \dfrac{1}{x}$. Hence, $\sigma_z^2 = \dfrac{\sigma_x^2}{\langle y \rangle^2} + \dfrac{\sigma_y^2}{\langle x \rangle^2}$ and, dividing both sides by $\langle z \rangle^2$,

$$\frac{\sigma_z^2}{\langle z \rangle^2} = \frac{\sigma_x^2}{\langle x \rangle^2} + \frac{\sigma_y^2}{\langle y \rangle^2},$$

which is therefore *identical* to (7.20)

Combing (7.19) and (7.21) we can conclude that, for a quantity obtained as a *rational function* of several uncorrelated variables,

$$\boxed{Z = \frac{X_1 X_2 \ldots X_r}{X_{r+1} X_{r+2} \ldots X_n} \implies \frac{\sigma_z^2}{\langle z \rangle^2} = \sum_{i=i}^{n} \frac{\sigma_{x_i}^2}{\langle x_i \rangle^2}} \qquad (7.23)$$

**Example 7.4** (*Range of a cannon shell*) Consider a cannon shell launched at an angle $\alpha$ with the horizontal with an initial speed $v_0$. Neglecting friction, the range $D$ of the shell is

$$D = \frac{v_0^2 \sin(2\alpha)}{g}.$$

Suppose we can set the initial speed and the firing angle with accuracies $\sigma(v_0)$ and $\sigma(\alpha)$, respectively. Since $v_0$ and $\alpha$ are not correlated, we have

$$\sigma_D^2 = \frac{\partial D}{\partial v_0} \sigma_{v_0}^2 + \frac{\partial D}{\partial \alpha} \sigma_\alpha^2 = \frac{4v_0^2}{g^2} \left[ \sin^2(2\alpha) \sigma_{v_0}^2 + \cos^2(2\alpha) \sigma_\alpha^2 \right].$$

Note that the maximum range (which is obtained for $\alpha = 45°$) is unaffected by a lack of accuracy in angle of fire. More precisely, this means that for $\alpha = 45° \pm \delta$ the first correction in the range is of order $\delta^2$.

**Example 7.5** (*Activity of a radioisotope*) Let the initial activity $N_0$ and the time constant $\tau$ of a radioactive source be known with an accuracy of 1%. We want to determine the uncertainty in the activity at a generic time $t$. From $N = N_0 \exp(-t/\tau)$ we have

$$\sigma_N^2 = \left( \frac{\partial N}{\partial N_0} \right)^2 \sigma_{N_0}^2 + \left( \frac{\partial N}{\partial \tau} \right)^2 \sigma_\tau^2 = \left[ \sigma_{N_0}^2 + \frac{N_0^2 t^2}{\tau^4} \sigma_\tau^2 \right] e^{-2t/\tau},$$

hence

$$\left(\frac{\sigma_N}{N}\right)^2 = \frac{\sigma_{N_0}^2}{N_0^2} + \left(\frac{t}{\tau}\right)^2 \frac{\sigma_\tau^2}{\tau^2} = 10^{-4}\left[1 + \left(\frac{t}{\tau}\right)\right].$$

Note that the two contributions to $\sigma_N/N$ become equal for $t = \tau$.

## 7.4  Significant Figures

In principle, what we have said so far should allow us to provide estimates and confidence intervals for both directly and indirectly measured quantities. Nonetheless, there is still a delicate issue to address. Our assessments make use the standard deviation $s_x$, which is the best estimate of $\sigma_x$. However, *how accurate* is this estimate? That is, how much can we trust the experimental value obtained for the standard deviation? This may seem more like a technicality than a real problem, but paying a little attention to the issue will give us a criterion to fix the number of significant figures in a result. Since we have, from (7.4),

$$s_x^2 = \frac{1}{N-1} \sum_{i=1}^{N} \delta_i^2,$$

$(N-1)s_x^2$ can be thought as the sum of the squares of $N$ i.i.d. variables $\delta_i$, each one with expectation $\langle \delta_i \rangle = 0$ and variance $\sigma^2(\delta_i)$ of course equal to $\sigma_x^2$. Hence,

$$\sigma^2(s_x^2) = \frac{1}{(N-1)^2} \sum_{i=1}^{N} \sigma^2(\delta_i^2). \tag{7.24}$$

At this point, we might think of applying noise rescaling once again and write $\sigma^2(\delta_i^2) = 4\,\langle \delta_i \rangle^2\,\sigma^2(\delta_i)$, but this *cannot* work, because $\langle \delta_i \rangle = 0$ for all $i$! However, this is just a failure of the small deviation approximation, so we can write, as in Remark 7.1,

$$\sigma(\delta_i^2) = 2\sigma^2(\delta_i).$$

Noticing that $\sigma^2(s_x^2) = 4\,\langle s_x \rangle^2\,\sigma^2(s_x) = 4\sigma_x^2\sigma^2(s_x)$, we obtain

$$4\sigma_x^2\sigma^2(s_x) = \frac{4N\sigma_x^4}{(N-1)^2}.$$

Neglecting the small difference between $N$ and $N-1$,

$$\sigma(s_x) \simeq \frac{\sigma_x}{\sqrt{N}},$$

that can be written, estimating $\sigma_x$ with $s_x$,

$$\sigma(s_x) \simeq \frac{s_x}{\sqrt{N}}. \tag{7.25}$$

*The precision of the standard deviation is therefore almost equal to that of the mean.* Since

$$\sigma[s(\bar{x})] = \sigma\left(\frac{s_x}{\sqrt{N}}\right) = \frac{\sigma(s_x)}{\sqrt{N}},$$

the uncertainty on the standard deviation of the mean is then

$$\sigma[s(\bar{x})] \simeq \frac{s_x}{N}. \tag{7.26}$$

The value of $\sigma[s(\bar{x})]$ is precisely what we need to fix the number of significant figures, because it tells us what is the degree of reliability of the confidence interval for $X$ that we have chosen using $s(\bar{x})$.

As an application, for the simple example given in Sect. 7.2.4 we obtain $\sigma[s(\bar{x})] = 1.24 \times 10^{-3}$. Hence, the confidence interval is correct up to the third decimal place, so it makes sense to report the result as we did, while stating $x = 3.5650 \pm 0.0124$ would not be justified.


## 7.5  Weighted Averages


In Chap. **??**, while discussing statistics by classes, we introduced the concept of *weighted average*. The same idea can be applied to a series of measurements of the same physical quantity made in different experimental conditions. As a simple example, suppose that two experimental devices $A$ and $B$ give, for the same quantity $X$ two different results $X = x_A$ and $X = x_B$, and that the precision of the two measurements, estimated by testing the instruments, are $\sigma_A$ and $\sigma_B$. If, for example, $\sigma_B = 2\sigma_A$, to obtain with the $B$ the same precision that is obtained with the $A$ we have to perform four times more measurements, because the precision in the estimate of $X$ increases with the square root of the number of measurements. Therefore, the measurement made with $A$ needs to be weighted *four times more* than the measurement made with $B$.

To obtain a correct estimate of $X$ with multiple measurements, we must then take into account the relative weight of each result $x_i$, associating with it an "effective number of measurements" equal to $1/\sigma_i^2$. So, if we have obtained $N$ outcomes $x_1, x_2, \dots x_N$ with different accuracy values $\sigma_1, \sigma_2, \dots, \sigma_N$, we must use the weighted average

$$\boxed{\bar{x}_w = \frac{\sum_{i=1}^{N} \left(x_i/\sigma_i^2\right)}{\sum_{i=1}^{N} \left(1/\sigma_i^2\right)}} \tag{7.27}$$

By introducing the *weighted variance* $\sigma_w^2$ as

$$\sigma_w^2 = \left( \sum_{i=1}^{N} \frac{1}{\sigma_i^2} \right)^{-1}, \tag{7.28}$$

the standard deviation of the weighted average is then given by

$$\boxed{\sigma^2(\bar{x}_w) = \sum_{i=1}^{N} \frac{\sigma_i^2}{\sigma_i^4} = \sigma_w^2} \tag{7.29}$$

Summing up, when we take multiple measurements of the same quantity with different precision, we can use the weighted average (7.27) as an estimate of the value, with a precision equal to the weighted variance (7.28). The assumption of the weighted average as the best estimate of the expectation is rigorously proved in the next chapter.

## 7.6  Small Samples

We have seen that the mean has a Gaussian distribution centered around the expected value, which implies that the standard variable

$$Z = \frac{\overline{X} - \langle x \rangle}{\sigma_x}$$

is normally distributed about the origin with unit variance. However, as we said, we often have no way of fixing $\sigma_x$, and we must rely on its best estimate $s(\bar{x})$. So we actually use the variable[12]

$$t = \frac{x - \bar{x}}{s(\bar{x})} = \sqrt{N}\, \frac{x - \bar{x}}{s_x}. \tag{7.30}$$

To correctly establish a confidence interval, however, we must know what is the probability distribution the variable $t$, which is proportional to the ratio between the Gaussian variable $(\bar{x} - \langle x \rangle)$ and the standard deviation, whose probability distribution is discussed in the next chapter.

For now, let me just point out that, in general, $t$ is *not* normally distributed, in particular when $s_x$ is obtained from a small sample of measurements. The shape of this distribution was derived, using some of the methods we developed in Chap. 4, by

---

[12] We normally use capital letters for the variable, but in this case $t$ is the standard, universally used notation. Thus, in this section we revert to lowercase.

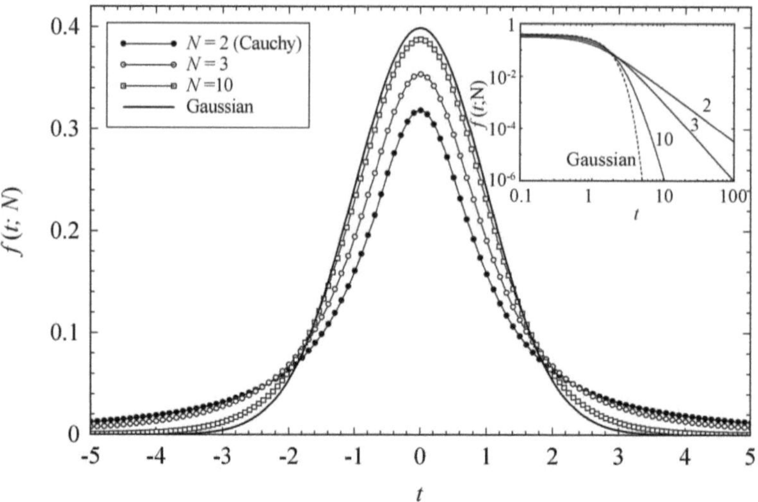

**Fig. 7.6** Probability density of the Student's $t$ for several values of $N$. The inset shows the power-law behaviour of the tails of $f(t; N)$ (only for $t > 0$, because of the double-log scale)

W. S. Gossett, who wrote under the pseudonym "Student",[13] and is therefore known as the *Student's t-distribution*. Its probability density is

$$f(t; N) = C_N \left(1 + \frac{t^2}{N-1}\right)^{-\frac{N}{2}} \quad (N \geq 2) \tag{7.31}$$

where $C_N$ is a constant that depends on $N$, explicitly given in the appendix, ensuring correct normalization. Figure 7.6 shows Student's distribution for a few values of $N$, compared to a normal distribution.For $N = 2$, obviously the smaller value of $N$ allowing $t$ to be defined

$$f_2(t) = \frac{1}{\pi(1 + t^2)},$$

namely, we obtain a Cauchy distribution. As $N$ increases, the Student distribution "interpolates" between a Cauchy distribution and a normal distribution, with a central

---

[13] Gossett was an employee of the Guinness breweries in Dublin and was forced to use a pseudonym to avoid suspicion of divulging industrial secrets (although it seems difficult to associate Guinness with "small samples"!).

region that approximates better and better a unit-variance Gaussian. Indeed, for $t \ll \sqrt{N}$ we have

$$\ln\left[\left(1 + \frac{t^2}{N-1}\right)^{-N/2}\right] = -\frac{N}{2}\ln\left(1 + \frac{t^2}{N-1}\right) \simeq -\frac{Nt^2}{2(N-1)} \xrightarrow[N\to\infty]{} -\frac{t^2}{2},$$

hence $f_N(t) \propto \exp(-t^2/2)$. As shown in the figure inset, the tails of the distribution retain for all values of $N$ a power-law decay, $f_N(t) \propto t^{-N}$, so $f_N(t)$ decreases much more slowly than a normal distribution. Nevertheless, the variance of the distribution is defined for all $N > 2$, and is given by

$$\sigma_t^2 = \frac{N-1}{N-3}.$$

Differences between the Student's distribution and the normal distribution are significant for wide confidence intervals. Using table A.3 in the appendix, you can for instance check that a 90% confidence interval estimated with the Student's distribution $f(t; 5)$ is approximately 30% wider than the one estimated with the normal distribution.

**Example 7.6** The apparent angular diameter $\vartheta$ of the Sun is measured using an instrument with a resolution $s_\vartheta \simeq 0.02°$. How many measurements must be made so that a 95% confidence interval is less than $0.02°$ wide?
If we assumed a Gaussian distribution, we would write for $s(\bar{\vartheta})$

$$1.96\, s(\bar{\vartheta}) = \frac{1.96 s_\vartheta}{\sqrt{N}} \le 0.02°,$$

hence $N \ge 4$. However, this result indicates a very small number of measurements is needed, so we better try with the Student's distribution. We are asking to have, with a confidence of at least 95%, $\bar{\vartheta} - 0.02 \le \vartheta \le \bar{\vartheta} + 0.02$. Therefore the value $t_{95}$ that gives a confidence interval of 95% must be less or equal than $\sqrt{N}$. However, table A.3 gives, for $N = 4$, $t_{95} = 1.592\sqrt{N}$, which is too much, whereas an acceptable value $t_{95} \le \sqrt{N}$ is obtained only for $N \ge 7$. Therefore, no less than seven measurements are needed to correctly evaluate a 95% confidence interval.

# Chapter 8
# Data Analysis and Model Testing

*With four parameters I can fit an elephant,*
*and with five I can make him wiggle his trunk*

*J. von Neumann*

Main goal of this chapter is assessing how "good" a set of data we have acquired is, a general objective that of course needs to be tailored to the specific issue we will be confronted with. This will require us to face typical "inverse problems", such as deducing from a limited set of data the probability distribution for a physical quantity that would be obtained in the limit of an infinite repetition of measurements, or comparing a set of data for two or more random variables to a theoretical prediction for the functional relation between them. Are there general guidelines for how to proceed? An answer to this basic question is given by the principle of maximum likelihood.

## 8.1 The Maximum Likelihood (ML) Principle

All along the journey we are embarking on, we will have as a lodestar the principle of *maximum likelihood*. To put it simply, we will assume that a dataset is better the higher the joint probability that we have of obtaining all these results.[1] This vague statement will become much clearer by specifying it to the individual tasks we plan to undertake, but the following example will already show you that the ML principle we have just introduced can be used to derive many results that we have already obtained.

---

[1] Remember that the joint probability density of a set of $N$ *mutually independent* measurements is simply $f(x_1, x_2 \ldots, x_N) \prod_{i=1}^{N} p(x_i)$.





R. Piazza, *An Invitation to Probability and Data Analysis for Physicists*, UNITEXT for Physics, https://doi.org/10.1007/978-3-031-83856-9_8

**Example 8.1** (*Some simple applications of the ML principle*) Suppose that all the data we collected have the same Gaussian probability density

$$f(x_i) = \frac{1}{\sigma\sqrt{2\pi}}\exp\left[-\frac{(x_i - \langle x\rangle)^2}{2\sigma^2}\right].$$

Then, if the data are the results of a series of mutually independent measurements, their joint probability density is simply

$$f(x_i; \langle x\rangle, \sigma) = \frac{1}{\sigma^N(2\pi)^{N/2}}\exp\left[-\frac{\sum_{i=1}^{N}(x_i - \langle x\rangle)^2}{2\sigma^2}\right],$$

where the expectation $\langle x\rangle$ and the standard deviation $\sigma$ are parameters that, according to the ML principle should be chosen so to maximize $f(x_i; \langle x\rangle, \sigma)$. Equivalently, we can maximize the logarithm of $f$, or minimize the *likelihood function*

$$\mathcal{L} = -\ln P = \frac{N}{2}\ln(2\pi) + N\ln\sigma + \frac{1}{2\sigma^2}\sum_{i=1}^{N}(x_i - \langle x\rangle)^2.$$

The minimum[2] is found by setting

$$\frac{\partial\mathcal{L}}{\partial\langle x\rangle} = -\frac{1}{\sigma^2}\sum_{i=1}^{N}(x_i - \langle x\rangle) = 0$$

$$\frac{\partial\mathcal{L}}{\partial\sigma} = \frac{N}{\sigma} - \frac{1}{\sigma^3}\sum_{i=1}^{N}(x_i - \langle x\rangle) = 0,$$

which give

$$\langle x\rangle = \frac{1}{N}\sum_{i=1}^{N}x_i = \bar{x}$$

$$\sigma^2 = \frac{1}{N}\sum_{i=1}^{N}(x_i - \langle x\rangle)^2.$$

Thus, we come across a result that we already found, i.e., the best choice for the $\langle x\rangle$ and $\sigma_x^2$ are the experimental mean and by the sum of the squares of deviations from $\langle x\rangle$ (in turn estimated by the adjusted standard deviation).

---

[2] Of course, this will not be maximum, since $f(x_i; \langle x\rangle, \sigma)$ can be made as small as we like by choosing $\langle x\rangle$ and $\sigma$ badly enough.

If instead we have obtained a set of data $k_1, k_2, ..., k_N$ of a quantity described by a discrete variable that we expect to follow a Poisson distribution, $P(k; a) = a^k e^{-a}/k!$, the likelihood function is

$$\mathcal{L} = -\ln\left[\frac{a^{k_1+k_2\ldots+k_N}e^{-Na}}{k_1!k_2!\ldots k_N!}\right] = Na - \ln a \sum_{i=1}^{N} k_i + \ln\left(\prod_{i=1}^{N} k_i\right),$$

so that the best value for $a$ is given by

$$\frac{\partial \mathcal{L}}{\partial a} = N - \frac{1}{a}\sum_{i=1}^{N} k_i = 0 \Longrightarrow a = \bar{k}.$$

Finally, suppose that $x_1, \ldots, x_N$ are still expected to follow a Gaussian distribution with the same expectation $\langle x \rangle$, but with a standard deviation that can vary, for example because the data have been taken with different precision. What is the best value for $\langle x \rangle$? The joint probability density is now

$$f(x_i; \langle x \rangle, \sigma) = \frac{1}{(2\pi)^{N/2} \prod_i \sigma_i} \exp\left[-\frac{1}{2}\sum_{i=1}^{N}\left(\frac{x_i - \langle x \rangle}{\sigma_i}\right)^2\right],$$

which corresponds to the likelihood function

$$\mathcal{L} = \frac{N}{2}\ln(2\pi) + \sum_{i=1}^{N}\ln(\sigma_i) + \frac{1}{2}\sum_{i=1}^{N}\left(\frac{x_i - \langle x \rangle}{\sigma_i}\right)^2.$$

Thus, the requirement $\partial \mathcal{L}/\partial \langle x \rangle = 0$ yields

$$\langle x \rangle = \frac{\sum_{i=1}^{N} x_i/\sigma_i^2}{\sum_{i=1}^{N} 1/\sigma_i^2},$$

that is, the best estimate of the expectation value is, as we had already established somewhat empirically in the previous chapter, the weighted average.


## 8.2  Chi-Squared Test

When the data have a Gaussian distribution, the principle of maximum likelihood provides a significant quantitative test. Let us indeed compare two series of measurements $\{x_i\}_A$, $\{x_i\}_B$ of the variables $X_1, X_2, \ldots X_r$, associated to $r$ (possibly different) quantities. Thus $(x_i)_A$ and $(x_i)_B$ are the results of two independent measurements of the same variable $X_i$ that has a normal distribution with expectation $\langle x_i \rangle$ and variance

$\sigma_i^2$. If we introduce the standard variables $Z_i$, centered at the origin and with unit variance, the joint probability for each series of measurements can be written

$$P_A[(z_1)_A, (z_2)_A, \ldots, (z_r)_A] = \frac{1}{(2\pi)^{r/2}} \exp\left[ -\frac{\sum_{i=1}^{N}(z_i)_A^2}{2} \right]$$

$$P_B[(z_1)_B, (z_2)_B, \ldots, (z_r)_B] = \frac{1}{(2\pi)^{r/2}} \exp\left[ -\frac{\sum_{i=1}^{r}(z_i)_B^2}{2} \right].$$

The second series of data will then be "worse" than the first if and only if

$$\sum_{i=1}^{N}(z_i)_A^2 > \sum_{i=1}^{N}(z_i)_B^2$$

We then define a new random variable,

$$\chi^2(z_1, \ldots, z_r) = \sum_{i=1}^{r} z_i^2 \tag{8.1}$$

called $\chi^2$ (*chi-squared*), given by the sum of the squares of the $r$ standard variables $Z_i = \sigma_i^{-1}(X_i - \langle x_i \rangle)$, i.e., of, the squared deviations of the variables $X_i$ with respect to their expectation, weighted by the reciprocal of their variance. Then, the ML principle implies that, to assess the quality of normally-distributed data, we can use the following $\chi$-*squared test*:

> The probability of obtaining a result worse than the one we actually obtained is equal to the probability $P(\chi^2 > \chi_0^2)$ of obtaining a value of $\chi^2$ greater than the value $\chi_0^2$ calculated from the measured values.

What do we gain from this different way of looking at the ML principle? Something really important. In fact, if we manage to find the probability density $f(\chi^2)$, the $\chi$-squared test can be quantitatively run by evaluating

$$P(\chi^2 > \chi_0^2) = \int_{\chi_0^2}^{\infty} f(\chi^2)\mathrm{d}(\chi^2). \tag{8.2}$$

### 8.2.1   Degrees of Freedom

Before looking for the probability distribution of $\chi^2$, we take a brief detour to discuss a subtle but important aspect of the analysis of a data sample, which we have neglected so far. A standard measurement consists of collecting a generic number

$N$ of independent data. However, we often need to relate these data to each other to compare the experimental results with a model.

To give a very simple example, suppose we want to compare the number of results $n_k$, obtained for a certain value $k$ of a variable that can assume $r$ distinct values, with the value predicted using the a theoretical distribution $P(k)$. This requires to evaluate $N P(k)$. However, the total number of data, $N = \sum_{k=1}^{r} n_k$, is obtained as a linear combination of the $n_k$'s, which are not anymore independent. Indeed, any $n_k$ can be obtained from $N$ and the other $r - 1$ data $n_{k' \neq k}$.

In order to obtain $\chi^2$, several of the problems that we plan to discuss require using the expectation or the variance of the data. If there is no model providing a theoretical value for these parameters, they have to be estimated by minimizing $\chi^2$, as we will see in the following examples. Once again, this entails one or more relationships between the data. In general, the number of really independent data decreases by one when we introduce a functional relationship between the results in order to determine a parameter or minimize a quantity. So if $N$ data $x_i$ must satisfy $m$ constraints $f_m(x_1, x_2, \ldots, x_n) = 0$, the number of data that are truly independent, called the number of *degrees of freedom*, is $\nu = N - m$.

The concept of degrees of freedom makes more transparent a result derived rather formally in the last chapter, namely that the standard deviation turns out to be the best estimate of the variance only if we introduce a correction factor. Indeed, to calculate the standard deviation we need the mean, $\bar{x} = (\sum_i x_i)/N$, which linearly relates the $N$ data. The number of really independent data is then $N - 1$. Thus, to assess the width of the distribution, we should not use $N$, but by the number of degrees of freedom $\nu = N - 1$.

Recalling that the mean is also the value that minimizes the sum of the squared deviations, we can also think that the reduction of the number of effective data is the counterpart we have to pay for making $s_x$ as small as possible. Similar considerations are also at the origin of the factor $N - 1$, and not $N$, in the Student's probability density (7.31).

### 8.2.2 Distribution of the Chi-Squared

The distribution of $\chi^2$ for $\nu$ degrees of freedom is more easily found by evaluating the probability density $f(\chi; \nu)$ for its square root,

$$\chi(z_1, \ldots, z_\nu) = \left( \sum_{i=1}^{\nu} z_i^2 \right)^{1/2},$$

where $z_1, \ldots, z_\nu$ are standard Gaussian variables, which can be seen as the magnitude of the position vector in a space of coordinates $\{(z_1, \ldots, z_\nu)\}$. We can use Eq. (A.21) in the appendix to find the surface area of a sphere of radius $\chi$ in a $\nu$-dimensional space,

$$S(\chi) = \frac{2\pi^{\nu/2}}{\Gamma(\nu/2)} \chi^{\nu-1}.$$

Therefore, for a spherical shell of radius $\chi$ and thickness $d\chi$,

$$f(\chi; \nu)d\chi = \frac{e^{-\chi^2/2}}{(2\pi)^{\nu/2}} S(\chi)d\chi = \frac{e^{-\chi^2/2}}{2^{\nu/2-1}\Gamma(\nu/2)} \chi^{\nu-1}d\chi.$$

Hence, since $f(\chi^2; \nu) = f(\chi; \nu)/(2\chi)$, we finally obtain

$$\boxed{f(\chi^2; \nu) = \frac{1}{2^{\nu/2}\Gamma(\nu/2)} (\chi^2)^{\nu/2-1} \exp\left(\frac{-\chi^2}{2}\right)} \tag{8.3}$$

Comparing with Eq. (3.92), we see that the probability density for the $\chi^2$ is nothing but a *gamma distribution* with $x_0 = 2$ and $\alpha = \nu/2$. Therefore, from (3.94) and (3.95),

$$\boxed{\langle\chi^2\rangle = \nu \ , \ \chi^2_{max} = \nu - 2 \ , \ \sigma^2(\chi^2) = 2\nu \ , \ \gamma(\chi^2) = 2\sqrt{\frac{2}{\nu}}} \tag{8.4}$$

Figure 8.1 shows $f(\chi^2; \nu)$ for some values of $\nu$. For $\nu = 1$, the distribution coincides with the probability density for the square of a standard Gaussian that we have found in (4.5). As $\nu$ increases, we expect $f(\chi^2; \nu)$ to approach a normal distribution, because of the CLT. However, Eq. (8.4) shows that the asymmetry of the distribution decreases only as $\nu^{-1/2}$. Thus the approach to a Gaussian shape is quite slow, as better appreciated from the distribution of the scaled variable $\chi^2_\nu = \chi^2/\nu$, shown in the figure inset.

We can then outline a general procedure that uses the $\chi^2$ test to judge the "goodness" of a set of $N$ data:

1. If $m$ parameters must first be fixed, we choose their values so to minimize $\chi^2$, whose value $\chi^2_0$ is then calculated from the experimental data using these parameter values;
2. Using the specific distribution $f(\chi^2; \nu)$, where $\nu = N - m$, we calculate the probability of obtaining a result which is "worse" than the one we actually found.

The table in Section (A.3) of the appendix gives percentage values of the probability $P(\chi^2 > \chi^2_0)$ for $\nu$ between 1 and 10, with $\chi^2_0/\nu$ ranging between 0.1 and 4. As a general criterion, we can state that the agreement between the data and the theoretical model is definitely weak when $P(\chi^2 > \chi^2_0)$ is less than 10%. On the other end, a value as high as $P(\chi^2 > \chi^2_0) = 0.95$ is suspect too. We may have been very lucky, but it could have resulted from a too conservative assessment of the uncertainties, as we will shortly see.

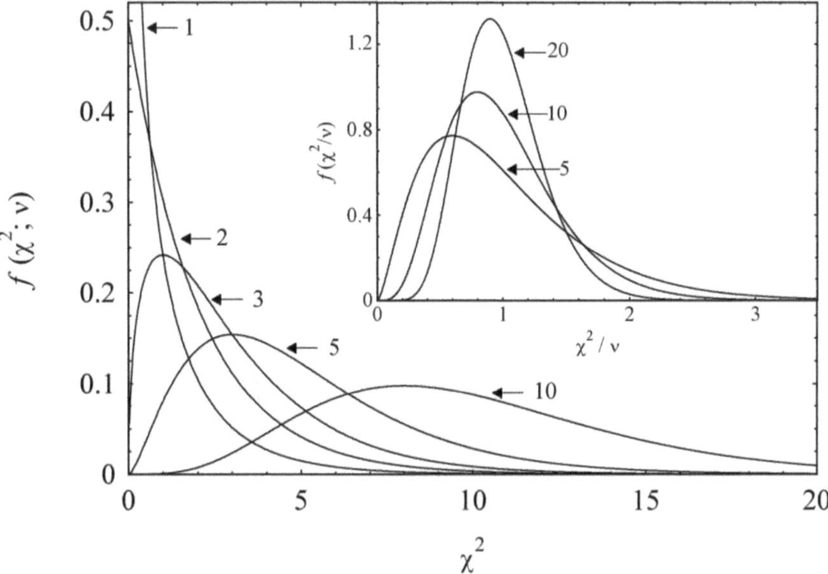

**Fig. 8.1**  Distribution of $\chi^2$ and of the scaled variable $\chi^2/\nu$ for several values of the number of degrees of freedom $\nu$, shown to the side of each curve

## 8.3  Chi-Squared Test for a Distribution

Let us apply the concepts we have just developed to determine how well a theoretical probability density $f(x)$ for a continuous variable $X$ fits a set of $N$ experimental data. Once we have constructed a histogram, made of into subintervals of width $\ell$, we have to compare the number of experimental data $n_k^e$ ranging in the $k$-th subinterval with the theoretical value $n_k^t = Nf(x_k)\ell$, where $x_k$ is a value within that range. Of course, if we repeat the measurement, $n_k^e$ will not be exactly the same: for independent measurements, we rather expect $n_k^e$ to fluctuate around the theoretical value $n_k^t$ according to a Poisson distribution of variance $n_k^t$.

For moderately large expectation values, however, the Poisson distribution is alike a normal distribution, so the variables $n_k^x$ are approximately Gaussian. We can then exploit the chi-squared test to examine whether $f(x)$ adequately describes the data, putting $\sigma_k^2 = n_k^t$ in Eq. (8.1),

$$\chi^2 = \sum_{i=1}^{N} \frac{(n_k^e - n_k^t)^2}{n_k^t} = N\ell \sum_{i=1}^{N} \frac{[f_k) - f(x_k)]^2}{f(x_k)}, \tag{8.5}$$

where we have introduced the (normalized) relative frequencies $f_k = n_k^e/(N\ell)$. Notice that we have been able to attribute a value to the variance of the individual data only because we *know* that the number of experimental data that range in a certain interval of $X$ has a Poisson distribution.

We can apply the same method to a discrete quantity $k$ if we regard it as a continuous variable with a probability density concentrated around the values $k_i$ it takes,

$$f(k) = \sum_i P(k_i)\delta(k - k_i).$$

In this case, using the bare relative frequencies $f(k_i) = n_i^s/N$, the expression for $\chi^2$ becomes

$$\chi^2 = N \sum_{i=1}^{N} \frac{[f(k_i) - P(k_i)]^2}{P(k_i)}. \tag{8.6}$$

**Example 8.2** (*Testing a coin and an old friend with* $\chi^2$) If we flipped a coin 200 times and got 110 heads and 90 tails, what is the probability that it was a fake coin? The calculated value of $\chi^2$ is

$$\chi_0^2 = \frac{(n_H^e - n_H^t)^2}{n_H^t} + \frac{(n_T^e - n_T^t)^2}{n_T^t} = \frac{(110 - 100)^2}{100} + \frac{(90 - 100)^2}{100} = 2.$$

Since the only constraint is $n_H^t + n_T^t = N$, we have $\nu = 2 - 1 = 1$ degrees of freedom. The table in (A.3) gives, for $\nu = 1$,

$$P(\chi^2/\nu > 2) = 15.73\%.$$

Although rather small, this probability value does not allow us to exclude that the coin is fair. However, had we obtained 5 heads more (and 5 tails less), which gives $\chi_0^2 = 4.5$, our conclusions would arguably be different, since[3]

$$P(\chi^2/\nu > 4.5) \simeq 3\%,$$

Let us now pay another visit to our faithful friend $\pi$, trying to decide whether the frequency distribution we obtained in Chap. 1 by analyzing $10^4$ digits is consistent with a uniform probability distribution. Using the last row in table 1.2, we have

$$\chi_0^2 = \frac{10^4}{0.1} \sum_{k=0}^{9} (f_k - 0.1)^2 = 9.1.$$

Here we have $10 - 1 = 9$ degrees of freedom, hence $\chi_0^2/\nu \simeq 1$, which yields $P(\chi/\nu > 1) \simeq 43\%$. Not bad, but not exceptional either.

---

[3] Just do a simple linear extrapolation using the results for $\chi_0 = 3.75$ and $\chi_0 = 4$, as my generation had to do in high school, when there were not pocket calculators, not to say smartphones!.

**Example 8.3** (*Statistics of road accidents*)  On a high-traffic road, the number of days, $n_k$, in which $k$ car accidents take place is given by the following table:

| $k$   | 0  | 1  | 2  | 3 | 4 |
|-------|----|----|----|---|---|
| $n_k$ | 42 | 36 | 14 | 6 | 2 |

If these accidents are independent events, we expect them to follow a Poisson distribution with an expectation that is estimated by the average number of accidents per day,

$$\bar{k} = \frac{1}{N} \sum_{k=0}^{4} k n_k = 0.9.$$

We can then assume as a test distribution

$$P(k; 0.9) = \frac{0.9^k \exp(-0.9)}{k!},$$

estimating the expected days with $k$ accidents as the integer that is closest to $n_k^t = N P(k)$. This yields

$$\chi^2 = \frac{(42 - 41)^2}{41} + \frac{(36 - 37)^2}{37} + \frac{(14 - 16)^2}{16} + \frac{(6 - 5)^2}{5} + \frac{(2 - 1)^2}{1} \simeq 1.5$$

Here we have only 3 degrees of freedom because, besides of the normalization condition $\sum_k n_k = N$, the 5 data are also related by the estimate $\langle k \rangle = \bar{k}$. For $v = 3$, the table in (A.3) gives

$$P(\chi^2 > 1.5) = P_3(\chi^2 v > 0.5) \simeq 68\%,$$

which is a very good result. However, the procedure we followed is not entirely correct. Indeed, 4 accidents were observed only in 2 days, which is a bit too low to approximate the Poisson distribution of $n_4$ with a Gaussian distribution. A more reliable estimate is obtained by grouping the data for $k = 3$ and $k = 4$. Thus we have a total of $n_3 + n_4 = 8$ data points in the interval $3 \le k \le 4$, to be compared with $N[P(3; 0.9) + P(4; 0.9)] \simeq 7$. This yield

$$\chi_0^2 = \frac{(42 - 41)^2}{41} + \frac{(36 - 37)^2}{37} + \frac{(14 - 16)^2}{16} + \frac{(8 - 6)^2}{6} \simeq 0.97$$

with just 2 degrees of freedom, so that $\chi_0^2 / v = 0.485$. From the table in (A.3)

$$P(\chi_0^2/v > 0.4) = 67.03\% \, ; \ P(\chi_0^2/v > 0.5) = 60.65\%.$$

A linear interpolation yields $P(\chi_0^2/v > 0.485) \simeq 61.6\%$, which is slightly less than the result we obtained previously.

**Example 8.4** (*The Rutherford-Chadwick-Ellis Experiment*) Let's revisit the landmark experiment on radioactive decays that was discussed in Example 3.12, whose results seem to be very well accounted for by a Poisson distribution with an average number of detected events $\bar{k} = 3.87$ in a time interval of 7.5 s. The number of times $n_k^e$ in which $k$ events were detected on a total of $N = 2608$ decays is compared in the following table to the theoretical expectation,

$$n_k^t = N P(k; \bar{k}) = 2608 \frac{(3.87)^k \exp(-3.87)}{k!},$$

approximated to the nearest integer value:

| $k$ | 0 | 1 | 2 | 3 | 4 | 5 | 6 | 7 | 8 | 9 | 10 | 11 | 12 | 13 | 14 |
|---|---|---|---|---|---|---|---|---|---|---|---|---|---|---|---|
| $n_k^e$ | 57 | 203 | 383 | 525 | 532 | 408 | 273 | 139 | 45 | 27 | 10 | 4 | 0 | 1 | 1 |
| $n_k^t$ | 54 | 210 | 407 | 525 | 508 | 394 | 254 | 140 | 68 | 29 | 11 | 4 | 1 | 1 | 1 |

Treasuring what we have learnt in the previous example, we discard the data with $k > 11$ for which $n_k^e < 4$, so that the number of degrees of freedom is $v = 12 - 2 = 10$. We have

$$\chi_0^2 = \sum_{i=1}^N \frac{(n_k^e - n_k^t)^2}{n_k^t} \simeq 12.9,$$

so that $\chi_0^2/v \simeq 1.29$, which, interpolating again between the values 1.2 and 1.3 in the table in (A.3), gives $P(\chi_0^2/v > 1.29) \simeq 23\%$, which is a very good result.

### 8.3.1 Maximum Likelihood or Maximum Entropy?

If you recall the applications of statistical entropy presented in Chap. 4, you may feel a bit perplexed when comparing the principle of maximum entropy to the ML method we have just introduced. Let's clarify the issue. We used the principle of maximum likelihood to assess, using the chi-squared test, the reliability of a specific distribution assumed for a set of experimental data, and to provide an estimate of its parameters. Determining the "best" probability distribution suggested by the data would be a much more ambitious task. For example, considering a discrete quantity that can take $r$ values, this is certainly hopeless if the number $N$ of data points is less than $r$, but in fact any reliable method would actually require $N \gg r$, because data suffers from noise. For a continuous variable $X$, an exact determination of the density $f(x)$ from a finite number of data is clearly mission impossible.

The principle of maximum entropy cleverly bypasses this complex inverse problem by stating *a priori* that the most "reasonable" probability distribution which satisfies some requirements must have a maximal entropy, that is, it is the distribution that can be obtained in the most diverse ways among those compatible with these constraints. On the other hand, the ML principle, by providing an assessment on the likelihood of an assumption about a distribution, has a "Bayesian scent" too. Is there no relationship at all between the likelihood function and the statistical entropy?

As a matter of fact, the two criteria are not as different as they seem. Consider indeed a random variable $k$ that can take $r$ values $k_j$, whose probability distribution, $P(k; a)$, depends on a parameter $a$ that we want to estimate. We can write the joint probability of obtaining set of $N$ independent results as

$$P(k_1, k_2 \ldots, k_r; a) = \prod_{j=1}^{r} P(k_j; a)^{n_j},$$

where $n_j$ is the number of times the value $k_j$ has been obtained,[4] with $\sum_j n_j = N$. According to the ML principle, the best estimate of $a$ is found by minimizing the likelihood function $\mathcal{L}(a) = -\ln[Pk_1, k_2 \ldots, k_r; a)]$, or, equivalently,

$$\Sigma_N(a) = \frac{\mathcal{L}(a)}{N} = -\frac{1}{N} \sum_{j=1}^{r} n_j \ln P(k_j; a) = -\sum_{j=1}^{r} f(k_j) \ln P(k_j; a).$$

In the limit $N \to \infty$ the experimental frequencies $f(k_j)$ approach the probability values $P(k_j; a_0)$, where $a_0$ is the value of $a$ that best fits the data,

$$\Sigma(a) = \lim_{N \to \infty} \Sigma_N(a) = -\sum_{j=1}^{r} P(k_j; a_0) \ln P(k_j; a).$$

Subtracting from $\Sigma$ the entropy of the distribution, $S = -\sum P(k_j; a_0) \ln P(k_j; a_0)$, evaluated for the best parameter estimate $a = a_0$, we have

$$\Sigma(a) - S = -\sum_{j=1}^{r} P(k_j; a_0) \left[\ln P(k_j; a) - \ln P(k_j; a_0)\right] = -\sum_{j=1}^{r} P(k_j; a_0) \ln \frac{P(k_j; a)}{P(k_j; a_0)}.$$

Observing that

$$-\ln \frac{P(k_j; a)}{P(k_j; a_0)} \geq 1 - \frac{P(k_j; a)}{P(k_j; a_0)},$$

since $\ln(x) \leq x - 1$ for every $x > 0$ (just check it on a graph), and using the normalization condition,

$$\Sigma(a) - S \geq -\sum_{j=1}^{r} P(k_j; a_0) - \sum_{j=1}^{r} P(k_j; a) = 1 - 1 = 0.$$

---

[4] This amounts to do "statistics by classes", as defined in Chap. 1.

Thus, $\Sigma(a) \geq S$ for every $a$, i.e., the minimum of $\Sigma(a)$ is $a = a_0$, where $\Sigma(a) = S$. At least for large data set, therefore, *the distributions estimated by the ML and the maximum entropy principles coincide.*

## 8.4   Testing, Comparing, or Guessing a Relationship

The problem we want to address is figuring out a functional relationship between two or more physical quantities whose values are measured simultaneously. There could actually be a variety of instances.

A. Theory may suggest that two quantities $X$ and $Y$ are linked by a definite functional relationship $f(x, y; \mathbf{p}) = 0$, where $\mathbf{p}$ is a set of parameters the function depends on. Our goal will then be finding those values $p_j$ that "fit better" with the experimental results. For example, if the theory predicts that $y = A \exp(-X/x_0)$, we will look for those values of $A$ and $x_0$ that, for the measured values of $X$, provide values of $Y$ closer to those experimentally obtained.

B. It may be the case that two or more distinct theoretical models provide *different* answers about the relationship between $X$ and $Y$, for example $f_1(x, y) = 0$ versus $f_2(x, y) = 0$. Here the target is finding a strategy that allows us to compare the various alternatives proposed by the theory and discriminate between them.

C. Finally, it may even be that there is *no* model that allows us to predict a relationship between the investigated quantities. All we can try to do is finding an *empirical relation* between $X$ and $Y$, hopefully by means of a fairly simple functional link, suggested by the experimental results, which any good theoretical model should be able to justify.

The three situations we have considered present an increasing degree of difficulty. We will see that problem A can be rigorously solved, at least in principle. Case B can be dealt with "decisional tests" that only provide a degree of confidence in a model. The final instance generally requires a certain amount of intuition by the experimentalist, which is hard to quantify. Nevertheless, in all three case we will deal with:

1. $N$ pair of values $(x_i, y_i)$ for $X$ and $Y$, measured in the same experimental conditions (that is, all the factors that determine the values of $X$ and $Y$ must be identical).

2. The uncertainties $\sigma(x_i)$, $\sigma(y_i)$ in each measurements of both $X$ and $Y$, These can either be *directly assessed*, by measuring each pair $(x_i, y_i)$ several times, or *estimated* from the specs of the experimental setup, which is the most common case.

3. One or more fit functions $f(x, y; \mathbf{p})$ we want to test, or whose parameters we wish to work out. Notably, as we will see, the latter task can be accomplished even when the uncertainties $\sigma(x_i)$, $\sigma(y_i)$ are *not* known, at the cost, however, of not being able to estimate how much the chosen relation $f(x, y; \mathbf{p})$ is reliable.

In many cases, however, one of the two variables, for example $X$, is regarded, at least in a first approximation, as an error-free quantity. This means that measurements of $X$ are considered so much more accurate than those of $Y$ that all uncertainties are put on the $y_i$'s, and simply indicated with $\sigma_i$.

## 8.5  The Method of Least Squares

As before, we can view each result $y_i$, corresponding to the value $x_i$ set for $X$, as a specific realization of a random variable $Y_i$ that, still assuming Gaussian noise, has an approximately normal distribution. If the $Y$ is known to be related to $X$ by a function $y = f(x; \mathbf{p})$, the estimated value of $y_i$ will be given by $\hat{y}_i = f(x_i; \mathbf{p})$. The variance $\sigma_i^2$ of the distribution of $y_i$ may instead depend on the $y_i$ and be in general different for different values $y_i$. Our plan is to follow a work program that consists of two tasks

$a_0$) Firstly, we want to find a set of values $\hat{\mathbf{p}}$ of the parameters $p_j$ such that $f(x, \hat{\mathbf{p}})$ is the function that "best describes" our data;

$b_0$) Once the best set $\hat{\mathbf{p}}$ of parameter values has been determined, we want to find a way to judge the "goodness" of the fit obtained.

We know that determining how much the results are "good" requires calculating the chi-squared value

$$\chi^2(\mathbf{p}) = \sum_{i=1}^{N} \frac{(y_i - \hat{y}_i)^2}{\sigma_i^2} = \sum_{i=1}^{N} \frac{[y_i - f(x_i; \mathbf{p})]^2}{\sigma_i^2} \tag{8.7}$$

which of course depends on the values assigned to the parameters $p_j$. Our work program can then be restated as follows

(a) We first determine the set of parameter values $\hat{\mathbf{p}}$ that minimizes $\chi^2(\hat{\mathbf{p}})$;

(b) Then, will judge the goodness of the fit by evaluating the probability $P(\chi^2 > \chi^2(\hat{\mathbf{p}}))$ of obtaining a value of $\chi^2$ larger than the one obtained from the experimental data and the fitted parameters.

To put it simply, we want to minimize the sum of the squared deviations of $Y$ with respect to the values calculated using $y = f(x; \mathbf{p})$, weighting each deviation with the uncertainty $\sigma(y_i)$. For this reason, the fit procedure we are introducing is called the method of *Least Squares*.

Let's start by working out the first step of our program. If we deal with a function that depends on a single parameter $p$, condition a) simply becomes

$$\left. \frac{d\chi^2(p)}{dp} \right|_{\hat{p}} = 0, \tag{8.8}$$

where we have specified that the derivative must be evaluated in $p = \hat{p}$. When $f$ depends on $r$ parameters, Eq. (8.8) is generalized by the system of equations

$$\left. \frac{\partial \chi^2(\mathbf{p})}{\partial p_j} \right|_{\hat{\mathbf{p}}} = 0 \quad (j = 1, 2, \ldots, r) \tag{8.9}$$

### *8.5.1   Linear Relationships (or Reducible to Them)*

We start by exploiting the Least Square method to find the best estimates for the slope $a$ and the intercept $b$ of a straight line, $y = ax + b$, that interpolates $N$ experimental data $(x_i, y_i)$, assuming that the values $x_i$ can be assumed as error-free. What we look for is sketched in Fig. 8.2, where the vertical bars centered on each $y_i$, usually called "error bars", have a length $2\sigma_i$. Note that the line does not necessarily cut across all the bars. If the distribution of $y_i$'s is Gaussian, we may indeed expect that the fraction of bars intersecting the line is about 2/3.

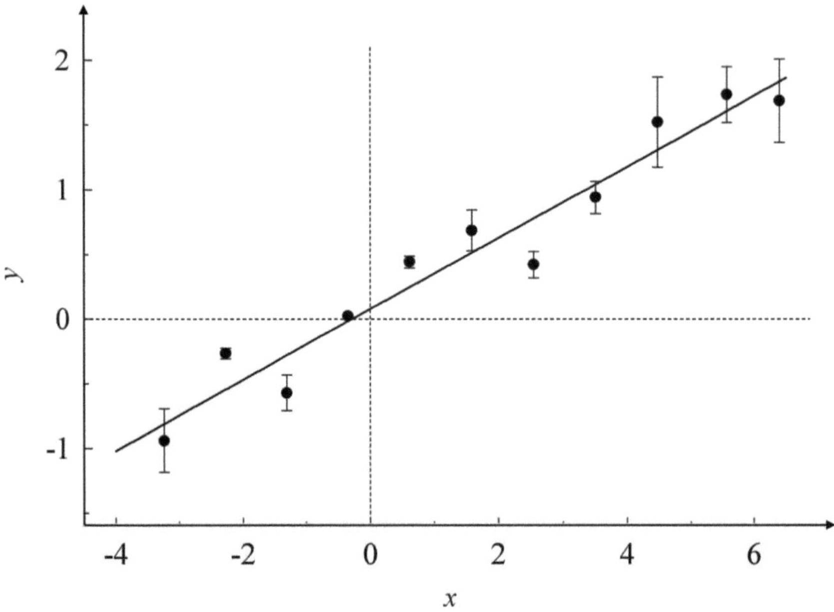

**Fig. 8.2**  Example of a linear best fit. The deviations of the individual points $(x_i, y_i)$ from an exact rectilinear trend have been obtained by sampling a Gaussian distribution with standard deviation $\sigma_i$ equal to half the corresponding error bar

#### 8.5.1.1  Best Line with Equal Uncertainties for All Data

When all the data can be assumed to have the same uncertainty $\sigma$, the expression for $\chi^2$ becomes

$$\chi^2(a, b) = \frac{1}{\sigma} \sum_{i=1}^{N} (y_i - ax_i - b)^2. \tag{8.10}$$

To estimate $\hat{a}$ e $\hat{b}$ we set

$$\left. \frac{\partial \chi^2(a, b)}{\partial a} \right|_{\hat{a},\hat{b}} = -\frac{2}{\sigma^2} \sum_{i=1}^{N} x_i \left( y_i - \hat{a}x_i - \hat{b} \right) = 0$$

$$\left. \frac{\partial \chi^2(a, b)}{\partial b} \right|_{\hat{a},\hat{b}} = -\frac{2}{\sigma^2} \sum_{i=1}^{N} \left( y_i - \hat{a}x_i - \hat{b} \right) = 0, \tag{8.11}$$

which gives

$$\sum_{i=1}^{N} x_i y_i - \hat{a} \sum_{i=1}^{N} x_i^2 - \hat{b} \sum_{i=1}^{N} x_i = 0$$

$$\sum_{i=1}^{N} y_i - \hat{a} \sum_{i=1}^{N} x_i - N\hat{b} = 0.$$

These equations can be written more compactly by putting $\sum_{i=1}^{N} x_i^n y_i^m = N\overline{x^n y^m}$ and dividing both them by $N$. Notice however that the quantities $\overline{x^n y^m}$ are *not* true statistical means, since $X$ is not a random variable, but just arithmetic averages involving products of the values that we have *chosen* for $X$ and the corresponding values obtained for $Y$. Using this convention, we have

$$\begin{cases} \overline{xy} - \hat{a}\overline{x^2} - \hat{b}\bar{x} = 0 \\ \bar{y} - \hat{a}\bar{x} - \hat{b} = 0 \end{cases}$$

wherefrom, solving the system

$$\boxed{\begin{aligned} \hat{a} &= \frac{\overline{xy} - \bar{x}\bar{y}}{\overline{x^2} - \bar{x}^2} \\ \hat{b} &= \bar{y} - \hat{a}\bar{x} \end{aligned}}$$

$$(8.12a)$$

$$(8.12b)$$

or, using the original sums,

$$\hat{a} = \frac{N \sum_{i=1}^{N} x_i y_i - \sum_{i=1}^{N} x_i \sum_{i=1}^{N} y_i}{N \sum_{i=1}^{N} x_i^2 - \left(\sum_{i=1}^{N} x_i\right)^2} \tag{8.13a}$$

$$\hat{b} = \frac{1}{N} \left(\sum_{i=1}^{N} y_i - \hat{a} \sum_{i=1}^{N} x_i\right). \tag{8.13b}$$

How precise are $\hat{a}$ and $\hat{b}$? Writing Eq. (8.12a) as

$$\hat{a} = \frac{1}{N} \sum_{i=1}^{N} \frac{x_i - \bar{x}}{\overline{x^2} - \bar{x}^2} y_i,$$

the slope we obtained can be seen as a linear combination of the values $y_i$. Therefore,

$$\sigma^2(\hat{a}) = \frac{1}{N^2} \sum_{i=1}^{N} \left(\frac{x_i - \bar{x}}{\overline{x^2} - \bar{x}^2}\right)^2 \sigma_i^2 = \frac{\sigma^2}{N^2 \left(\overline{x^2} - \bar{x}^2\right)^2} \sum_{i=1}^{N} (x_i - \bar{x})^2.$$

Thus, since $\sum_{i=1}^{N} (x_i - \bar{x})^2 = N(\overline{x^2} - \bar{x}^2)$,

$$\sigma^2(\hat{a}) = \frac{\sigma^2}{N(\overline{x^2} - \bar{x}^2)}. \tag{8.14}$$

With a similar calculation, one easily finds

$$\sigma^2(\hat{b}) = \frac{\overline{x^2}\sigma^2}{N(\overline{x^2} - \bar{x}^2)}. \tag{8.15}$$

Note that both $\sigma_{\hat{a}}$ and $\sigma_{\hat{b}}$:

1. are proportional to $\sigma$;
2. decrease as $\sqrt{N}^{-1/2}$;
3. decrease as $\overline{x^2} - \bar{x}^2$, which is larger the wider the measurement interval of $X$ (and, remember, has nothing to do with the "variance of $X$").

We can also wonder whether $\hat{a}$ and $\hat{b}$ are independent or if they are correlated. Using noise rescaling, (8.11) yields

$$\sigma^2(\bar{y}) = \bar{x}^2 \sigma^2(\hat{a}) + \sigma_{\hat{b}}^2 + 2\bar{x}\sigma_{\hat{a}}\sigma_{\hat{b}}\rho_{ab}.$$

which, since $\sigma^2(\bar{y}) = \sigma^2/N$, becomes

$$2\bar{x} \frac{\sigma^2 \sqrt{\overline{x^2}}}{N(\overline{x^2} - \bar{x}^2)} \rho_{ab} = -\frac{\sigma^2}{N}.$$

the correlation coefficient between slope and intercept is then given by

$$\rho_{ab} = -\frac{\bar{x}}{\sqrt{\overline{x^2}}} \tag{8.16}$$

*When the center of the measurement interval lies on the negative (positive) x-axis, the slope and intercept values are then positively (negatively) correlated.*

Once the best fit parameters have been determined, we can estimate the value $\hat{y}_i = \hat{a}x_i + \hat{b}$ of $Y$ that we expect to obtain for a value $x_i$ of $X$. The precision of this estimate can be obtained using again noise rescaling and expression (8.15) for $\rho_{ab}$,

$$\sigma^2(\hat{y}_i) = |x_i|^2\sigma^2(\hat{a}) + \sigma^2(\hat{b}) - \frac{2|x_i|\bar{x}}{\sqrt{\overline{x^2}}}\sigma_{\hat{a}}\sigma_{\hat{b}} \tag{8.17}$$

The last term, which can give a large contribution, vanishes when $\bar{x} = 0$. Therefore, to obtain better estimates of the values $\hat{y}_i$, it is advisable to pick a set of values for $x$ that is as centered around the origin as possible.

The expressions (8.14) and (8.15) for the errors on the parameters are however correct only for a sufficiently large number of pairs of experimental data. That $\sigma(\hat{a})$ and $\sigma(\hat{a})$ have a well-defined value even for $N = 2$ sounds rather strange, because two point define a unique straight line with no uncertainties in its slope or intercept! Once again, this inconsistency is removed by replacing $N$ with the number $\nu$ of degrees of freedom. Since we have introduced two relationships[5] between the data pairs of data $(x_i, y_i)$, we have $\nu = N - 2$. Calling $s(\hat{a})$, $s(\hat{b})$ the "corrected" precisions on $\hat{a}$ and $\hat{b}$, we have:

$$s^2(\hat{a}) = \frac{\sigma^2}{(N-2)(\overline{x^2} - \bar{x}^2)} \tag{8.18a}$$

$$s^2(\hat{b}) = \frac{\overline{x^2}\sigma^2}{(N-2)(\overline{x^2} - \bar{x}^2)}. \tag{8.18b}$$

### 8.5.1.2 Linear Dependence and Correlation Coefficient $r_{xy}$

What conditions must be met to obtain a *perfect* linear relationship between $x$ and $y$? This implies that, using the fitted values for the slope and intercept,

$$\chi^2(\hat{a}, \hat{b}) = \frac{1}{N}\sum_{i=1}^{N}(y_i - \hat{a}x_i - \hat{b})^2 = 0.$$

---

[5] In general, the degrees of freedom reduce to $N - r$ for a fit function with $r$ free parameters.

Substituting $a$ and $b$ from Eq. (8.12) one obtains, after some algebra,

$$(\overline{y^2} - \bar{y}^2)(\overline{x^2} - \bar{x}^2) - (\overline{xy} - \bar{x}\bar{y})^2 = 0,$$

which implies that the correlation coefficient $r_{xy}$ must satisfy

$$\boxed{r_{xy}^2 = \left(\frac{s_{xy}}{s_x s_y}\right)^2 = 1}$$

Therefore, as anticipated in Chap. 1, *there is an exact linear relationship between x and y if and only is they are fully correlated or anticorrelated.* Notice also that Eq. (8.12a) can be written

$$\hat{a} = r_{xy}\frac{\Delta y}{\Delta x}, \tag{8.19}$$

where $\Delta x = (\overline{x^2} - \bar{x}^2)^{1/2}$ and $\Delta y = (\overline{y^2} - \bar{y}^2)^{1/2}$ estimate the width of intervals of values for $X$ and $Y$.

**Example 8.5** (*Temperature profile along a heated bar*)  This example may teach us how delicate it can be to assign uncertainties correctly, even in simple cases. Consider a metal bar of length $L = 1\,\text{m}$ that connects hot water at an unknown temperature $T_h$ with a reservoir containing melting ice at $T_c = 0°\text{C}$. The entire system is thermally isolated from the external environment. In the bar are inserted five temperature probes, with a stated accuracy of $\pm 0.5°\text{C}$, that measure the local temperature at fixed distances $d_i$ from the hot source. When the temperature profile along the bar reaches a steady state, the probes give the following readings

| $d$ (m) | 0.2 | 0.35 | 0.5 | 0.65 | 0.8 |
|---------|------|------|------|------|------|
| $T$ (°C) | 54.5 | 45.5 | 35.5 | 25.0 | 16.0 |

If the bar is homogeneous, we expect the steady-state temperature profile to be linear, $T(d) = \nabla T d + T_h$, where $\nabla T = (T_h - T_c)/L$. Using the table,

$$\bar{d} = 0.5\,\text{m}\ ;\ \overline{d^2} = 0.295\,\text{m}^2\ ;\ \overline{T} = 35.3°\text{C}\ ;\ \overline{dT} = 14.725°\text{C}\,\text{m}$$

Therefore, the best estimates for the temperature gradient and for the temperature of the hot source, respectively the slope $a$ and the intercept $b$ of the linear fit, are

$$\hat{a} = -65.0°\text{C}\,\text{m}^{-1}\ ;\ \hat{b} = 67.8°\text{C}.$$

For $s(\hat{a})$ and $s(\hat{b})$, with $\nu = 5 - 2 = 3$, we obtain

$$s(\hat{a}) = 1.36°\text{C}\,\text{m}^{-1}\ ;\ s(\hat{b}) = 0.74°\text{C}$$

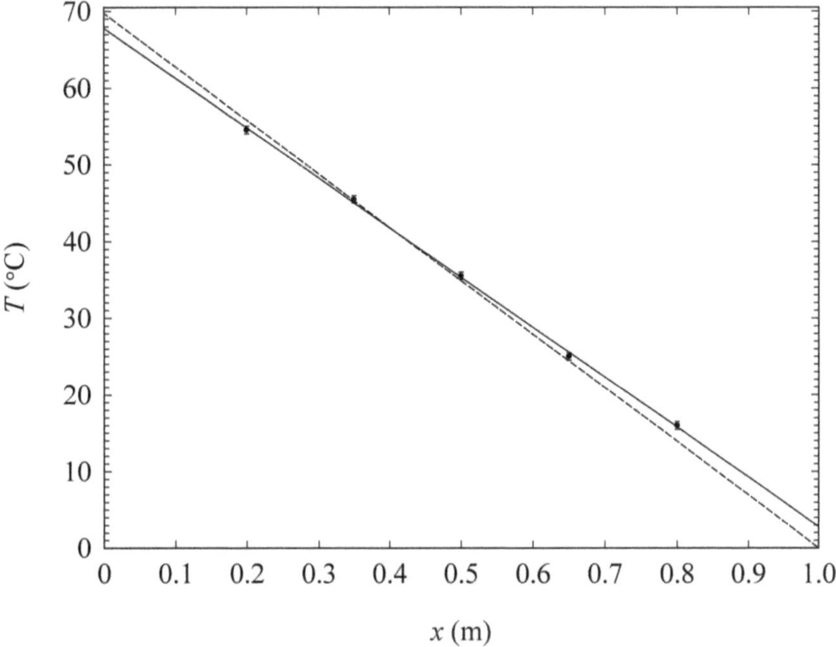

**Fig. 8.3** Best fit to the temperature profile along rod heated at one end and in contact with melting ice at the other using a straight line (full line) or a simple proportion (broken line)

so we can estimate

$$\nabla T = -65 \pm 1\,^{\circ}\mathrm{C\,m^{-1}}\ ;\ \ T_h = 67.8 \pm 0.7\,^{\circ}\mathrm{C}.$$

The full line in Fig. 8.3 seems to indicate that the fit is very good. Using $\hat{a}$ and $\hat{b}$, however, the temperature of the cold source is estimated to be $T_c = 2.8\,^{\circ}\mathrm{C}$, which is highly suspicious. The water/ice mixture acts in fact as a temperature marker, a physical information that we *cannot* disregard.

A straightforward way to take it into account is to impose that the straight line passes through the point $(d = 1\,\mathrm{m}, T = 0\,^{\circ}\mathrm{C})$, which is equivalent to choose as a fit function a simple proportion, $T = ax$, where $x = 1 - d$ is the distance from the *cold* source, and $a = \nabla T$ again. Minimizing $\chi^2$ with respect to the single parameter $a$, one easily finds (and you may want to check)

$$a = \frac{\overline{xT}}{\overline{x^2}}\ ;\ \ s^2(a) = \frac{\sigma^2}{\nu\overline{x^2}},$$

where now $\nu = N - 1 = 4$. We then obtain $a = 69.75$, $s(a) = 0.46$, and therefore $T_h = 69.8 \pm 0.5$, which is $2\,^{\circ}\mathrm{C}$ higher than the previous estimate.

The fit with a simple proportion, shown in Fig. 8.3 as a broken line, looks definitely poorer than the straight line fit, but it suggests the occurrence of deviations that are correlated with the position along the bar. Indeed, the temperatures measured close to the hot/cold end of the bar are respectively lower/higher than predicted by the fit. A possible explanation is the presence of a thermal resistance between the probes and the bar due to an imperfect thermal contact, which may indeed produce a biased reading with these features (try to figure out why). Thus, it may not have been wise to assume the accuracy of all the probes was $\pm 5\,°C$, without considering the potential spurious effects from their assembly.

### 8.5.1.3   Straight Line with Variable Measurement Accuracy

In laboratory practice, it is often the case that the larger the measured value, the greater its uncertainty, which means that standard deviations of the data are not constant. Therefore, the conditions (8.11) have to be modified as

$$
\begin{cases}
\sum_{i=1}^{N} \dfrac{x_i(y_i - \hat{a}x_i - \hat{b})}{\sigma^2(y_i)} = 0 \\[2ex]
\sum_{i=1}^{N} \dfrac{y_i - \hat{a}x_i - \hat{b}}{\sigma^2(y_i)} = 0,
\end{cases}
$$

which makes the calculations a bit more tedious. It is not hard to see, however, that the results (8.12) for $\hat{a}$ and $\hat{b}$ remain *unchanged* provided that all the quantities $\overline{x^n y^m}$ are regarded as weighted averages, and that in the expressions for $\sigma(\hat{a})$ and $\sigma(\hat{b})$ we put $\sigma^2 = N[\sum_i 1/\sigma^2(y_i)]^{-1}$.

### 8.5.1.4   Fit Functions that Can Be Converted to a Linear Form

The results we have obtained can be extended to the more general class of fit functions that can be converted to a linear relationship with a transformation of variables. Let us consider the two simplest and most common cases.

- When the test function is a *power law*, $y = Ax^\alpha$, we can write

$$
\ln y = \alpha \ln x + \ln A.
$$

  In other words, new variables $\ln X$ and $\ln Y$ are linearly related. The slope of the straight line is the exponent of the power law, and its intercept is the logarithm of the amplitude $A$. Changing variable from $Y$ to $\ln Y$, however, the uncertainties of the data points are modified too. To estimate the precision of the "transformed" experimental points $(\ln x_i, \ln y_i)$ we have indeed to set

$$\sigma^2(\ln y_i) = \frac{1}{y_i^2}\,\sigma^2(y_i).$$

So, if the original data have all the same uncertainty, this is not the case for their logarithms. Finally, we have $\sigma(\hat{A}) = |A|\sigma(\hat{b})$, with $\sigma(\hat{b})$ given by (8.15).

- For an *exponential* fit function, $y = A\exp(\pm x/x_0)$, taking again the logarithms of both members we have

$$\ln y = \pm\frac{x}{x_0} + \ln A,$$

which is still a linear relationship between the variables $X$ and $\ln Y$ with a slope $x_0^{-1}$ and an intercept given by the logarithm of the amplitude $A$.

In both cases, the particular form for the uncertainties of the new variable $\ln(Y)$ often simplifies the treatment. If, for example, the measurement precision $\sigma(y_i)$ is approximately *proportional* to $y_i$, maybe because it is proportional to the full scale of an instrument, then the accuracy on $\ln Y$ is *constant*. Hence, we can use the results (8.13) with no need for weighted averages.

**Example 8.6** (*Brownian diffusion in particle mixtures*) This example, taken from a real experiment, may teach us how useful can be a change of variable that transform a nonlinear function into a linear relationship. We have seen in Chap. 5 that the diffusion coefficient of a colloidal particle in a simple fluid (the "solvent"), which can be accurately measured using Dynamic Light Scattering (DLS),[6] can be written as $D_0 = k_B T/f$. In particular, for a spherical particle of radius $a_t$ the friction coefficient is given by $f = 6\pi\eta_0 a_t$, where $\eta_0$ is the fluid viscosity coefficient.

But what happens if the solvent is itself a suspension of spherical colloidal particles of radius $a_h$? The diffusion coefficient $D$ of the "tracer" particle is reduced with respect to $D_0$ by an amount that depends on the fraction $\phi$ of the total volume occupied by the "host" particles. At first order in $\phi$ one has indeed $D = D_0(1 - \kappa\phi)$, where $\kappa$ depends only on the size ratio $r = a_t/a_h$.

Two limiting values for $\kappa$ are easily spotted. Indeed, when the host particles are very small compared to the tracer, i.e., when $r$ is very large, we can reasonably assume that, as for an additive of molecular size their effect is just increasing the bare solvent viscosity $\eta_0$ according to a very general formula obtained by Einstein,[7] $\eta = \eta_0(1 + 2.5\phi)$, which, as you see, does not depend on the host particle size but only on their volume fraction. In the opposite limit $r \ll 1$, the host particles can be viewed as a macroscopic porous medium with huge cavities, allowing for a unhindered diffusion of the tracer, thus $D = D_0$.

An accurate result, $\kappa = 1.832$, can also be found for $r = 1$, which corresponds to the self–diffusion of a tracer particle in a "sea" of identical particles. Finding the value of $\kappa$ for a generic value of $r$, however, is a very challenging task that was carried out by some of the best theorists in particle hydrodynamics. Although no analytical

---

[6] See B. J. Berne and R. Pecora, *Dynamic Light Scattering*, Wiley-Interscience (1976).

[7] A. Einstein, *Ann. Phys.* **339**, 591 (1911).

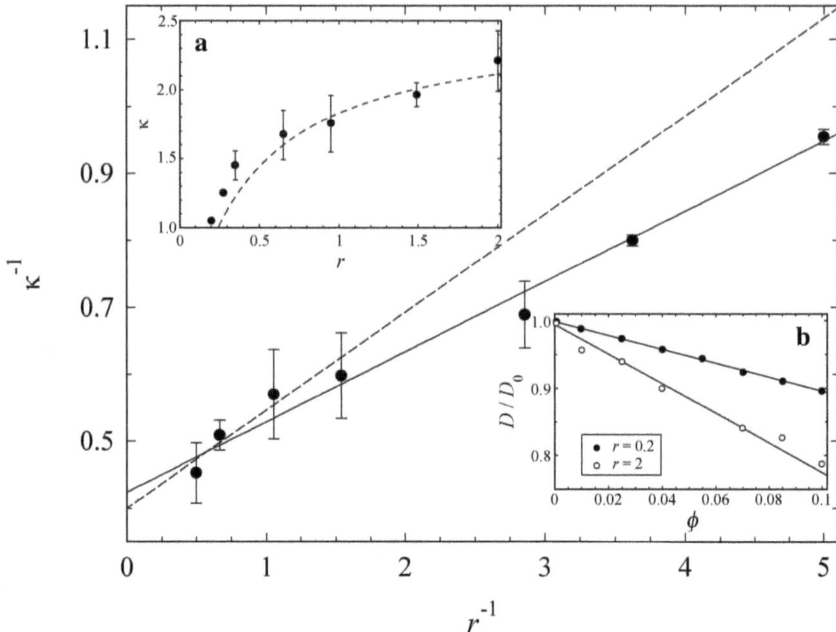

**Fig. 8.4** Inset A: Experimental dependence of the coefficient $\kappa$ on the size ratio $r$ compared with Eq. (8.20) (broken line). The same data are re-plotted in the figure body after the transformation into a linear relation described in the text. The full line is best linear fits to the data. Inset B shows two examples of the linear fits yielding the values for the coefficient $\kappa$

form for $\kappa(r)$ can be found, the numerical results seem to be quite well fitted by the empirical expression

$$\kappa(r) = \frac{c_1}{1 + c_2/r},\tag{8.20}$$

where $c_1 = 2.5$, to comply with Einstein's result, and $c_2 \simeq 0.37$.

Measuring $D$ by DLS is at least as challenging as obtaining its value analytically or numerically. A major issue is distinguishing the scattering contribution of the tracers, which is extremely weak, from that of the much more concentrated host particles. This is particularly hard for small values of $r$, since the scattered intensity grows as the *sixth power* of the particle radius.

Recently,[8] this strong limitation was overcome by using as host fluid a suspension of colloidal particles whose refractive index can be accurately matched with that of the solvent, so that they behave as "ghost" particles that scatter practically no light. This allowed the whole range $0.2 \leq r \leq 2$ to be investigated. Inset B in Fig. 8.4 shows that a linear fit to the dependence of $D/D_0$ on the host volume fraction $\phi$ holds up to

[8] V. Ruzzi, S. Buzzaccaro, P. Moretti, and R. Piazza, *J. Chem. Phys.* **161**, 121102 (2024).

the largest investigated value $\phi = 0.1$, and the measured trend of the coefficient $\kappa$ versus $r$ is compared to Eq. (8.20) in inset A using the theoretical values for the coefficients $c_1$ and $c_2$.

From the plot, however, it is not easy to assess if the observed differences are just quantitative (different values for $c_1$ and $c_2$) or whether the functional form (8.20) does not satisfactorily matches the experimental evidence. One could try a nonlinear fit with Eq. (8.20) using one of those routines we will mention in Sect. 8.5.3, which would arguably give a better agreement with the data points. But would this convince you that (8.20) is really the *best* fit function? An easier comparison is obtained by transforming Eq. (8.20) as

$$\kappa^{-1} = \frac{1}{c_1}(1 + c_2 r^{-1}),$$

which is a *straight line* of $y = \kappa^{-1}$ versus $x = r^{-1}$. The body of Fig. 8.4 shows that the experimental data lie reasonably well on a straight line, although the experimental for the slope is about 30% lower than the theoretical prediction. On the other hand, the intercept gives $c_1 = 2.36 \pm 0.07$, which is very close to Einstein's exact value. While it is not easy to spot the origin of the 30% discrepancy between the numerical and the experimental results for $c_2$, one can be rather confident that the empirical function (8.20) yields a rather good description of the dependence of $\kappa$ on $r$.

### 8.5.2  Regression

There is a profound difference between the data collected in the measurement discussed in the example 8.5 and, for instance, the approximately linear relation between height and body weight in Fig. 1.12.[9] In the case of the heated bar, by increasing the precision of the temperature probes and improving the thermal contact between them and the bar, we can expect the experimental points to move closed exact linear behavior predicted by the heat equation. Conversely, the uncertainty of football players height and weight is insignificant, but the data are fluctuating by their very nature: there is no way to lower their spread, which derives from their statistical distribution. To put it differently, instrumental noise can be minimized, but *intrinsic* noise cannot.

Estimating relationships between two quantities that are not consequences of deterministic laws, but simply emerge as common statistical trends is usually known as *regression analysis*. So, finding the best straight line becomes in economic and social sciences a *linear regression* fit. The origin of the word "regression" is curious. This term was originally used by Francis Galton, who observed that stature of children

---

[9] For the latter, the best fit is $w = (0.83 \pm 0.03)\,h - (76 \pm 5)$. It is worth pointing out that, although the data spread is large, the uncertainty in the slope is less than 4% due to the large number of data points.

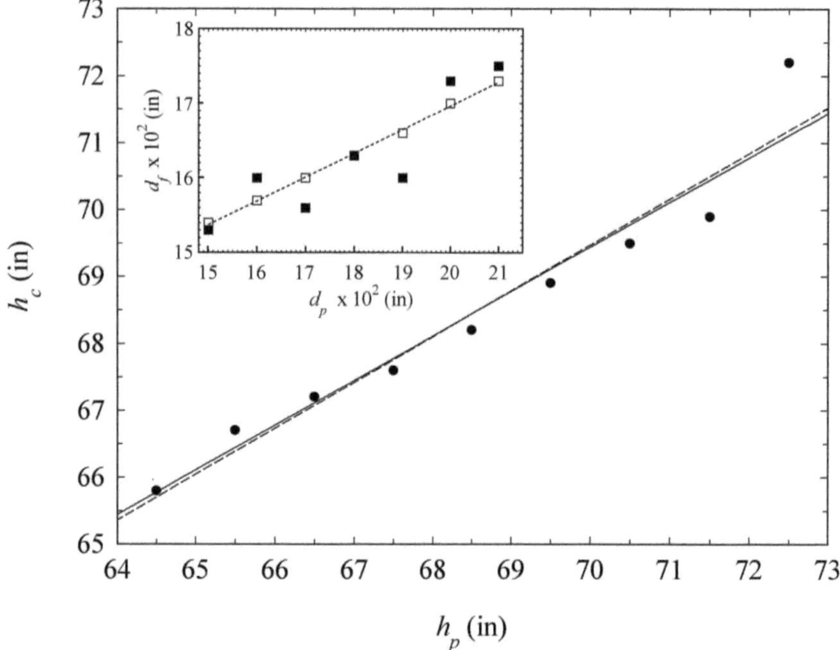

**Fig. 8.5** Galton's data for the median height of the children height $h_c$ versus their mid-parents stature $h_p$. The broken line is the best linear fit to the data, whose slope $s \simeq 0.69$ is pretty close to Galton's suggestion $s = 2/3$ (full line). The inset shows the data for sweet peas filial vs. parent seeds (full squares), together with Galton's linear "smoothing" (open squares). Note that, in both cases, the investigated range is so limited that a high value of the linear correlation coefficient is not surprising

deviate less from the average than that of their parents. Of course, tall/short parents typically beget children that are taller/shorter *than the average*, but Galton pointed out that these children are at the same time statistically shorter/taller *than their own parents*. Galton called this tendency "regression to the mediocrity", although today it is more appropriate to call it *regression to the norm* or *to the mean*.

But what has this to do with linear fits? Well, reading carefully the original paper by Galton[10] one can see that the a linear relation between the medians of children's and parents' height is actually shown (although rather implicitly) in Plate IX a, which Galton properly names "Rate of Regression in Hereditary Stature". Galton's conclusion that "the deviations from the average of the children's height are to the those of their mid-parents as 2 to 3" seems to be pretty well confirmed by the data in Fig. 8.5. In the same paper, Galton also discusses the results of an experiment on sweet peas that his friends grew for him. In Table VI, where he compares the

---

[10] F. Galton, *J. Anthropol. Inst. Great Britain and Ireland* **15**, 246 (1886). Actually, Galton uses what he calls "mid-parents stature", calculated as the average of the parents' height, with the mother's height multiplied by a correction factor of 1.08.

size of the parent seeds, $d_p$, with that of their produce, $d_f$, Galton explicitly uses a linear approximation to provide "smoothed values" for the mean diameter of the filial seeds (see the inset in Fig. 8.5). But it was Karl Pearson, the father of statistics applied to biological sciences, who explicitly pointed out the connection between regression rate and slope of a linear trend and introduced the correlation coefficient $r_{xy}$, showing its deep connection with a linear relationship expressed by Eq. (8.19).[11] Person's general concept of linear regression is at the origin of the use in economic and social sciences of the expression "regression analysis" to indicate any statistical method for estimating the functional relationship between a dependent and one or more (error-free) independent variables.

### 8.5.3 *Polynomial and Nonlinear Fits*

The least squares method we used to determine the best straight line is also applicable to a polynomial relationship of degree $r$, $y = \sum_{k=0}^{r} a_k x^k$. The reason why this approach still works is that, although the *functional relation* between $X$ and $Y$ is not linear, the *fit parameters* still appear linearly in the function. Consequently, minimizing the value of $\chi^2(a_0, a_1, \ldots, a_r)$ requires to solve a system of $r$ *linear* equations that, if soluble, has a *single* solution.

Consider, for example, a parabolic fit function, $y = ax^2 + bx + c$. Assuming equal uncertainties for all points and imposing that the derivatives of $\chi^2(a, b, c)$ vanish, one finds that $\hat{a}$, $\hat{b}$, and $\hat{c}$ have to satisfy

$$\begin{cases} \sum_i x_i^2 y_i - \hat{a} \sum_i x_i^4 - \hat{b} \sum_i x_i^3 - \hat{c} \sum_i x_i^2 = 0 \\ \sum_i x_i y_i - \hat{a} \sum_i x_i^3 - \hat{b} \sum_i x_i^2 - \hat{c} \sum_i x_i = 0 \\ \sum_i y_i - \hat{a} \sum_i x_i^2 - \hat{b} \sum_i x_i - \hat{c} = 0 \end{cases} \tag{8.21}$$

Of course, the higher the degree of a polynomial, the better we can approximate the experimental data.[12] but it is also evident that it becomes increasingly difficult to give a precise meaning to the coefficients of the resulting fit function. Moreover, the more we increase the degree of the polynomial, the more sensitive the fit becomes to small variations in the data. The continuous line in Fig. 8.6 shows, for example, the best fit with a 5th-degree polynomial of 10 experimental points, while the dashed

---

[11] The connection between linear relationship and regression towards the mean can also be seen by noticing that, using Eq. (8.19 and writing $\hat{y}_i = \hat{a}x_i + \hat{b}$, $\bar{y} = \hat{a}\bar{x} + \hat{b}$ for the means, one obtains

$$\frac{\hat{y}_i - \bar{y}}{\Delta y} = r_{xy} \frac{x_i - \bar{x}}{\Delta x}.$$

Since $|r_{xy}| \leq 1$, this means that the fitted values $\hat{y}_i$ are closer to their mean $\bar{y}$ than the $x_i$'s to $\bar{x}$, provides that the differences are scaled to the widths of the intervals $\Delta x$ and $\Delta y$.

[12] Obviously, there is always a polynomial of degree $N - 1$ that passes *exactly* through each one of $N$ data pairs!.

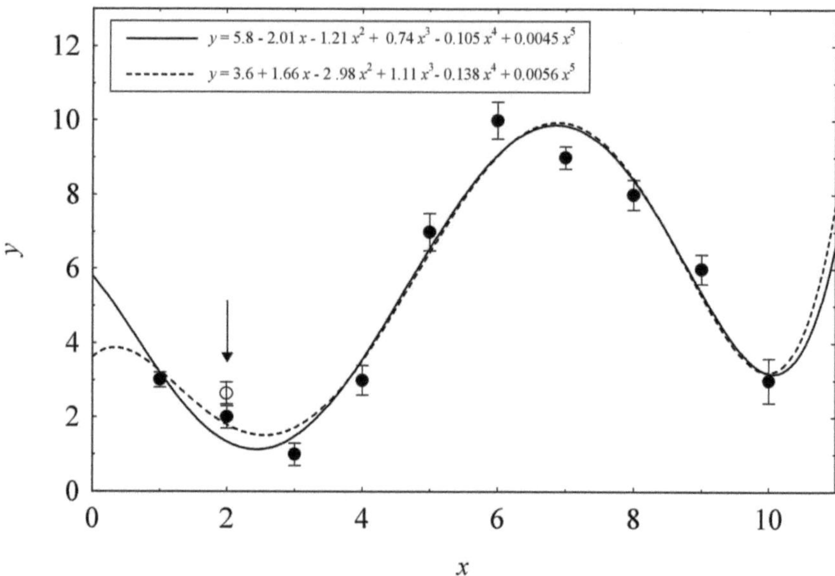


**Fig. 8.6** Effect on the parameters of a 5th degree polynomial fit due to change of the *single* data point shown by the arrow

curve is the fit function obtained by altering only the result shown by the arrow. While the two fits are not dramatically different, the values of the fit coefficients are *very* different, even of opposite sign. Therefore, to obtain meaningful fits, it is worth limiting as much as possible the the number of fit parameters. Unless you are John von Neumann, of course.

We only mention the more general problem of nonlinear fits, that is, attempts to fit a set of experimental data with test functions such as $y = \sin(ax)\exp(-bx)$, where some parameters appear nonlinearly. This gives rise to complications of different nature that make the issue very challenging, because a) the equations obtained by minimizing $\chi^2(\mathbf{p})$ are *nonlinear*, hence generally solvable only numerically, and b) more importantly, $\chi^2(\mathbf{p})$ usually shows *more than a single minimum* in the parameter space. Usually, the numerical methods used to solve linear equations are based on iterative approximations, starting from an initial estimate $\mathbf{p}_0$ of parameter values. However, if $\chi^2(\mathbf{p}_0)$ is close to a secondary minimum, the iterated solution can remain trapped within the "attraction basin" of this minimum, and the fitting algorithm does not spot the absolute minimum of $\chi^2(\mathbf{p})$. Luckily, several effective numerical methods allow these problems us to be overcome, at least when the when a plausible initial estimate of the parameter values is made. Among them, it is worth mentioning the Levenberg-Marquardt algorithm, a method on which most of the nonlinear fitting programs are based.

## 8.6  Chi-Squared Test for a Fit

### 8.6.1  Benefits and Limits of the $\chi^2$ Test

Once we have found the best estimates for the parameter values of the test function, we can ask ourselves how "good" is the fit we made, which means calculating $\chi_0^2$ using the parameter estimates we have obtained and evaluating then the probability of obtaining $\chi^2 > \chi_0^2$. However, this is a bit trickier than comparing an experimental data histogram to a theoretical distribution, where we can we quantitatively estimate the Poisson fluctuations of the number of points that range within an interval of the histogram. Here, even if we expect a normal distribution for the values $y_i$, the width of the Gaussian can be estimated only by repeating the measurement of each pair $(x_i, y_i)$ many times, something which is not usually done.

More commonly, as we already mentioned, the uncertainties $\sigma(y_i)$ are *estimated* from the specs of the used instrumentation. Unfortunately the value of $\chi_0^2$ *crucially* depends on the values $\sigma(y_i)$, which appear in the denominator of the expression that defines $\chi^2$. Increasing the estimated uncertainties, cause a decrease of $\chi_0^2$, so the fit "looks better". Two different attitudes can be adopted:

- A pessimistic or conservative scientist may overestimate the uncertainties. This yields less precise values for the fit parameters, but $\chi_0^2$ will be lower than it should be, and the scientist will be more confident about that goodness of the fit function.
- An optimistic or self-confident scientist can conversely be prone to overestimating the precision of the results by blindly trusting the instrumentation used and the skill of their coworkers. This yields accurate estimates for the fit parameters, but the high value obtained for $\chi_0^2$ may undermine the scientist's confidence in the test function they used.

A proper use of the chi-squared test requires therefore an accurate estimate of the experimental uncertainties. Yet, this does required if our goal is *comparing* two different functional relations between $X$ and $Y$. Using the same uncertainties, the absolute values of $\chi_0^2$ may be wrong for both fits, but these two values can still be compared, and the best functional relation selected.

### 8.6.2  Reversed Chi-Squared Test

The purpose of this section is solving a little mystery. You may already be so accustomed to use a spreadsheet that you may have failed to notice that its software can calculate the best straight line trough a set of data points *without* requiring the uncertainties of the data. And the algorithm does not only provide you with the line slope and intercept, but also with their *accuracy*.

Is this pure magic? No, it is not. Clearly, if we cannot say anything (or anything that can be trusted) about data uncertainties, we won't be able to gather all the information we want about the quality of a fit. But we can still estimate the fit parameters and, in part, their accuracy, provided that we are willing to pay a price: giving up the possibility of deciding whether the fit function we choose is good or not. Namely, we must blindly trust our choice.

As worrying as it may seem, this bold assumption is often largely justified. Suppose, for example, that you want to determine the value of the gravity acceleration $g$ by measuring the oscillation period $T$ of an almost ideal pendulum of length $L$. Evidently, no one will ever convince you to use a function other than $g = 4\pi^2 L/T^2$. Unless goblins or gremlins are around,[13] there is no reason to doubt it. However, if we have a reasonable degree of confidence about $f(x, \mathbf{p})$, we *know* from (8.4) that the expected value of $\chi^2$ is $\nu$. That is, if we assume equal uncertainties for all data and call $E^2(\mathbf{p}) = \sum_i [y_i - f(x_i, \mathbf{p})]^2$ the sum of the quadratic deviations, we expect

$$\chi^2(\mathbf{p}) = \frac{E^2(\mathbf{p})}{\sigma^2} \simeq \nu,$$

We can then give an a posteriori estimate of the uncertainties on the data by summing the experimental quadratic deviations and choosing

$$\sigma \simeq \frac{E^2}{\nu} \tag{8.22}$$

value that can then be used to estimate the accuracy of the parameters. It is clear that this "reversed $\chi^2$ test works only if all the data have approximately the same precision. Besides, this is just a rough estimate, since $\chi^2$ is not a deterministic quantity with a single value $\chi^2 = \nu$, but a random variable with a rather wide distribution.


## 8.7   Hypotheses Non Fingo

By presenting you with the $\chi$-squared method for testing the goodness of a fit, I have sneakily taken you into the tangled forest of hypothesis testing. Following the Prince of Physics—who actually made many of what we call today "hypotheses"— I'd really do without it. Nevertheless, I feel it is my duty to spend a few words about this important branch of statistics, at least from the perspective of a physicist.

If you have taken a look at Chap. 6, you already know that this subject is eminently Bayesian. In the abstract-measure probability theory, there is no room for estimating "how good" is a conjecture, which is a meaningless question. In the Bayesian approach, conversely, evaluating the level of our confidence in an interpretation of the experimental data, or comparing two different assumptions, is a major issue. Once we remember that conditional probability can be considered as a geometric

---

[13] Or a mischievous teacher, who has concealed a lodestone under the table over which the pendulum iron blob oscillates.

projection, without any reference to an individual's information, this is not necessarily the realm of subjectivists. However, as extensively discussed in Chap. 6, the "hypotheses space" is a rather fuzzy concept. So, we better proceed cautiously.

The first thing to note is that there is no way to prove that a conjecture is true beyond any doubt.[14] The only thing we can rigorously do is to *disprove* a hypothesis, which according to Popper is the distinctive trait of scientific laws. But can we gauge our level of confidence in an assumption, even if we cannot guarantee it is absolutely true?

The first thing to point out is that hypotheses can either be simple or composite. A simple hypothesis is, for example, stating that a random variable $X$ is such that $\langle x \rangle = 0$. This is a self-contained statement: nothing else has to be specified. On the contrary, $f(x) = \lambda \exp(-\lambda x)$ is a composite hypothesis, because the decay rate $\lambda$ is not specified: it is just a general assumption on the functional form probability density for $X$. Simple hypotheses are much easier to deal with, but they have limited practical value. However, hypothesis testing is built on tests for simple hypotheses. And there is a simple, I'd say elementary test that is crucial in science.

### 8.7.1 Is There a Signal? The Null Hypothesis

In statistics, the best way to gain confidence in a hypothesis is to first assume the opposite and then show that it is not true. Namely, to try and show that something is there, you have to try, and fail, to assert the opposite. This is called testing the *null hypothesis*, usually indicated with $H_0$, and is particularly useful for dichotomic statements—either this or that, *tertium non datur*. As we said, however, there is no way to be sure that your assumption is absolutely right, but a high confidence level can still be reached if $P(H_0) \ll 1$.

For example, suppose that you suspect that a coin is biased, and that you flip it ten times to test. Your hypothesis is that the probability $p$ of heads (or tails) in a single trial is not $1/2$, therefore $H_0$ means assuming $p = 1/2$. Using the binomial distribution we find that the probability of obtaining 8 or more heads is $P(k \geq 8) = 2^{-10} \left[ \binom{10}{8} + \binom{10}{9} + \binom{10}{10} \right] \simeq 0.099$, and the same of course for the probability of obtaining 8 or more tails. Therefore, if you get 8 or more results of the same kind, the probability that the coin is fair is less than 20%, so you can argue that, to this confidence level, it is a trick coin.

In many cases, however, there is not a dichotomic choice between two options, so you cannot state that a specific assumption is valid. However, if the null hypothesis turns out to be, to a high degree of confidence, false, then you can claim that *something* is there, even if you don't know what it is. Suppose for instance that a region of the Milky Way is imaged by a telescope on the sensor of a CCD camera. Each pixel

---

[14] This seems to be crucial for a jury to deliver a guilty verdict, which makes me question why anyone has ever been convicted.

of the camera detects an average background noise $\bar{\epsilon}$ of 25 photons/second. You observe that one specific pixel gives a reading of 40 photons/second. Is this just a fluctuation, or indicates the presence of a feeble light source? Photon detection is a Poisson process, so we expect $\sigma_\epsilon = \sqrt{25} = 5$ photons/second, so the observed fluctuation exceeds the average background noise $\langle\epsilon\rangle$, estimated as $\bar{\epsilon}$, by $3\sigma_\epsilon$. The value of $\bar{\epsilon}$ is large enough for using a Gaussian approximation, hence the chance that the observed anomaly is just a fluctuation is only $P(\epsilon > \langle\epsilon\rangle + 3\sigma_\epsilon) = 0.6\%$.

## *8.7.2   The Neyman-Pearson Test*

The simplest way to decide whether we can trust a simple hypothesis $H_1$ is testing it against the null hypothesis $H_0$. Then, if $H_1$ is true, $H_0$ is false, and vice versa. Since the decision is based on a finite sample and not on the entire population, however, there is always the possibility of making a wrong decision. We can actually make two kinds of errors.[15] The first is rejecting $H_0$ if $H_1$ is false. This is called a *Type I error*. We will indicate its probability as[16]

$$\alpha = P(\bar{H}_0 \mid \bar{H}_1) \quad \text{(Type I)},$$

which is called the *significance level* of the test. But we can conversely accept $H_0$ if $H_1$ is true. This is called a Type II error, and we write its probability as

$$\beta = P(H_0 \mid H_1) \quad \text{(Type II)},$$

The probability of correctly *rejecting* $H_0$ when $H_1$ is true, which is of course

$$P(\bar{H}_0 \mid H_1) = 1 - \beta,$$

is called the *power* of the test against $H_1$. The goal of testing $H_0$ versus $H_1$ is *maximizing the power of the test subjected to a significance level of the test of at most $\alpha$*, that is, maximizing $P(\bar{H}_0 \mid H_1)$ keeping $P(\bar{H}_0 \mid \bar{H}_1)$ less or equal to a preset value $\alpha$, typically chosen of the order of a few percent.

This "constrained optimization" problem may seem to be rather abstract. To make it more practical consider a data sample $\mathbf{x} = (x_1, x_2, \ldots, x_n)$ for the continuous variable $X$, and suppose that the hypotheses to compare are

$H_0$: $X$ is distributed with the joint p.d.f. $f_0(\mathbf{x}) = f_0(x_1, x_2 \ldots, x_n)$
$H_1$: $X$ is distributed with the joint p.d.f. $f_1(\mathbf{x}) = f_1(x_1, x_2 \ldots, x_n)$.

---

[15] Here we can safely speak of errors, since *errare humanum est*.

[16] This is not the standard notation used in professional statistics, but it is more in tune with what we have done so far, if by it we mean that is the probability of rejecting $H_0$ (that is, accepting $\bar{H}_0$) if $H_1$ is false ($\bar{H}_1$ true).

If we define the *rejection region* for the test as the region $\mathcal{R} \in \mathbb{R}^n$ such that we reject $H_0$ when $\mathbf{x} \in \mathcal{R}$, the two goal of our test is maximizing

$$\int_R f_1(\mathbf{x})\mathrm{d}x_1 \ldots \mathrm{d}x_n$$

with the constraint

$$\int_R f_0(\mathbf{x})\mathrm{d}x_1 \ldots \mathrm{d}x_n \leq \alpha.$$

How does this define the rejection region $R$? Clearly, we want to increase the power of the test as much as possible, but at the same time have the smallest possible increase of Type I errors. If you think about it, this can be obtained b choosing those points $\mathbf{x}$ that give the smallest value of the quantity

$$L(\mathbf{x}) = \frac{f_0(\mathbf{x})}{f_1(\mathbf{x})},$$

which is called the *likelihood ratio*. A fundamental and general result obtained by Jerzy Neyman and Egon Pearson (son of Karl Pearson) is that a test based on the likelihood ratio is the most powerful test of $H_1$ against $H_0$. More precisely, the *Neyman-Pearson lemma*, which applies to *any* two simple hypotheses $H_0$ (which does not need to be the null hypothesis) and $H_1$, states the following

Suppose that the likelihood ratio test which rejects $H_0$ when $L(\mathbf{x}) < c$ has significance level $\alpha$. Then any other test of $H_1$ against $H_0$ with significance level less or equal to $\alpha$ has at most the power of this likelihood ratio test.

**Example 8.7** (*Likelihood ratio for Gaussians*) Let us give a simple example of a likelihood ratio test. Suppose that we want to contrast two hypothesis on a continuous variable $X$,

$H_0$: $f(x) = g(x; 0, 1)$, i.e., $X$ is a Gaussian with $\sigma = 1$ and $\langle x \rangle = 0$;
$H_1$: $f(x) = g(x; 2, 1)$, i.e., $X$ is a Gaussian with $\sigma = 1$ and $\langle x \rangle = \mu$,

testing them with a data set $\mathbf{x} = (x_1, x_2, \ldots, x_n)$ that we have collected. Here $\mu$ is a *given* value, so $H_0$ and $H_1$ are simple hypotheses because both distributions have no free parameters. The joint p.d.f. of the data under hypothesis $H_0$ is

$$f_0(x_1, x_2, \ldots x_n) = \left(\frac{1}{\sqrt{2\pi}}\right)^n \exp\left(-\frac{x_1^2 + x_2^2 + \ldots + x_n^2}{2}\right),$$

while, under hypothesis $H_1$, we have

$$f_1(x_1, x_2, \ldots x_n) = \left(\frac{1}{\sqrt{2\pi}}\right)^n \exp\left[-\frac{(x_1 - \mu)^2 + (x_2 - \mu)^2 + \ldots + (x_n - \mu)^2}{2}\right],$$

Taking the ratio, expanding the squares, and simplifying, we have

$$L(x_1, x_2, \ldots x_n) = \frac{f_0(x_1, x_2, \ldots x_n)}{f_1(x_1, x_2, \ldots x_n)} = e^{n\mu(\mu/2 - \bar{x})},$$

where $\bar{x}$ is the sample mean. If $\mu < 0$, $L(x_1, x_2, \ldots x_n)$ is a strictly decreasing function of $\bar{x}$, so rejecting $H_1$ when $L(x_1, x_2, \ldots x_n)$ is too small is equivalent to do it when $\bar{x}$ is larger than a upper threshold $c$.

But how do we choose $c$, if we want that, under the hypothesis $H_0$, the significance level of the test is $\alpha$? We know that, if $X$ is distributed according to $f_0(x_1, x_2, \ldots x_n)$, the distribution $f_{\bar{x}}$ of the mean is still a Gaussian centered at the origin, but with a standard deviation reduced by a factor $\sqrt{n}$, so we must choose

$$c = f_{\bar{x}}(\bar{x} = \alpha) = \frac{1}{\sqrt{2\pi}} \exp\left(-\frac{\alpha^2}{2n}\right).$$

When $\mu < 0$, one should conversely choose $x < -f_{\bar{x}}(\alpha)$.

Although it applies just to the comparison of two simple hypotheses, The Neyman-Person lemma provides the bases for more advanced methods. Our necessarily brief survey of the rather complex art of hypothesis testing, however, must come to an end. After all, this is primarily a Bayesian domain, so I would rather refer you to a good textbook like the one by Sivia mentioned in the suggested readings. As the Ecclesiastes and the Byrds say, there is a time for everything, and this is the time to bring this book to an end.

# Suggested Readings


**Entry Level**

- D. Huff, *How to lie with statistics*, W. W. Norton & Co., New York, 1993.
  *An antidote against the often inaccurate, sometimes thoughtless, almost always dangerous use of statistics by journalists, politicians and, above all, influencers (whatever this means).*
- M. J. Moroney, *Facts from figures*, Penguin Books, Harmondsworth, 1990.
  *In the Internet era, a book that requires pen, paper, and a pocket calculator to be followed may look a bit updated. All the way around: it is still and by far one of the best introductory books on statistics.*
- Harold W. Lewis, *Why Flip a Coin? The Art and Science of Good Decisions*, J. Wiley & Sons, New York, 1993.
  *Probably the most effective introduction to decision theory, written, for once, not by a philosopher, an economist, or a social scientist, but by a physicist. In this little gem, Hal Lewis discusses issues such as voting, judging, investing, gambling, using a nontechnical but thoughtful approach.*
- L. Tarasov, *The world is built on probability*, Mir Publishers, Moscow, 2023 (freely available on the web at http://mirtitles.org).
  *Another nice and funny book, aimed to convince the readers that the random world begins directly in their own living room because, in fact, all modern life is based on probability.*
- M. R. Schroeder, *Fractal, Chaos, Power Laws: Minutes from an Infinite Paradise*, W. H. Freeman & Co, New York, 1991.
  *Although somewhat dated, this book remains in my opinion one of the best introductions to scale invariance and self-similarity.*

**Intermediate Level**

- W. Feller, *An Introduction to Probability Theory and its Applications*, Vol. 1, 3rd ed., John Wiley & Sons, New York, 1968.
  *A masterpiece, which cannot be missed on the bookshelves of anyone who wants to seriously deal with probability theory.*



© The Editor(s) (if applicable) and The Author(s), under exclusive license to Springer Nature Switzerland AG 2025

R. Piazza, *An Invitation to Probability and Data Analysis for Physicists*, UNITEXT for Physics, https://doi.org/10.1007/978-3-031-83856-9

- B. V. Gnedenko, *Theory of Probability*, 6th ed., CRC Press, 2020.
  *Generations of Soviet physicists have learned probability on Gnedenko's reference textbook, first published in 1950. It is the Russian counterpart of Feller (but simpler). Hard to choose between them!*
- A. Papoulis, Probability, Random Variables and Stochastic Processes, 4th ed., McGraw-Hill Europe, 2002
  *A comprehensive introduction to the fundamental concepts of probability theory and stochastic processes, and a great classic for applied physicists and engineers.*
- S. Miller and D. Childers, *Probability and Random Processes*, 2nd ed. Academic Press, 2012.
  *A very good introduction to random processes manly addressing students in communication engineering, but surely very useful for physicists too.*
- E. Milotti, *The Physics of Noise*, IOP Concise Physics Morgan & Claypool Publ., 2019.
  *A little, splendid book on the physics and mathematics of noise.*
- K. Jacobs, *Stochastic Processes for Physicists*, Cambridge Univ. Press, Cambridge, 2010.
  *A very good and readable introduction to stochastic calculus.*
- E. T. Jaynes, Probability Theory: the Logic of Science, Cambridge Univ. Press, Cambridge , 2003.
  *As you may have guessed from Chap. 6, Jaynes' approach is not really my thing, but I still think that this book, the result of a life's effort, is surely worth reading.*
- R. B. Griffiths, *Consistent Quantum Theory*, Cambridge University Press, Cambridge, 2008.
  *A deep but mathematically simple introduction to modern quantum mechanics, essential for anyone who wants to understand quantum probability.*
- J. Orear, *Notes on statistics for physicists*, Laboratory for Nuclear Studies, Cornell University
  *A concise, but very useful introduction to data analysis written by a student and coworker of Enrico Fermi. With the benefit of being freely available on* http://nedwww.ipac.caltech.edu/level5.
- R. J. Barlow, *Statistics: A guide to the Use of Statistical Methods in the Physical Sciences*, John Wiley & Sons, Chichester, 1989. *In my opinion, the best available basic textbook on experimental data analysis, particularly for applications to physics.*
- G. Cowan, *Statistical Data Analysis*, Clarendon Press, Oxford, 1998.
  *Slightly more advanced than Barlow's book, this is a classic reference in data analysis. Particle physics provides the majority of the examples, but the discussed tools and methods are applicable to all other fields of physics.*
- D. S. Sivia (with J. Skilling), *Data analysis, a Bayesian tutorial*, Oxford Univ. Press, Oxford, 2006.
  *If you wish to take a look at Bayesian estimation, which I admit I snubbed too much, this is a very well-made introduction.*

## Advanced Level

- W. Feller, *An Introduction to Probability Theory and its Applications*, Vol. 2, 2nd ed., John Wiley & Sons, New York, 1971.
  *If you need to deal with continuous distributions, or look for a rigorous theoretical approach to probability, Feller's second volume is a must-have. But is much harder to digest than the first one...*
- V. Balakrishan, *Elements of Nonequilibrium Statistical Mechanics*, Springer Nature (with ANS Books Pvt.), 2021.
  *Apparently off-topic, but actually a great and deep introduction to random processes in physics.*
- A. Amir, *Thinking Probabilistically: Stochastic Processes, Disordered Systems, and Their Applications*, Cambridge Univ. Press, Cambridge 2021.
  *One of the best books providing a physicist's application-driven approach to random processes. Very nice, but be careful: it's not easy at all.*
- J. M. Bernardo and A. F. M. Smith, *Bayesian Theory*, John Wiley & Sons, 2000.
  *An authoritative books on Bayesian inference and modeling, treated at a pretty advanced level.*
- B. R. Frieden, *Probability, Statistical Optics, and Data Testing*, 3rd ed., Springer-Verlag, Berlin, 2001.
  *An original and thorough treatment of probabilistic methods and data analysis, using a Bayesian approach, with numerous applications in statistical optics.*

# Appendix A
# Mathematical Notes

## A.1 Useful Functions (and False Friends)

### A.1.1 The Gamma Function

The gamma function was first introduced in 1730 by Leonhard Euler with the purpose of generalizing factorials to non integer values. It is defined as[1]

$$\Gamma(x) = \int_0^\infty t^{x-1} e^{-t} dt \tag{A.1}$$

In fact, for $x > 0$, a simple integration by parts yields

$$\Gamma(x+1) = \int_0^\infty t^x e^{-t}\, dt = x \int_0^\infty t^{x-1} e^{-t}\, dt - \left[ t^x e^{-t} \right]_0^\infty = x\Gamma(x), \tag{A.2}$$

which generalizes the usual recursive relation for factorials. Hence, when the argument of the gamma function is an integer $n$ we have

$$\Gamma(n) = (n-1)!,$$

---

[1] Actually, Euler originally defined the gamma function as

$$\Gamma(x) = \int_0^1 (-\ln z)^{x-1} dz,$$

but a simple change of variable $t = -\ln z$ yields the usual form (A.1), which is due to Legendre. Putting instead $\ln z = -t^2/2$ gives the equivalent expression

$$\Gamma(x) = \frac{1}{2^{x-1}} \int_0^\infty t^{2x-1} e^{-t^2/2} dt,$$

a result that we shall find useful to evaluate the moments of a Gaussian function.

© The Editor(s) (if applicable) and The Author(s), under exclusive license to Springer Nature Switzerland AG 2025
R. Piazza, *An Invitation to Probability and Data Analysis for Physicists*, UNITEXT for Physics, https://doi.org/10.1007/978-3-031-83856-9

because, of course, $\Gamma(1) = 1$.

It is also easy to find the values of $\Gamma(x)$ for half–integer numbers. Indeed, using Footnote 1, we have

$$\Gamma(1/2) = \sqrt{2} \int_0^\infty e^{-t^2/2}\, \mathrm{d}t = \sqrt{\pi},$$

where the last equality comes from the value of the integral of a normalized Gaussian (see Sect. A.2), while all the other values follow from the recursive relation (A.2). We have, for instance,

$$\Gamma(3/2) = \sqrt{\pi}/2 \;\; ; \;\; \Gamma(5/2) = 3\sqrt{\pi}/4 \;\; ; \;\; \Gamma(7/2) = 15\sqrt{\pi}/8,$$

and in general, for any positive integer,

$$\Gamma\left(n + \frac{1}{2}\right) = \frac{1 \cdot 3 \cdot 5 \cdots (2n - 1)}{2^n} \sqrt{\pi}.$$

Using Eq. (A.2), written in the form $\Gamma(x) = \Gamma(x + 1)/x$, the definition of $\Gamma(x)$ can be extended to the whole negative axis, with the exception of the negative integers, where it diverges (because $\Gamma(x) \xrightarrow[x \to 0^+]{} +\infty$).[2] A graph of $\Gamma(x)$ for small values of its argument is shown in Fig. A.1. For $x \gg 1$, the Stirling approximation we have introduced for $n!$ is also valid for $\Gamma(x)$,

$$\Gamma(x + 1) \simeq \sqrt{2\pi x}\, x^x e^{-x}. \tag{A.3}$$

### A.1.1.1    The Beta Function

A close relative of the gamma function is the *beta function* (also called the Euler integral of the 1st kind), a function of two positive variables[3] $x_1, x_2$ defined as

$$\mathrm{B}(x_1, x_2) = \int_0^1 t^{x_1 - 1}(1 - t)^{x_2 - 1}\mathrm{d}t, \tag{A.4}$$

which enters in the definition of one of the most important distributions for Bayesian statistics (and bears its name, see Chap. 3). The relation with the gamma function is evident because[4]

---

[2] For the more skilled in mathematics, the gamma function can also be extended to complex values $z$ by analytical continuation: $\Gamma(z)$ is analytic everywhere, except of course in the origin and for negative integer values, where it has simple poles.

[3] Or, more generally of two complex variables with a positive real part.

[4] See, for instance, E. Artin, *The Gamma Function*, Holt, Rinehart and Wiston, New York (1964).

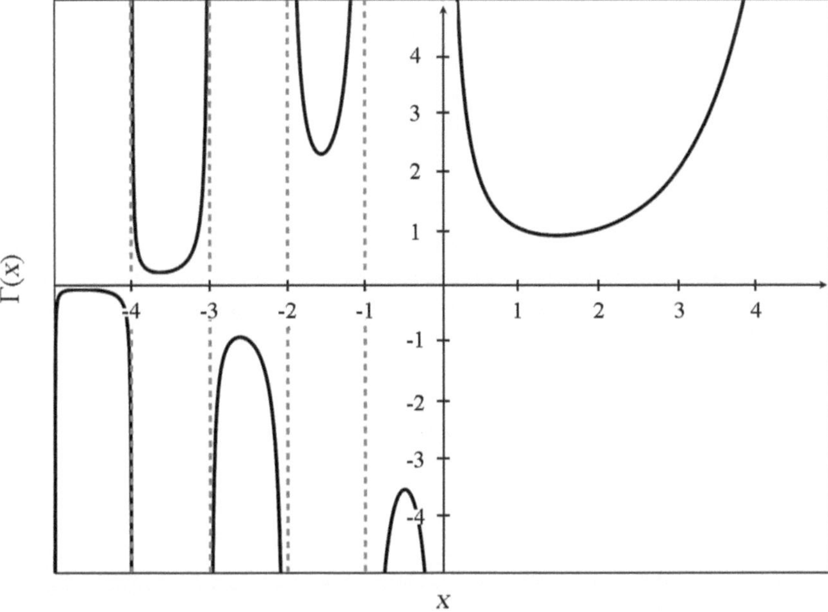

**Fig. A.1** Graph of the gamma function. The dashed lines indicate the vertical asymptotes of $\Gamma(x)$. On the positive axis, $\Gamma(x)$ has a single minimum for $x \simeq 1.462$, where $\Gamma(x) \simeq 0.886$ (very close to the value of $\Gamma(3/2)$)

$$B(x_1, x_2) = \frac{\Gamma(x_1)\Gamma(x_2)}{\Gamma(x_1 + x_2)}. \tag{A.5}$$

Note that, for $x_1 = x_2 = x$, $B(x, x) = \Gamma^2(x)/\Gamma(2x)$. When $x_1$ and $x_2$ are two integers, $n_1$ and $n_2$, we then have

$$B(n_1, n_2) = \frac{(n_1 - 1)!(n_2 - 1)!}{(n_1 + n_2 - 1)!}. \tag{A.6}$$

## A.1.2 The Dirac Delta

The Dirac "delta", which we extensively used in the text, is one of the most powerful mathematical tools not only in physics, but also in engineering, in particular in electronics, telecommunications, and structural mechanics. In fact, although it was already implicitly used by Joseph Fourier and Gustav Kirchhoff, it was an electrical engineer, Oliver Heaviside, who fully understood the extreme usefulness of this mathematical object. To model the response of an electrical circuit when a switch (a telegraph key) was suddenly connected, Heaviside represented the resulting voltage, which he called $Q$, as a quantity proportional to the step function $H(t)$, which he first introduced. Since the current is proportional to $\mathrm{d}Q/\mathrm{d}t$ ($pQ$, in his notation), Heaviside noticed that

> If, then, as in the present case $Q$ is zero before and constant after $t = 0$, $pQ$ is zero except when $t = 0$. It is then infinite. But its total amount is $Q$. That is to say, $pH(t)$ means a function of $t$ which is wholly concentrated at the moment $t = 0$, of total amount 1. It is impulsive so to speak.

Therefore, by stressing the usefulness of considering the response of a system to an "impulsive" input, Heaviside not only highlights the origin of $\delta(t)$ as the derivative of $H(t)$, but also implicitly admits that it is a mistake to call it a "function", because no true function can vanish everywhere except at a point, and have at the same time a finite integral. In fact, Heaviside's ideas were strongly opposed by the mathematicians of the Royal Society, who rejected the last part of his paper On Operators in Physical Mathematics, because of "total lack of rigor".[5] Hence, it is not a coincidence that Paul Dirac, who extensively used what we now call the "Dirac delta" to develop the theory of quantum mechanics, had been a student of Heaviside in Bristol. However, the true nature of this strange object remained obscure until Laurent Schwartz generalized the idea of function by introducing the concept of distributions. We shall briefly speak about them in the last section.

For the moment, we take a "gentler" approach, operationally regarding the Dirac delta $\delta(x - x_0)$ as a "sampling" tool that picks up the value of a function $f(x)$ of time or space at a point $x_0$. Let us first examine the "discrete" equivalent of $\delta(x - x_0)$, the simple *Kronecker* delta $\delta_{ij}$, which allows to extract a given term $f_i$ from a sequence $\{f_j\}$ of numbers,

$$\delta_{ij} = \begin{cases} 1 & \text{if } j = i \\ 0 & \text{if } j \neq i, \end{cases}$$

so that, for a series $\sum_j f_j$, we obtain $\sum_j f_j \delta_{ij} = f_i$. Yet, the definite integral of a function $f(x)$ of a real variable (either between finite extremes, or possibly over the entire axis) can be regarded as a "summation" where the discrete index $j$ is substituted by a *continuous* index $x$. To sample the value $f(x_0)$ of a function $f(x)$ defined in the domain $[a, b]$, the analogous of the Kronecker delta should then satisfy, for each $x_0 \in [a, b]$,

$$\int_a^b \delta(x - x_0) f(x) \, \mathrm{d}x = f(x_0). \tag{A.7}$$

Because $\delta(x - x_0)$ "sifts out" the value of $f$ in $x_0$, this basic definition of $\delta(x - x_0)$ is often called the *sifting property* of the delta function. In particular, choosing $x_0 =$

---

[5] Heaviside was so annoyed by the way his work was handled, that he later refused both to be funded by the Royal Society and to accept the prestigious Hughes Medal from this institution. However, although a brilliant scientist (just to give an example, it was Heaviside who recast Maxwell equations in the modern form), he was surely not an easy person, and was actually described by his best friends as a "first–rate oddity". He detested so much his opponents that he planned to entitle his autobiography (which he never finished) Wicked People I Have Known....

0, we can formally introduce a symbol $\delta(x)$ such that $\int_{-\infty}^{\infty} \delta(x) f(x)\, dx = f(0)$. Evidently, $\delta(x)$ cannot be a true function, because, applying this definition for the function $f(x) \equiv 1$, we should have

$$\int_a^b \delta(x)\, dx = \begin{cases} 1 & \text{if } 0 \in [a, b] \\ 0 & \text{if } 0 \notin [a, b]. \end{cases} \tag{A.8}$$

Thus, $\delta(x)$ should vanish for all $x \neq 0$, but have a unit integral over any interval $[a, b]$ containing the origin, no matter how small. Therefore, Eq. (A.7) should only be viewed as the formal definition of an operator,

$$\delta_{x_0}[f] : f \rightarrow f(x_0),$$

that associates a number, $f(x_0)$, to a function $f$, which is what in mathematics is known as a *functional*. For now, we can try and give an intuitive meaning to $\delta(x)$ by considering it as a kind of "improper" limit of a sequence of true functions $\delta_\epsilon(x)$ that depend on a parameter $\epsilon$. For example, consider the "rectangular" (or *unit boxcar*) functions

$$\delta_\epsilon(x) = \frac{1}{\epsilon} \operatorname{rect}(x/\epsilon) = \begin{cases} 1/\epsilon, & \text{if } |x| \leq \epsilon/2 \\ 0, & \text{if } |x| > \epsilon/2, \end{cases}$$

By decreasing $\epsilon$, the shape of $\delta_\epsilon(x)$ on a graph becomes "thinner" and "taller", but the area under the rectangle remains unitary. The same happens for a sequence of normalized Gaussians centered on the origin and of standard deviation $\sigma = \epsilon$,

$$\delta_\epsilon(x) = g(x; 0, \epsilon) = \frac{1}{\epsilon \sqrt{2\pi}} \exp\left(-\frac{x^2}{2\epsilon^2}\right).$$

When $\epsilon$ gets smaller and smaller, both these sequences of functions get more and more similar to $\delta(x)$. Yet, we know that there is no true limit for $\epsilon \rightarrow 0$, since $\delta(x)$ is not a function, but some kind of "alien" object: comparing with the definition of $\delta(x)$, what we really mean is that

$$\lim_{\epsilon \to 0} \int_{-\infty}^{+\infty} f(x) \delta_\epsilon(x)\, dx = f(0). \tag{A.9}$$

Actually, it is not even required that, when $\epsilon \rightarrow 0$, the interval where $\delta_\epsilon(x) \neq 0$ shrinks. For instance, one can show that the functions

$$\delta_\epsilon(x) = \frac{1}{\pi x} \sin\left(\frac{x}{\epsilon}\right)$$

satisfy (A.9), although, for all values of $\epsilon$ they oscillate over the whole real axis with an amplitude that grows for $x \rightarrow 0$.

A particularly useful representation of $\delta(x)$ is

$$\delta(x) = \frac{1}{2\pi} \int_{-\infty}^{\infty} e^{-iqx}\, dq = \frac{1}{2\pi}\left[\int_{-\infty}^{\infty} \cos(qx)\, dq - i\int_{-\infty}^{\infty} \sin(qx)\, dq\right], \quad (A.10)$$

which we can qualitatively justify as follows. The first integral at the right–hand side is a superposition of a continuous distribution of oscillating functions with all possible wavelengths $\lambda = 2\pi/q$. At any point $x_0 \neq 0$, the phases $kx_0$ of these oscillations are uniformly distributed in $[-\pi, \pi]$, and then their amplitude will be uniformly distributed in $[-1, 1]$. By superimposing a large number of contributions of positive and negative signs, we may then expect the total amplitude to vanish. However, for $x_0 = 0$, $\cos(qx_0) = 1$ for all values of $q$, so that all oscillations superimpose *in phase*, leading to a *diverging* total amplitude. The situation is not very different for the second integral, but in the origin $\sin(qx_0) = 0$: hence the imaginary part must vanish.[6] Note that Eq. (A.10) also means that $\delta(x)$ is a kind of "generalized" inverse Fourier transform of the constant function $f(q) \equiv 1$ (which is not integrable, and therefore does not admit an "ordinary" Fourier transform). Some particularly useful properties of the delta function are listed below.

**(a) Scaling.** If we scale the argument of the delta function by $a \neq 0$, we have

$$\delta(ax) = \frac{\delta(x)}{|a|}, \qquad (A.11)$$

which is obtained from Eq. (A.7) with a simple change of variable (examine the cases $a > 0$ and $a < 0$ separately). This means that $\delta(-x) = \delta(x)$, and that the delta function is a *homogeneous function of degree* $-1$.

**(b) Convolution.** A simple corollary of the sifting property is that the *convolution* of $\delta(x - x_0)$ with $f(x)$ is given by

$$f(x) * \delta(x - x_0) = \int_{-\infty}^{+\infty} f(x')\delta(x - x_0 - x')\, dx' = f(x - x_0), \qquad (A.12)$$

where we have used $\delta(-x) = \delta(x)$. Hence, the convolution of a generic function $f(x)$ with $\delta(x - x_0)$ just yields a translation of $f(x)$ by $x_0$. What is interesting is that this property holds not only for a "true" function $f(x)$, but also for the delta function itself. Thus we have

$$\int_{-\infty}^{+\infty} \delta(x - x_1)\delta(x - x_2)\, dx = \delta(x_2 - x_1) \qquad (A.13)$$

**(c) Composition** Suppose that $y = g(x)$ is a continuous and differentiable function of $x$, with $g'(x) = dy/dx$. We can define a "composed" delta function $\delta(y)$ by requiring that, for any regular function $f(y)$ of $y$, the integral

---

[6] The same conclusion can also be reached by simply observing that $\int_{-a}^{a} \sin(qx)\, dq = 0$, no matter how large $a$ becomes, because it is the integral of an odd function.

$$\int_{-\infty}^{+\infty} \delta(y) f(y) \mathrm{d}y = \int_{-\infty}^{+\infty} \delta[g(x)] \, f(x) \, |g'(x)| \, \mathrm{d}x,$$

sifts out all the values of $f$ where $y$ vanishes. Then, if $g(x)$ has $n$ distinct roots $x_i$, we must have

$$\delta(y) = \sum_{i=1}^{n} \frac{\delta(x - x_i)}{|g'(x_i)|}. \tag{A.14}$$

For instance, if $y = (x^2 - x_0^2)$, $\delta(y) = [\delta(x - x_0) + \delta(x + x_0)] \, / 2|x_0|$.

We have discussed the Dirac delta $\delta(x - x_0)$, which acts on a function of a single variable, but the operational definition (A.7) can be readily extended to more dimensions. If we consider a function $f(\mathbf{x})$, where $\mathbf{x}$ is a vector in $\mathbb{R}^n$, we then define $\delta(\mathbf{x})$ via the relation

$$\int_{\mathbb{R}^n} \delta(\mathbf{x} - \mathbf{x}_0) f(\mathbf{x}) \, \mathrm{d}\mathbf{x} = f(\mathbf{x}_0). \tag{A.15}$$

We shall mostly deal with functions of the position vector, $\mathbf{r} = (x, y, z) \in \mathbb{R}^3$, where the Dirac delta can be formally written as $\delta(\mathbf{r}) = \delta(x)\delta(y)\delta(z)$, so that, in analogy with Eq. (A.10),

$$\delta(\mathbf{r}) = \frac{1}{(2\pi)^3} \int e^{-i\mathbf{q}\cdot\mathbf{r}} \, \mathrm{d}^3 q \tag{A.16}$$

The properties we listed above are of course modified in more than one dimension. In particular, the scaling property becomes

$$\delta(a\mathbf{x}) = |a|^{-n} \delta(\mathbf{x}), \tag{A.17}$$

hence $\delta(\mathbf{x})$ is a homogeneous function of degree $-n$. The composition property is slightly more complicated, since, in an $n$-dimension in general, the points where a function $y = g(\mathbf{x}) = 0$ constitute a (hyper) surface $\Sigma_0$ with $(n - 1)$ dimensions. Then one can define $\delta(y)$ via the relation

$$\int_{\mathbb{R}^n} f(\mathbf{x})\delta[g(\mathbf{x})]\mathrm{d}\mathbf{x} = \int_{\Sigma_0} \frac{f(\mathbf{x})}{|\nabla g(\mathbf{x})|} \, \mathrm{d}\Sigma_0. \tag{A.18}$$

## A.2 Useful Integrals

**Gaussian Integrals**. We shall often need to evaluate the moments of a Gaussian function. It is therefore useful to calculate the integrals of the form

$$\int_{-\infty}^{\infty} x^n e^{-ax^2} \mathrm{d}x,$$

where $a > 0$ is related to the standard deviation $\sigma$ of the Gaussian distribution by $a = (2\sigma^2)^{-1}$. We plan to show that, if $r$ is a positive integer,

$$\int_{-\infty}^{\infty} e^{-ax^2}\,dx = \sqrt{\frac{\pi}{a}} = \sqrt{2\pi}\,\sigma \tag{A.19a}$$

$$\int_{-\infty}^{\infty} x^2 e^{-ax^2}\,dx = \frac{1}{2a}\sqrt{\frac{\pi}{a}} = \sqrt{2\pi}\,\sigma^3 \tag{A.19b}$$

$$\int_{-\infty}^{\infty} x^{2r-1} e^{-ax^2}\,dx = 0. \tag{A.19c}$$

**A.19a:** Let us make things apparently more complicated by evaluating the integral *squared*, namely, the double integral

$$I^2 = \left(\int_{-\infty}^{\infty} e^{-ax^2}\,dx\right)^2 = \int_{-\infty}^{\infty}\int_{-\infty}^{\infty} e^{-a(x^2+y^2)}\,dx\,dy.$$

The form of this integral suggests, of course, the use of polar coordinates $(r, \vartheta)$, with $r = x^2 + y^2$ and $dx\,dy = r\,dr\,d\vartheta$.[7] We obtain

$$I^2 = \int_0^{2\pi} d\vartheta \int_0^{\infty} r e^{-ar^2}\,dr = -\frac{\pi}{a}\int_0^{\infty} d\left(e^{-ar^2}\right) = \frac{\pi}{a}.$$

**A.19b.** Evaluating the second integral is now almost immediate, because

$$\int_{-\infty}^{\infty} x^2 e^{-ax^2}\,dx = -\int_{-\infty}^{\infty} \frac{\partial}{\partial a}\left(e^{-ax^2}\,dx\right) = -\frac{\partial}{\partial a}\left(\int_{-\infty}^{\infty} e^{-ax^2}\,dx\right) = -\frac{\partial}{\partial a}\sqrt{\frac{\pi}{a}},$$

wherefrom Eq. (A.19b) follows. The same method can of course be used to obtain any higher even moments of a Gaussian. For instance,

$$\int_{-\infty}^{\infty} x^4 e^{-ax^2}\,dx = -\frac{\partial}{\partial a}\left(\int_{-\infty}^{\infty} x^2 e^{-ax^2}\,dx\right) = \frac{3\sqrt{\pi}}{4}\frac{1}{a^{5/2}} = 3\sqrt{2\pi}\,\sigma^5.$$

**A.19c.** Since the integrand is antisymmetric with respect to the origin the contributions from $(-\infty, 0]$ and from $[0, +\infty)$ are equal and of opposite sign, hence the integral vanishes.

If the previous integrals must be evaluated only for *positive* values of $x$, you can use the original Euler's definition of the gamma function, given in footnote 1, to easily show that, for any integer $n \geq 0$,

---

[7] Rigorously, we should first evaluate the double integral between *finite* extremes $(-b, b)$, then observe that the area of the rectangle of integration is always between that of the inscribed circle of diameter $b$ and that of the circumscribed circle of diameter $\sqrt{2}b$, and finally take the limit $b \to \infty$.

$$\int_0^\infty x^n e^{-ax^2}\, dx = \frac{\Gamma\left(\frac{n+1}{2}\right)}{2a^{(n+1)/2}} = 2^{(n-1)/2}\,\Gamma\left(\frac{n+1}{2}\right)\sigma^{n+1}, \qquad (A.20)$$

This also gives us a general expression for the even moments of a Gaussian, which are just twice as large.[8]

**Example A.1** (*The volume of a hypersphere*) As a simple and useful application that involves both Gaussian integrals and the gamma function, let us calculate the volume of a sphere of radius $r$ in a $D$-dimensional space (which, for $D > 3$, is popularly known among science-fiction fans as a "hypersphere"). The sphere volume must be proportional to $r^D$, thus we can write $V(r) = C_D r^D$. To explicitly evaluate $C_D$, consider the integral

$$I_D = \int_{-\infty}^{\infty} dx_1 \cdots dx_D\, e^{-(x_1^2 + x_2^2 + \cdots + x_D^2)},$$

where $\{x_1, x_2, \ldots x_D\}$ are orthogonal Cartesian coordinates. Since this is the product of $D$ identical Gaussian integrals (A.19a) with $a = 1$, we simply have $I_D = \pi^{D/2}$. On the other hand, noticing that $x_1^2 + x_2^2 + \cdots + x_D^2 = r^2$, we can write $I_D$ in spherical coordinates as

$$I_D = \int_0^\infty dr\, S_D(r)\, e^{-r^2},$$

where $S_D(r) = D C_D r^{D-1}$ is the surface area of the sphere. Introducing $z = r^2$ and using (A.1), we have

$$I_D = \frac{D}{2} C_D \int_0^\infty dz\, z^{D/2-1} e^{-z} = \frac{D}{2} C_D \Gamma(D/2).$$

By equating these two expressions for $I_D$, we find

$$C_D = \frac{2\pi^{D/2}}{D\Gamma(D/2)} = \frac{\pi^{D/2}}{\Gamma(D/2 + 1)},$$

so that the volume and the surface area of the sphere are

$$V(r) = \frac{\pi^{D/2}}{\Gamma(D/2 + 1)} r^D \ , \quad S(r) = \frac{2\pi^{D/2}}{\Gamma(D/2)} r^{D-1} \qquad (A.21)$$

---

[8] Check that, for $n = 0, 2$, the results coincide with those we have previously found.

## A.3 Numeric Tables

**Normal distribution.** Cumulative distribution $G(z) = \frac{1}{\sqrt{2\pi}} \int_{-\infty}^{z} e^{-t^2/2} dt$ for percentage values of the normalized variable $0 \leq z \leq 3.5$. Entries should be read by crossing the line giving the first two digits with the column giving the last one. For example, $G(0.73) = 76.74\%$ (shown in bold). For negative values of $z$, one can simply use $G(-z) = 100\% - G(z)$.

| z | 0 | 1 | 2 | 3 | 4 | 5 | 6 | 7 | 8 | 9 |
|---|---|---|---|---|---|---|---|---|---|---|
| **0.0** | 50.00 | 50.41 | 50.82 | 51.22 | 51.62 | 52.02 | 52.41 | 52.81 | 53.21 | 53.61 |
| **0.1** | 54.01 | 54.40 | 54.80 | 55.19 | 55.59 | 55.98 | 56.38 | 56.77 | 57.16 | 57.56 |
| **0.2** | 57.95 | 58.34 | 58.73 | 59.12 | 59.50 | 59.89 | 60.28 | 60.66 | 61.05 | 61.43 |
| **0.3** | 61.81 | 62.19 | 62.57 | 62.95 | 63.33 | 63.70 | 64.08 | 64.45 | 64.82 | 65.19 |
| **0.4** | 65.56 | 65.93 | 66.29 | 66.66 | 67.02 | 67.38 | 67.74 | 68.10 | 68.46 | 68.81 |
| **0.5** | 69.16 | 69.51 | 69.86 | 70.21 | 70.56 | 70.90 | 71.24 | 71.58 | 71.92 | 72.26 |
| **0.6** | 72.59 | 72.92 | 73.25 | 73.58 | 73.91 | 74.23 | 74.55 | 74.87 | 75.19 | 75.50 |
| **0.7** | 75.82 | 76.13 | 76.44 | **76.74** | 77.05 | 77.35 | 77.65 | 77.95 | 78.24 | 78.54 |
| **0.8** | 78.83 | 79.11 | 79.4 | 79.68 | 79.97 | 80.24 | 80.52 | 80.80 | 81.03 | 81.34 |
| **0.9** | 81.60 | 81.87 | 82.13 | 82.39 | 82.65 | 82.90 | 83.16 | 83.41 | 83.65 | 83.90 |
| **1.0** | 84.14 | 84.38 | 84.62 | 84.86 | 85.09 | 85.32 | 85.55 | 85.78 | 86.00 | 86.22 |
| **1.1** | 86.44 | 86.66 | 86.87 | 87.08 | 87.29 | 87.50 | 87.70 | 87.91 | 88.11 | 88.30 |
| **1.2** | 88.50 | 88.69 | 88.88 | 89.07 | 89.26 | 89.44 | 89.62 | 89.80 | 89.98 | 90.15 |
| **1.3** | 90.33 | 90.50 | 90.66 | 90.83 | 90.99 | 91.15 | 91.31 | 91.47 | 91.63 | 91.78 |
| **1.4** | 91.93 | 92.08 | 92.22 | 92.37 | 92.51 | 92.65 | 92.79 | 92.93 | 93.06 | 93.19 |
| **1.5** | 93.32 | 93.45 | 93.58 | 93.72 | 93.83 | 93.95 | 94.07 | 94.18 | 94.30 | 94.41 |
| **1.6** | 94.52 | 94.63 | 94.74 | 94.85 | 94.95 | 95.06 | 95.16 | 95.26 | 95.35 | 95.45 |
| **1.7** | 95.55 | 95.64 | 95.73 | 95.82 | 95.91 | 96.00 | 96.08 | 96.17 | 96.25 | 96.33 |
| **1.8** | 96.41 | 96.49 | 96.56 | 96.64 | 96.71 | 96.79 | 96.86 | 96.93 | 97.00 | 97.06 |
| **1.9** | 97.13 | 97.19 | 97.26 | 97.32 | 97.38 | 97.44 | 97.50 | 97.56 | 97.62 | 97.67 |
| **2.0** | 97.73 | 97.78 | 97.83 | 97.88 | 97.92 | 97.98 | 98.03 | 98.08 | 98.12 | 98.17 |
| **2.1** | 98.21 | 98.26 | 98.30 | 98.34 | 98.38 | 98.42 | 98.46 | 98.50 | 98.54 | 98.57 |
| **2.2** | 98.61 | 98.65 | 98.67 | 98.71 | 98.75 | 98.78 | 98.81 | 98.84 | 98.87 | 98.90 |
| **2.3** | 98.93 | 98.96 | 98.98 | 99.01 | 99.04 | 99.06 | 99.09 | 99.11 | 99.13 | 99.16 |
| **2.4** | 99.18 | 99.20 | 99.22 | 99.25 | 99.27 | 99.29 | 99.31 | 99.32 | 99.34 | 99.36 |
| **2.5** | 99.38 | 99.40 | 99.41 | 99.43 | 99.45 | 99.46 | 99.48 | 99.49 | 99.51 | 99.52 |
| **2.6** | 99.53 | 99.55 | 99.56 | 99.57 | 99.59 | 99.60 | 99.61 | 99.62 | 99.63 | 99.64 |
| **2.7** | 99.65 | 99.66 | 99.67 | 99.68 | 99.69 | 99.70 | 99.71 | 99.72 | 99.73 | 99.74 |
| **2.8** | 99.74 | 99.75 | 99.76 | 99.77 | 99.77 | 99.78 | 99.79 | 99.79 | 99.80 | 99.81 |
| **2.9** | 99.81 | 99.82 | 99.83 | 99.83 | 99.84 | 99.84 | 99.85 | 99.85 | 99.86 | 99.86 |
| **3.0** | 99.87 | 99.87 | 99.87 | 99.88 | 99.88 | 99.89 | 99.89 | 99.89 | 99.90 | 99.90 |
| **3.0** | 99.87 | 99.87 | 99.87 | 99.88 | 99.88 | 99.89 | 99.89 | 99.89 | 99.90 | 99.90 |
| **3.1** | 99.90 | 99.91 | 99.91 | 99.91 | 99.92 | 99.92 | 99.92 | 99.92 | 99.93 | 99.93 |
| **3.2** | 99.93 | 99.93 | 99.94 | 99.94 | 99.94 | 99.94 | 99.94 | 99.95 | 99.95 | 99.95 |
| **3.3** | 99.95 | 99.95 | 99.95 | 99.96 | 99.96 | 99.96 | 99.96 | 99.96 | 99.96 | 99.97 |
| **3.4** | 99.97 | 99.97 | 99.97 | 99.97 | 99.97 | 99.97 | 99.97 | 99.97 | 99.97 | 99.98 |

**Student's $t$-distribution.** Values $t_P$ of the Student's $t$-variable giving a percentage cumulative distribution

$$P = C_N \int_{-t_P}^{+t_P} \left( 1 + \frac{t^2}{N-1} \right)^{-\frac{N}{2}} dt$$

as a function of the number of sample data (corresponding to $\nu = N - 1$ degrees of freedom). The explicit value of the normalization constant $C_N$ is

$$C_N = \frac{1}{\sqrt{\pi(N-1)}} \frac{\Gamma\left(\frac{N}{2}\right)}{\Gamma\left(\frac{N-1}{2}\right)}. \tag{A.22}$$

| P N | 50.0% | 66.3% | 90.0% | 95.0% | 50.0% |
|---|---|---|---|---|---|
| 2 | 1.000 | 1.838 | 6.314 | 12.71 | 63.65 |
| 3 | 0.817 | 1.322 | 2.920 | 4.303 | 9.925 |
| 4 | 0.765 | 1.197 | 2.353 | 3.182 | 6.841 |
| 5 | 0.741 | 1.142 | 2.132 | 2.776 | 4.604 |
| 6 | 0.727 | 1.111 | 2.015 | 2.571 | 4.032 |
| 7 | 0.718 | 1.091 | 1.943 | 2.447 | 3.707 |
| 8 | 0.711 | 1.077 | 1.895 | 2.365 | 3.499 |
| 9 | 0.706 | 1.067 | 1.860 | 2.306 | 3.355 |
| 10 | 0.703 | 1.059 | 1.833 | 2.262 | 3.250 |
| 11 | 0.700 | 1.053 | 1.812 | 2.228 | 3.169 |
| 12 | 0.697 | 1.048 | 1.796 | 2.201 | 3.106 |
| 13 | 0.695 | 1.044 | 1.782 | 2.179 | 3.055 |
| 14 | 0.694 | 1.040 | 1.771 | 2.160 | 3.012 |
| 15 | 0.692 | 1.037 | 1.761 | 2.145 | 2.977 |
| 16 | 0.691 | 1.035 | 1.753 | 2.131 | 2.947 |
| 17 | 0.690 | 1.033 | 1.746 | 2.120 | 2.921 |
| 18 | 0.689 | 1.031 | 1.740 | 2.110 | 2.898 |
| 19 | 0.688 | 1.029 | 1.734 | 2.101 | 2.878 |
| 20 | 0.687 | 1.027 | 1.729 | 2.093 | 2.861 |
| $\infty$ | 0.675 | 1.000 | 1.645 | 1.960 | 2.576 |

**Chi-squared distribution.** Percentage probability $P(\chi^2 > \chi_0^2)$ for a number of degrees of freedom $\nu$ between 1 and 10 and for a value of the scaled chi-squared $\chi_0^2/\nu$, calculated from the experimental data, ranging from 0.1 to 4.

| $\nu$ — $\chi_0^2/\nu$ | 1 | 2 | 3 | 4 | 5 | 6 | 7 | 8 | 9 | 10 |
|---|---|---|---|---|---|---|---|---|---|---|
| 0.1 | 75.18 | 90.48 | 96.00 | 98.25 | 99.21 | 99.64 | 99.83 | 99.92 | 99.96 | 99.98 |
| 0.2 | 65.47 | 81.87 | 89.64 | 93.85 | 96.26 | 97.69 | 98.56 | 99.09 | 99.43 | 99.63 |
| 0.3 | 58.39 | 74.08 | 82.54 | 87.81 | 91.31 | 93.71 | 95.41 | 96.92 | 97.50 | 98.14 |
| 0.4 | 52.71 | 67.03 | 75.30 | 80.88 | 84.92 | 87.95 | 90.29 | 92.12 | 93.57 | 94.74 |
| 0.5 | 47.95 | 60.65 | 68.23 | 73.58 | 77.65 | 80.89 | 83.52 | 85.71 | 87.55 | 89.12 |
| 0.6 | 43.86 | 54.88 | 61.49 | 66.26 | 70.00 | 73.06 | 75.65 | 77.87 | 79.81 | 81.53 |
| 0.7 | 40.28 | 49.66 | 55.19 | 59.18 | 62.34 | 64.96 | 67.22 | 69.19 | 70.96 | 72.54 |
| 0.8 | 37.11 | 44.09 | 49.36 | 52.49 | 54.94 | 56.97 | 58.72 | 60.25 | 61.63 | 62.88 |
| 0.9 | 34.28 | 40.66 | 44.02 | 46.28 | 47.99 | 49.36 | 50.52 | 51.52 | 52.41 | 53.21 |
| 1.0 | 31.73 | 36.79 | 39.16 | 40.60 | 41.59 | 42.32 | 42.89 | 43.35 | 43.73 | 44.05 |
| 1.1 | 29.43 | 33.29 | 34.76 | 35.46 | 35.80 | 35.94 | 35.98 | 35.95 | 35.86 | 35.75 |
| 1.2 | 27.33 | 30.12 | 30.80 | 30.84 | 30.62 | 30.28 | 29.87 | 29.42 | 28.97 | 28.51 |
| 1.3 | 25.42 | 27.25 | 27.25 | 26.74 | 26.06 | 25.31 | 24.56 | 23.81 | 23.08 | 22.37 |
| 1.4 | 23.67 | 24.66 | 24.07 | 23.11 | 22.06 | 21.02 | 20.02 | 19.06 | 18.16 | 17.30 |
| 1.5 | 22.07 | 22.31 | 21.23 | 19.92 | 18.60 | 17.36 | 16.20 | 15.12 | 14.13 | 13.21 |
| 1.6 | 20.59 | 20.19 | 18.70 | 17.12 | 15.62 | 14.25 | 13.01 | 11.89 | 10.88 | 9.96 |
| 1.7 | 19.23 | 18.27 | 16.46 | 14.68 | 13.08 | 11.65 | 10.39 | 9.28 | 8.30 | 7.44 |
| 1.8 | 17.97 | 16.53 | 14.47 | 12.57 | 10.91 | 9.48 | 8.25 | 7.19 | 6.28 | 5.50 |
| 1.9 | 16.81 | 14.96 | 12.72 | 10.74 | 9.07 | 7.68 | 6.51 | 5.54 | 4.72 | 4.03 |
| 2.0 | 15.73 | 13.53 | 11.16 | 9.16 | 7.52 | 6.20 | 5.12 | 4.24 | 3.51 | 2.93 |
| 2.25 | 13.36 | 10.54 | 8.03 | 6.11 | 4.66 | 3.58 | 2.75 | 2.12 | 1.64 | 1.28 |
| 2.5 | 11.38 | 8.20 | 5.76 | 4.04 | 2.85 | 2.03 | 1.44 | 1.03 | 0.74 | 0.53 |
| 2.75 | 9.72 | 6.39 | 4.11 | 2.66 | 1.73 | 1.13 | 0.74 | 0.49 | 0.33 | 0.22 |
| 3.0 | 8.32 | 4.97 | 2.93 | 1.74 | 1.04 | 0.62 | 0.38 | 0.23 | 0.14 | 0.09 |
| 3.25 | 7.14 | 3.87 | 2.08 | 1.13 | 0.62 | 0.34 | 0.19 | 0.11 | 0.06 | 0.03 |
| 3.5 | 6.14 | 3.02 | 1.48 | 0.73 | 0.36 | 0.18 | 0.09 | 0.05 | 0.02 | 0.01 |
| 3.75 | 5.28 | 2.35 | 1.05 | 0.47 | 0.21 | 0.10 | 0.05 | 0.02 | 0.01 | 0.00 |
| 4 | 4.55 | 1.83 | 0.74 | 0.30 | 0.12 | 0.05 | 0.02 | 0.01 | 0.00 | 0.00 |

# Index

MIX
Papier aus verantwortungsvollen Quellen
Paper from responsible sources
FSC® C105338

If you have any concerns about our products,
you can contact us on
ProductSafety@springernature.com

In case Publisher is established outside the EU,
the EU authorized representative is:
**Springer Nature Customer Service Center GmbH**
**Europaplatz 3, 69115 Heidelberg, Germany**

Printed by Libri Plureos GmbH
in Hamburg, Germany